D. Dietrich und
H. Schweinzer (Hrsg.)

Feldbustechnik
in Forschung, Entwicklung und Anwendung

Beiträge zur Feldbustagung FeT '97
in Wien, Österreich,
13.–14. Oktober 1997

Springer-Verlag Wien GmbH

Univ.-Prof. Dipl.-Ing. Dr. Dietmar Dietrich
Institut für Computertechnik
Ass.-Prof. Dipl.-Ing. Dr. Herbert Schweinzer
Institut für Elektrische Meßtechnik
Technische Universität Wien, Wien, Österreich

Gedruckt auf säurefreiem, chlorfrei gebleichtem Papier – TCF
Graphisches Konzept: Ecke Bonk

Mit 233 Abbildungen

ISBN 978-3-211-83062-8   ISBN 978-3-7091-6881+3 (eBook)
DOI 10.1007/978-3-7091-6881-3

Vorwort

Die Feldbustechnik hat sich etabliert. Dies gilt für den Bereich der Anwendungen, auf breiter Basis aber auch für Forschung und Entwicklung. Die Anzahl der Serienapplikationen, der großen Einzelprojekte, in die bis zu 20.000 und mehr Knoten integriert werden, und andererseits der wissenschaftlichen Veröffentlichungen ist sprunghaft gestiegen. Die zentralen Themen im Bereich der Feldbusse kreisen schon lange nicht mehr um Fragen der Protokolle; jetzt geht es um weit umfassendere, komplexere Aufgaben. Darüber hinaus eröffnen Feldbusse ungeahnte Möglichkeiten, wie die Resultate belegen: es lassen sich enorme Energiemengen einsparen, die Wartung von Systemen kann entscheidend vereinfacht werden, die Sicherheit in vielen Bereichen läßt sich steigern, aber auch viele einfache, routinemäßige Aufgaben lassen sich mit Hilfe der Feldbusse effizient automatisieren. Der Einsatzbereich ist in der Tat weit gespannt; überall, wo Daten anfallen, verarbeitet und Systeme gesteuert werden, sind Feldbusse einzusetzen: Kläranlagen, Kraftfahrzeuge, Gebäude, Wetterstationen, Bandstraßen, Büroräume, Raumfahrt, Flutlichtanlagen, Sicherheitssysteme, Heizkraftwerke, Skiliftanlagen sind Beispiele dafür. Feldbussysteme sind im Zusammenwirken mit LANs und WANs als Nervensystem anzusehen, das in alle Bereiche unseres Lebens eingebettet wird. Dadurch erlebt die Automatisierung zur Zeit einen Quantensprung, der eine ähnlich revolutionäre Wirkung zeigen wird, wie sie der Personal Computer verursachte. Darauf muß man sich einstellen und das auf vielen verschiedenen Ebenen.

Daß diese einsetzende Dynamik von vielen erkannt wurde, spiegeln auch die emsigen Aktivitäten der verschiedenen internationalen Normungsgruppen wider, wo teilweise hart um die Durchsetzung bestimmter Feldbustypen gekämpft wird. Es war von Anfang an klar, daß nur wenige Bussysteme von den vielen entwickelten übrig bleiben werden. Womit vor einigen Jahren nur wenige rechneten: man hat letztendlich, zumindest nach außen hin, akzeptiert, daß mehrere Feldbusstandards nebeneinander bestehen werden. Entsprechende Forschungsprogramme, Gremien und Arbeitsgruppen wurden zur Lösung dieses Problems ins Leben gerufen. Dort bemüht man sich um die Schaffung gemeinsamer Oberflächen, um busunabhängige Profile und um Systemübergänge. Man tut sich schwer, da es um große Märkte geht, doch ohne Einigung wird so rasch kein Fortschritt erzielbar sein.

Die Feldbustagung in Wien hat sich zum Ziel gesetzt, viele der aktuellen Fragen zu diskutieren. Gegenüber der ersten Veranstaltung dieser Art, der FeT 95, bei der Industrie- und Forschungsaktivitäten getrennt behandelt wurden, wurde jedoch die Vorgangsweise geändert. Das Programmkomitee entschied, für die zweitägige Tagung globale Fachschwerpunkte zu definieren, um eine optimal Zusammenführung zwischen Industrie und Hochschule zu erreichen. Das Programmkomitee hat sich auch dazu entschlossen, seinen Kreis um zusätzliche international bekannte Experten aus dem Bereich der Feldbustechnik und Automatisierung zu erweitern.

Da eine große Anzahl interessanter, viele Bereiche abdeckender Aufsätze eingereicht wurde, haben wir uns entschlossen, diesen Sammelband weitgehend unabhängig vom Ablauf der Tagung so zu gestalten, daß er auch unabhängig von der Tagung einen wesentlichen Beitrag zum Thema der Feldbustechnik darstellt. Die Anzahl der in den Sammelband aufgenommenen Beiträge ist etwa doppelt so groß, wie die bei der Tagung vorgetragenen Beiträge, und umfaßt die vom Programmkomitee best-bewerteten Aufsätze. Die Gliederung des Buches erfolgte nach Kriterien, die den Aufsätzen entnommen wurden. Damit ist gut abzulesen, welche Schwerpunkte gegenwärtig bearbeitet werden. Einleitend zu den Kapiteln werden die jeweiligen Schwerpunkte vorgestellt. Zur Tagung selbst wurden wiederum vier Vortragende eingeladen, um bestimmte Themenkomplexe zu unterstreichen und direkt anzu-sprechen. Wir freuen uns, daß alle vier Vortragende dazu bereit waren.

Wir möchten uns bei all denen bedanken, die sich mit großem Engagement für das Gelingen der Tagung eingesetzt haben. Zu nennen sind vor allem die Mitglieder des Organisationskomitees, insbesondere Herr Franz Raschbacher und Herr Thilo Sauter, aber auch das Programmkomitee, die Sekretärinnen des Instituts für Computertechnik, Frau Renate Wimmer und Frau Hilda Wurm, des Instituts für Elektrische Meßtechnik, Frau Gerda Breitmeyer und Frau Beatrix Lauinger, die Mitarbeiter des Institutes für elektrische Meßtechnik und die des Institutes für Computertechnik. Weiters möchten wir uns bei Frau Helga Stangl bedanken, der in Zusammenarbeit mit verschiedenen Assistenten der beiden Institute ein exzellenter Entwurf des Flyers 'Invitation to FeT 97' glückte. Nicht vergessen werden dürfen selbstverständlich die Autoren und Referenten selbst, die diese Tagung mit Leben erfüllten. Um die Ausgaben der Tagung weitgehend abzudecken, wäre eine drastische Anhebung der Tagungsgelder erforderlich gewesen. Das steht aber in krassem Widerspruch zu unserer Absicht. Unser Ziel ist es, ein möglichst breites Publikum ansprechen zu können, was nur durch eine zusätzliche finanzielle Unterstützung * erreicht wurde, wofür wir hiermit unseren Dank aussprechen. Es freut uns, daß sich das Komitee CENELEC TC65CX entschieden hat, seine Konferenz einen Tag nach der FeT 97 ebenfalls in Wien abzu-halten, so daß allen Mitgliedern des Komitees die Gelegenheit geboten wird, ebenfalls an der FeT 97 teilzunehmen. Letztendlich ist dem Springer-Verlag zu danken, der es ermöglicht hat, diesen Sammelband rasch und unkompliziert aufzulegen.

Wien, Oktober 1997

Dietmar Dietrich   Herbert Schweinzer

* Unterstützt von: Bundesministerium für Wissenschaft und Verkehr, A
                   Bundesinnung d. Elektrotechniker, Radio- u. Videoelektroniker, A
                   Phoenix Contact GmbH & Co., D
                   Siemens AG Österreich, A
                   Weidmüller ConneXt GmbH & Co., D

Preface

Fieldbus technology has established itself. This is true for the field of applications, but on a broad basis also for research and development. The numbers of off-the-shelf applications and singular large-scale projects with up to 20.000 and more nodes as well as those of scientific publications have increased dramatically. The focal topics in the area of fieldbus systems are no longer questions of protocols; today we deal with more comprehensive and complex tasks. As the results show, fieldbus systems open up undreamt of possibilities: enormous amounts of energy can be saved, the safety in many areas may be improved, but also simple everyday problems can be automated efficiently. The area of application is wide indeed; wherever data need to be acquired and processed or systems must be controlled, fieldbus systems may be employed: purification plants, cars, buildings, weather stations, assembly lines, offices, space research, floodlight systems, security systems, heating plants, ski-lifts are examples. In combination with LANs and WANs, fieldbus systems may be regarded as a nervous system that is being embedded in all spheres of our life. At present, automation experiences a quantum leap that will have a revolutionary effect similar to that of the Personal Computer. We will have to accommodate ourselves to this fact in many different ways.

The fact that this dynamical evolution has been recognized by many people is reflected in the busy activities of various international standardization committees with their partly fierce struggles for a breakthrough of certain types of fieldbus systems. It has been clear from the very beginning that only few of the many developed systems will remain in the end. But what only few expected some years ago is that it has been accepted, at least outwardly, that several fieldbus standards will exist in parallel. To tackle this problem, suitable research programmes, committees, and working groups have been established. They strive for the creation of common interfaces, fieldbus-independent profiles and systems interconnections. It is hard work in view of the large markets, but without agreement no progress will be made in a short time.

The goal of the fieldbus conference in Vienna is to discuss many of the up-to-date topics. In contrast to the first event of this kind, the FeT 95, where industrial and research issues had been treated separately, the procedure has been changed this time. The programme committee decided to define global subjects in order to bring together industry and universities in an optimal way. Furthermore, the programme committee was enlarged by additional renowned experts on fieldbus systems and automation.

Since a large number of interesting papers covering many areas was submitted, we decided to make these proceedings largely independent of the conference itself, such that they represent an independent, essential contribution to fieldbus technology. The number of printed articles (those which attained the highest ratings by the programme committee) is roughly twice as large as that of the papers presented at the conference. The compilation followed criteria taken from the articles, which makes the current

VIII

working fields evident. These subjects are presented in the introductions of the respective chapters. For the conference, four speakers were invited to address and emphasize certain topics. We are glad that all four followed our invitation.

We would like to thank all those who devoted strong efforts to the success of the conference. These are above all the members of the organizing committee, in particular Franz Raschbacher and Thilo Sauter, but also the programme committee, the secretaries of the Institut für Computertechnik, Renate Wimmer and Hilda Wurm, and of the Institut für Elektrische Meßtechnik, Gerda Breitmeyer and Beatrix Lauinger, as well as the staff of the institutes. Furthermore, we would like to thank Helga Stangl, who in co-operation with several assistants of the two institutes succeeded to make the excellent design of the flyer 'Invitation to FeT 97'. We must not forget, of course, the authors and speakers who filled the conference with life. To cover the expenses of the conference, a drastic increase of the conference fees would have been necessary. This, however, contradicts our intention. Our aim is to address a broad audience. This became possible only with the support of sponsoring organizations [*], whose financial support we gratefully acknowledge. We are also glad that the committee CENELEC TC65CX took the decision to hold its meeting in Vienna the day after the FeT, so that the members of the committee have the opportunity to participate in the FeT. Finally, we thank Springer-Verlag for the quick and uncomplicated printing of the proceedings.

Vienna, October 1997

Dietmar Dietrich   Herbert Schweinzer

[*] Supported by: Bundesministerium für Wissenschaft und Verkehr, A
                   Bundesinnung der Elektrotechniker, Radio- u. Videoelektroniker, A
                   Phoenix Contact GmbH & Co., D
                   Siemens AG Österreich, A
                   Weidmüller ConneXt GmbH & Co., D

Inhaltsverzeichnis
Table of contents

Tagungsleitung
Head of conference

Prof. Dr. H. Schweinzer, TU Wien, A

Vorsitz des Programmkomitees
Chairman of programme committee

Prof. Dr. D. Dietrich, TU Wien, A

Programmkomitee
Programme committee

Dr. M. Adams, Dr. Seufert GmbH, D
Prof. Dr. K. Bender, TU München, D
Dipl.-Ing. R. Bent, Phoenix Contact GmbH & Co., D
Dr. W. J. Borst, Borst Automation, D
Prof. Dr. J. Böttcher, Univ. der Bundeswehr München, D
Prof. J.-D. Decotignie, EPFL-LIT, CH
Prof. Dr. D. Dietrich, TU Wien, A
Prof. Dr. W. Dilger, TU Chemnitz-Zwickau, D
Prof. Dr. K. Etschberger, Steinbeis-Transferzentrum, D
Prof. P. Fischer, FH Dortmund, D
Prof. DDr. W. A. Halang, Fernuniversität Hagen, D
Dr. J. W. Hertel, ECHELON, D
J. Johansen, Proces-Data, DK
Dr. K. P. Judmann, TU Wien, A
Prof. Dr. K. Kabitzsch, TU Dresden, D
Prof. Dr.-Ing. K.-M. Koch, Carinthian Tech Research Institute, A
Dr. M. Merx, Fa. Weidmüller ConneXt GmbH & Co, D
Ing. W. Morrenth, Siemens AG Österreich, A
Prof. Dr. P. Neumann, Institut für Automation und Kommunikation e.V., D
Prof. Dr. G.-H. Schildt, TU Wien, A
Dr.-Ing. W. Schruttke, Fa. Weidmüller GmbH & Co, D
Dipl.-Ing. D. Schultschik, LeGrand Österreich GmbH, A
Prof. Dr. H. Schweinzer, TU Wien, A
Dr. R. Sommer, Siemens AG., D
Prof. Dr. H. U. Steusloff, Fraunhofer-Institut, D
Prof. Dr. F. Thuselt, FH Pforzheim, D
H. K. Tronnier, EIBA, B
Dr. Gerd-Ulrich Vack, SYSMIC GmbH. Dresden, D
H. Zeltwanger, CAN-in-Automation e.V., D

Organisationskomitee
Organization committee

Dipl.-Ing. Christoph Mittermayer
Dipl.-Ing. Franz Raschbacher
Dipl.-Ing. Thilo Sauter
Dipl.-Ing. Hans-Jörg Schweinzer

Verzeichnis der Autoren
Index of Authors

Teil 1 Beiträge aus Forschung und Entwicklung

Contributions from research and development

1 Industrieautomation

Industrial automation

Die Aufsätze spiegeln sehr gut die derzeit aktuellen Schwerpunkte im Bereich der Industrieautomation wider: die netzübergreifenden Aktivitäten in den Normungsausschüssen, die Normungsaktivitäten im IEC, die Echtzeitproblematik bei Feldbussen und natürlich die zunehmenden Aktivitäten in der Prozeßautomation.

The articles reflect precisely the current major topics in industrial automation: the network-independent activities of the standardization committees, the standardization activities in the IEC, the real-time problem of fieldbus systems, and of course the growing activities in process automation.

Fieldbus Standardization,
the European Approach and Experiences

Guido H. Gürtler

SIEMENS AG
Technical Regulations and Standardization
Siemens AG, 81730 München
Tel. ++49 89 636 40700
Fax ++49 89 636 40705
Guido.Guertler@mchp.siemens.de

Abstract. Focusing on the purpose of standardization in general, reflecting on the historical developments of Fieldbus Standardization and taking into account the actual challenges given by progress of technologies, the assumption is justified that traditional standardization structures and procedures are not able to lead in time to market accepted results.

The European Approach is an experiment with promising results because it takes due account of the real market needs: it has led to an accepted European Standard EN 50170 and the Common Application Interface between bus and peripheral equipment (NOAH Project) is under development.

It is a great pleasure for me to present some reflections on Fieldbus Standardization. Looking at the developments of Fieldbus Standardization from a fair distance it may not be a surprise that conclusions are drawn which may sound provocative.

I will first concentrate on some essentials of standardization, then look a bit into history, give the important details on the European Approach and the experiences made, and finally dare some perspectives.

1 Standardization, Aim and Purpose

Standardization is understood as the work in consensus-based recognized standards organizations like ISO and IEC with their member organizations nominating the participants in technical work.

Standardization has to serve the market interests, to bring economic benefits to all parties involved, has definitely to keep pace with the progress in technologies because otherwise the finally accepted standards are out-dated quickly standardization, has to bring the different and diverging interests, in good time, to consensus, and finally, by its economic impact, standardization has to give certainty to investors like users and manufacturers.

Having said this, we must also take into account, which type of standard we are talking about because the impacts of competition are quite different.

First everybody knows the standardized units like „meter", as an example for a standard for mutual understanding. Here the relevance of competition is really low and it is more often difficult to get the experts to work: everyone gains benefits from these standards and asks in the same moment why he should invest manpower.

The second type are horizontal standards like EMC (Electromagnetic Compatibility) or mechanical or electrical Safety, standards which have the advantage that otherwise the same regulations would have to be drafted repeatedly in many product standards: this would lead to slight differences for the same technical subject because the experts would be different and this would cause in consequence a lot of confusion, would lead to multiple testing etc. Therefore, it is a wise decision by IEC and ISO to have given the so-called „Pilot-Function" to many horizontal standards, i.e. the product standards refrain from defining the same subjects but make reference to the horizontal standards. Nevertheless, it must be stated that not all Technical Committees obey this rule. You will understand that such horizontal standards have a relatively neutral effect on competition.

This is very different if we look into the third category of product standards like fieldbus. Here the goal is to bring diverging business interests to a common position: the technical work is usually combined with time consuming and tough negotiations. Normally a contribution reflects the business interest of the contributors and suspicions of the others are on hand.

The fourth type are standards for rationalisation, i.e. for example the reduction of screwtypes in order to gain economic advantages in purchase, production and maintenance. Here the different interests of manufacturers and users of such material have to be brought to consensus and it is obvious that the latter ones normally gain the greater benefits from standardization.

2 Reflections on History and Developments

Let's have this in mind and look a bit into history. Fieldbusses are a child of digitalization in information technology. They are in practical implementation already for some 20 years. Many different types of products have been developed and found their market. Just to name a few:

CAN, EIB, PROFIBUS, BACNET, INTERBUS-S, FIP or the Fieldbus Foundation's Specification. IBM and Daimler announced last year to develop a bus for automotive control systems and this can also be seen, technologically, as a fieldbus, driven by its own intelligence together with actors and sensors.

Besides these names, and I am very sure that all of you know many more, an increasing number of proprietary solutions, let's say company standards, emerge and do co-exist in the market.

The observation can be made that all these solutions are market- driven, find their implementations, i.e. customers are ready to pay for, but they differ in many respects, e.g. function and specialized purpose, price, speed of data transmission, interfaces to peripheral equipment, security concepts, network management, centralized or decentralized intelligence, noise immunity, the kind of implementation in hardware and/or software, the concept of wired or wireless installation etc. etc. This makes the whole fieldbus scenery really complex but also fascinating.

But this all justifies on the other hand the question whether one single standardized fieldbus will ever be able to fulfil the different application needs. One rule seems to be logic:

> **Each single solution can not serve all applications!**

In consequence it is, at first glance, a worth-while attempt to get as many application needs as possible covered by one single standard. But such an approach has implicitly the disadvantage that each single application does not need all functionality offered and may be too costly for the dedicated applications.

But an even more important standardization rule and need is to take into account the course of time: different solutions emerge at different points of time and customers do not wait for getting their specific problem solved, e.g. in process automation, by a general purpose solution. They want to *run an efficient production line now!*

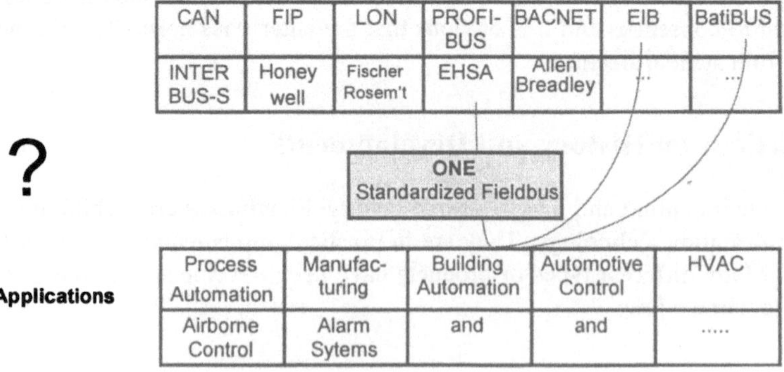

Fig. 1. Is ONE Standard possible?

Another observation, looking into recent history, is a definite change in relations between traditional standardization and innovation. Everybody knows that the life-cycle of innovative products has come down to months. Well, if a technology like fieldbus belongs to the range of innovative products, it is not a surprise that traditional structures and procedures, being additionally subject to *public* consensus, have lost attractiveness, compared with consortia and fora. In addition, it can be stated that the higher the complexity of such innovative solutions is, the earlier they contain intellectual property rights of the originators. These, again, are a real obstacle to standardization.

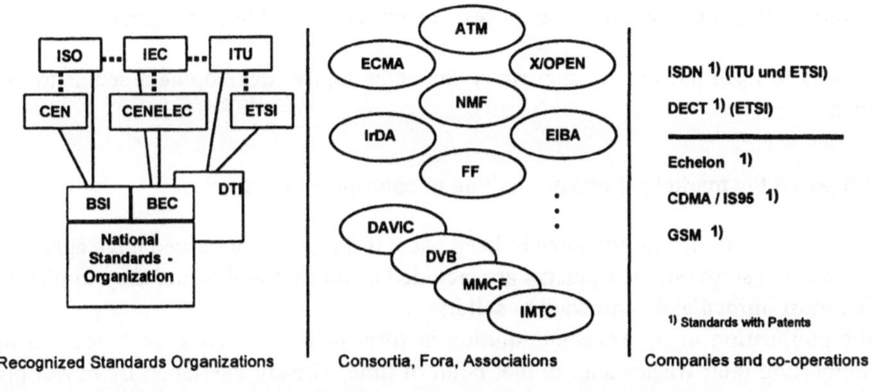

Fig. 2. Lost Attractiveness

Further observations can be made: All developments have their *specific time constraints*; this applies particularly to standardization items which are basically driven by technology, like fieldbus. In detail I would like to state:

1. Fieldbus-Standardization is now going on for over 10 years, at IEC

2. to publish a Standard takes still 3 years *after* having finished technical work at Technical Committee level

3. micro-electronics are penetrating applications, alowing the creation of completely new concepts

4. communication technologies are offering new possibilities

5. design and development use these possibilities and this is nothing else but market-driven

6. new concepts emerge, using de-centralized intelligence

7. Standards Organizations slowly open their processes, e.g. for PASs[1]

8. Standards Organizations hesitatively invite other contributers

9. the *real* specification and test work remains in consortia and user organizations

10. the market needs primarily efficient, not universal solutions.

After these observations I would like to come back to the earlier mentioned aim and purpose of standardization and draw the conclusion that the traditional approach, as followed by IEC, has not led to the success the market would like to expect.

Especially to keep pace with progress in the technological development seems to be a problem:

* supposed the market players are willing to contribute effectively,

* the procedures are still too slow to keep pace: if the parties involved have agreed on a technical solution, their patents are included as far as possible and they implement it almost immediately and want to sell it;
  the publication of the relevant solution in form of a standard after 3 years is not supporting their wishes and, in that point of time, already overtaken by newer products.

3 The European Approach and Experiences

Therefore, CENELEC tried to find other ways which fulfil these market needs in a better way. When the parties concerned met and discussed the possible evolution in the next few years, the concept of a multi-solution standard was found, in combination with the development of a Common Application Interface between the various buses and the peripheral equipment.

It was a great event to see how the interested parties met in CENELEC/TC 65CX and agreed on this concept. The selection criteria have been specified and agreed on and, on this basis, later the decision was taken to put PROFIBUS, FIP and P-NET together in the EN 50170. At the same time TC 65CX agreed on its scope that the ultimate aim is to take the IEC Standard 61158 when it is mature, has proven its applicability, covers the needed functionality and provides efficiency in use.

---

[1] Publicly Available Specifications

A multi-solution standard: EN 50170

- defining selection criteria

- qualify candidates

- publish the qualified ones in one standard

in combination with

the development of **one Application Interface,**

common to the qualified solutions:

- providing adequate independence between bus systems and peripheral equipment

Fig. 3. The European Approach

In CENELEC it was also agreed to get the NOAH Project started, aiming at a „Common Application Interface". This is to be developed and will be brought in line with the application layer of the IEC 61158.

CENELEC as a very innovative organization agreed to offer the interested parties a new scheme of co-operation: the so-called "Standards Projects with Reference Implementation". NOAH is the first project where this approach is practised. A characterization can be the following:

1. Input, i.e. technical specifications, can come from any competent source

2. the project participants have to qualify, i.e. to fulfil some conditions, but they are nominated by the Standards Organizations.

3. the qualified participants work together in a consortium-like project, put their specifications together, test them and

4. come up with tested results for TC consolidation: this is the point of time for normal voting procedures of CENELEC to apply again.

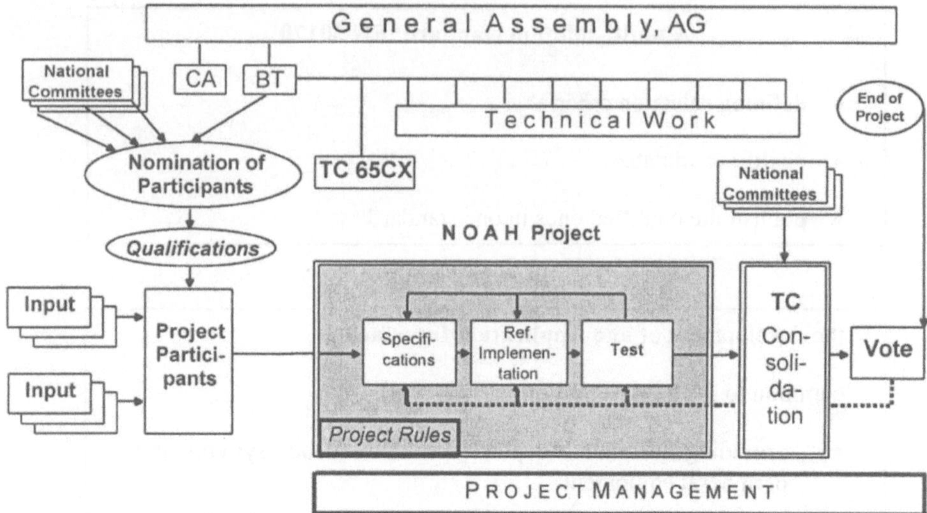

Fig. 4. The NOAH Project

The qualification criteria, as agreed by CENELEC BT (Technical Board) were the following: A candidate shall

1. gather relevant experience in field communication systems

2. be nominated by the National Committees

3. give a commitment in writing to actively support the projects' goals, underlined by a minimum capacity of 6 men-months per year

4. indicate to which work item he is willing to contribute, describing their relevant experience and a planning proposal how best to perform this specific work item

5. describe briefly the initial contribution he can offer

6. express his agreement to refrain from further participation if the necessary constructive support is not visible to the other participants any more.

Particularly the 6th condition will ensure a real productive work of all participants. What are the experiences so far?

For the EN 50170 this can be stated as a good success:

- the approval was given by CENELEC Member Bodies (BT)

- the EN gives certainty to users and manufacturers and the market appreciates this for the foreseeable future

- concerns arose from non-European partners, expressed particularly at IEC level: the EN 50170 would not be a real standardization. But one should not forget that also in IEC some other multi-solution standards exist and nobody opposes!

- an endeavour of other candidates to join is welcomed by CENELEC, e.g. the Fieldbus Foundation: their approach, through the BSI channel, to be accepted as 4th part in EN 50170 is at present under negotiation in TC 65CX and this action may be completed early 1998.

In summary, the EN 50170 is a good solution.

The ongoing work at CENELEC/TC 65 CX shows also into the right direction, again taking due account of actual and future market needs. Next to the "General Purpose Field Communication System" of EN 50170 work has been started on "High Efficiency Communication Systems for Small Data Packages" and will lead to an EN 50254, comprising those solutions which qualified, i.e. which fulfil the qualification criteria as specified in consensus by TC 65 CX .

Concerning the NOAH Project it was amazing to see that the conditions to participate were all accepted during the inaugural meeting in February 1996. It was not a surprise for anybody that non-European interests wanted to join this project but this was not possible due to statutory reasons: CENELEC can't accept project participants from non-European countries without provoking IEC. But the offer was officially made that those interested parties from consortia outside CENELEC could bring their contributions into the NOAH Project through one of their European members being at the same time a NOAH Participant.

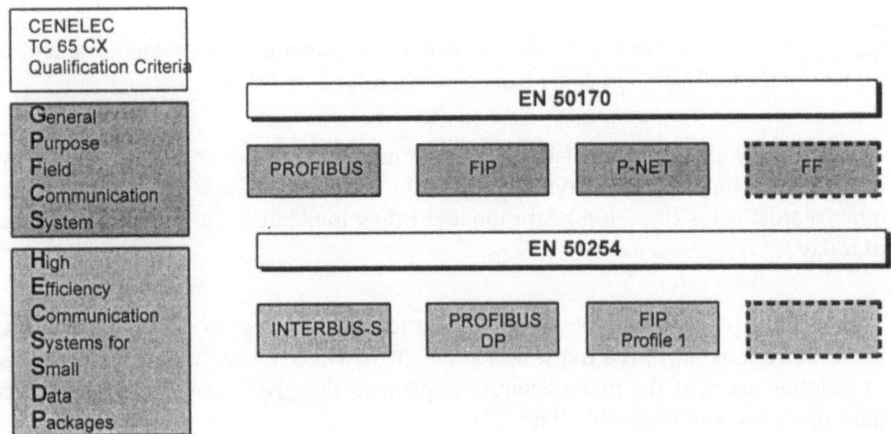

Fig. 5. The Market Solution in Progress

The first rough time schedule for the NOAH work summarized at 2 to 3 years. The work is actually in progress.

If this CENELEC approach is now compared with the earlier mentioned aims and purpose of standardization, reference is made again to the same picture, adding the columns for IEC and CENELEC:

Aim and Purpose	at IEC	at CENELEC
- to serve the market interests	?	YES
- to bring economic benefits to all parties involved	?	YES
- to keep pace with progress in technologies	NO	YES
- to bring the different interests, in time, to consensus	?	YES
- to give certainty to investors (users and manufacturers)	?	YES

Fig. 6. Aims and Purpose of Standardization, Judgement

You may be sure that I am well aware that such a comparison is not liked by several readers. But economic realities and visible developments should not be neglected, even if they deem, at first glance, not to be comfortable.

4 Perspectives for the next 3 to 7 Years

It is always difficult to predict the future! But a few statements are rather sure and it would be my great pleasure to come back on them after 3 to 7 years:

1. Market reality shows that existing and emerging solutions are accepted, when they contribute technically and economically to the efficiency of use in applications; in other words: users (investors) earn money if they implement the fieldbus solutions of today.

2. The technical progress will continue, particularly related to micro-electronics, communications, infra-red transmission etc. This will, rapidly, bring more benefits to fieldbus users, if the manufacturers implement the given new possibilities into their products (and they will all do so!).

   Whether these new possibilities are standardized or not, is a very second question. It is justified to ask what the added value of genuine standardization (defining one single fieldbus standard for the most of applications) would be, or whether all such efforts are rather a destruction of values.

3. The IEC work on the standard 61158 seems to be in good progress, but the question is, see 2 above, whether this result will not be immediately overtaken by technical progress which is gaining increasing speed. At least there is this great risk!

4. The EN 50170 is, and will remain, in practical use because it reflects the actual market situation, together with EN 50524.

The Common Application Interface which is under development in the NOAH Project will surely enjoy greater market acceptance because it brings economic advantages particularly to the users of fieldbus applications.

Ein einheitliches Programmierinterface für Feldbusapplikationen

J. Quade

Softing GmbH, Richard-Reitzner-Allee 6
D-85440 Haar
quade@softing.com

Abstract. Im Rahmen des europäischen Verbundprojekts RACKS wurde ein einheitliches Programmierinterface für die Feldbusse der Europäischen Norm und des Foundation Fieldbus entwickelt. Während ein Low-Level Interface die Portierung bestehender Applikationen erleichtert, hat der Programmierer bei Verwendung des Comfort-Interfaces einen sehr einfachen Zugriff auf das Kommunikationssystem mit seinen Objekten, ohne sich spezielle Kommunikationskenntnisse aneignen zu müssen.

Abstract. A Uniform Fieldbus Programming Interface (UFI) for the fieldbus systems of the European standard EN 50 170 and the Foundation Fieldbus has been specified and implemented in the joint European project RACKS (Reusable Application interface for Communicating real time KernelS). The specification is based on a virtual fieldbus model, to which the real fieldbus has to be mapped. The functionality of the virtual RACKS fieldbus can be accessed by both a low-level and a high-level interface. The low-level interface simplifies the portation of existing fieldbus applications, whereas the high-level comfort interface enables applications easy access to the communication system, without being burdened with the communication itself.

1 Einführung

Das durch die europäische Union geförderte Projekt RACKS (Reusable Application interface for Communicating real-time KernelS) hat ein einheitliches Programmierinterface (Uniform Fieldbus Interface, UFI) für die drei Feldbusse der Europäischen Norm EN50170 PROFIBUS, P-Net, WorldFIP und dem Foundation Fieldbus spezifiziert und implementiert. Das Interface ermöglicht es, eine Applikation zu erstellen, die unabhängig vom verwendeten Feldbussystem ist.

Der Programmierer kann dabei über zwei unterschiedliche Zugriffsmethoden die Funktionalität eines "virtuellen" Feldbusses verwenden. Reale Feldbusse lassen sich direkt auf das Modell dieses virtuellen Feldbusses mappen, ohne daß dabei Modifikationen am Protokoll selbst vorgenommen werden müssen, damit die Interoperabilität mit bestehenden Devices gewährleistet ist.

RACKS bietet zum Zugriff auf die Funktionalität eine Low-Level (harmonized) und eine High-Level (comfort) Schnittstelle an. Während sich die Low-Level Schnittstelle stark an die bestehenden Programmierinterfaces anlehnt, bietet die High-Level Schnittstelle einen für die Feldbuswelt neuartigen objektorientierten Zugang, der den Programmierer beinahe vollständig von den Kommunikationsaspekten entlastet und dabei dennoch effizient ist.

Im folgenden soll zunächst das virtuelle Feldbusmodell vorgestellt werden, das dem Uniform Fieldbus Interface zugrunde liegt. Die beiden Varianten des Interface sind Gegenstand des darauffolgenden Abschnitts.

2 Das virtuelle Feldbusmodell

Der virtuelle RACKS-Feldbus stellt einer Applikation Objekte und Dienste auf diese Objekte zur Verfügung, die sich in die sogenannten Basis- und die Zusatzobjekte beziehungsweise Dienste unterteilen lassen (core and additional objects and services). Die Verwendung der Basisfunktionalität garantiert die Portabilität zwischen den unterschiedlichen Feldbussen, die Möglichkeit, zusätzliche Funktionalität verwenden zu können, stellt sicher, daß eine Applikation sämtliche Möglichkeiten eines spezifischen Feldbussystems ausnutzen kann.

Die Objekte und die Dienste weisen eine Verwandtschaft zu MMS auf. Als Objekte wurden das Virtual Device-Objekt, das Variablen-Objekt, Domain-Objekt, Program Invocation-Objekt, das Variablenliste-Objekt und das Event-Objekt mit entsprechenden Diensten festgelegt. Für den Zugriff auf gepufferte Objekte, wie sie beispielsweise bei PROFIBUS-DP oder MPS vorkommen, gibt es spezielle Lese- und Schreibdienste. Darüber hinaus ist das Objektmodell noch um Systemobjekte (z.B. Rootobjekt, Netzwerkobjekt, Segmentobjekt oder Stationsobjekt) erweitert, die die gesamte Netzstruktur für die Applikation repräsentieren.

Zusätzlich zu den Objekten und den Kommunikationsdiensten sind auch die Dienstsequenzen spezifiziert worden. Bevor eine Applikation auf ein Objekt zugreifen kann, muß sie eine Verbindung zum Virtual-Device-Objekt öffnen, das für das Objekt zuständig ist (im virtuellen RACKS Feldbus gibt es nur verbindungsorientierte Kanäle, auch wenn es sich real um einen verbindungslosen Kanal handelt). Der Versuch, eine offene Verbindung erneut zu öffnen, führt nicht dazu, daß die Verbindung wieder abgebaut wird.

3 Die Applikationsschnittstelle

Eine Anwendung hat prinzipiell zwei Möglichkeiten, auf die Funktionalität des virtuellen RACKS-Feldbus zuzugreifen. Am Low-Level Interface, das durch den sogenannten H-Layer implementiert wird, sind die diversen Kommunikationsdienste noch sichtbar. Dagegen ist am High-Level Interface, das durch den sogenannten C-Layer implementiert wird, die Kommunikation versteckt. Die Applikation kann die (Kommunikations-) Objekte direkt lesen und schreiben.

Prinzipiell eröffnet das Uniform Fieldbus Interface die Möglichkeit, innerhalb einer Applikation auf mehrere Kommunikationssysteme (Netzwerke) des gleichen oder auch unterschiedlichen Typs zuzugreifen, wobei der Anlauf (z.B. Laden von Firmware auf den Kommunikationscontroller, Einstellen von Busparametern, Übergeben der Verbindungsparameter an den Stack) für ein Netzwerk durch die Interface-implementierung durchgeführt wird.

Die von den Kommunikationssystemen her bekannten Dienstkennungen *Indication* und *Confirmation* sind sowohl am H-Layer als auch am C-Layer Interface auf bekannte Zugriffsarten abgebildet. Für die Bearbeitung von *Indications* beispielsweise kann die Applikation Callback-Funktionen zur Verfügung stellen, die von der Schnittstelle aufgerufen werden. Über die Beendigung eines Kommunikationsdienstes (*Confirmation*) kann die Applikation ebenfalls per *Callback* informiert werden, sie kann aber ebenso auf die Beendigung warten (blockierender Zugriff) oder - anstatt zu warten - selbst überprüfen, ob der Kommunikationsdienst abgeschlossen wurde oder nicht (nicht blockierender Zugriff). In jedem Fall wird sichergestellt, daß ein von der Applikation gestellter Kommunikationswunsch auch in einer endlichen Zeit - positiv oder negativ - beendet wird (timeout).

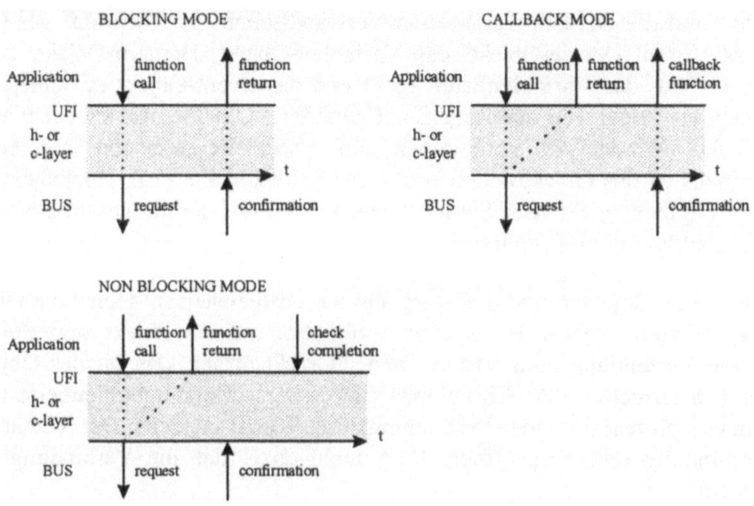

Fig. 1. Zugriffsarten

Ein wesentliches Merkmal sowohl der H-Layer als auch der C-Layer Schnittstelle ist, daß diese Schnittstellen Betriebs- oder Runtimesystemunabhängig sind.

Das virtuelle Feldbusmodell spezifiziert weder die Konfiguration noch Konfigurationsobjekte. Dafür aber stellt das Uniform Fieldbus Interface eine Funktion zur Verfügung, um die busspezifische Konfiguration zu laden und damit die jeweilige Kommunikationshardware, Kommunikationssoftware und auch das Kommunikationssystem selbst zu konfigurieren.

PROPERTIES	H-LAYER	C-LAYER
object oriented	limited	yes
multi network support	yes	yes
blocking access mode	yes	yes
callback access mode	yes	yes
non blocking access mode	yes	yes
platform independent	yes	yes
operating system independent	yes	yes
symbolic addressing scheme	no	yes
data conversion	no	yes
interoperability	yes	yes
client and server	yes	yes
communication oriented	yes	no
application oriented	no	yes
access to network objects	no	yes
same object access for client and server	limited	yes

Fig. 2. Interface-Eigenschaften

3.1 Der Harmonization-Layer

Das H-Layer Interface ähnelt existierenden Interfaces, so daß eine bestehende Applikation ohne strukturelle Änderung portiert werden kann.

Es bietet Funktionen an, mit deren Hilfe Kommunikationsaufträge an den realen Stack übergeben werden. Diese Kommunikationsaufträge (Services) werden in einer Datenstruktur (exchange block) beschrieben, die neben den Daten gleichzeitig Informationen darüber enthält, wie der Auftrag abgewickelt werden soll und welchen Zustand die Auftragsabwicklung hat.

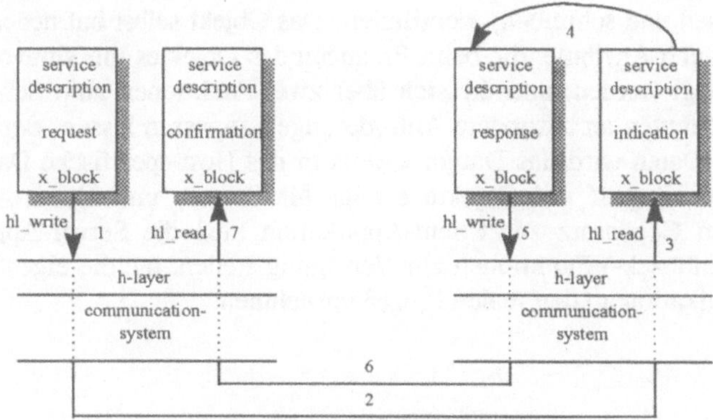

Fig. 3. Datenfluß im H-Layer

Objekte werden am H-Layer über die Verbindung und über einen Index adressiert, wobei die Verbindung durch einen lokalen und einen entfernten (remote) Endpunkt gekennzeichnet ist. Soll eine Applikation daher mit verschiedenen Feldbussen zusammenarbeiten, sind entweder diese Adreßangaben anzupassen, oder aber die Bussysteme identisch zu konfigurieren.

H-LAYER FUNCTION	FUNCTION PARAMETER	FUNCTION RETURN VALUE
hl_open	configuration	h-layer instance handle
hl_close	h-layer instance handle	
hl_read	exchange block	success/failure
hl_write	exchange block	success/failure
hl_select	array of exchange blocks, timeout	completed requests
hl_get_error_string_function	function return code	error string
hl_get_error_string_service	exchange block	service error string

Fig. 4. H-Layer-Funktionen

3.2 Der Comfort-Layer

Während die Low-Level Schnittstelle nur den Zugang zu den direkten Kommunikationsobjekten gibt, ermöglicht die High-Level Schnittstelle den Zugriff auf das gesamte Netzwerk. Das sogenannte Comfort-Interface bietet Objekte an, die selber wieder Objekte beinhalten können (ähnlich den Directories eines Dateisystems). Dabei ist das Kommunikationssystem über einen Objektbaum zugänglich. Das Root-Objekt beinhaltet die Informationen über die einzelnen Netzwerk-Objekte. Diese wiederum verweisen auf die zugehörigen Node-Objekte, die für die eigentlichen Kommunikationsobjekte verantwortlich sind.

Ähnlich wie der Zugriff auf Dateien vollzieht sich auch der Zugriff auf die über das Feldbussystem zur Verfügung gestellten Kommunikationsobjekte. Ein Kommunikationsobjekt wird über einen symbolischen Namen ausgewählt und erzeugt. Dabei bekommt der Programmierer ein Handle zurück, das das Objekt für die weiteren Zugriffe (z.B. lesen und schreiben) identifiziert. Das Objekt selbst hat neben den eigentlichen Daten auch Attribute, die beim Erzeugen des Objektes mit sinnvollen Default-Werten vorbelegt werden, und die sich über zwei Funktionen zum Setzen und zum Lesen von Attributen an besondere Anforderungen anpassen lassen. Beim Lesen und Schreiben der Daten wird das Datum jeweils in das Host-spezifische Datentypformat gewandelt. Der Zugriff auf Objekte erfolgt für Server- und Client-Applikationen gleichartig. Im Gegensatz zur Client-Applikation muß die Server-Applikation nur zusätzliche (Callback-) Funktionen zur Verfügung stellen, die die eigentliche Anbindung der Applikation an den realen Prozeß vornehmen.

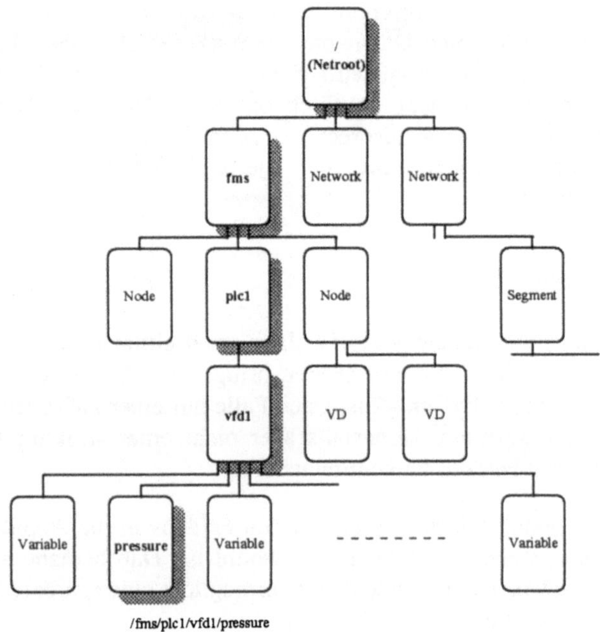

Fig. 5. Objektbaum

Die Verwendung von objektorientierter Technologie ermöglicht eine effiziente Implementierung der Schnittstelle. Da die Objekte, bevor sie in der Applikation verwendet werden können, einmalig angelegt werden müssen, können zeitaufwendige Aktionen (z.B. die Zuordnung von symbolischen Namen zur realen Busadresse) zu diesem Zeitpunkt stattfinden. Durch die Möglichkeit, im Interface Zustandsinformationen vorzuhalten, kann der eigentliche Zugriff auf das Objekt ohne weitere Berechnungen direkt erfolgen.

C_LAYER FUNCTIONS	FUNCTION PARAMETER	RETURN VALUE
cl_init_c_layer	configuration	object handle to /
cl_exit_c_layer		
cl_create	object name	object handle
cl_delete	object handle	
cl_read	object handle, memory buffer, length of buffer	bytes read
cl_write	object handle, memory buffer, length of buffer	bytes written
cl_set_attribute	object handle, memory, attribute name	success/failure
cl_get_attribute	object handle, memory, attribute name	success/failure
cl_get_object	object name	object handle
cl_get_error_string	object handle	error string

Fig. 6. C-Layer Funktionen

Anders als beim H-Layer sind kommunikationsspezifische Funktionen der Applikation verborgen. Eine notwendige Datenkonvertierung zwischen dem hostspezifischen und dem busspezifischen Datentyp wird ebenso transparent durchgeführt, wie das gesamte Verbindungsmanagement. Dadurch beschränkt sich der Aufwand für eine Applikation, verteilte Objekte anzusprechen - nach der Initialisierung - nur darauf, das Objekt lokal anzumelden und eventuell - je nach Zugriffsmode - eine Callback-Funktion bereitzustellen.

4 Zusammenfassung

Die Existenz eines einheitlichen und vor allem auch einfach zu verwendenden Zugangs zu unterschiedlichen Feldbussystemen bringt dem Anwender erhebliche Vorteile. Seine Applikation bedarf im günstigsten Falle nur einer auf einen anderen Feldbus angepaßten Konfiguration, keinesfalls aber mehr einer strukturellen Änderung, um über einen anderen Feldbus zu kommunizieren.

Insbesondere der Comfort-Zugang integriert den Feldbus in die Applikation, wie der Programmierer es von anderer Peripherie gewohnt ist. Daß beinahe sämtliche kommunikationsspezifischen Teile versteckt sind, ermöglicht eine spürbare Produktivitätssteigerung bei der Erstellung der Applikation.

Neben der Spezifikation existieren erste Implementierungen des Interfaces für PROFIBUS FMS, für WorldFIP subMMS, P-Net und Fieldbus Foundation.

Literatur

1. General Purpose Field Communication Systems. CEN CENELEC standard EN50170, 1996

2. Szymanski, J., Quade, J., Johansen, J.: RACKS - European Initiative for Fieldbus Interface Harmonisation. Tagungsband European Conference on Integration in Manufacturing, Dresden, 1997.

3. Patz, M.: Schluß mit dem Feldbus-Wirrwar? Produktion Nr. 28, Seite 3, 1996.

Pre-Requisites for End User Exploitation of Fieldbus in the Process Industries

Iain G Brownlie

Eutech Engineering Solutions Ltd
PO Box 99, Belasis Hall Technology Park,
Billingham, Cleveland, TS23 4YS
e-mail iain.brownlie@gbeut02.eutech.com

Abstract. Fieldbus technology is well established in the Manufacturing Industries but has yet to make a significant impact in the Process Industries. This paper reviews the requirements for distributed automation in the process sector, and identifies the key benefits to be realised from deploying fieldbus. A systematic approach which enables distributed applications to be easily constructed from different manufacture's components is required. The critical importance of fieldbus standards and the key role of the "User Layer" is discussed. Technology is slowly becoming available which enables most of the benefits to be realised, although a single international standard remains elusive. As End Users begin to consider deploying fieldbus, there are several practical issues surrounding the usability of the technology which need to be considered.

1 Requirements for Distributed Automation

1.1 Business Drivers

The Process Industries face a constantly changing business environment with strong shareholder pressures on performance. The industry is undergoing considerable reorganisation to establish companies which can compete effectively on a global basis. Many of the core process technologies are reaching maturity at a time when new competitors are emerging from Asia Pacific. The commoditisation of many products requires new capabilities to be developed to become more customer focused, and responsive to changes in market demands.

As a consequence of these pressures there is a strong focus towards maximising the return on investment, and this is resulting in many end users looking not just at initial capital costs, but also the implications on longer term operation and maintenance costs. Asset Management throughout the plant lifecycle provides an opportunity for generating real competitive advantage.

In addition to the normal business pressures, the risks associated with processing hazardous materials under difficult reaction conditions require the Process Industries to operate to the very highest standards in safety, health and environmental performance.

1.2 Importance of Process Control

Process Control systems are responsible for ensuring that plants continue to produce consistent product conforming to specification. Discrete logic is required to control motors, pumps, and on-off valves, whereas temperatures, pressures, levels and flows require analogue controls. Although there is no absolute separation between these types of control, Programmable Logic Controllers (PLC) are often used where there is a high proportion of discrete control, whereas Distributed Control Systems (DCS) are widely used for the analogue controls needed by the Process Industries.

Distributed control has been important since the earliest implementation of feedback control strategies using mechanical instrumentation. Each control loop was physically independent resulting in a high degree of fault tolerance. The principle of single loop integrity was kept when the control moved into electronic single loop controllers. Computer control systems based on mini-computers centralised control. The evolution of the current generation of DCS Systems based on controller modules with local I/O capability has begun to move control back towards the field. The partitioning of the application across the controllers represents a critical aspect of system design, and can significantly affect the overall system integrity, reliability and performance.

Many large scale process plants were designed to operate continuously for long periods. Such plants were often started and shut down manually, and controls were switched to automatic only after the plant had stabilised at the normal operating conditions. Business drivers are forcing end users to adopt much more sophisticated and flexible strategies for control, particularly for multi-product batch plants which require different conditions and flexible reconfiguration strategies.

The Function Block paradigm is widely supported by the vendors of DCS and PLC systems and is capable of solving a large number of common control problems, using a relatively small set of basic algorithms. Many processes exhibit non linear dynamics or multi-variable interactions which require more sophisticated techniques such as model based predictive control. At the business level, product prices and raw material costs can significantly affect the optimal control strategy. For many plants in the Process Industry profitable operation relies on successful integration of a hierarchy of systems for planning, optimisation, advanced control which operate above the basic controls.

Control systems have an important role in ensuring plant safety. The first level of integrity is to maintain the plant at its designed operating point. Failure of equipment or operator error can lead to a potentially hazardous incidents, so alarms are required to identify where operator intervention may be needed. The control system may itself take some level of protective action in response to an alarm. The most serious hazards which could result in an acceptable danger to life or the environment require an automatic shutdown protection system which will trip part or all of the plant. It is a general requirement to ensure clear separation of trip and control functions. IEC 1508 [1] is being introduced to provide a framework for the functional safety of electrical, electronic and programmable safety related systems and will have a significant impact on the design of future control applications.

2 Vision for Fieldbus

2.1 Which Fieldbus ?

The IEC efforts to create a single fieldbus standard have taken much of the last decade and has proved to be a particularly difficult task. At the same time a number of products have come to market which manifestly deliver real benefits to their users. Within Europe, the EN50170 [2] standard offers end users a choice of three established technologies, or a three headed monster, depending on your viewpoint.

Anyone who still believes that a universal fieldbus architecture will solve all user needs is probably going to be disappointed. Connecting low cost devices in building automation, controlling equipment in explosive atmospheres, and synchronising high speed drive systems all have significantly different requirements. There are good reasons why specific technologies offer advantages in certain classes of application.

Even within the Process Industries area, there are requirements for sophisticated protocols to exchange information with complex intelligent instrumentation, and different requirements to communicate with comparatively simple discrete devices where a low cost simpler technology could offer advantage.

There are probably three or four current fieldbus technologies which have a suitable architecture for application in the Process Industry. The key problem which needs to be solved is not the details of the communication protocol, but the higher level "User Layer" functions which are required to build interoperable applications.

2.2 Direct Cost Savings

A key advantage of all fieldbus architectures is the ability to connect several instruments to a single fieldbus cable. Current design technologies require large heavy multi-core cables to be run to field junction boxes where single cores are then run to each instrument. In the centre, each core must be patched to the correct DCS I/O card terminal. The cost of engineering design, commissioning and loop testing is enormous.

With fieldbus, a single segment carries the information from several instruments back to the centre. The order of connection is not significant resulting in much simpler installation, particularly at the control room. The principle of ensuring single loop integrity on a single bus segment will probably mean that many end users will only actually connect around 4-6 instruments on a fieldbus used for control, as against a technology constraint of 32 devices. This limits the savings potential, but we still expect 25% of the instrumentation and control system costs can be saved in this way.

Direct cost savings are particular important where there are many simple devices where bus transactions are limited to read/write of current values and status. Many of the existing fieldbus technologies do little more than this, but can nevertheless generate significant savings for their owners. This has proved particularly effective for integrating remote I/O into PLC systems in the Manufacturing sector.

2.3 Smart Instrumentation

The real value of fieldbus for many end users is not the direct cost savings but the increased plant efficiency and maintenance benefits which will arise through using the additional information available from smart instrumentation, and from the integration of this information into the higher level decision support systems.

- secondary measurements (e.g. flow + temperature) reduce the number of instruments required and the number of process connections

- ability to run control function blocks within the instrumentation will have an important benefit in reducing the complexity of the centralised DCS system

- remote programmable configuration facility and on-line instrument datasheets will simplify define calibration, commissioning and ongoing support

- maintenance diagnostics will detect many instrument and process faults, resulting in better availability, repeatability and accuracy of the measurement data

Many users already use smart HART devices with 4-20mA connection to the control system. This has demonstrated important improvements in reliability and performance using intelligent devices. A key problem is the hand held configurator required to interrogate devices. Digital integration through fieldbus offers a solution to this issue.

Many of these benefits will be application specific, and the savings can be difficult to quantify in advance. Users are reluctant to pay a premium price for smart devices.

2.4 Safety and Reliability

Fieldbus and smart instrumentation offer a number of distinct advantages to improve the safety and reliability of process plant.

At the device level, better diagnostics will lead to an improvement in the reliability and availability of the primary measurements. The use of digital communication methods will improve accuracy and repeatability through elimination of the analogue to digital conversion process.

Users may have concerns about electrical or magnetic interference. In fact, the basic communications are very robust, and any data corruption will be detected before it affects the control action, so integrity will be better than conventional wiring.

The introduction of standard function block definitions will greatly assist the control engineer who will no longer need to learn the idiosyncrasies of many different proprietary algorithms. This should result in fewer configuration errors reaching the field, particularly those leading to unexpected behaviour under fault conditions.

The migration of control algorithms to the field will ensure that basic control is not lost with a failure of the central control system. This will eliminate the need to provide much of the expensive redundancy in the higher level systems. The comparative reliability and simplicity of instrumentation means that overall security will be maintained at significantly lower cost.

The system design must consider the impact of power supply failure when the fieldbus is used to power the devices. End users should design for "single loop integrity" by ensuring that only one control loop is implemented on each bus segment, and configure the actuators to move to appropriate positions on loss of power or communication.

There are established principles and supporting legislation covering the installation of electrical equipment where there is potential for explosive atmospheres and/or dust hazards. Local legislative requirements may differ from country to country.

Intrinsically safe (Ex'i') devices store insufficient energy to cause a means of ignition. All devices including power supplies, barriers, and instrumentation must be certified according to the gas type and temperature classification of the process fluid. Cable type and lengths are important as they affect the inductance and capacitance of the system. Intrinsically safe installations are mandatory where there is a continuous presence of explosive materials. Fieldbus allows several devices to be installed in both the safe and hazardous areas. Detailed calculations and appropriate documentation are required to prove that the electrical characteristics of the whole system are within the specification for I.S. A significant advantage is that maintenance work can be carried out on live instruments, although any modifications requires detailed assessment/certification.

Many users prefer alternative design strategies in plant areas which have significantly reduced risks. Other methods offer protection using construction materials which ensure that any electrical fault is contained within the instrument (e.g. flameproof enclosures (Ex'd'), powder filling (Ex'p')). Such installations still require certified devices but have fewer design constraints allowing bus power, cable lengths and the number of instruments per segment to be increased. With these installations, it is not possible to open live equipment for maintenance. Removing power from the whole bus may be unacceptable as it would affect all the other instrumentation on the segment. There is a need for bus coupling technology which will facilitate safe disconnection of instruments at the field junction box, possibly within the hazardous area.

One of the fundamental reasons why fieldbus has not had the same impact in the process sector as it has had in manufacturing is that there are very few suitably certified devices available from the vendors. This situation is now beginning to change.

While many control systems are safety related, instrumentation is also used in safety critical Protection Systems which safeguard human life and the environment. These systems must have the highest levels of integrity. The role which fieldbus devices could play in such systems is not clear. The increased reliability of instrumentation is beneficial, but failure modes will need to be considered against the Safety Integrity Levels as defined in IEC1508. Considerable practical experience on operating performance is needed before this topic can be addressed properly. Users are advised to design conventional independent plant protection systems for the foreseeable future.

3 Importance of Standards

3.1 Application Areas and Fieldbus Architectures

Many end users in the Process Industries have been slow to adopt fieldbus, but this is certainly not through a lack of choice. There are perhaps as many as 40 technologies available. The EN50170 standard supports three leading technologies and there are plans to extend this. The proposed prEN50254 [3] standard promises yet more options for "High Efficiency Communication Subsystems for Small Data Packages".

This confusion has caused many end users to wait and see what happens before making a decision to deploy fieldbus. Meanwhile, many of the instrumentation suppliers are waiting to see which technologies are successful before committing to the development costs. The prospect of a single international fieldbus standard seems as elusive as ever.

There are reasons why particular technologies offer distinct advantages for specific applications. The fieldbus market is segmented according to the fundamental application requirements. Many end users may need more than one type of fieldbus. The vendors are responding by developing bridges between different architectures.

3.2 What do we Mean by Integration ?

It is true that many technologies available today have the potential to generate benefit for the end user, particularly savings on plant wiring and other installation costs.

Higher added value comes not from the fieldbus technology, the smart instruments, or the new maintenance information, but from the new opportunities to reduce costs over the whole plant lifecycle, covering Design, Construction, Operation and Maintenance.

These enhanced benefits require a framework for integration of information at all levels in the business. The ability to easily construct systems from components is an essential tool to enable the agile enterprise to succeed in the future. Different fieldbus technologies can be classified according to the following levels of integration [4].

Incompatibility	Devices fail to work together due to differences in data, functions, communication mapping or protocol
Compatibility	Devices can operate independently on the same network, or work together using unique purpose built applications
Interconnectability	Devices can operate using the same protocols and user layer parameter access methods, although different parameters may be supported in similar devices from different vendors.
Interworkability	Two devices will support transfer of device parameters between devices having the same parameter definitions and communication mappings.
Interoperability	Two devices can support distributed applications using the same parameters and application function concepts (e.g. function blocks, modes, scheduling, parameter status, events) but may have different dynamic responses.
Interchangeability	Two devices support the same parameters, functions and dynamic response so as to allow replacement without needing re-design or re-configuration.

3.3 Implications for the Fieldbus Standards

Any level of integration beyond incompatibility requires devices to be designed according to a set of publicly available specifications, ideally in the form of an international standard. Most fieldbus architectures are defined in terms of a reduced ISO 7 layer communication model illustrated below.

1	Physical	Physical media, connectors, and signalling conventions
2	Data Link	Secure communication of messages between devices
7	Application	Application interface (e.g. Read / Write data)

This level of definition is generally sufficient to allow Interconnectable devices, and this is achieved (at least) by most fieldbus architectures.

The scope of the fieldbus specification must be extended to include the "User Layer" functionality to achieve the higher levels of integration. Device Profiles, Function Blocks and Scheduling are three important concepts to be defined within the User Layer to allow the degree of Interoperability required by the Process Industries.

Device Profiles should define the parameters of a specific class of device covering the information relating to measurement technology, linearisation, calibration and other operational data (e.g. "SENSOR_VALUE" might hold the current measurement value). Some form of device profile is a requirement of Interworkability.

The Function Block reference model defines properties of algorithms, modes, alarms, parameters, and interfaces required to implement distributed applications. Analogue Input (AI) and Analogue Output (AO) blocks can then be defined to support measurement and actuator functions. A richer set of algorithms (e.g. PID Controller) will enable users to migrate control from the central systems into field devices.

The third requirement of the User Layer is to provide a scheduling model. This must address the scheduling of algorithms within a device, and the scheduling of communications between devices. The degree of determinism inherent in the scheduling process can be affected by the underlying communication architecture. Many applications in the Process Industry are not too demanding, but care would be required in the design of some time critical loops (e.g. compressor anti-surge).

3.4 Maturity of the International Standards

It is unfortunately the case that the existing international standards in the fieldbus arena do not address the critically important User Layer. In practice, only the IEC1158-2 [5] Physical Layer standard has been generally accepted. Individual technologies may provide solutions to some of the User Layer requirements but these are not formally part of the standard, and the user is advised to tread carefully.

WorldFIP Companion Standards [6] provide device profile data, as do the Profibus-PA Profiles [7]. The latter also define a Function Block model although only basic AI/AO blocks are defined. Fieldbus Foundation [8] have the most sophisticated User Layer which most closely matches the aspirations of many in the Process Industries.

Interworking or Interoperability are achievable in principle with several technologies, although the level of conformance for the current generation of actual devices needs to be carefully checked. Interchangeability is a gold standard which may be impossible to achieve in practice.

There is important work being done to harmonise the User Layer, notably in IEC SC65C/WG7 [4]. Within Europe, the RACKS [9] and NOAH [10] initiatives are trying to establish a common user layer interface across the three architectures within EN50170, and there are developments within the domain of specific technologies.

The last few years have seen considerable progress. Many vendors are announcing products for delivery in 1998, which meet the definition for Interoperability, although the set of supported function blocks may be somewhat limited.

4 Usability

When the End User first commits to using fieldbus there will be a number of significant issues which need to be addressed. Some of these should be taken into account when selecting the most appropriate fieldbus technology.

4.1 Training

There are many new concepts to be learned and training courses will be required ranging from an overview of the technology, design principles for distributed applications, device configuration and troubleshooting. There are some formal courses available but they tend to be technology or device specific (e.g. Sira [11] offer a multi-vendor training facility for Fieldbus Foundation product).

Many fieldbus groups and individual vendors offer documentation explaining the key aspects of their technology, case studies, and general advice.

4.2 Design Philosophy

There will be important implications for the design process. There will be an increasing need to consider the implications of smart instrumentation and highly distributed control at an earlier stage in the plant design. The disciplines of process, control and instrument engineering will need to be much more integrated with fieldbus. The information required to be captured will change and this will affect the symbology used on engineering drawings and datasheet systems. New systems will be required to capture network topology and function allocation schemes. Instrument datasheet systems will need to be extended to cope with the advanced requirements for intelligent maintenance and other operational data.

A Design Philosophy should be developed covering basic principles to be used. This should describe basic rules for cable type, topology, distances, and power supplies. The methods for connecting devices should be defined including mechanisms for isolation of individual devices for maintenance. Guidance should be given on allocation of devices to bus segments and allocation of control functions to devices. Change control procedures need to be developed which ensure the overall system integrity in a distributed environment. The effect of fieldbus failure needs to be considered during the hazard and operability studies.

While emphasis is naturally focused on the instrumentation, the detailed design of the instrument connections and field junction boxes to include proper termination, isolation switches and fuses, will greatly enhance the integrity of the fieldbus installation, by ensuring faults are not propagated through the system.

4.3 Tools

Many of the changes in the design process can be supported by Computer Aided Software Engineering Tools. The ACORN 1479 [12] project is being partially funded by the European Commission under the Brite EuRam Workprogramme to develop a toolkit for the design of highly distributed process control systems based on fieldbus.

System vendors are developing specific tools addressing device and network configuration of specific technologies. Asset Management systems to centrally manage device configuration and maintenance data are becoming increasingly available.

4.4 Availability

Device availability is likely to be an important constraint for early adopters, particularly for installation in hazardous areas. End users would be well advised to discuss the specific plans for product developments with their preferred suppliers, paying close attention to the level of conformance with the underlying specification.

4.5 Higher Level Systems

Most of the debate relating to fieldbus technology has been concentrated on the communication protocols, or at the device level. End Users need a solution for the total system problem. It is essential that fieldbus devices are integrated with higher level central control systems. These will continue to provide the user interface, sequence control for batch processes together with more advanced control and optimisation functions. Many vendors are introducing new low cost PC based platforms for control.

5 Conclusions

This paper has confirmed that fieldbus technology promises to bring many benefits to end users in the process industries. There are many quick savings that could be easily realised by implementing one of the available technologies.

Many users believe that the strategic importance of fieldbus is to allow the new information available from smart instrumentation will be delivered to those who need to know on a timely basis. This can only happen if there is a fully functional User Layer which defines the necessary device profiles and functions required to produce truly interoperable distributed applications. This requires not only well defined standards, but widespread support by the vendors in general, and the end user's preferred suppliers in particular.

Safety is of paramount importance to end users, and there are particular requirements for design and certification of devices for installation in hazardous areas. Fieldbus technology can be applied, and indeed offers some advantages in increased safety and reliability. The limited availability of suitably certified available devices has been an important reason why fieldbus has not yet been widely deployed in the process sector.

While the basic specification must meet certain fundamental needs, it is vitally important that the vendors support the standard by providing the range of instrumentation, control system interfaces and configuration tools required to implement the whole application. Most vendors will not support fieldbus interfaces to older generation DCS systems.

Most of the technological problems have been solved, with several technologies emerging in the market capable of meeting the requirements for interoperability. End users can make a rational choice based on analysis of the functionality offered by the specification, the technology supported by their preferred suppliers of instrumentation and control systems, and the availability of training and engineering support in their local area.

Once the technology question has been answered, there are important practical implications for the methods used for design, commissioning, operating and maintenance of fieldbus applications. Software engineering tools such as being developed by the ACORN project and other third party offerings will greatly improve the management of design information.

Some End Users may still wait to see which technology succeeds in the market before committing. Others may seek to get competitive advantage from rapid early deployment of the technology. Many will attempt small scale applications to build knowledge and experience.

References

1   ISO/IEC 1508, Functional Safety : Safety Related Systems.

2   CENELEC EN50170, General Purpose Fieldbus Communication Systems, EN50170-1 P-NET, EN50170-2 Profibus, EN50170-3 WorldFIP.

3   CENELEC prEN50254, High Efficiency Communication Subsystems for Small Data Packages.

4   IEC SC65C WG7 Distributed Function Block Application Requirements (draft).

5   IEC 1158-2 Fieldbus : Physical Layer ("Wire Medium").

6   WorldFIP, 2 rue de Bône, 92160 Antony, France.

7   Profibus International, Haid-und Neu-str 7, D-76131 Karlsruhe, Germany.

8   Fieldbus Foundation, 9390 research Boulevard, Suite II-250, Austin, Texas, 78759-9780, USA.

9   ESPRIT Project 20468, RACKS, Reusable Application Interface for Communicating Real Time Kernels.

10 CENELEC Project NOAH, Network Oriented Application Harmonisation based on General Purpose Field Communication Systems.

11 Sira Test & Certification, South Hill, Chislehurst, Kent, UK.

12 Brownlie et al, ACORN 1479 - Design Tools for Fieldbus, FeT '97, *ibid*

Engineering von Verteilten Automatisierungssystemen

René Simon, Peter Neumann, Christian Diedrich

Institut für Automation und Kommunikation e.V. (ifak)
Steinfeldstraße 3 (IGZ), D-39179 Barleben
Tel.: 039203-81060, Fax: 03920381100, Email: rsi@ifak.fhg.de

Kurzfassung. Neben der funktionalen Betrachtung von Verteilten Automatisierungssystemen rückt in zunehmendem Maße ihr Engineeringprozeß in den Mittelpunkt der Betrachtungen. Es ist in zunehmendem Maße erforderlich, ein durchgängiges Engineering zu erreichen. Dieses zeichnet sich dadurch aus, daß zwischen den einzelnen Phasen des Engineeringprozesses eine verlustfreie Übertragung der notwendigen Informationen gewährleistet wird. Es wird vorgeschlagen, das Verteilte Automatisierungssystem als Produkt im Sinne von ISO 10303 (STEP) zu betrachten. Damit steht STEP als Methode und mit der bereits aufgebauten Infrastruktur vollständig zur Verfügung, das Problem der durchgängigen Beschreibung von Verteilten Automatisierungssystemen wird auf das Problem der durchgängigen Beschreibung von Produktdaten zurückgeführt. Im Projekt „Advanced Control Network" (ACORN) wird dieser Ansatz einer Validierung unterzogen.

Abstract. Besides the functional view on distributed control systems, the engineering aspect becomes more and more important. It will be essential to integrate the whole engineering process. Here the transport of information without losses from one engineering phase to an other must be provided. It is proposed here, to see the distributed control system as product as defined in ISO 10303 (STEP). Then all the methods and the already built infrastructure of STEP is available. The problem of describing the distributed control system is put down to the problem of describing product data (solved in STEP). The project „Advanced Control Network" (ACORN) validates this approach.

1 Verteilte Automatisierungssysteme

Die klassischen Meß-, Steuer- und Regelgeräte beruhten meistens auf einfachen physikalischen Wirkprinzipien (mechanisch, hydraulisch, pneumatisch, elektrisch) und wurden überwiegend als Einzelgeräte für eine relativ abgeschlossene Automatisierungsaufgabe eingesetzt. Mit der Einführung der Mikroprozessortechnik und deren sprunghaften Verbreitung erfolgte eine deutliche Verlagerung des Fokusses vom Einzelgerät zum Geräteverbund. Dieser Verbund von Automatisierungsgeräten einschließlich der dazu notwendigen Kommunikationsmedien wird als Leitsystem bezeichnet.

Prozeßleitsysteme (PLS) der 1. Generation sind durch eine zentrale Struktur sowie Unterschiede in der Fertigungstechnik und in der Verfahrenstechnik gekennzeichnet [1]. Bei Prozeßleitsystemen findet zur Zeit der Übergang zur 2. Generation statt, die eine dezentrale Struktur (explizite Programmierung der Kommunikationsmechanismen - z.B. mittels Kommunikations-FB nach IEC 1131-5) besitzt. Weiterhin ist eine Verschmelzung der fertigungs- und verfahrenstechnischen Lösungen zu beobachten. Wesentliche Merkmale einer zukünftigen 3. Generation von PLS werden ein durchgängiges Engineering sowie Lösungen sein, bei denen durch den Anwender lediglich die Anwendungsverbindungen anzugeben sind, nicht aber die Kommunikation explizit zu programmieren ist (implizite Kommunikationsmechanismen). Diese PLS werden hier als Verteilte Automatisierungssysteme (VAS) bezeichnet.

Neben der funktionalen Betrachtung von Prozeßleitsystemen rückt in zunehmendem Maße ihr Engineeringprozeß in den Mittelpunkt. Dieser wird ständig komplexer, da die einzusetzenden Geräte und Werkzeuge komplexer und vielfältiger werden. Das folgende Bild verdeutlicht, welche Faktoren (aus Sicht des Automatisierungsgerätes) diese Komplexität beeinflussen:

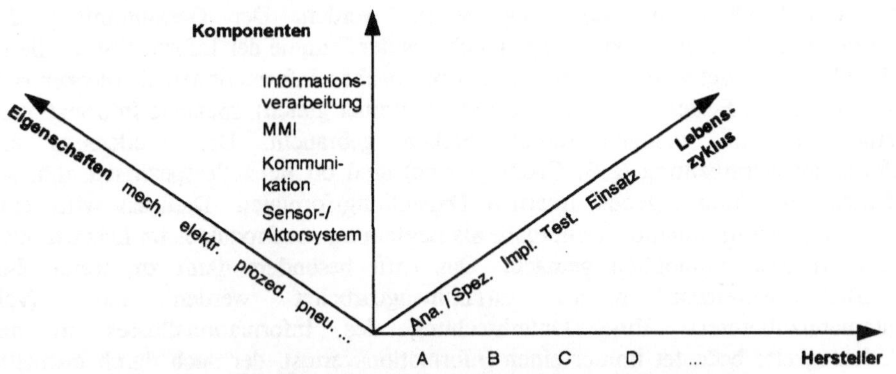

Abb. 1. Komplexität des Engineeringumfeldes aus Gerätesicht

In der Automatisierungstechnik kommt eine Vielzahl von Geräten zum Einsatz, die zur Realisierung der verschiedenen Komponenten des Leitsystems verwendet werden. Für das Engineering des verteilten Automatisierungssystems sind jedoch neben der Kenntnis der prozeduralen Eigenschaften dieser Geräte viele weitere Informationen notwendig (z.B. elektrische Anschlußwerte und mechanische Abmessungen eines Motors für ein Stellglied). Die Klassifizierung dieser Informationen wird in obiger Abbildung durch die Aufteilung einer Achse nach verschiedenen Eigenschaften angedeutet. Die aufgeführten Eigenschaften (elektrisch, mechanisch, prozedural, pneumatisch) lassen sich ergänzen (z.B. faseroptisch) und weiter unterteilen. Der Engineeringprozeß wird außerdem durch die zunehmende Verwendung von Gerätetechnik verschiedener Hersteller in einer Anlage und der funktionalen Vielfalt der Gerätetechnik (Komponenten) gekennzeichnet. Der Lebenszyklus der Automatisierungsgeräte ist in vielfältiger Weise mit dem des Systems verbunden.

Während der Engineeringprozesse wird ein Gerät unter verschiedenen Sichten (Punkten, Linien, usw., die in obiger Abbildung einzutragen wären) betrachtet.

Es ist in zunehmendem Maße erforderlich, ein durchgängiges Engineering des Systems zu erreichen. Dieses zeichnet sich dadurch aus, daß zwischen den einzelnen Phasen des Engineeringprozesses eine verlustfreie Übertragung der notwendigen Informationen gewährleistet wird. Für diese Übertragung muß notwendigerweise eine einheitliche Transfersyntax vereinbart sein, weiterhin muß die Semantik der Information erhalten bleiben. Das zur Zeit übliche Interfaceengineering (kleinere und größere Soft- und Hardwareadapter) sowie aufwendige Inbetriebnahmeprozesse (trial-and-error) werden durch ganzheitliche Betrachtungen (Modellbildung, formale/formalisierte Beschreibungen, automatische Implementierung) abgelöst.

2 ISO 10303 - STEP

Ein zu fertigendes Produkt durchläuft nacheinander verschiedene Phasen seines Lebenszyklusses (z. B. Design, Prototyping, Fertigung, Wartung). In jeder dieser Phasen müssen dem Bearbeiter (und seinem Werkzeug) eine Reihe von Informationen über das Produkt zur Verfügung gestellt werden. Der Gesamtumfang der Informationen über ein Produkt ergibt sich aus der Summe der Informationen, die in jeder Phase benötigt werden. Diese phasenspezifischen Informationen überdecken sich nur teilweise (z. B. der Name des Produkts ist immer gleich), spezielle Informationen werden nur an wenigen (einer) Stellen gebraucht. Die Werkzeuge zur Informationsverarbeitung (z.B. CAD-Systeme) sind oft herstellerspezifisch, d.h. sie arbeiten mit ihren eigenen internen Darstellungsformaten. Dadurch wird eine kontinuierliche Informationsweitergabe als Begleitung des Produktes im Lebenszyklus erschwert bzw. unmöglich gemacht. Das trifft besonders dann zu, wenn über Unternehmensgrenzen hinaus zusammengearbeitet werden muß (vgl. Automobilzulieferer). Eine Unterbrechung des Informationsflusses in der Werkzeugkette bedeutet immer einen Informationsverlust, der auch durch manuelle Neueingabe nur bedingt ausgeglichen werden kann (zwangsweise Inkonsistenzen). Dieses Problem kann nur dann gelöst werden, wenn eine hersteller- und lebenszyklusunabhängige Produktbeschreibung durchgesetzt werden kann.

Im Bereich CAD tauchte zuerst das Problem auf, Produktdaten austauschen zu müssen. Daher existieren hier mehrere Ansätze, diesen Austausch zu vereinheitlichen (IGES, SET, VDAFS). Diese Aktivitäten wurden dann unter dem Dach der ISO in TC184 SC4 vereint, dessen Arbeiten im Standard ISO 10303 "Industrial automation systems - Product data representation and exchange" [2] resultieren. ISO 10303 wird durch die DIN übernommen.

Der Standard ISO 10303 läßt sich grob in 4 Abschnitte unterteilen. Der erste Abschnitt (Beschreibungsmethoden) umfaßt neben einer Einführung und der Erläuterung der fundamentalen Prinzipien vor allem die Spezifikation der Datenbeschreibungssprache EXPRESS und ihrer Varianten (z.B. EXPRESS-G).

Darauf aufbauend werden im zweiten Abschnitt (Implementierungsmethoden) die notwendigen Implementierungsmethoden beschrieben. Dazu gehören der Fileaustausch und ein einheitliches Datenbankzugriffsinterface (SDAI). Der dritte Abschnitt (Testung) dient dazu, Umsetzungen der ersten beiden Abschnitte in realen Produkten (Softwarewerkzeugen) hinsichtlich ihrer Konformität zum Standard zu überprüfen. Zusätzlich zu den nicht-applikationsspezifischen Festlegungen der ersten drei Abschnitte werden im vierten Abschnitt Daten für häufig benötigte Aktivitäten / Anwendungsgebiete spezifiziert. Diese Spezifikationen sind hierarchisch aufgebaut. Die Basismodelle bilden die unterste Schicht. Es gibt integrierte Ressourcen z.B. für Material oder Formeigenschaften. Darauf aufbauend existieren die Anwendungsressourcen z.B. für Kinematik oder Zeichnen. Diese Anwendungsressourcen werden durch Application Protocols (APs) konkretisiert bzw. eingeschränkt (ähnlich Profile bei Feldbussen). Es existieren zum Beispiel Application Protocols für 2D-CAD und den Schiffbau [3].

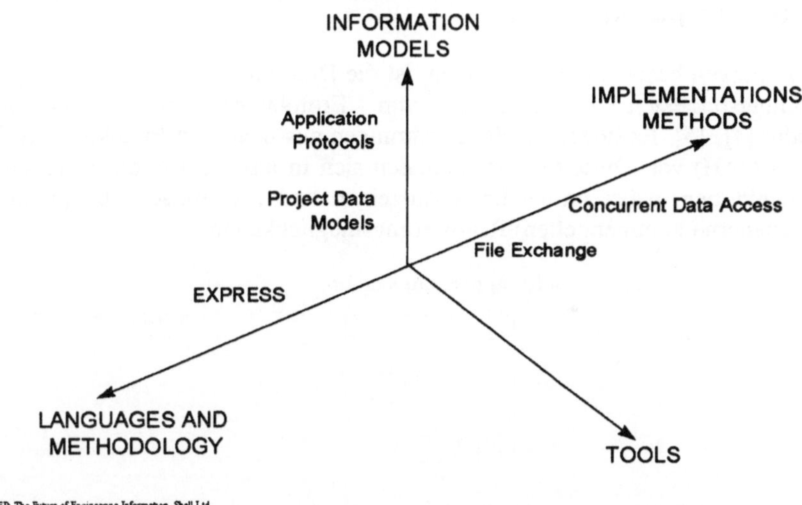

Abb. 2. STEP-Aspekte

Es wird vorgeschlagen, das Verteilte Automatisierungssystem als Produkt im Sinne von ISO 10303 zu betrachten. Damit steht STEP als Methode und mit der bereits aufgebauten Infrastruktur vollständig zur Verfügung, das Problem der durchgängigen Beschreibung von Verteilten Automatisierungssystemen wird auf das Problem der durchgängigen Beschreibung von Produktdaten zurückgeführt. Dieses Problem der durchgängigen Beschreibung von Produktdaten in deren Lebenszyklus ist mit dem Standard ISO 10303 und der dazu aufgebauten Infrastruktur grundsätzlich gelöst, wenn auch die umfassende praktische Umsetzung zur Zeit noch aussteht. Folgende Fragen sind zu beantworten:

1. Ist die vorgeschlagene Betrachtungsweise gerechtfertigt, d.h. lassen sich die STEP-Methodiken auf die Behandlung von Verteilten Automatisierungssystemen anwenden?

2. Welche inhaltlichen Erweiterungen, z.B. hinsichtlich der vorhandenen Application Protocols, müssen am STEP-Standard durchgeführt werden?

3. Welche methodische Erweiterungen, z.B. hinsichtlich der spezifizierten Kommunikationsmechanismen, müssen am STEP-Standard vorgenommen werden?

4. Welche zusätzlichen Werkzeuge, z.B. Modifikation bestehender Werkzeuge aus beiden Welten, müssen entwickelt werden?

5. Welche Schritte müssen unternommen werden, um die Migration existierender Lösungen sicherzustellen?

3 Advanced Control Network

Um diese Fragen beantworten zu können, ist die Durchführung von Forschungs- und Entwicklungsprojekten einschließlich von Erprobungen beim Endanwender notwendig [4], [5]. Es liegen bereits Erfahrungen aus mehreren Projekten (ACORN, MOVA, NOAH) vor. Diese Projekte befinden sich in unterschiedlichen Phasen. Der Schwerpunkt liegt dabei immer beim Aufzeigen und Nachweisen von praktischen (technischen und kommerziellen) Realisierungsmöglichkeiten.

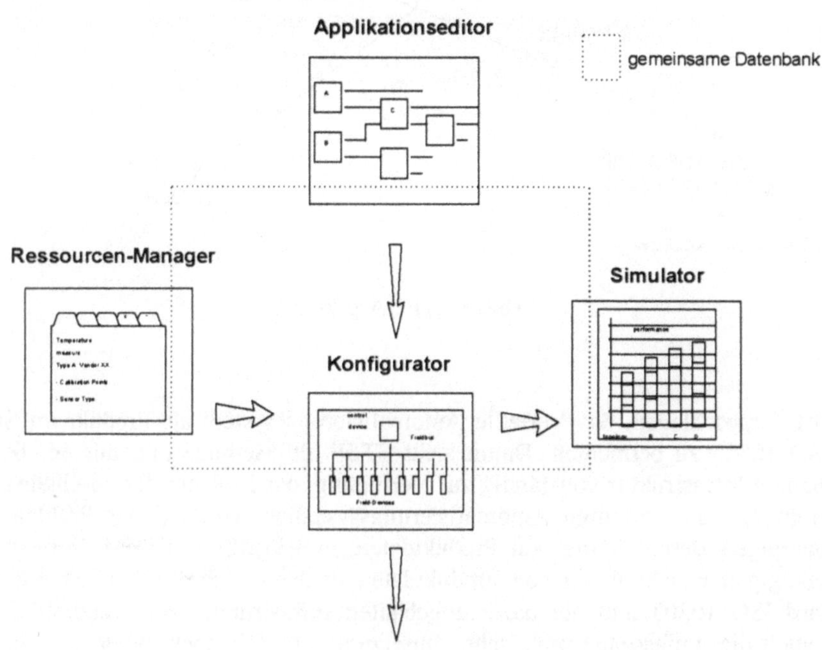

Abb. 3. ACORN-Toolkit

Das Verbundprojekt ACORN 1479 wird von der EU im Rahmen des F&E-Programmes Brite EuRam gefördert. Hauptziel ist die Entwicklung eines Toolkits, welches für das Engineering von Verteilten Automatisierungssystemen auf Basis der Funktionsbaustein- und Feldbus-technologien eingesetzt wird. Abb. 3 gibt einen Überblick über das ACORN-Toolkit.

Der Konfigurator erzeugt aus den durch den Applikationseditor beschriebenen Applikationen unter Verwendung der durch den Ressourcen-Manager erstellten Ressourcen Konfigurationen, die sowohl zur Simulation (off-line) oder zur Inbetriebnahme des Automatisierungssystems verwendet werden. Voraussetzung für die Funktionstüchtigkeit dieser Struktur ist neben dem semantischen Aspekt die Schaffung der Infrastruktur für eine heterogene Umgebung der Plattformen.

Es ist zu beachten, daß durch diese Werkzeuge eine Trennung von funktionaler Beschreibung des Automatisierungssystems (Applikationen) und der Ressourcenzuordnung (im Konfigurator) stattfindet. Dies stellt eine neue Qualität im Engineering dar, die nur durch die Offenheit der Beschreibungen erreicht werden.

Die Implementierung dieses Toolkits erfolgt in einer PC-Umgebung (Windows95 / NT). Als Datenbank kommt Oracle zum Einsatz, dabei wird eine Datenbankstruktur implementiert, die dem STEP Application Protocol 212 „Electrotechnical design and installation" entspricht [5].

Das Projekt ACORN befindet sich zur Zeit in der Implementierungsphase, d.h. bis zum Herbst 1997 werden die Prototypen der verschiedenen Werkzeuge vorliegen. Gleichzeitig werden 3 Pilotanlagen vorbereitet (1*Chemie, 2*Kraftwerk), in denen die Tools für das Engineering von feldbus-basierten Automatisierunsgssystemen (PROFIBUS, WorldFiP, FF) eingesetzt und damit validiert werden.

Literatur

1. Polke, M.: Prozeßleittechnik. Oldenbourg Verlag München. 1994.

2. STEP: Industrial Automation Systems - Product Data Representation and Exchange, ISO 10303

3. West, M.: STEP: The Future of Engineering Information, SHELL Ltd., 1995

4. Svemar, B.; Koechlin, J.-B.; Simon, R.; Zubimendi, E.; John, K.-H.; Brandl, Th.; Dahlström, L. E.; Gustavsson, L.; Salfati, R.: SIMPLE: Standardization of Interfaces between multiple Programming and Control Environments, Proposal for the Domain of Integration Manufacturing, ESPRIT, 1995

5. Hörger, J.; Simon, R.; Nerke, R.: Field Architecture Reference Model, Brite EuRam Project BE95-479 ACORN, Deliverale D03.1, 1996

Integration of Open Distributed Systems with IEC 61499

James H. Christensen

Rockwell Automation
1 Allen-Bradley Drive
Mayfield Heights, OH 44124-6118 USA

Abstract. A full hypertext (HTML) version of this paper, incorporating the most recent research results and standardization status, will be made available on the Internet following the FeT'97 conference at the URL ftp://ftp.cle.ab.com/stds/ iec/tc65wg6/liaison/fet97.htm)

1 The challenges of 21st century manufacturing

21st-century manufacturing, including both continuous and discrete process industries, will face an environment of rapid and radical change on a global scale [1]. This will include new consumer demands for higher quality design, customized/personalized product functions at lower cost per function, and shorter delivery cycles for these "made-to-order" products. In the face of these demands, manufacturers will also have to deal with globalization and regionalization of markets, manufacturing operations, and standards for both product and process quality and environmental impact.

Significant barriers to meeting these challenges will be raised in the advanced industrial economies by the aging of the work force, combined with the preference of younger workers for jobs in the tertiary (service) industries, which are perceived as being less "dirty, dark and dangerous".

A major technological barrier to be overcome will be the integration of "islands of automation" which have been constructed on an ad hoc basis without standardized interfaces with each other or the rest of the manufacturing enterprise. This barrier must be overcome in order the achieve a high degree of product customizability and reduced design/delivery cycles of "mass customized" products.

Combined with this lack of standardized interfaces is a lack of systematization of manufacturing technology as a whole. This leads to an insufficient base of easily reusable technology to support the required globalization and rapid reconfiguration of manufacturing operations.

2 Control system requirements

The agile enterprise [2] which will be able to survive and prosper on the challenges enumerated above must be capable of bringing out totally new products quickly, assimilating field experience and technological innovations easily into continual modification of product offerings.

Such enterprises will impose new challenges to industrial-process measurement and control systems, in particular *minimizing the costs of change* due to continuous reconfiguration on the factory floor [3].

Most communication standards to date, e.g. Fieldbus [4], have addressed the issues of reducing the cost of wiring via bus technologies, and reducing the cost of controllers through the use of distributed microprocessor-based control. However, additional work needs to be done in order to address the true major cost issues identified above, namely the *costs of continuous change*.

3 The IEC 61499-1 standard

The draft IEC 61499 standard attacks the costs of change head on through an explicit model for encapsulation, reuse and dynamic reconfiguration of functionality in distributed industrial process measurement and control systems (IPMCSs).

The draft IEC 61499-1 [5] defines a system architecture consisting of a minimal number of functional unit types (devices, resources, applications and function blocks). By dealing explicitly with the relationships among events, data and communication in IPMCSs, this architecture is applicable to both static and dynamic configuration, integration and re-configuration of open distributed IPMCSs for both continuous process control and manufacturing automation. (NOTE: Examples of distributed system configurations will be given in the on-line documentation).

The draft IEC 61499-2 [6] builds upon this simplified system architecture to define the principles for interoperability of life cycle support tools for open distributed control systems.

4 The future

In order to meet future demands for flexibility and reconfigurability, control systems will need to acquire substantial ability to re-configure themselves as the result of negotiated agreements to cooperate in the accomplishment of dynamically changing tasks. IEC 61499 provides a solid basis for future architectures which can meet these requirements, such as those based on autonomous cooperative agents (holons) [7].

References

1. IROFA, "Joint International Research Programs into an Intelligent Manufacturing System," Ver.3, Tokyo, January 1990.

2. Iacocca Institute, "21st Century Manufacturing Enterprise Strategy: An Industry-Led View," 1991, p. 7.

3. Chrysler/Ford/GM, "Requirements of Open, Modular Architecture Controllers for Applications in the Automotive Industry, Version 1.1," Dec.13, 1994, pp. 2-3.

4. IEC 61158, various Parts and drafts.

5. IEC 65/216/CD, Function blocks for industrial-process measurement and control systems -- Part 1: Architecture, 1997-07-04.

6. IEC 65/217/CD, Function blocks for industrial-process measurement and control systems -- Part 2: Life-cycle support model, 1997-07-04.

7. J.H.Christensen, M.Thouati and O.J.Struger, "Les Systemes Holoniques," Terrain 7 (Jan-Mar 1996), CiMax, Nanterre, France, 26-31.

OPEN CONTROL - Der Standard für PC-basierende Automatisierungstechnik

Alfredo Baginski

Manager Marketing International INTERBUS
PHOENIX CONTACT, 32825 Blomberg, Deutschland

Abstract. Der Industrie-PC gewinnt in der Steuerungstechnik zunehmend an Bedeutung. Maschinen- und Anlagenbauer können sich schon heute ihre Automatisierungslösungen aus den Hard- und Softwarekomponenten unterschiedlicher Hersteller selber zusammenstellen. Damit eröffnet sich ihnen neue Möglichkeiten, herstellerunabhänige und damit kostengünstige Automatisierungslösungen zu realisieren. Damit in solchen offenen Lösungen keine Schnittstellenprobleme auftreten und sich so die Engineeringkosten erhöhen., definiert die Open Control-Nutzergruppe die CALL Interfaces. Die CALL Interfaces stellen sowohl während der Entwurfsphase als auch während der Laufzeit einer Automatisierungslösung den einheitlichen Hersteller sicher. Zusätzlich sichert CALL den Zugriff auf dezentral angeordnete Komponenten über ein Feldbus-System wie INTERBUS oder andere.

Abstract. The industrial PC is playing an increasingly significant role in the field of control technology. These days, machine and system builders can compile their own automation solutions using hardware and software components from a variety of manufacturers. This opens up a whole new world of options for implementing manufacturer-independent, cost effective automation solutions, the Open Control User Group has defined the CALL interfaces. These CALL interfaces enable simple data exchange, not only during the design phase, but also during the runtime of an automation solution. In addition, CALL enables access to distributed components via a fieldbus system such as INTERBUS (or other).

1 Anforderungen an zukunftsorientierte Automatisierungslösungen

Heutige Automatisierungslösungen müssen ohne Änderungen auf dem Weltmarkt eingesetzt werden können. Unter ökonomischen Gesichtspunkten ist es für Maschinen- und Anlagenbauer nicht mehr vertretbar, Anpassungen an die lokalen SPS-Hersteller mit ihren geschlossenen Hard- und Softwarelösungen vorzunehmen. Die Lösung für dieses Problem ist der Einsatz von offenen herstellerunabhängigen Technologien.

Der erste Schritt in diese Richtung ist der Einsatz offener Bus-Systeme wie z. B. INTERBUS oder Sercos Interface. Die nächsten Schritte zur Herstellerunabhängigkeit sind der Einsatz der PC-Architektur als Alternative zur herstellerspezifischen SPS-Lösung und die Verwendung von Standard-Software (z. B. nach IEC 1131-3) zur Programmerstellung.

Um hier eine einfache Austauschbarkeit zwischen den Produkten unterschiedlicher Hersteller sicherzustellen, ist die Definition von offenen Schnittstellen zwischen den einzelnen Komponenten einer kompletten Automatisierungslösung notwendig. Dies ist das Ziel der Open Control-Nutzergruppe.

Die Open-Control-Nutzergruppe wird weltweit von mehr als 100 Anwendern und Herstellern unterstützt. Dazu gehören Firmen wie IBM, Microsoft, PHOENIX CONTACT, Mercedes-Benz oder Indramat. Das Ziel der Open-Control-Nutzergruppe ist es, auf Basis existierender Standards (z. B. PC-Architektur, Windows, OLE, OPC, IEC 1131-3, INTERBUS oder Sercos) einen internationalen Industriestandard für PC-basierende Automatisierungslösungen zu etablieren. Open Control setzt als Automatisierungshardware auf die offene PC-Architektur. Der PC bietet den Vorteil der weltweiten Verfügbarkeit sowohl was die Hardware betrifft als auch das weltweit verfügbare Know-how. Durch den Wettbewerb der verschiedenen Hersteller untereinander sind marktgerechte Preise und Innovationen für die Zukunft gesichert.

Vergleicht man unter Kostengesichtspunkten eine existierende SPS-Lösung mit einer IPC-Lösung, so erkennt man sehr schnell, daß durch den Einsatz von Komponenten, die im PC-Bereich zum Standard gehören, enorme Einsparungen erreicht werden; ein Beispiel ist hier eine Ethernet-Karte. Im PC-Bereich gehört sie zum Standard und wird von vielen Herstellern in verschiedenen Ausführungen angeboten. Am Markt herrscht Wettbewerb und entsprechend ist das Preis-/Leistungsverhältnis. Eine Ethernet-Karte für eine SPS hat eine spezielle Bauform und wird in dieser Ausführung nur vom jeweiligen SPS-Hersteller angeboten. Ein freier Wettbewerb findet nicht statt. Entsprechend ist das Preis-/Leistungsverhältnis.

Die Akzeptanz für PC-basierende Automatisierungstechnik wächst heute in Nordamerika genauso wie in Asien oder Europa. Der PC ist heute in unterschielichen Ausführungen vom vollausgestatteten Industrie-PC bis zum Embedded PC verfügbar. Damit lassen sich je nach Applikationsanforderung bei einem durchgängigen Konzept individuell skalierbare Automatisierungslösungen realisieren.

Im ersten Ansatz also nur Vorteile. Durch die Möglichkeit auf der offenen Hardwareplattform des PCs Soft- und Hardwarekomponenten unterschiedlicher Hersteller gleichzeitig einzusetzen, ergibt sich im Vergleich zur klassischen geschlossenen SPS-Lösung eine neue Herausforderung. Sie besteht darin, daß im Gegensatz zur total integrierten (= geschlossenen) Lösung die Schnittstellen zwischen den einzelnen Paketen untereinander nicht abgestimmt sind. Damit ist der erforderliche Aufwand zum Austausch von Informationen zwischen den Softwarepaketen ungleich höher.

2 Die Open Control Lösung

Um diesen Zustand zu ändern, ist die Open-Control-Nutzergruppe ins Leben gerufen worden. Ziel der Open-Control-Nutzergruppe ist es, einen Baukasten unterschiedlicher Hard- und Softwarekomponenten zu schaffen, bei dem die einzelnen Elemente ohne Kompatibilitätsprobleme zu einer individuellen Automatisierungslösung zusammengestellt werden können. Durch die gemeinsame Festlegung der Software-Schnittstellen entfallen die erhöhten Engineeringaufwände. Damit sinken die Kosten einer Gesamtlösung im Vergleich zu einer klassischen SPS-Lösung besonders im Engineeringbereich bis zu 40 %.

2.1 Die Open Control Architektur

Unter dem Namen CALL (Control Applacation Link Layer) definiert Open Control offene Software-Schnittstellen. Nicht nur aus Kommunikationssicht sind in einem Projekt die Engineeringphase und die Laufzeit bezüglich ihrer Anforderungen zu unterscheiden. Open Control definiert deshalb unterschiedliche Schnittstellen CALL-E (Engineering) und CALL-R (Runtime). Die Kommunikation mit dezentralen Feldgeräten erfolgt bei einer PC-basierenden Automatisierungslösung über ein Bus-System. Die offene Kommunikation in diesem Bereich stellt dem Open-Control-Konzept CALL-P (Periphery) sicher.

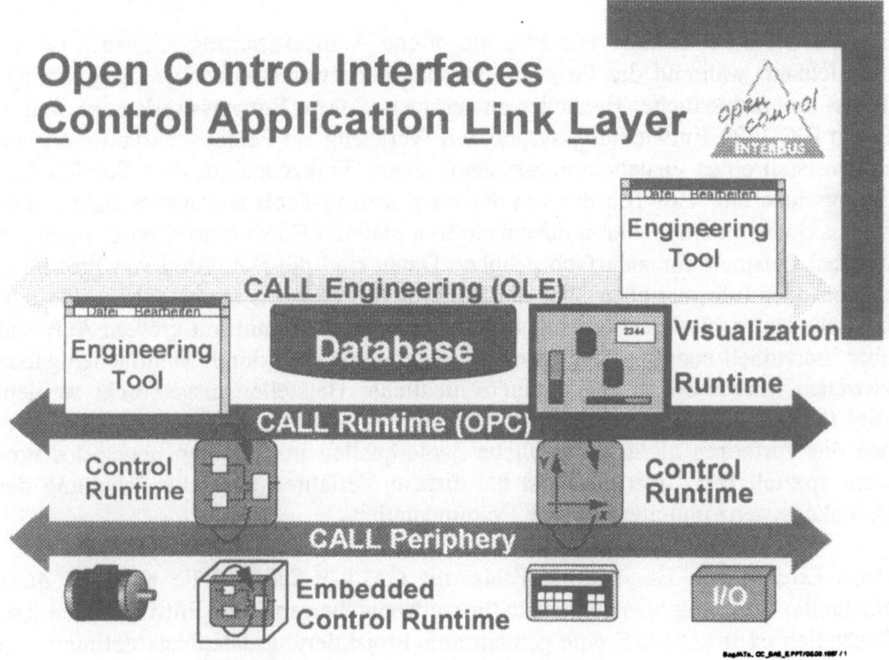

Abb.1. Architektur Control Application Link Lauer

42

CALL-E definiert unter Nutzung des OLE-Standards den durchgängigen Datenaustausch zwischen verschiedenen Entwurfswerkzeugen. Dazu enthält CALL-E u.
a. Definitionen zur Nutzung einer gemeinsamen Projektdatenbank und Richtlinien
zur Gestaltung von Bedienoberflächen des Engineering-Tools. Weitere Bestandteile
von CALL-E sind die Definition von Schnittstellen zur Initialisierung und Koordination gemeinsam aktiver Werkzeuge und die Festlegung gemeinsamer Installati-
onsvorschriften. CALL-E integriert damit singuläre Engineering-Tools in ein ho-
mogenes Gesamtsystem mit einheitlichem Aussehen gemeinsamer Datenhaltung.

CALL-R definiert die Software-Schnittstellen zwischen den Laufzeitsystemen.
Dazu gehören z.B. Programmierungs- sowie Visualisierungs- oder Konfigurationswerkzeuge. CALL-R beinhaltet die von Microsoft unterstützte OPC-
Spezifikation (OLE for Process Control). Nur die zusätzlichen Anforderungen in
industriellen Anwendungen werden von Open Control neu definiert.

CALL-P stellt den Datenaustausch zwischen einem Laufzeitsystem und dezentralen
Komponenten über einen Feldbus in Echtzeit sicher. Außerdem ermöglicht CALL-
P den schnellen Datenaustausch zwischen zwei Laufzeitsystemen. Dies ist besonders bei dezentralen Steuerungen, bei denen einzelne Teilaufgaben autark bearbeitet werden können, eine wichtige Kommunikationsart.

2.2 Die CALL-E Schnittstelle

Ein Vorteil beim Einsatz des PCs als offene Automatisierungshardware ist die
Möglichkeit, während der Projektierungsphase Entwurfswerkzeuge (Engineering-
Tools) unterschiedlicher Hersteller einzusetzen. Solche Entwurfswerkzeuge sind z.
B. ein IEC 1131-Entwicklungssystem, ein Werkzeug zur Feldbuskonfiguration, der
Entwurfsteil eines Visualisierungssystems sowie Werkzeuge aus dem Bereich Fertigungssteuerung. Ein Teil der von den Engineering-Tools angelegten Daten müssen auch von einem weiteren oder mehreren anderen Entwurfswerkzeugen genutzt
werden. Beispiele für mehrfach genutzte Daten sind die Definition von Prozeßvariablen oder Informationen über die Struktur des Gesamtsystems. Ohne eine einheitliche Datenschnittstelle können solche Informationen nur mit großem Aufwand
über individuell erstellte Konvertierungsprogramme oder durch Mehrfacheingaben
zwischen Entwurfswerkzeugen unterschiedlicher Hersteller ausgetauscht werden.
Dies führt zu erhöhten Aufwänden und damit zu erhöhten Kosten. Zusätzlich bergen die Verfahren nicht unerhebliche Fehlerquellen in sich. Ein ungelöstes Problem speziell bei Änderungen ist bei diesem Verfahren auch die Erhaltung der
Datenkonsistenz und die benötigte Dokumentation.

Beim Einsatz von Engineering-Tools mit CALL-E-Schnittstelle entfallen diese
Nachteile. CALL legt Datenformate für werkzeugübergreifende Entwurfsdaten fest.
Zusätzlich ist in CALL-E eine gemeinsame Projektierungsdatenbank definiert, auf
die sämtliche Entwurfswerkzeuge zugreifen. Aufbauend auf der Projektierungsdatenbank entstehen Werkzeuge für die einheitliche generische Projektierung und

Dokumentation. Damit entfallen die oben beschriebenen Nachteile, die sonst durch die fehlende Kopplung von Entwurfswerkzeugen entstehen.

CALL-E gliedert sich in drei Bereiche: FRAMEWORK, COMPONENT und OB-JECT LAYER. Mit FRAMEWORK LAYER wird das übergreifende Werkzeug bezeichnet, mit dessen Hilfe die Integration und Koordination der einzelnen Engineering-Tools innerhalb eines Open-Control-Systems sichergestellt wird. Die Steuerung der aktiven Engineering-Tools erfolgt dabei über eine FRAMEWORK-Management-Schnittstelle der einzelnen Werkzeuge. Darüber hinaus definiert CALL E gemeinsame Richtlinien zur Gestaltung von Bedienoberflächen der Engineering-Tools.

Der Open-Control-FRAMEWORK LAYER hat die Aufgabe, die verschiedenen Engineering-Tools und den Datenaustausch zu koordinieren. Dazu gehört das Management der gemeinsamen Datenbasis, wie z. B. Öffnen und Speichern des Projektes, Benachrichtigung der Tools bei Änderungen in der Datenbasis usw., sowie die Initialisierung des Open-Control-Systems z.B. durch das Einleiten von Programmdownloads.

Als COMPONENT LAYER werden die unterschiedlichen Engineering-Tools zur Projektierung einer Anlage bezeichnet. Sie nutzen die gemeinsame Datenbasis zur Archivierung der eigenen Daten und nutzen über sie die Information, die von anderen Engineering-Tools im Verlauf der Projektierung erstellt wurden. Damit Engineering-Tools in einer Open-Control-Lösung integriert werden können, müssen die Tools um eine Toolmanagement-Schnittstelle zum FRAMEWORK LAYER und zur gemeinsamen CALL-E-Datenbasis erweitert werden.

Als OBJECT LAYER wird die gemeinsame CALL-E-Datenbasis bezeichnet. Sie dient zum Datenaustausch zwischen den Engineering-Tools. Die Archivierung der Daten erfolgt getrennt nach Daten, die nur für das erzeugende Tool relevant sind (private Daten) und Daten, die für zwei oder mehrere Tools bestimmt sind. Die gemeinsam benutzte Datenbasis enthält die Beschreibung der Eigenschaften aller Komponenten eines Automatisierungssystems, die für verschiedene Engineering-Tools relevant sind. Die Beschreibung der Eigenschaften erfolgt in Form einer Objekthierarchie. Alle Beschreibungsdaten eines Automatisierungssystems werden zu einem Projekt-Objekt mit hierarchisch unterlagerten Beschreibungsobjekten zusammengefaßt. Diese als Projekt-Objekt-Baum bezeichnete Struktur wird in der CALL-E-Datenbasis abgelegt. Über die gemeinsame Schnittstelle kann von allen Engineering-Tools darauf zugegriffen werden.

2.3 Die CALL-R Schnittstelle

CALL-R ist die Kommunikationsplattform für Applikationsprogramme (z. B. Steuerungsprogramme oder Visualisierungspakete) zur Laufzeit. In heutigen Applikationen ist die Kommunikation zwischen zwei Laufzeitsystemen im Fertigungsumfeld recht aufwendig, da die Semantik der übertragenen Daten nicht einheitlich festgelegt ist. Eine Notlösung ist die Kommunikation auf Bitebene in Form von 24 V-Binärsignalenoder der Datenaustausch über individuell angepaßte Kommunikationsprotokolle. Dieses Verfahren geht zu Lasten der Anwender, die individuelle Treiber und Anpassungen entwickeln oder einsetzen müssen. Die Erstellung und Pflege dieser Anpassung erweist sich als sehr aufwendig und stößt bei Änderungen, die sich naturgemäß an sehr vielen Stellen auswirken, oder im Wartungsfall sehr schnell auf ihre Grenzen.

CALL-R schafft es, über eine einheitliche Semantik den Datenaustausch zu vereinfachen. Damit ist CALL-R eine offene Kommunikationsplattform für den Datenaustausch zwischen CALL-R-Client und CALL-R-Server zur Laufzeit.

CALL-R integriert die im wesentlichen von US-amerikanischen Firmen entwikkelte und von Microsoft unterstützte OPC-Spezifikation (OLE for Process Control). Zur Erfüllung der Anforderungen in der Automatisierungstechnik an dieser Stelle ergänzt CALL-R die OPC-Spezifikation um weitere Funktionalitäten. Dazu gehören z. B. die Definition von Programmobjekten oder Funktionen, die besonders die Verwaltung und Wartung größerer Systeme erleichtern wie dynamsiche Konfiguration, netzwerkweite, symbolische Namen und die Verbindung mit der CALL-E-Datenbasis.

Die CALL-R-Spezifikation definiert die Schnittstelle zu einem OPC-Server in Form von Funktionen, die der OPC-Server einem OPC-Client zur Verfügung stellt. Die Basis für den Informationsaustausch bildet der OLE/(D)COM-Mechanismus.

Das CALL-R-Architekturbild zeigt eine Implementierung auf Basis der OPC-Spezifikation. Der CALL-R-Server enthält mindestens eine vollständige OPC-Server-Implementierung alle Objekte und Interfaces, die bei OPC als mandatory gekennzeichnet sind. Damit kann ein CALL-R-Server auch von jedem OPC-Client aus angesprochen werden.

Ein CALL-R-Client enthält einen OPC-Handler und unterstützt damit sowohl das OPC-Automation-Interface wie auch das OPC-Custom-Interface. Die zusätzliche CALL-R-Funktionalität wird über den CALL-R-Handler zugänglich.

2.4 Die CALL-P Schnittstelle

CALL-P dient zum echtzeitfähigen Datenaustausch zwischen Steuerungsanwendungen und der Peripherie. Zusätzlich können über CALL-P auch zwei Steuerungsanwendungen über ein entsprechendes Netzwerk untereinander Daten austauschen.

In industriellen Anwendungen werden über die CALL-P-Schnittstelle mehrere Datenklassen parallel übertragen. Dazu gehören zyklische Prozeßdaten (Soll-/Ist-Werte, Synchronisationsdaten ...), bedarfsorientierte Geräteparameter oder Programmvariablen. Engineering-Tools können über einen transparenten Datenkanal herstellerspezifische Funktionalitäten der unterlagerten Steuerungssysteme nutzen.

Bei der Kommunikation von dezentralen Steuerungsprogrammen untereinander über CALL-P kommt dem verwendeten Bus-System eine besondere Bedeutung zu. Es müssen zeitgleich folgende Anforderungen erfüllt werden:

1. Zyklischer Austausch von zeitkritischen Prozeßdaten zwischen Steuerungseinheit und Peripheriekomponenten.

2. Bedarfsorientierter Austausch von Programmen und Prozeßdaten zwischen intelligenten Netzwerkteilnehmern.

3. Zeitkritischer Austausch von Prozeßdaten wie z. B. Synchronisation oder Verriegelungsdaten zwischen zwei Netzwerkteilnehmern ohne Umweg über ein zentrales Steuerungsprogramm.

CALL-P arbeitet kanalorientiert. Für die Verwaltung der Kommunikationskanäle stellt CALL-P entsprechende Managementdienste zur Verfügung. Über einen geöffneten Kanal können Informationen bidirektional übertragen werden. Die verwendeten Lese- und Schreibbefehle sind dabei geschwindigkeitsoptimiert, um die Anforderungen an ein Übertragungssystem in der Fertigungstechnik zu erfüllen.

CALL-P minimiert die Entwicklungsaufwendungen für Hersteller von PC-basierten Steuerungssystemen. Die Vereinheitlichung der Peripheriezugriffe führt zur Verkürzung von Entwicklungszeiten und senkt so die Kosten und steigert die Modularität und Attraktivität von Programmpaketen.

CALL-P setzt nicht zwingend ein Feldbussystem voraus. CALL-P kann auch zum Datenaustausch mit PC-Einsteckkarten für analoge oder digitale Signale verwendet werden. Weiterhin kann CALL-P zur echtzeitfähigen Kommunikation zwischen Steuerungsprogrammen eingesetzt werden, die auf demselben Rechner ablaufen. Ein Beispiel dafür ist ein Soft-PLC-Programm, das mit einer CNC-Anwendung Daten austauscht.

3 Zusammenfassung

PC-basierende Architekturen stellen als neutrale und herstellerunabhängige Hardwareplattform die Basis für die Zukunft der Automatisierungstechnik dar. In Verbindung mit den CALL-Interfaces der Open-Control-Nutzergruppe läßt sich eine vollständige Entkopplung von Hard- und Software erreichen. Das bedeutet Investitionsschutz und langfristige Updatemöglichkeiten von Maschinen und Anlagen. Unabhängig davon, ob man einen Hochleistungsrechner auf Basis eines Pentium-Prozessors oder einen kleinen Embedded-Controller für dezentrale Steuerungsaufgaben verwendet, läßt sich das gleiche Steuerungskonzept verwenden. Auch in der Wahl der anderen Automatisierungskomponenten erlangt der Anwender mit Open Control die Freiheit, die jeweils passende Teillösung einzusetzen, z. B. kann ein komfortables Visualisierungspaket für einen Leitstand eingesetzt werden und ein anderes angepaßtes Paket für eine kleine Visualisierungsaufgabe direkt an eine Maschine.

Die skalierbaren Hardware- und Softwarelösungen in Verbindung mit den CALL-Interfaces bilden einen offenen Automatisierungsbaukasten, mit dessen Elementen unter technischen und wirtschaftlichen Gesichtspunkten individuell angepaßte Lösungen realisiert werden können.

Die Open-Control-Nutzergruppe garantiert über ein entsprechendes Zertifizierungsverfahren problemloses Zusammenspiel aller Komponenten einer offenen Automatisierungslösung.

Literatur

1. INTERBUS-S Club e. V. (Hrsg.):
   Spezifikation Open Control Interface
   Baden Baden, April 1997

2. Alfredo Baginski, Martin Müller
   INTERBUS-S Grundlagen und Praxis
   Hüthig Verlag, Heidelberg 1994

3. Roland Bent, Wolfgang Blome
   Offene Automation nur mit dem PC?
   Elektronik, No. 7, April 1996

4. Roland Bent
   Reale Zukunftsvisionen
   iee, 41. Jahrgang, No. 8, August 1996

Open Control:
Der Weg zum Network Computing

Peter Fröhlich [1], Thomas Speidel [2]

[1] IBM Deutschland InformationssystemePascalstr. 100, D 70569 Stuttgart
peter_froehlich@de.ibm.com

[2] Fritz Electronic GmbHSchönbergstr. 45, D 73760 Ostfildern

Abstract. Der Open-Control-Arbeitskreis spezifiziert Schnittstellen für modulare und offene Automatisierungssysteme. Die industrielle Realisierung des Network Commputers ermöglicht Automatisierungssysteme auf Basis dieser Spezifikationen und der Web-Technologie.

Abstract. The Open Control user group specifies interfaces for modular and open automation systems. The industrial implementation of the Network Computer is the basis for automation systems building on these specifications and web technology.

1 Control Application Link Layer

Der PC hält Einzug in die Steuerungstechnik. Die Open-Control-Nutzergruppe hat seit Ihrer Gründung im April 1996 eine umfassende Schnittstellenspezifikation vorgelegt, genannt "CALL", *Control Application Link Layer* [1]. Auf Basis allgemein akzeptierter Soft- und Hardware-Standards des PC- und Feldbus-Umfelds bietet CALL drei Schnittstellen:

1. CALL-E (Engineering): Die Entwurfs- und Konfigurationswerkzeuge für Feldbus, Steuerung, Visualisierung und MES greifen über die standardisierte CALL-E-Schnittstelle auf eine gemeinsame Projektdatenbank zu. Neben der Topologie des Automatisierungssystems werden dort zentral alle Datenobjekte mit wichtigen Attributen beschrieben. Über die Zugriffsschnittstelle, die z.B. unter Windows NT auf OLE aufbaut, gleichen die Einzelwerkzeuge ihre internen Datenbanken bidirektional mit der Projektdatenbank ab. Dadurch kann der Projektierungsaufwand stark reduziert werden. Vor allem aber ist die Datenkonsistenz nicht nur bei der Erstellung, sondern auch bei späteren Änderungen gewährleistet

2. CALL-R (Runtime) schafft eine Kommunikationsplattform, die den Datenaustausch zur Laufzeit mit einer einfachen, aber prägnanten Semantik vereinheitlicht. Die Schnittstellen zu den Applikationen sind auch in heterogener Umgebung, bei kleinen lokalen Systemen wie in großen vernetzten Anlagen absolut identisch. CALL-R integriert bereits die von US-amerikanischen Automatisierungsherstellern in Zusammenarbeit mit Microsoft entwickelte OPC-Spezifikation (OLE for Process Control) [2], die eine Kommunikationsschnittstelle für Prozeßvariablen unter Verwendung von OLE beschreibt. CALL-R verfügt

außerdem über eine C-Aufrufschnittstelle für dieselben Funktionen, die die Plattformunabhängigkeit gewährleistet.

3. CALL-P (Periphery) schließlich ergänzt CALL-R um den Echtzeittransfer zwischen Steuerung und Peripherie, sowie zwischen den Steuerungen untereinander. Ohne besondere technische Eigenschaften bestimmter Feldbusse zu beschränken, werden die wichtigsten Funktionen zur Kommunikation über den Feldbus passend zum durchgängigen CALL-Datenmodell vereinheitlicht. Die Hersteller PC-basierter Steuerungssysteme sparen Entwicklungskosten und Zeit bei der Erstellung von Feldbustreibern. Der Anwender profitiert von der dadurch erlangten Modularität und Aktualität.

Die Basis für CALL ist grundsätzlich der PC, der mit seinem breiten Angebot an Hard- und Software einen unschlagbaren Kostenvorteil gegenüber althergebrachten, proprietären Automatisierungsumgebungen bietet. Dem Bedürfnis des Anwenders an Verfügbarkeit und Zuverlässigkeit wird auch mit PC-basierten Lösungen Rechnung getragen. Je nach Anforderung besteht die Wahl zwischen Einprozessor- und Koprozessor-Lösungen. Letztere erlauben skalierbare Leistung bei gleicher Zuverlässigkeit wie entsprechende SPS-Lösungen.

2 Der Network Based Controller

Mit der Orientierung am breiten Bürotechnik-Markt ist die Automatisierungstechnik auf dem Weg zu offenen und kostengünstigen Lösungen, insbesondere was die zunehmende Kommunikationsvernetzung anbelangt. Stolperstein für viele Anwender sind derzeit noch die Konfigurations- und Installationszeiten von Automatisierungssystemen. Nach einem Hardwareausfall muß die korrekte Softwareumgebung auf dem Ersatzgerät wieder hergestellt werden. Im Extremfall verursacht dies Produktionsausfallzeiten.

Im Büroumfeld werden solche Konfigurations- und Installationszeiten in der "Total Cost of Ownership" berücksichtigt, die nach unabhängigen Schätzungen zwischen \$6.144 und \$13.270 pro Jahr betragen. *Network Computer* schaffen hier Abhilfe. Ihr international festgeschriebenes Referenzprofil wird u.a. von Apple, IBM, Netscape, Oracle und SUN getragen. Die Daten und Programme erhalten NCs direkt aus dem Intranet. Sie führen Java-Applikationen aus, arbeiten als Webbrowser und intelligente Multistandard-Grafikterminals.

Zusammen mit der Firma Fritz electronic GmbH entwickelt IBM aus dem NC der Bürowelt [3] den Network Based Controller (NBC), der als universeller Grundbaustein intranetgestützter Automatisierungssyteme dient. Der NBC verbindet die Grundfunktionen von Automatisierungssystemen mit den Standards von Intra- und Internet:

v Echtzeitgerechte und zuverlässige Verarbeitung von Prozeßsignalen.
v Bedienung, Visualisierung und Diagnose (HMI: Mensch-Maschine-Schnittstelle).
v Aufbereitung von Prozeßdaten.
v Aufbau vernetzter, dezentraler Strukturen.

Um diese Funktionen mit dem NBC zu realisieren, sind vier mögliche Szenarien denkbar, die in Abb. 1 dargestellt sind:

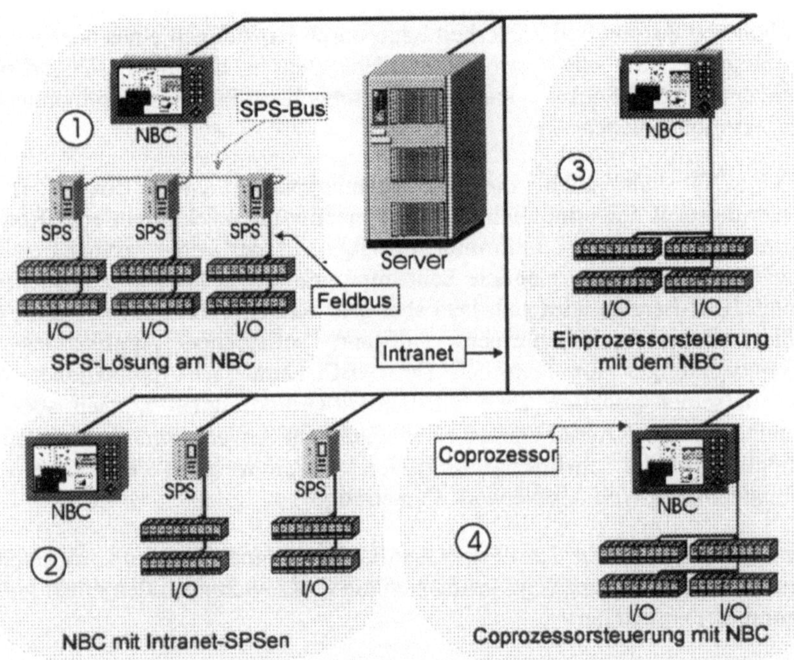

Abb. 1. Szenarien

1. **Klassischer Einsatz des NBC als Zellenrechner zur direkten Kopplung mit gängigen SPS-Systemen. Der NBC fungiert dabei als Server, um die CALL-R-Schnittstelle der Open-Control-Spezifikation für klassische SPS zu realisieren und im Netzwerk zur Verfügung zu stellen. Gleichzeitig bereitet er die Daten auch für direkten Zugriff von Webbrowsern aus dem Intranet auf. Er realisiert also sowohl eine Daten- als auch eine grafische Schnittstelle. Mit dem integrierten Webbrowser greift der NBC auch als Client auf diese grafische Schnittstelle zu, um HMI-Funktionen zu realisieren.**

2. **Zukünftige SPS-Generationen werden bereits von sich aus TCP/IP-fähig sein und ihre Daten entsprechend aufbereiten und zur Verfügung stellen. In diesem Fall übernimmt der NBC gegebenenfalls die Umsetzung von CALL-R-Datenzugriffen auf proprietäre Zugriffe. Als Client realisiert er wie unter 1. HMI-Funktionen.**

3. **Äußerst effizient ist die direkte Prozeßsteuerung mit dem NBC. Hierzu wird die Laufzeitumgebung einer SPS im NBC mit einem Echtzeitbetriebssystem hergestellt. Die Lösung unterscheidet sich von einer PC-basierten Soft-SPS, da die Kompatibilität zum nicht-echtzeitfähigen Windows oder OS/2 nicht gewahrt**

werden muß. Der NBC wird also zu einer SPS mit integriertem Network Computer. Alle MMI-Dienste sowie die Aufbereitung der SPS-Daten für das Netz werden wie oben ebenfalls vom NBC wahrgenommen.

4. Noch höhere Leistung und Sicherheit kann durch Hinzufügen eines oder mehrerer Coprozessoren zum NBC erreicht werden. Der in der IBM Networkstation vorhandene PMC-Sockel stellt die schnelle PCI-Bus-Verbindung zur Coprozessorkarte sicher.

Mit dem NBC hält auch die Programmiersprache Java Einzug in die Automatisierungswelt. Sie ermöglicht die Implementierung objektorientierter Konzepte und ist gleichzeitig absolut plattformunabhängig. Mit der heute verfügbaren Java-Technologie ist es noch nicht möglich, Echtzeitprogramme in Java zu implementieren. Im gesamten HMI-Bereich wird sich Java aber sehr schnell durchsetzen können. Dazu tragen die weltweit mit enormem Aufwand betriebenen Applikations- und Werkzeugentwicklungen für Java bei. Der NBC kann Java-Applikationen direkt ausführen. In seiner Funktion als Multistandard-Grafikterminal können aber auch herkömmliche Windows-Applikationen *remote* auf dem Server ausgeführt werden. Der NBC ermöglicht so die schrittweise Migration von proprietären Plattformen in die plattformunabhängige Welt des Network Computing.

An dieser Stelle zeigt sich die Stärke der Open-Control-Standards. Von vorneherein auf Plattformunabhängigkeit ausgelegt, lassen sich die CALL-Schnittstellen direkt auf Java und ins Intranet übernehmen.

3 Zusammenfassung

Der NBC ist die konsequente Weiterführung der Grundsätze von Open Control, die Modularität und Offenheit von Automatisierungssystemen auf der Basis verbreiteter und technologisch fortschrittlicher Standards zu gewährleisten. Das firmeninterne Intranet bietet erstmals die Möglichkeit, Automatisierungsdaten und andere Unternehmensdaten über standardisierte Zugriffsmechanismen unternehmensweit zugänglich zu machen. Da der Zugriff bei entsprechender Absicherung auch über das Internet erfolgen kann, sind Fernwartung oder Datenaufbereitung im globalen Unternehmensverbund ohne zusätzliche Kosten realisierbar. Open Control hat mit den Standards für PC-basierte Steuerungslösungen jetzt schon Meilensteine gesetzt. Die Web-Technologie wird darauf aufbauend ganz neue Perspektiven eröffnen.

Literatur

1. Open Control: http://www.interbusclub.com

2. OPC: http://www.microsoft.com

3. IBM Network Station: http://www.ibm.com/Products/netcomputers.htm

Feldbusse und IEC 1131 -
Symbiose zum Nutzen des Anwenders

Martin Wollschlaeger

Institut für Prozeßmeßtechnik und Elektronik (IPE),
Otto-von-Guericke-Universität Magdeburg,
PF 4120, D-39016 Magdeburg, Germany

Abstract. Die Flexibilität moderner feldbusgestützter Automatisierungssysteme basiert auf der Modularität ihrer Komponenten, die durch geeignete Konfigurierung und Programmierung mittels unterschiedlicher, herstellerspezifischer Werkzeuge an die automatisierungstechnischen Erfordernisse angepaßt werden müssen. Die Einführung der Norm IEC 1131 hat im SPS-Bereich zu einer herstellerunabhängigen Betrachtung von SPS-Systemen geführt. Die in der IEC 1131 definierten Konzepte lassen sich auch auf Feldbussysteme anwenden und erlauben die Realisierung von universellen, busunabhängigen Werkzeugen für Feldbussysteme. Aspekte der Integration von Feldbussen und spezifischen Werkzeugen in die IEC 1131 werden beschrieben und Ergebnisse für Anwender sowie künftige Entwicklungen diskutiert.

Abstract. The flexibility of modern automation and control systems using a fieldbus is based on the modular structure of fieldbusses. Their components have to be adapted to the specific demands of an automation task by configuring and programming themselves using different, vendor-dependent tools. In the area of PLCs, the introduction of the international standard IEC 1131 has lead to a vendor-independent view to PLC-based systems. The concepts defined in IEC 1131 may be applied to fieldbus systems, too. They allow the introduction of universal, bus-system independent tools. Aspects of the integration of fieldbusses as well as of existing tools into IEC 1131, users' benefits and future trends are discussed.

1 Motivation

Die Einführung der in der Norm IEC 1131 definierten Konzepte im Bereich der SPS-Programmierung hat eine Reihe von Vorteilen mit sich gebracht. Sie erlaubt die Behandlung von Gerätesystemen unterschiedlicher Hersteller und mit unterschiedlicher Funktionalität nach einem durchgängigen Konzept und hat wesentlich zu einer einheitlichen Betrachtung von SPS-Systemen beigetragen. Es entstand ein anwenderfreundliches System, das die Erstellung wiederverwendbarer Programme sehr gut unterstützt und damit zu Kostensenkungen in der Programmierung von Anwendungssoftware sowie zu Einsparungen bei der Einsatzvorbereitung und im Servicebereich geführt hat.

In zunehmendem Maße bestehen Automatisierungssysteme neben SPS-Strukturen aus dezentralisierten, feldbusgestützten Komponenten. Bedingt durch die differenzierten Einsatzcharakteristika und unterschiedlichen Leistungsparameter von Feldbussen und deren Komponenten ist ihre Handhabung in den verschiedenen Phasen des Lebenszyklus einer Automatisierungsanlage unterschiedlich komplex und derzeit fast ausschließlich mit bussystemspezifischen Werkzeugen zu lösen. Demgemäß erfordert der Einsatz von Feldbussen umfangreiche systemspezifische Kenntnisse beim Anwender und verursacht entsprechende Kosten in der Einsatzvorbereitung. Letztlich sind die Probleme der Normung mit dafür verantwortlich, daß beim Anwender immer noch eine Hemmschwelle hinsichtlich des Einsatzes von Feldbussen existiert.

2 Integration von Feldbussen in die IEC 1131

Die durch den Einsatz der IEC 1131 [1] erreichte anwenderfreundliche, universelle Betrachtung von SPS läßt sich auch auf Feldbusse ausdehnen. Eine geeignete Beschreibung der funktionellen und topologischen Eigenschaften von Feldbussen und ihren Komponenten bildet die Basis für deren Integration in Entwicklungsumgebungen auf Basis von IEC 1131. Die Programmierung von SPS nach IEC 1131 [2] folgt einem in der Norm definierten Programmiermodell. Neben der Spezifikation von Datenformaten und von Anweisungen zur Typ- und Variablendeklaration werden globale organisationsbeschreibende Elemente wie Ressource, Task und Programm eingeführt. Darüber hinaus werden die Gültigkeitsbereiche von Variablen und der Mechanismus des Datenaustauschs zwischen verschiedenen Funktionselementen definiert. Von wesentlicher Bedeutung für IEC 1131-konforme Systeme sind die in der Norm definierten Programmiersprachen. Dabei lassen sich entsprechend der Form ihrer Notation grafische und textuelle Sprachen unterscheiden. Zu den grafischen Sprachen gehören Kontaktplan (Ladder Diagram), Funktionsplan (Function Block Diagram) und Ablaufdiagramm (Sequential Function Chart), während als textuelle Sprachen die Anweisungsliste (Instruction List) sowie Strukturierter Text (Structured Text) benutzt werden. Die Sprachen sind beliebig miteinander kombinierbar. Die Rückübersetzbarkeit zwischen den verschiedenen Programmiersprachen ist nicht zwingend erforderlich, wenngleich sie anzustreben ist.

Feldbusse und ihre Komponenten lassen sich mit den Konzepten der IEC 1131 beschreiben. Die in der Norm definierten Organisationsobjekte lassen sich ebenso wie die Programmiersprachen auf verteilte Ressourcen wie feldbusgestützte Systeme anwenden. Die Integration der Feldbusse in die IEC 1131 ist ein momentan ablaufender Prozeß [3, 4], der die Beschreibung der unterschiedlichen Feldbuskonzepte und -systeme nach einheitlichen Richtlinien voraussetzt. Dazu ist es notwendig, die Bestandteile der Feldbusprotokolle mittels geeigneter Interfaces so aufzubereiten, daß sie nahtlos in bestehende IEC 1131-konforme Entwicklungsumgebungen integriert werden können (Fig. 1) [5]. Neben dem Mapping des Softwaremodells sind dabei auch das Kommunikationsmodell sowie das Feldbusmanagement abzubilden. Die in Feldbussystemen verwendeten Datentypen entsprechen weitestgehend den Definitionen in der IEC 1131. Um eine Variable nach IEC 1131 der entsprechenden Datenstruktur im Feldbussystem zuzuordnen, muß sie auf ein geeignetes Objekt abgebildet werden. Ein solches Objekt kann beispielsweise ein Kommunikationsobjekt bei PROFIBUS, ein Identifier in der verteilten Datenbasis bei CAN oder eine Softwire-Number beim P-NET sein (Fig. 2). Abhängig von der Lokalisation des Datenobjekts müssen Zugriffspfade definiert werden. Die anderen organisatorischen Elemente der IEC 1131 wie Konfigurationen, Ressourcen, Programme, Tasks, Funktionsblöcke und Funktionen besitzen Entsprechungen in den jeweiligen Bussystemen. So kann ein Funktionsblock z.B. als P-NET-Channel oder als SDS-Device bei CAN betrachtet werden. Konfigurationen repräsentieren die Module eines Feldbussystems, während ein Programm nach IEC 1131 mit einem über MMS-Services gesteuerten Programm in einem Busmodul verglichen werden kann.

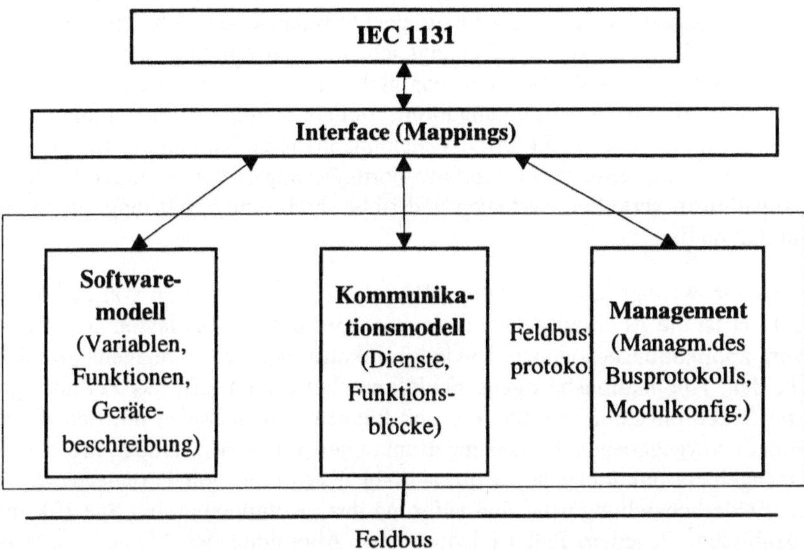

Fig. 1. Abbildung eines Feldbusprotokolls auf die IEC 1131

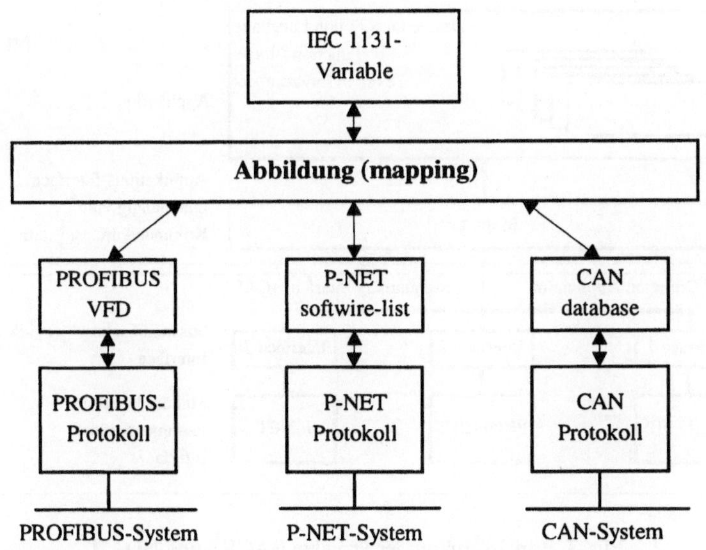

Fig. 2. Abbildung einer IEC 1131-Variablen auf Kommunikationsobjekte

Die durch das Busprotokoll zur Verfügung gestellten Kommunikationsfunktionen lassen sich auf Kommunikationsfunktionsblöcke nach IEC 1131-5 abbilden. Sie definieren die Nutzung von systemweiten Transportmechanismen zum Datenaustausch zwischen Programmen, Funktionen und Funktionsblöcken in den einzelnen Feldbusmodulen unter Nutzung des Feldbusprotokolls. Die in der Norm definierten Kommunikationsfunktionsblöcke müssen demzufolge mit Hilfe der durch das jeweilige Busprotokoll zur Verfügung gestellten Anwendungsfunktionen realisiert werden. Abhängig vom Bussystem (Verfügbarkeit von confirmed bzw. unconfirmed services) ist ihre Implementierung unterschiedlich aufwendig. Wenngleich die Anwendung der Konzepte der IEC 1131 in vielen Fällen die Nutzung von globalen oder direkt repräsentierten Variablen begünstigt (implizite Kommunikation), stellen Kommunikationsfunktionsblöcke einen sinnvollen Ausgangspunkt für bussystemunabhängige Kommunikationsfunktionen dar. Ihre Anwendung ermöglicht explizite Formulierungen von kommunikationsbezogenen Abläufen in verteilten Systemen und stellt somit eine Ergänzung zur impliziten Kommunikation dar.

Ein komplexer, wenngleich notwendiger Prozeß bei der Integration von Feldbussen in die IEC 1131 ist die Abbildung des Feldbusmanagements. Dabei lassen sich die Abbildung von applikationsbezogenen sowie protokollbezogenen Managementfunktionen unterscheiden. Applikationsbezogene Funktionen beinhalten z.B. das Zeitmanagement sowie die Synchronisation der Module und können üblicherweise mit den durch das Busprotokoll vorgegebenen Managementfunktionen realisiert werden. Protokollbezogene Managementfunktionen wie Adressierung, Definition von Parametern des Bus-Timings, Fehlerbehandlung usw. sind aufgrund ihrer Komplexität und Spezifik schwieriger abzubilden. In jedem Fall ist jedoch eine Abbildung des Managements unverzichtbar für eine durchgängige Integration der Feldbusse in die IEC 1131.

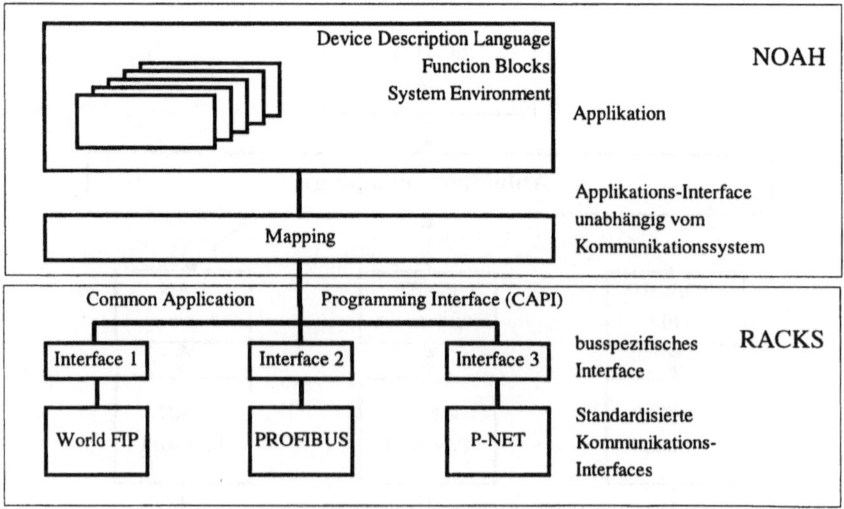

Fig. 3. Prinzipstruktur der Projekte RACKS und NOAH

Neben der Definition von busspezifischen Mappings bieten aktuelle Projekte interessante Ansatzpunkte für die Realisierung universeller Mappings. Das Projekt RACKS (Reusable Application Interface for Communicating Real-Time Kernels) definiert auf der Basis der drei in die europäische Feldbusnorm EN 50170 eingeflossenen Bussysteme P-NET, PROFIBUS und World-FIP ein einheitliches Programmierinterface (CAPI) (Fig. 3) [6], auf das Anwendungsprogramme wie Werkzeuge zurückgreifen können. Einen weiteren Schritt in Richtung auf universelle Schnittstellen geht NOAH (Network Application Harmonisation), dessen Ziel ein einziges, bussystemunabhängiges Application Interface für beliebige Applikationen ist. Letztlich steht das aus der PC-Welt bekannte Prinzip des „plug & play" als Grundgedanke hinter NOAH. Hierbei ist die Nutzung von IEC 1131-Prinzipien vorgesehen. Damit schaffen RACKS und NOAH Voraussetzungen für die Einführung funktionell orientierter, busunabhängiger Programmierungskonzepte.

3 Integration von Werkzeugen in IEC 1131-Umgebungen

3.1 Programmierung von Busmodulen

Die Module eines feldbusgestützten Automatisierungssystems lassen sich hinsichtlich ihrer Programmierung in zwei Gruppen einteilen. Die erste Gruppe bilden vorkonfektionierte Module, die bereits über als Firmware implementierte Anwendungsfunktionen verfügen. Solche Funktionen sind typischerweise im Bereich der Sensorik/Aktorik angesiedelt, lassen sich aber auch verarbeitenden Funktionen zuordnen. So können vorkonfektionierte Busmodule beispielsweise über Anwendungsprogramme zur Analogwerterfassung und -aufbereitung, zur Binär-Ein-/Ausgabe oder zur PID-Regelung verfügen. Die implementierten Algorithmen lassen sich durch entsprechende Parameter beeinflussen und somit an den konkreten Anwendungsfall anpassen. Die Programmierung derartiger vorkonfektionierter Module reduziert sich demzufolge auf die Parametrierung der Anwenderprogramme. Die zweite Gruppe von Modulen bilden freiprogrammierbare Busmodule. Sie werden z.B. als Controller oder als SPS mit entsprechender Anschaltbaugruppe eingesetzt und enthalten ein Betriebssystem sowie Kommunikationsfunktionen, meist jedoch keinerlei vorprogrammierte Anwendungsfunktionalität. Die eigentlichen Anwendungsprogramme müssen während Projektierung und Konfigurierung erzeugt und in die Module implementiert werden. Sie sind typischerweise wesentlich komplexer als Anwendungsprogramme in vorkonfektionierten Modulen. Zur Erstellung der Anwenderprogramme stehen spezielle Entwicklungsumgebungen (Compiler, Debugger) oder bereits IEC 1131-basierte Programmierwerkzeuge (z.B. für viele speicherprogrammierbare Steuerungen) zur Verfügung.

3.2 Integrationskonzepte

Zur Handhabung von Feldbussen und Busmodulen existieren zahlreiche Werkzeuge. Sie werden vielfach von Feldbus-Herstellern entwickelt und mit den Busmodulen angeboten. Die Werkzeuge sind demzufolge busspezifisch und repräsentieren zumeist eine herstellerspezifische, topologisch orientierte Sichtweise. Neben Werkzeugen, die spezifisch bezüglich der einzelnen Betriebsphasen eines Bussystems sind, existieren durchgängige Lösungen, wie beispielsweise Open Control [7].

Moderne Werkzeuge für Feldbussysteme stellen sich dem Anwender als Set von einzelnen, miteinander informationell verkoppelten Komponenten dar. Die einzelnen Komponenten realisieren dabei betriebsphasenspezifische Aufgaben (z.B. Konfigurator, Monitor, Diagnosetool). Die Grundlage des Datenaustauschs der Komponenten bildet eine gemeinsame strukturierte Datenbasis, in der neben Beschreibungsdaten der einzelnen Module auch die jeweiligen Anwenderprojekte abgelegt werden. Der Zugriff der Werkzeugkomponenten auf die Datenbasis erfolgt mit standardisierten Abfragen, etwa auf der Basis von SQL. Neben dem Interface zur Datenbasis können die einzelnen Werkzeuge über standardisierte Interfaces verfügen, mit deren Hilfe sie aus anderen Anwendungen heraus angesprochen werden können. Ein typisches Beispiel hierfür stellt das P-NET-Werkzeugsystem auf der Basis von VIGO (Virtual Interface Global Objects) dar, das über eine OLE2-Automation-Schnittstelle verfügt.

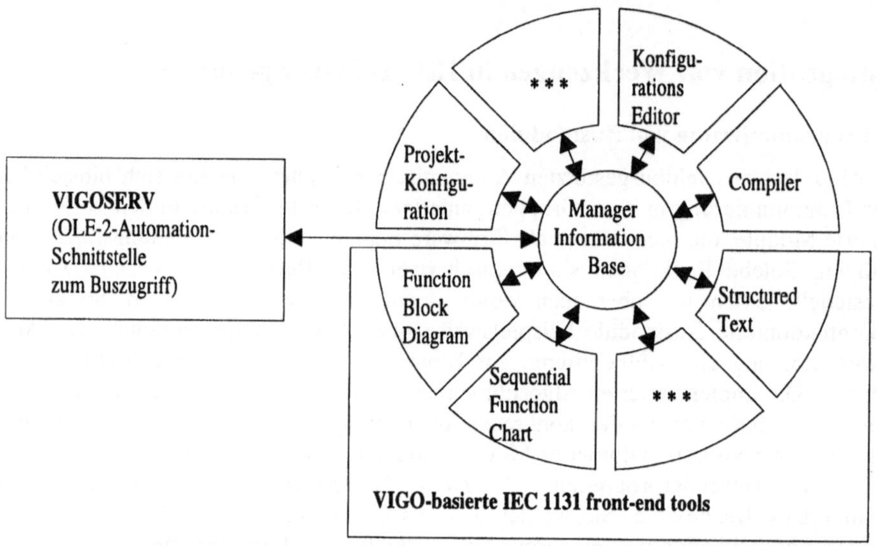

Fig. 4. Erweiterung des P-NET-Werkzeugsets um IEC 1131-Entwicklungswerkzeuge

Die zentrale Datenbasis der Werkzeugsets stellt einen hervorragenden Ansatzpunkt zur Integration existierender Werkzeuge in IEC 1131-Programmierumgebungen dar. Sie erlaubt den Zugriff der IEC 1131-Werkzeuge (z.B. Compiler, Projektverwaltung, Bibliotheken) auf die bussystemspezifischen Daten. Darüber hinaus ist die Nutzung der vorhandenen spezifischen Werkzeuge innerhalb der IEC 1131-Entwicklungsumgebung möglich, indem die Schnittstellen der Werkzeuge angesprochen werden und diese die übergebenen Daten weiter bearbeiten können. Es erfolgt somit eine Erweiterung der bestehenden Werkzeugsets durch IEC 1131-konforme Komponenten, wie sie am Beispiel des P-NET in Fig. 4 unter Nutzung von VIGO dargestellt ist. Die vom VIGO zur Verfügung gestellten Funktionen des OLE2-Automation-Interfaces können beispielsweise genutzt werden, um Einträge für neue Module, neue Projekte oder Codefiles anzulegen. Die Funktionen erlauben ebenso die projektspezifische Eintragung von Daten (Parametern) in die Initialisierungswerte der Module. Abhängig vom Typ der anzusprechenden Daten (Projekt, Modul, Channel oder Softwire-Number) können mittels des Funktionsaufrufs spezielle Algorithmen angesprochen werden, von Konvertierungsfunktionen über channel-spezifische Komponenten bis hin zur Einstellung von Moduladressen oder Compileraufrufe für die Controllerprogrammierung.

4 Ergebnisse für den Anwender

Die Integration von Feldbussen in die IEC 1131 erlaubt die Behandlung von Feldbussystemen und ihren Komponenten als verteilte E/A-Baugruppen bzw. verteilte Steuerungen. Dadurch ist die durchgängige Programmentwicklung für Steuerungen wie Feldbussysteme nach den gleichen Konzepten und mit einem identischen Programmiermodell realisierbar. Konfigurierung und Programmierung von vorkonfektionierten wie freiprogrammierbaren Feldbusmodulen sind ebenso wie ihre Handhabung aus IEC 1131-basierten Softwareentwicklungsumgebungen für SPS heraus möglich. Feldbuskomponenten wie komplette Feldbussysteme lassen sich demzufolge durch Integration in die IEC 1131 als das betrachten, was sie letztlich sind - integrale Bestandteile von Automatisierungssystemen.

Für den Anwender bedeutet die Handhabung von Steuerungen und Feldbussystemen nach einem einheitlichen, durchgängigen Systemkonzept eine drastische Senkung des Einarbeitungsaufwands. Es erlaubt ihm die gewünschte funktionelle Sicht auf den Feldbus als Kommunikationsmedium des Automatisierungssystems. Die eigentlichen Kommunikationsvorgänge werden gegenüber dem Anwender gekapselt. Dadurch ist eine Orientierung auf die zur Lösung des automatisierungstechnischen Problems notwendigen Programmier- und Konfigurierungshandlungen möglich. Eine detaillierte Beeinflussung des Feldbussystems mit spezifischen Werkzeugen wird durch die Integration nicht verhindert, sondern durch Bereitstellung zusätzlicher Informationen unterstützt. Die Wiederverwendung von existierenden Automatisierungslösungen wird wesentlich erleichtert. Die einfache Änderbarkeit einer installierten Anlage zur Anpassung an wechselnde Einsatzanforderungen ist ebenfalls gegeben. Nicht zuletzt verringert die modulare Struktur eines derartigen Systems wesentlich den Aufwand für Wartung, Diagnose und Fehlerbehandlung. Für den Hersteller von Feldbussystemen und -komponenten liegt der Vorteil IEC 1131-gestützter Programmierumgebungen vorrangig in der Akzeptanzverbesserung beim Anwender. Die durch Definition von Mappings und Schnittstellen für front-end tools spezifizierten Interfaces ermöglichen eine Verbesserung der Modularität und damit Wiederverwendung und Kostensenkung bei der Erstellung von neuen Komponenten, Werkzeugen und Testumgebungen.

5 Ausblick

Die prognostizierten Entwicklungen der PC-Technik werden in zunehmendem Maße auch den Bereich der Programmierung von Automatisierungssystemen und Anwendersoftware beeinflussen. Wesentliche diesbezügliche Trends stellen Komponentensoftware und PC-Control dar. Der in der Entwicklung moderner Informationstechnologien erkennbare Trend zur Einführung von komponentenbasierter Software erlaubt die Implementation modularer, skalierbarer Entwicklungsumgebungen für die Programmierung von Steuerungen wie Feldbuskomponenten. Die durchgängige Einbeziehung verteilter Informationssysteme, etwa für das Management von Bibliotheken und Spezifikationen, läßt Steigerungen der Effektivität in der Anwendungsprogrammerstellung erwarten. Der Bereich des PC-Control bietet in Kombination mit kostengünstigen Anschaltbaugruppen für Feldbussysteme eine ernstzunehmende Alternative zu SPS für viele Steuerungsaufgaben. Hervorzuheben sind vor allem sehr günstige Möglichkeiten für Simulation, Debugging und Dokumentation sowie die Integration von kostengünstiger Standardsoftware aus dem Bereich der Büro- und Netzwerktechnik. Darüber hinaus werden anwenderfreundliche, über breite Akzeptanz verfügende Lösungen für die Steuerungs- und Automatisierungstechnik verfügbar. Die betriebssystemimmanenten Mechanismen zum Datenaustausch finden auch im Bereich der Automatisierungstechnik zunehmend Akzeptanz, wie die Entwicklung von OPC (OLE for Process Control) [8] zeigt. Die beschriebenen Integrationskonzepte für Feldbusse und IEC 1131 mit dem Ziel der einheitlichen Programmierung von Steuerungen und Feldbussystemen stehen mit den zu erwartenden Entwicklungen im Einklang und werden von ihr gefördert. Sie erschließen letztlich die Vorteile von feldbusgestützten Systemen für den Anwender, ohne detaillierte Kenntnisse über interne Details konkreter Busprotokollimplementierungen vorauszusetzen, und tragen in hohem Maße zu einer einheitlichen Betrachtungsweise von Feldbussen sowie zur Verbesserung ihrer Akzeptanz bei.

Literatur

1. n.n.: DIN IEC 1131 - Speicherprogrammierbare Steuerungen. Beuth Verlag GmbH Berlin, 1992.

2. Neumann, P.; Grötsch, E.; Lubkoll, C.; Simon, R.: SPS-Standard: IEC 1131. Programmierung in verteilten Automatisierungssystemen. R. Oldenbourg Verlag München Wien, 1995.

3. Sperber, M.: Distributed IEC 1131-programming and CANopen. ICP'96 1st international PLCopen conference on Industrial Control Programming. Paris, 01.-02. Oktober 1996, Tagungsband S. 87-90.

4. Wollschlaeger, M.: Integration of P-NET into IEC 1131. 4th International conference on the P-NET Fieldbus system. Porto, 02.-03. Mai 1996, Tagungsband.

5. Wollschlaeger, M.: Aspects Of the Integration Of Fieldbusses Into IEC 1131. ICP'96 1st international PLCopen conference on Industrial Control Programming. Paris, 01.-02. Oktober 1996, Tagungsband S. 81-85.

6. Johansen, J.: P-NET in the RACKS project. 4th International conference on the P-NET Fieldbus system. Porto, 02.-03. Mai 1996, Tagungsband.

7. n.n.: Open Control Interface. InterBus-S Club e.V., Baden Baden, 1996.

8. n.n.: OLE for Process Control specification v 1.0. OPC task force, 29.08.1996.

Simulation of Function Block Networks in Distributed Field Applications

Reinhard Herzog and Kym Watson

Fraunhofer-Institut für Informations- und Datenverarbeitung (IITB),
Fraunhoferstr. 1, D-76131 Karlsruhe

Abstract. Field applications in process control and manufacturing are experiencing a trend towards modularization and distribution of functionality across available computing and communication resources. Function blocks have become a widely accepted technique to describe distributed applications. The paper defines an abstract function block model which supports common execution policies. The model is described from the view point of simulation tools which are a valuable aid in assessing the dynamic behaviour of a distributed system. The main characteristics of the simulation tool OOST of the Fraunhofer Institute IITB are mentioned.

1 Introduction

International efforts to define digital field communication systems have led to a number of fieldbus solutions. Recent efforts have been increasingly dedicated towards the specification of function blocks to describe the application modules running over the communication platform provided by the fieldbus itself. The most common function blocks include analogue input and output and PID control. Function block methodology is used to define applications in industrial process measurement and control systems. Such systems can be viewed as a collection of devices interconnected by field communication networks. The applications are modelled as function block networks, whereby each function block is assigned to a resource of a device. The function blocks are linked via their input and output channels, which may represent a remote link over a fieldbus as required. A function block is broadly characterised by its inputs, outputs and states and by its algorithm (the function of the block which calculates output values based on the input data and state, including internal parameters). A function block model must also include rules on the execution of the blocks in a distributed system with (limited) resources. The model covers the flow of data and events through the function block network and together with the algorithms allows measurement and control systems to be described. The standardisation group IEC TC65 WG6 has defined one of the most notable generic function block models [1]. We will follow the terminology used there. Specific function block specifications available are those of the Fieldbus Foundation [4] and PROFIBUS [5].

A network of function blocks can in the first instance be conceived independently of its mapping onto the available resources, i.e. devices and field communication networks. This opens up the possibility of creating distributed applications and offers the application programmer a high degree of flexibility in structuring his application.

The mapping of function blocks onto resources, however, is an essential task in a field automation system as typical field devices can run only certain function blocks and the communication bandwidth between devices is limited. This task is further complicated by the need to schedule when function blocks may use the available resources in order that the resulting dynamic behaviour of the function blocks meets user requirements (such as on control algorithms). A further significant aspect arises when considering the dynamic behaviour of a system in abnormal conditions (e.g. response to stale measurement values, or to an alarm situation). This raises the need for tools to assess the dynamic behaviour of function blocks, both before and after the mapping onto the resources. Ideally, such tools should be integrated into the complete design process of a distributed automation system as in the approach taken by the BRITE-EURAM project ACORN [2].

This paper will first define a general function block model relevant to modern automation systems based on fieldbus. The description will apply an abstract approach in order to identify those model features essential for simulation.

The paper will then outline a function block simulation system based on the simulator OOST (Object Oriented Simulation System) of the Fraunhofer Institute IITB. The simulation system can be used to assess the dynamic behaviour of an automation system, both in the early design phase and after the function blocks have been assigned to resources. Observation points such as the time to execute a given control loop can be defined and monitored. The loading of the fieldbus segments and resource utilisation can be assessed. An integrated animation of the process values allows the impact on the process to be visualised and the control strategy to be evaluated for different scenarios.

2 An abstract function block model

In contrast to the various function block specifications which leave certain aspects up to the specific implementation, we require a precise model in order to be able to simulate it. We first define the concept of a function block, then a network of function blocks, followed by the assignment of function blocks to resources and an execution (or scheduling) model. The model is in part an abstraction and, where necessary, a specialisation of the function block architecture defined in [1]. The specification of function block types for specific purposes is not of concern here, although a simulation tool would in practice use the same types as defined by or used by other tools in the system engineering process.

We consider a function block to be an automaton with data inputs I_d, event inputs I_e, data outputs O_d, event outputs O_e and states Z. The states are used to model the various modes of function blocks such as out of operation, manual and automatic. States also cover the notion of parameters internal to the function block. The state space Z can typically be written as $P \times Z'$, where P is the internal parameter space and Z' is a set of function block modes. An input (x, y) with $x \in I_d$ and $y \in I_e$ of a function block is a pair of finite vectors x_i, i=1,...,n and y_i, i=1,...,m of data and event inputs respectively corresponding to the input channels. These input variables can be general structures (for example a process value together with an associated status value). Similar statements apply to the outputs. The execution of a function block is a transition

$$I_d \times I_e \times Z \rightarrow O_d \times O_e \times Z.$$

That is, for a given inputs $x \in I_d$ and $y \in I_e$ and state $z \in Z$, the function block executes an algorithm to compute the outputs $x' \in O_d$ and $y' \in O_e$ and the new state $z' \in Z$. We assume that the values of the input variables at the beginning of execution are taken by the algorithm (snap on input values at execution begin). The invocation of the execution will be described below.

Function blocks are linked by connecting the outputs to inputs, whereby each data input is connected to exactly one data output and each event input is connected to exactly one event output. An output may remain unconnected. On the other hand, an output may be connected to several inputs. An input or output may be connected to an external instance representing for example the technical process / user interaction or the communication system. In concrete fieldbus function block specifications the link to the process is often achieved via special "transducer blocks".

Function blocks are assigned to "resources", which represent facilities responsible for their execution (such as a CPU or communication medium). We assume as in [1] that each function block is assigned to exactly one resource, although generalisations such as the assignment of a function block to several resources may yield better device models (this applies to composite function blocks; e.g. that m of n given resources must be simultaneously available). Such generalised models can be handled easily by a simulation tool. In any case, a device may contain several resources.

Each resource possesses a "scheduling function" which determines when a function block is executed after it has been invoked. The invocation is triggered by an event input. The event input may come either from a connected function block (including the same function block) or from the scheduling function. In the latter case the scheduling function could generate events based on a time schedule (as [4, 5]). A wide range of execution schemes can be realised with these mechanisms, ranging from as fast as possible to clock driven. We assume in our model that every event input is notified to the scheduling function which processes the events in a first come first served policy (various generalisations are conceivable, but unlikely in practice). The event is processed by executing the function block algorithm with the input variables and state current at the start of processing. The (new) output variables and successor state are activated upon completion of the algorithm. The time for the resource to execute the algorithm is a resource-specific parameter (constant or given via a probability distribution) which has to be set in the simulation model. It is assumed that there is no delay in transferring the new output variables to the linked inputs. Thus in this model event inputs are queued for processing, whereas data inputs are overwritten by subsequent inputs (only the current values are sampled). However, it is apparent that the model allows "data" to be transferred with the events.

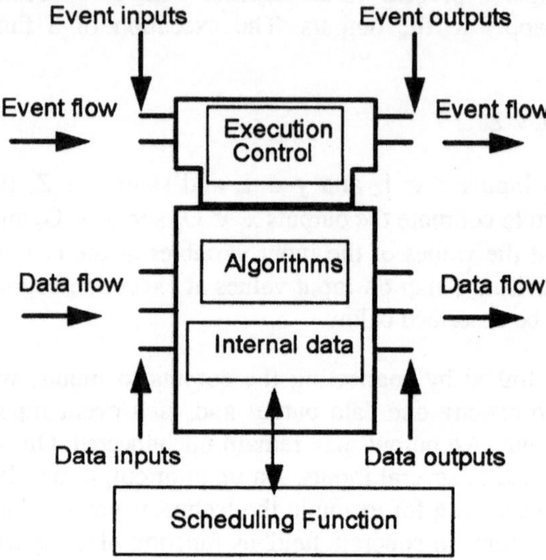

Fig. 1. Function Block Model [1]

3 Simulation

3.1 Objectives

There are three main areas where simulation can be applied: visualisation of dynamic behaviour, operator training and quantitative assessment of system performance. These are features of the simulator OOST (Object Oriented Simulation System) of the Fraunhofer Institute IITB.

An important aid in verifying the dynamic behaviour of the system is the ability to animate selected parts of the function block network by visualising the dynamic behaviour of function block variables.

By incorporating the interface to the user one can extend the functionality of the simulation to training applications. Training scripts can be executed and the user reaction evaluated.

Simulation has been and is often used to gather statistical data which can be taken to evaluate performance metrics. In the context of function block networks it is important to assess the execution rate of function blocks (average and fluctuation), the time to transfer data / events through a part of the function block network and the utilisation of resources. The former is relevant to assessing the quality of control loops and other control operations. The latter indicates where bottle necks have arisen / are likely to arise.

The first two types of simulation can be conducted on the function block network before it has been assigned to resources: that is in the application design phase. The affect of individual unit operations in the control system (for example associated with a phases of a production process or following reconfiguration after faults) can be tracked though the network.

3.2 Approach

The simulation of a distributed field application covers not only the simulation of the function blocks itself, but also the simulation of the process which is connected to the function blocks as well as the user actions interacting with the application.

When modelling a system for simulation several levels of detail are possible. The level chosen depends mainly on the available system information and the amount of time which can be invested in the modelling.

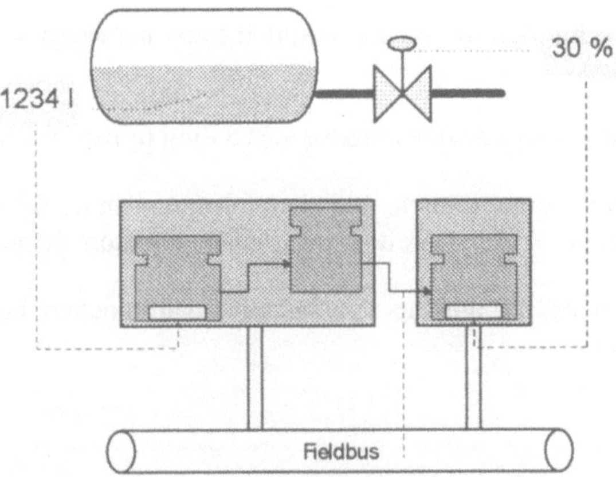

Fig. 2. Scope of Simulation

Where simulation systems are integrated into the actual design process, the deepest level of simulation can be chosen for the function blocks, which is the emulation of function blocks. It is feasible to directly execute the actual code of the function blocks in the simulation system (i.e. a function block emulation), so that in this case no abstraction from the real system is needed. In the ACORN toolkit [2] the function block code is given in the standardized language IEC 1131-3 [3], which is the related language definition for [1].

Beside the function blocks themselves the environment of the application must also be modelled and simulated. This environment covers the controlled process, the configuration of resources on which the application is running and the user interaction.

64

System engineers often study the process by using a simulation model. Then it is important to have an open simulation solution in order to be able to integrate existing models running in third party simulation systems.

The function blocks define only the logic of the application. In order to execute the application real resources such as devices, CPUs and fieldbus networks are required. Here, additional information is required, e.g. device models, communication protocols and performance measurements. In sophisticated simulation systems this is typically available as model components which are automatically configured according to the configuration information delivered by the integrated configurations system [2].

4 Influence of the communication platform

The function block model as described above is independent of the fieldbus architecture. Nevertheless it is obvious that the existing fieldbus concepts will have different solutions for the realisation of such function blocks. Important influencing factors are:

- how does the fieldbus handle events? (explicit event messages or hidden as flags in normal messages)

- how is scheduling realised? (event based scheduling or time based)

- timing behaviour? (deterministic and therefore good basis for the calculation of jitter / asynchronous therefore better performance but more difficult to handle)

- flexibility? (is a device able to dynamically load function bocks?; is dynamic linking possible)

References

1. IEC TC65/WG6(PT1ACD)1, Committee Draft - Function Blocks for Industrial Process Measurement and Control Systems, June 5 1997.

2. Brownlie, I., Herzog, R. and Simon, R.: Automatisierung mit Verteilten Systemen, atp 6/97, pp46-55.

3. Programmable controllers - Part 3: Programming languages, IEC 1131-3, 1993.

4. Fieldbus Foundation: User Layer Technical Specification (Function Block Application Process Parts 1 and 2), Order No. FF-002-1.0-1-1, 1997.

5. PROFIBUS Nutzerorganisation: PROFIBUS-PA Profile for Process Automation Devices, Order No.: 3.042, 1996.

Analyse von Profibus-Netzwerken durch Simulation

Georg Marschall

Universität der Bundeswehr Hamburg
Fachbereich Elektrotechnik
Holstenhofweg 85   D-22043 Hamburg

Kurzfassung. Die Leistungsdaten eines PROFIBUS-Netzwerkes, insbesondere die Systemreaktionszeit, hängen nicht nur von der benutzten Datenübertragungsrate ab, sondern werden auch maßgeblich von der Netzwerk-Konfiguration und den verwandten Profibusimplementierungen beeinflußt. Oftmals sind daher für komplexere Netzwerke bereits vor der eigentlichen Inbetriebnahme Versuchs- und Testaufbauten notwendig, um die Leistungsfähigkeit eines dezentralen Automatisierungskonzeptes beurteilen zu können.
Die Simulation konkreter Profibus-Netzwerke liefert entsprechende Ergebnisse bei wesentlich geringerem Geräte- und Zeitaufwand. Darüber hinaus ermöglicht sie auch ein Abschätzen des Optimierungspotentials der simulierten Anwendung.
Im vorliegenden Beitrag wird ein solches Simulationsmodell basierend auf höheren Petri-Netzen für dezentrale Automatisierungssysteme beschrieben und anhand konkreter Profibus-FMS-Netzwerke im Mono- und Multi-Master-Betrieb angewandt.

Abstract. This paper introduces a high level Petri Net model of distributed Automation facilities on the basis of a Profibus-FMS network, which enables analysis of the application under user aspects by simulation.
The model is based on the developments described in [9] and considers now mono- and multi-master systems, so that Profibus networks consisting of several PLC's can be simulated in general. The application is based on a typical PLC-architecture and Profibus-implementation. The functionality of the system is discussed and modelled in a hierarchical net structure, which allows an easy modification of the model by exchanging or extending single subnets.
The model is verified by different aspects indicating the efficiency of automation facilities.

1 Problemstellung und Abgrenzung

Automatisierungssysteme lassen sich in die drei Hauptbereiche "Regel- bzw. Steueralgorithmus", "Automatisierungseinrichtung" und "Technischer Prozeß" aufteilen, wobei der Algorithmus über die Automatisierungseinrichtung auf den Prozeß zugreift, um ein gewünschtes Prozeßverhalten zu erzielen. Bei dezentralen Automatisierungssystemen besteht die Automatisierungseinrichtung aus mehreren Automatisierungsgeräten in denen der Algorithmus als Programm abgelegt ist, sowie aus verschiedensten Peripheriegeräten zur Instrumentierung des Prozesses. Alle Geräte werden zur Kommunikation über einen Feldbus miteinander verbunden (Fig. 1).

66

Die Automatisierungseinrichtung umfaßt damit sowohl das verwandte Kommunika-
tionssystem, als auch die Implementierung der Kommunikationsschnittstellen, die
Architektur der Steuerungsgeräte sowie die Funktionalität zugehöriger Schnittstellen.
Der Steueralgorithmus und der technische Prozeß gehören hingegen nicht zur
Automatisierungseinrichtung.

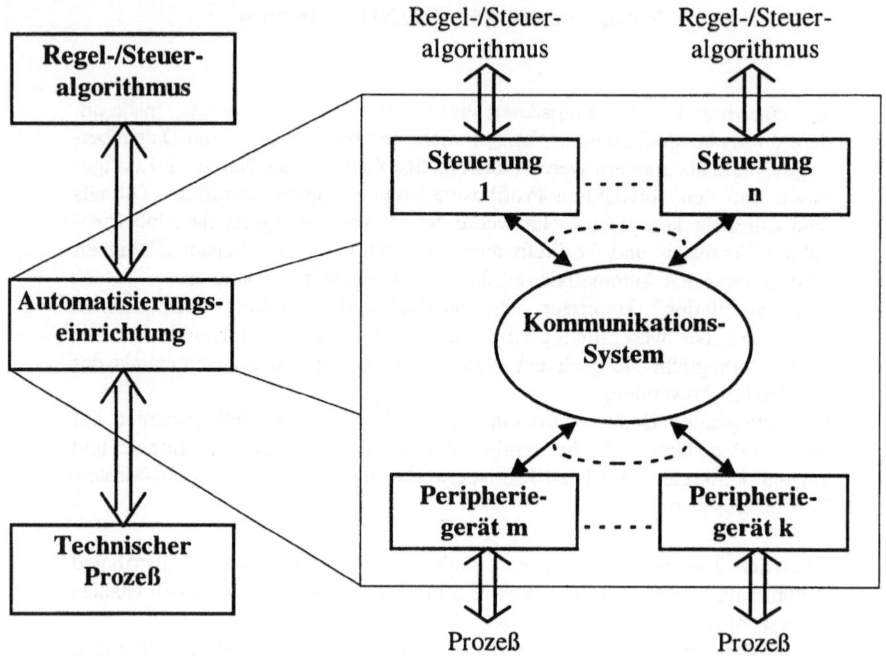

Fig. 1. Modell eines Automatisierungssystems und seiner Automatisierungseinrichtung

Für die Leistungsfähigkeit eines kompletten Automatisierungssystems wird als typi-
scher Indikator oftmals die Systemreaktionszeit angeführt, d.h. die Zeit vom Auslösen
einer Aktion durch den Algorithmus bis zur Auswirkung dieser Aktion an der
Prozeßschnittstelle. Diese Zeit wird nun maßgeblich von der oben beschriebenen
Automatisierungseinrichtung bestimmt, da sie sowohl durch die technischen Daten des
Kommunikationssystems als auch durch die Konfiguration sowie die Integration der
Kommunikationsschnittstellen innerhalb der Automatisierungs- und Peripheriegeräte
bestimmt wird.

Die Simulation einer solchen Automatisierungseinrichtung ermöglicht daher unabhängig von der eigentlichen Anwendung die Leistungsanalyse konkreter Feldbusnetzwerke für verschiedene Applikationen, da die Modellteile Algorithmus und
Prozeß, in denen sich die eigentliche Anwendung widerspiegelt, nicht berücksichtigt
werden brauchen.

2 Entwicklung des Simulationsmodells

Für die Modellierung dezentraler Automatisierungseinrichtungen eignen sich insbesondere höhere, farbige Petri-Netze, wie etwa in [6] beschrieben, da sich Kommunikationsnetzwerke durch ein hohes Maß an Nebenläufigkeit, Komplexität und Parallelität auszeichnen. Diese sind analytisch oft nur schwer zu beschreiben, so daß sich eine Analyse auf Simulationsbasis, wie z.B. auch in [4] angewandt, anbietet. Zur Programmentwicklung wurde ein kommerzielles Werkzeug [5], basierend auf erweiterten, zeitbewerteten Petri-Netzen [3], verwandt. Neben der Simulation mit Hilfe von Petri-Netzen finden sich auch Modellierungen von Feldbussystemen aus informationstheoretischer Perspektive, z.B. basierend auf Warteschlangensystemen, wie etwa in [7, 8].

Die Modellierung eines Profibus-Master, z.B. einer Speicherprogrammierbaren Steuerung (SPS), ist aufgrund der hohen Funktionalität von besonderem Interesse und soll im folgenden kurz vorgestellt werden. Prinzipiell stellt sich dabei zum einen die Aufgabe der Modellierung der Masterfunktionalität des Kommunikationssystems, während die Modellierung der Systemarchitektur des Automatisierungsgerätes die zweite große Komponente darstellt. Ein Automatisierungsgerät mit Kommunikationsinterface arbeitet dabei typischerweise als Zweiprozessorsystem [10]. Dabei muß ein Hauptprozessor in einer Multitaskumgebung neben dem eigentlichen Steuerprogramm auch weitere Tasks bearbeiten, während davon unabhängig ein eigener Kommunikationsprozessor das Protokoll abarbeitet. Der Datenaustausch zwischen dem Programm und dem Kommunikationssystem wird dabei von einer der Tasks des Hauptprozessors durchgeführt, dem sogenannten Application Layer Interface (ALI) (Fig. 2).

Aus dieser Architektur ergeben sich bereits wichtige Parameter für die Leistungsfähigkeit der Automatisierungseinrichtung, nämlich die Zykluszeit der Multitaskumgebung, die Operationszeit der ALI-Task und die „Breite" des Datentransfers zwischen ALI und Kommunikationsschnittstelle, die alle im Modell als Systemparameter Eingang finden.

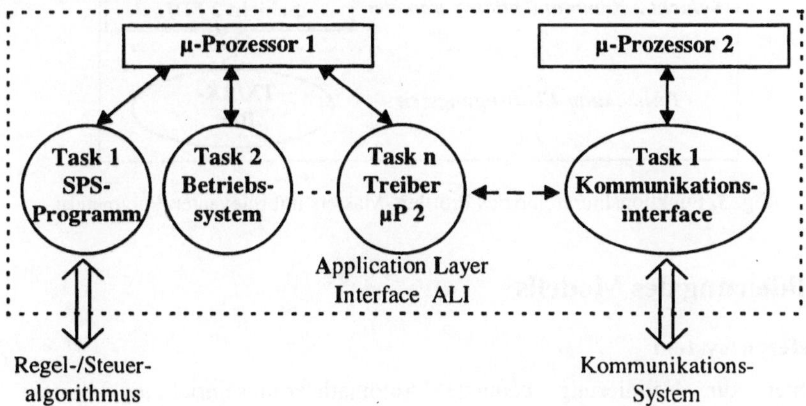

Fig. 2. Systemarchitektur eines Automatisierungsgerätes mit Kommunikationsinterface

Das Modell des Kommunikationssystems orientiert sich prinzipiell am ISO/OSI-Modell. Das Profibus-Modell enthält dabei die Schichten 7 (FMS) und 2 (FDL), wobei die Erstellung der verschiedenen Profibusfunktionen als Petri-Netz aufgrund der entsprechenden schriftlichen Beschreibung dieser Funktionen, etwa in [1] und [2], durchgeführt werden muß. Jede Funktionalität wird dabei als eigenes Unternetz modelliert, so daß sich eine hierarchische Netzstruktur ergibt. Dies erleichtert die Erweiterung des Modells um zusätzliche Funktionen und schafft eine transparente Netzstruktur. Bei der Modellierung der einzelnen Schichten des Profibus sind insbesondere auch die unterschiedlichen Kommunikationsparameter, wie Token Target Rotationtime (T_{TR}), Station Delay Initiator (T_{SDI}), Slottime (T_{SL}) etc. zu berücksichtigen, welche meist als Verzögerungszeit für das Feuern einer Transition in das Petri-Netz Eingang finden.

Jede Komponente der Automatisierungseinrichtung (SPS, E/A-Modul, ...) wird so zuerst als mehrschichtiges Funktionsmodell entwickelt. Anschließend werden die einzelnen Funktionalitäten in Petri-Netze umgesetzt, welche zusammen dann ein hierarchisches Petri-Netz für jede Komponente aufbauen. In Abbildung 3 sind die obersten Funktionsebenen des Profibus-Masters mit einigen zugehörigen Zeit-parametern dargestellt. Eine ausführliche Beschreibung der Funktionsschichten und der Modellierung ist in [9] aufgeführt.

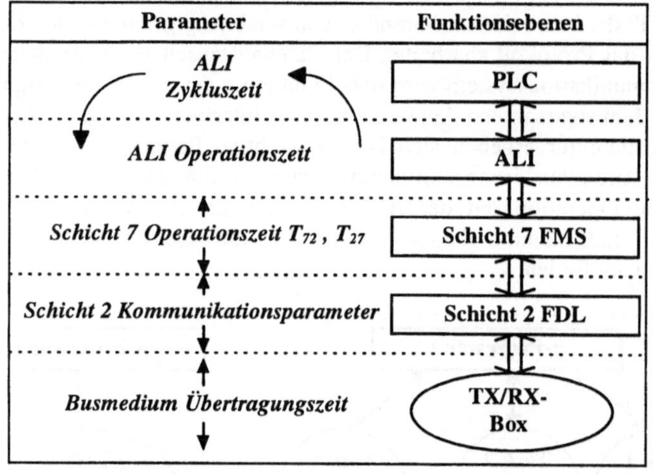

Fig. 3. Funktionshierarchie des Profibus-Masters mit relevanten Parametern

3 Validierung des Modells

3.1 Referenzsystem

Die hier zur Validierung benutzte Automatisierungseinrichtung besteht aus Speicherprogrammierbaren Steuerungen und E/A-Modulen der Firma Kuhnke GmbH. Bei der SPS handelt es sich um ein Zweiprozessorsystem mit im allgemeinen drei Tasks des Hauptprozessors. Für eine genauere Beschreibung der Prozessorkopplung und der Funktionalität des ALI sei auf [9] verwiesen.

Die Profibus-Implementierung der SPS sieht azyklische Master-Slave-Verbindungen und Master-Master-Verbindungen vor. Demzufolge werden Master-Slave-Dienste auf einer niederprioren SRD-Verbindung (Send and Request Data, SRD_{low}) und Master-Master-Dienste auf einer SDA-Verbindung (Send and Request Data with Acknowledge, SDA) ausgetauscht.
Das Validierungssystem stellt damit also eine typische Profibus-Systemanbindung und -Implementierung dar und kann daher als repräsentativ angesehen werden.

3.2 Simulationsergebnisse

Die Modellvalidierung muß prinzipiell für verschiedene Netzwerk-Konfigurationen durchgeführt werden, um die Verläßlichkeit des Modells nachzuweisen, wobei insbesondere der Betrieb von Mono- und Multi-Master-Anwendungen zu berücksichtigen ist. Da es sich bei dem Simualtionsmodell um eine mikroskopische Simulation handelt, also jedes Telegramm einzeln abgearbeitet wird, bietet sich als Validierungskriterium einerseits die Wiederholrate bzw. Häufigkeit bestimmter Telegrammtypen an, welche meßtechnisch mittels eines Busmonitors und anschließender Auswertung einfach zu ermitteln ist. Andererseits ist aber auch eine anwendungsorientierte Validierung sinnvoll, um zu brauchbaren Simulationsergebnissen für Applikationen zu gelangen. Die bereits erwähnte Systemreaktionszeit stellt einen solchen Aspekt dar. Als Indikator für diese Systemreaktionszeit bietet sich die Ausführungszeit eines Schreibbefehls innerhalb der Automatisierungseinrichtung an. Die Ausführungszeit ist dabei die Zeit vom Setzen eines dezentralen Ausgangs innerhalb des SPS-Programms (Master) bis zum physikalischen Signalwechsel diese Ausgangs auf dem entsprechenden Teilnehmer (Slave). Da diese Zeit variiert, werden jeweils die Minimal-, Maximal- und Mittelwerte der Ausführungszeit über 500 Messungen (Schreibbefehle) ermittelt. Größere Stichproben wirken sich auf den Mittelwert nur geringfügig aus, ergeben jedoch für Minimum und Maximum bessere Deckungen.
Mit der Telegrammverteilung und der Ausführungszeit eines Write-Service stehen also zwei verschiedene charakteristische Merkmale der Automatisierungseinrichtung zur Validierung bereit. Jede Testkonfiguration wird mit verschiedenen Einstellungen des Kommunikationsparameters T_{SDI} gemessen bzw. simuliert, da dieser als "Master-Wartezeit" unmittelbaren Einfluß auf die Validierungskriterien hat. Bei Profibus-Netzwerken mit Adreßlücken wird zusätzlich der Parameter T_{SL} variiert, da dieser bei jedem Aufruf einer solchen Adreßlücke durch den Master zum Tragen kommt und sich deshalb ebenfalls in beiden Kriterien widerspiegelt.
In den Abbildungen 4 bis 8 ist exemplarisch die Validierung eines Mono-Master-Systems bestehend aus einem Master und fünf Slaves und einem Multi-Master-Netzwerk bestehend aus zwei Mastern und je drei Slaves dargestellt. Änderungen der Kommunikationsparameter beziehen sich dabei stets auf alle vorhandenen Master. Grundsätzlich muß bei der Validierung von Multi-Master-Systemen mit höheren Abweichungen als bei Mono-Master-Systemen gerechnet werden, da aufgrund einer höheren Anzahl von Mastern sich auch Modellvereinfachungen im Master stärker auswirken. Dieser Effekt ist mit wachsender Slavezahl nicht so ausgeprägt, da deren Struktur weniger komplex ist und somit weniger Modellvereinfachungen vorgenommen wurden.

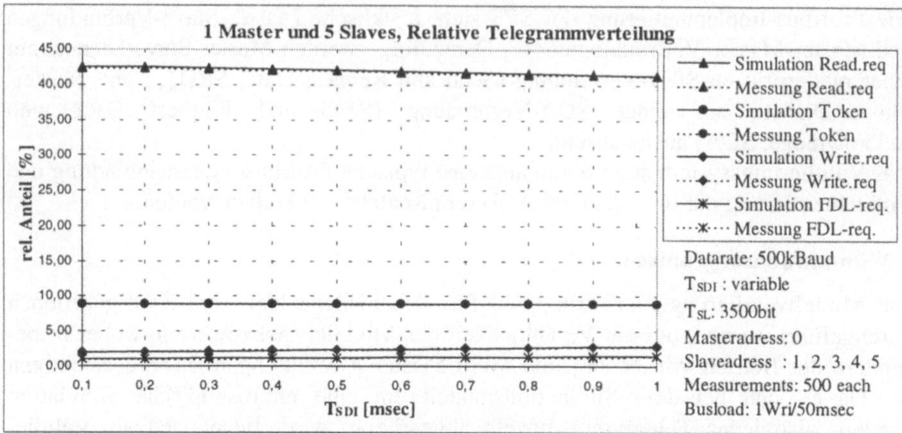

Fig. 4. Mono-Master-System, Relative Telegrammverteilung über T_{SDI}

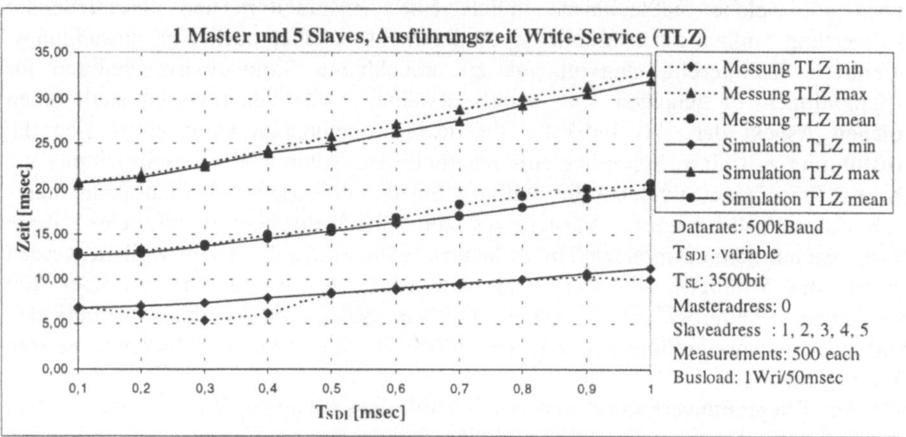

Fig. 5. Mono-Master-System, Ausführungszeit eines Write-Service über T_{SDI}

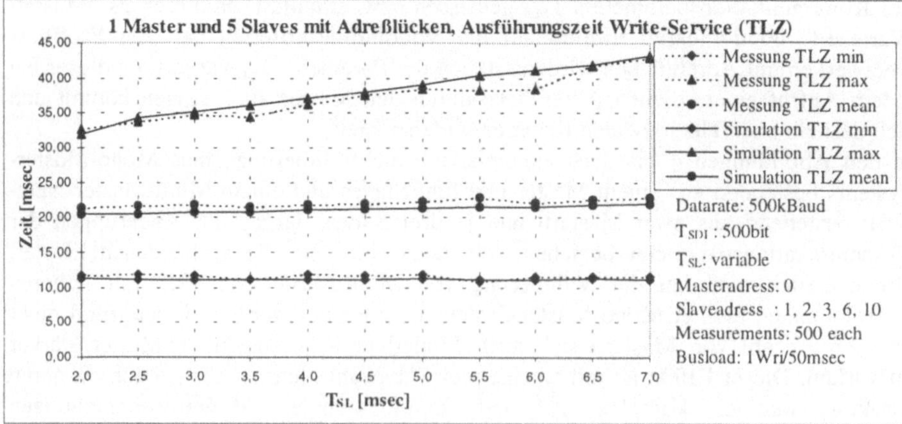

Fig. 6. Mono-Master-System mit Adreßlücken, Ausführungszeit eines Write-Service über T_{SL}

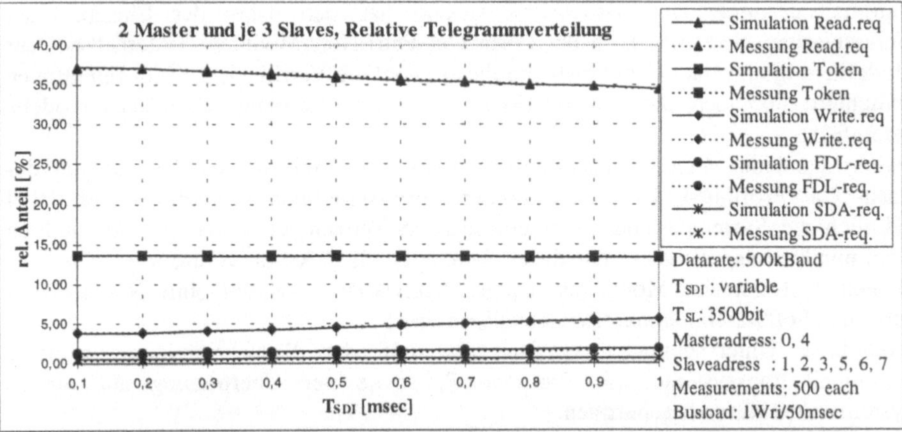

Fig. 7. Multi-Master-System, Relative Telegrammverteilung über T_{SDI}

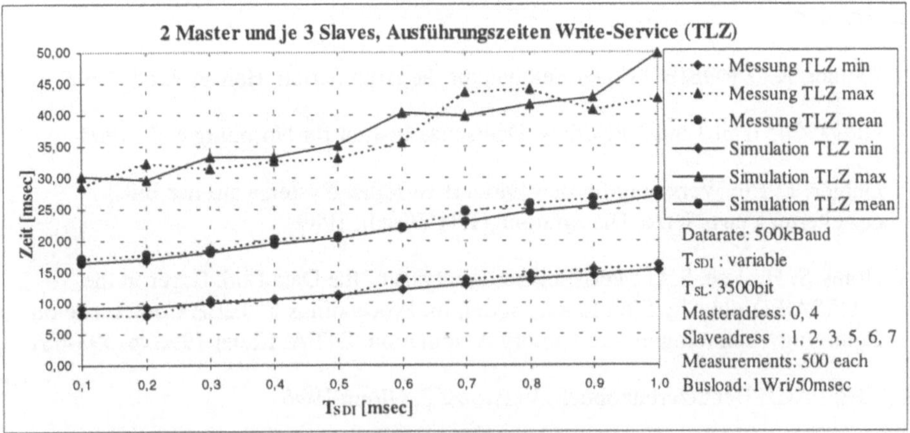

Fig. 8. Multi-Master-System, Ausführungszeit Write-Service über T_{SDI}

4 Zusammenfassung

Die Validierung des Modells anhand der beschriebenen Testaufbauten zeigt eine gute Approximation des Systems durch die Simulation in verschiedenen Betriebsbedingungen sowohl hinsichtlich des Kommunikationsprotokolls, als auch bezüglich des Gesamtverhaltens. Abweichungen sind vor allem auf Vereinfachungen im Modell der Multitaskumgebung der SPS als auch im Modell der Schicht 7 der Profibus-Schnittstelle zurückzuführen. Die so erreichte Reduzierung des Modellierungsaufwandes rechtfertigt diese Vereinfachungen jedoch in vollem Umfang, da das Ziel der Simulation in der anwendungsorientierten Unterstützung von Projektierung und Analyse dezentraler Automatisierungssystemen und nicht in der Verifizierung einzelner Protokollfunktionen liegt. Die beschriebene Modellierung und Simulation dezentraler Automatisierungseinrichtungen mit Hilfe erweiterter Petri-Netze liefert so verläßliche Aussagen über die Leistungsfähigkeit eines darauf basierenden

Automatisierungssystems. Besonders bewährt hat sich dabei der Einsatz einer hierarchischen Netzstruktur in der Form von Petri-Netz-Modulen (Unternetzen) zur Modellierung einzelner Funktionen, da dies sowohl die Verifikation dieser Funktionen erleichtert, als auch die spätere Ergänzung oder Änderung des Gesamtmodells vereinfacht.

Das vorhandene Modell berücksichtigt Mono- und Multi-Master-Systeme auf der Basis von Profibus-FMS mit beliebiger Adreßverteilung. Grundlage sind dabei lediglich die Systemparameter der eingesetzten Teilnehmer sowie die eingestellten Kommunikationsparameter. Da diese Parameter als bekannt vorausgesetzt werden können, läßt sich das Modell auf andere Netzwerke prinzipiell ohne Kenntnis der einzelnen Softwareimplementierungen übertragen.

Zukünftige Untersuchungen werden sich mit der Berücksichtigung weiterer Kommunikationssysteme wie Profibus-DP, sowie der Übertragung auf andere Systemarchitekturen beschäftigen.

Literatur

1. Bender, K.: PROFIBUS: Der Feldbus für die Automation, Hanser 2. ed. 1992

2. DIN 19245 Teil 1und 2 Profibus, Deutsches Institut für Normung e.V., 1991

3. Dähler, J.: Ein Werkzeug für den Entwurf verteilter Systeme auf der Basis erweiterter Petri-Netze, Dissertation, ETH Zürich, 1989

4. Hong, S. H., Lee, S. G.: Performance Analysis of the Data Link Layer in the IEC/ISA Fieldbus by Simulation Model, In: Proceedings 5th IEEE Conference on Emerging Technologies and Factory Automation, ETFA, Kauai 1996, p. 593-601

5. IBE: PACE Benutzerhandbuch, Version 2.3, Glonn 1996

6. Jensen, K.: Coloured Petri Nets, Vol. 1, Springer 2. ed. 1996

7. Klehmet, U.: Leistungsbewertung und Optimierung von zeitbegrenzten Polling Systemen am Beispiel des Feldbusses PROFIBUS, Dissertation, VDI-Fortschrittberichte Reihe 10, Nr.371, VDI-Verlag 1995

8. Li, Y.: Bewertung der Echtzeitfähigkeit von Feldbus-Systemen, Dissertation, VDI-Fortschrittberichte Reihe 10, Nr.235, VDI-Verlag 1993

9. Marschall, G.: Petri Net Simulation of a Distributed Automation System Based on Profibus, In: Troch, I., Breitenecker, F. (eds): Proceedings IMACS Symposium on Mathematical Modelling, 2nd MATHMOD, (ARGESIM Report No. 11) Wien 1997, p.241-246

10. Schnieder, E.: Prozeßinformatik, Vieweg 2. ed. 1993

Echtzeitkopplung zwischen Maschinenemulatoren und Maschinensteuerungen

Joachim Schullerer

Lehrstuhl für Informationstechnik im Maschinenwesen, Prof. Dr. Klaus Bender, TU-München,

Abstract: To execute the softwaretest of machine-control-systems parallel to the development of the machine (simultaneous engineering), the machine is simulated (*virtual machine*) in real-time. During the simulation, the virtual machine is connected to the real machine-control-system by a high performance distributed real-time communication system, called I/O-Emulator. The Architecture of the I/O-Emulator is presented in the following article.

Zusammenfassung: Um den Steuerungstest parallel zur Maschinenentwicklung durchzuführen (Simultaneous Engineering), wird ein Simulationsmodell der Maschine auf einem Emulatorrechner ausgeführt. Der Emulatorrechner ist über ein Koppelsystem - der E/A-Emulation - mit den realen Maschinensteuerungen verbunden. Die E/A-Emulation integriert verschiedene Feldbussysteme und andere E/A-Systeme zu einem echtzeitfähigen und homogenen Kommunikationssystem. Zusätzlich unterstützt sie die Instrumentierung des Emulationsaufbaus mit Testwerkzeugen zum Test der Steuerungs-Software.

1 Motivation

Ein Hauptkennzeichen moderner Maschinen und Anlagen ist der hohe und weiter steigende Durchdringungsgrad mit Informationstechnik. Man spricht in diesem Zusammenhang auch von der *intelligenten Maschine*.

Dabei ist der Produktionsprozeß zur Herstellung intelligenter Maschinen und Anlagen weiterhin weitgehend sequentiell [1]. An die mechanische Konstruktion schließt sich die Elektrokonstruktion und die mechanische Fertigung an. Erst nach der Montage und der mechanischen Inbetriebnahme der Maschine erfolgt die elektrische Inbetriebnahme mit der Steuerungsinbetriebnahme und dem Softwaretest. Inzwischen beansprucht die elektrische Inbetriebnahme bereits bis zu 90% der gesamten Inbetriebnahmezeit [2] mit weiter steigender Tendenz.

Die Kapitalbindung ist in der Inbetriebnahmephase am höchsten. Fehler oder Mängel, die erst während der Inbetriebnahme entdeckt werden, sind aufwendig und teuer zu beheben. Die Inbetriebnahme erfolgt zunehmend beim Kunden, Fehler oder Verzögerungen zu dieser Zeit gehen mindestens mit einem Imageschaden einher, im schlimmsten Fall drohen Vertragsstrafen wegen Terminüberschreitungen.

Seitens der Industrie besteht daher ein großer Bedarf, die elektrische Inbetriebnahme während der kritischen Inbetriebnahmephase wesentlich zu verkürzen. Dies kann durch vorziehen von Aktivitäten erreicht werden, Stichwort hierfür ist der Begriff *Simultaneous Engineering*. Auf die Steuerungsinbetriebnahme bezogen heißt das Ziel, den Steuerungsaufbau und die Steuerungs-Software *parallel zur Maschinenfertigung* zu testen und in Betrieb zu nehmen.

2 Entwicklung intelligenter Maschinen

Am Lehrstuhl für Informationstechnik im Maschinenwesen (*itm*) der TU-München wird das Problemfeld *Steuerungsinbetriebnahme* im Rahmen des Projektes *SEMI* (Simultaneous Engineering bei der Entwicklung intelligenter Maschinen) bearbeitet [3].

Der im SEMI-Projekt gewählte Ansatz besteht darin, den Steuerungstest mit Hilfe einer Echtzeitsimulation der zu entwickelnden Maschine (*Maschinenemulation*) bereits vor der Maschineninbetriebnahme durchzuführen. Die realen Maschinensteuerungen sind über die Sensor-Aktor-Schnittstelle als *Hardware-in-the-Loop* an die *virtuelle Maschine* angeschlossen. Die Übertragung der Prozeßsignale zwischen Maschinensteuerung und virtueller Maschine wird als *E/A-Emulation* bezeichnet.

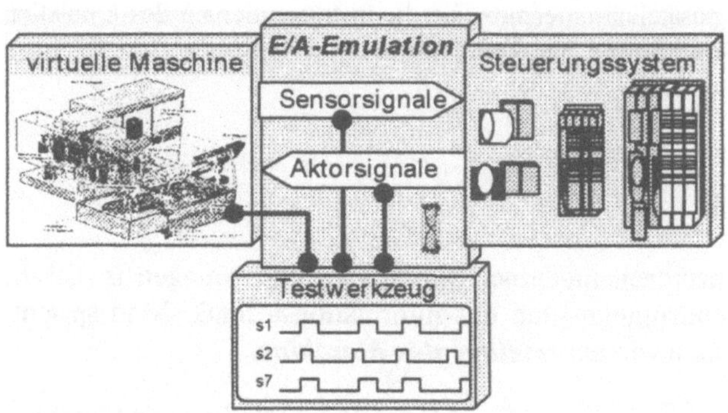

Abb. 1: Closed-Loop-Anordnung der Maschinenemulation

Zur Emulationszeit liest die virtuelle Maschine die Ausgangssignale der Maschinensteuerung als Aktorwerte ein, berechnet daraus und aus dem momentanen Zustand der virtuellen Maschine neue Sensorwerte und gibt diese als Eingangssignale an die Maschinensteuerung zurück, die wiederum neue Ausgangssignale berechnet, usw. Diese *Closed-Loop*-Anordnung ergibt die harten Echtzeitanforderungen an die Maschinenemulation und an die E/A-Emulation (Abb. 1).

Die Projektpartner und das *itm* verfolgen im SEMI-Projekt das Ziel, die Zeit für die elektrische Inbetriebnahme um 60 % zu reduzieren. Die gesamte Inbetriebnahmezeit soll um 40 % reduziert werden. Der vorliegende Beitrag behandelt den Schwerpunkt der E/A-Emulation.

3 Problemstellung und Ziel

Die E/A-Emulation hat die Aufgabe, die virtuelle Maschine über die Sensor-Aktor-Schnittstelle an die realen Maschinensteuerungen (*Feldgeräte*) und an physikalisch bereits vorhandene Maschinenkomponenten anzubinden. Da die Software der Feldgeräte getestet werden soll, darf das System unter Test - also die Maschinensteuerung - nicht verändert werden. Alle bei der Automatisierung von Maschinen und Anlagen gängigen Steuerungssysteme sowie Testwerkzeuge sollen angekoppelt werden können.

Durch die geringen Stückzahlen im Maschinen- und Anlagenbau hat die effiziente und wirtschaftliche Wiederverwendbarkeit sowie die leichte Handhabbarkeit der E/A-Emulation einen nicht zu unterschätzenden Stellenwert. Daraus resultieren hohe Anforderungen an die Konfigurierbarkeit, Parametrierbarkeit, Skalierbarkeit und Portierbarkeit der Lösung.

Die Randbedingungen für die E/A-Emulation bestimmen im wesentlichen der modellierte technische Prozeß mit seinen Signalen, die Modellierungsmethode, der Emulatorrechner und nicht zuletzt die Maschinensteuerungen.

3.1 Technischer Prozeß und Prozeßsignale

Der technische Prozeß definiert Art, Anzahl und Frequenz der Signale an der Prozeßschnittstelle. Anzahl und Frequenz der Signale können je nach Maschinentyp sehr stark variieren.

Über 95 % der Prozeßsignale sind *Binärsignale* (Endschalter, Ventile, usw.) mit Signalfrequenzen zwischen 1 Hz und 1 kHz auf. *Analogsignale* mit Signalfrequenzen von 10 Hz bis 40 Hz stellen den restlichen Anteil der Signale. Vereinzelt kommen *höherfrequente Signale* vor (Inkrementalgeber: 1 kHz bis ca. 100 kHz).

Stehen Signale in engem semantischen Zusammenhang, so enstehen hohe Anforderungen an die Übertragungspräzision. Die Übertragung hat dann nicht nur *rechtzeitig* und *isochron* sondern sogar zu einem definierten Zeitpunkt zu erfolgen.

3.2 Virtuelle Maschine und Emulatorrechner

Für die Maschinenemulation wird mit der objektorientierten Modellierungsmethode ROOM [4] ein Modell des technischen Prozesses erstellt: die virtuelle Maschine. Ein Echtzeit-Ablaufsystem führt das daraus automatisch erzeugte Programm mit Hilfe eines Echtzeit-Betriebssystems auf einer leistungsfähigen Multiprozessorarchitektur aus.

3.3 Feldgeräte

Feldgeräte sind kompakte oder modulare Steuerungen sowie feldbusfähige integrierte oder intelligente Sensoren/Aktoren [5]. Alle Feldgeräte verfügen i. d. R. über eine interne Speicherdarstellung der E/A-Signale, das *Prozeßabbild*. Die Kopplung an den technischen Prozeß erfolgt über *Signalklemmen* und zunehmend über dezentrale Feldgeräte. Modular aufgebaute Steuerungen bieten die Möglichkeit, Baugruppen in den *Rückwandbus* einzustecken. Integrierte und intelligente Sensoren/Aktoren haben als Schnittstelle zum technischen Prozeß physikalische Größen, die Feldbusschnittstelle dient der Anbindung an übergeordnete Steuerungen.

4 Architektur der E/A-Emulation

Die vielfältigen und zum Teil konträren Anforderungen definieren im Detail die von der E/A-Emulation zu lösende Aufgabe. Das zur Lösung implementierte Kommunikationssystem wird im folgenden als *I/O-Manager* vorgestellt (Abb. 2).

Auf einer PC- oder VME-Bus-Multiprozessorarchitektur mit dem Echtzeitbetriebssystem VxWorks übernimmt das Ablaufsystem (μRTS: microRunTimeSystem) die Aufgabe, die virtuelle Maschine unter Echtzeitbedingungen auszuführen. Teil dieses Ablaufsystems ist der I/O-Manager, dessen Anwendungsschicht als *Signal-Manager* und dessen Transportschicht als *Device-Manager* realisiert ist [6]. Die Anschaltung der Feldgeräte und der Testumgebung übernehmen sogenannte Stecker (*I/O-Connector, Test-Connector*).

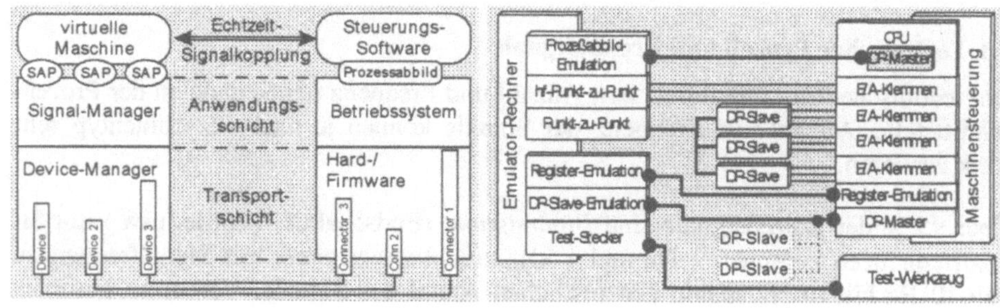

Abb. 2: Architektur und Aufbau der E/A-Emulation

4.1 Anwendungsschicht: Signal-Manager

Der *Signal-Manager* stellt der virtuellen Maschine für die Ende-zu-Ende-Kommunikation mit der Maschinensteuerung oder Testumgebung logische Kommunikationskanäle über Service Access Point (SAP) zur Verfügung (Abb. 2).

Der nicht echtzeitfähige *TestChannel* dient der Anschaltung der Testumgebung über einen Teststecker mit Funktionen für das Monitoring und Forcen von E/A-Signalen. Der *RtIoChannel* garantiert die Echtzeitübertragung typischer Prozeßsignale mit Frequenzen von 1 Hz bis 1 kHz zwischen Maschinensteuerung und virtueller Maschine. Der *HfRtIoChannel* übernimmt die Echtzeitübertragung höherfrequenter Signale mit Frequenzen ab 1 kHz bis 50 kHz.

Da die virtuelle Maschine ereignisorientiert arbeitet, die Steuerungssignale dagegen meist zyklisch abgetastet werden, übernimmt der Signal-Manager das Umschalten zwischen ereignisorientierter (virtuelle Maschine) und zyklischer (Steuerungen) Welt.

4.2 Transportschicht: Device-Manager

Der *Device-Manager* bildet die Transportschicht des I/O-Managers auf den ISO-OSI-Schichten 1 bis 4 und stellt die transparente Echtzeitverbindung zwischen Signal-Manager und Feldgeräteanschaltungen sicher. Er integriert als Treiber-Kernel die *Emulator-Devices* und stellt dem Signal-Manager deren Funktionalität über eine einheitliche Schnittstelle zur Verfügung.

Die Emulator-Devices sind die eigentlichen E/A-Geräte und bestehen aus einer Treiber-Software und aus einer Anschaltungs-Hardware. E/A-Geräte sind - z. T. intelligente - I/O-Boards oder Busanschaltungen (Profibus-DP, CAN, ASi, Ethernet, usw.). Die Koppelarten *Punkt-zu-Punkt-Kopplung, Prozeßabbild-Emulation, Feldgeräte-Emulation, Register-Emulation* und *MMI-Kopplung* sind als Emulator-Devices implementiert. Aus Platzgründen werden hier nur die Koppelarten erläutert, die neue Feldbusanwendungen beinhalten.

4.2.1 Feldgeräte-Emulation: Profibus-DP-Slave-Emulator

Hat die Maschinensteuerung ihre Prozeßperipherie mit einem Feldbus verteilt, so erfolgt die Kopplung über den *Feldgeräte-Emulator*, der die dezentrale Peripherie der Steuerung am Feldbus emuliert. Der am *itm* entwickelte Emulator für Profibus-DP kann 32 DP-Slaves bei 500 kBaud mit nur einer Schicht 1-Anschaltung emulieren.

Eine intelligente Profibus-PC-Karte mit zwei Mikrocontrollern bildet die Hardware-Basis. Ein Prozessor führt das erweiterte Profibus-Protokoll der Schicht 2 aus. Die Erweiterung um eine Slave-Datenbasis ermöglicht es, online so zwischen Profibus-Slaves umzuschalten, daß am Bus eine korrekte Reaktion auf Master-Dienste erfolgt.

Der zweite Prozessor bedient die Schnittstelle zum Device-Manager und führt die für die Slave-Emulation modifizierte DP-Schicht aus. Für jeden emulierten Slave existiert ein DP-Slave-Zustandsautomat, der durch die Destination-Address des empfangenen Telegramms ausgewählt wird und der die entsprechende DP-Slave-Aktion ausführt. Dabei wird die Slave-Datenbasis der Schicht 2 aktualisiert.

4.2.2 Prozeßabbild-Emulation

Die *Prozeßabbild-Emulation* implementiert eine Speicher-Speicher-Kopplung des Prozeßabbildes zwischen Maschinenemulator und Feldgeräten über einen Feldbus. Speziell bei Kompaktsteuerungen und bei intelligenten Sensoren/Aktoren spielt sie eine wichtige Rolle, da bei diesen Feldgeräten das Prozeßabbild sonst nur über Punkt-zu-Punkt-Kopplung (Kompaktsteuerungen) oder gar nicht (intelligenter Sensor/Aktor) manipuliert werden kann.

In Feldgeräten führen Betriebssystemfunktionen den Datenaustausch normalerweise zwischen Prozeßabbild und E/A-Baugruppen aus. Für die Prozeßabbild-Emulation führen die Betriebssystemfunktionen den Datenaustausch stattdessen zwischen Prozeßabbild und Maschinenemulator aus. Dies erfordert im Feldgerät die neue Betriebsart *Emulationsbetrieb*.

Am *itm* wurde ein Funktions-Prototyp für Siemens Steuerungen der Reihe S7-300/400 realisiert. Derzeit erfolgt der Ausbau des Prototypen um die nötige Bandbreite und Echtzeitfähigkeit zu gewährleisten. Die Kopplung erfolgt über Profibus-DP, da dieser Feldbus heute bei den meisten Steuerungen verfügbar ist und die Anforderungen der Maschinenemulation bzgl. Bandbreite und Echtzeitfähigkeit erfüllt.

4.2.3 Register-Emulation

Bei modular aufgebauten Steuerungen wird die Prozeßperipherie in Form von E/A-Baugruppen über Steckerleisten in den Rückwandbus gesteckt. Die Peripherie-Baugruppe verfügt neben der Rückwandbusanschaltung über ein Datenregister zur Zwischenpufferung der E/A-Daten (= Prozeßperipherie) sowie über die Ein-/Ausgabe-Elektronik zur Anschaltung der E/A-Signale an den technischen Prozess. Die Steuerungs-CPU greift auf die Datenregister der Peripheriebaugruppen über spezielle Backplane-Protokolle zu.

Die *Register-Emulation* emuliert alle in der Steuerung projektierten Peripherie-Baugruppen mit ihren Datenregistern über *eine* Rückwandbus-Anschaltung. Parallel dazu überträgt sie Signalveränderungen über einen leistungsfähigen Feldbus zwischen Maschinenemulator und Steuerung. Sie kann für modulare Steuerungen mit bekannter oder offengelegter Backplane-Architektur realisiert werden.

Am *itm* wurde die Register-Emulation für Siemens-Steuerungen der Baureihe S5 (95U, 115U-155U) mit dem Feldbus CAN integriert.

4.3 Feldgeräteanschaltung: I/O-Connector und Test-Connector

Der *I/O-Connector* realisiert die Anschaltung des I/O-Managers an die Feldgeräte. Einen Überblick über die verschiedenen Ausprägungen verschafft Tab. 1.

Koppelart	Ausprägung des *I/O-Connectors* in der Maschinensteuerung
Punkt-zu-Punkt	Vorhandene Steuerungsklemmen werden genutzt.
Feldgeräte-Emulation	Vorhandener Feldbus wird unverändert genutzt.
Prozeßabbild-Emulation	Emulationsbetriebsart, OPC-Server, Feldbuskopplung.
Register-Emulation	Modulare SPS: intelligente Backplane-Anschaltung mit Feldbus.

Tab. 1: I/O-Connector

Die Aufgabe des *Test-Connectors* kann mit der Aufgabe des Servicesteckers bei PKW verglichen werden. Er bietet Testwerkzeugen die Schnittstelle zum Test des gesamten Emulatorsystems *virtuelle Maschine - E/A-Emulation - Maschinensteuerung*. Der Test-Connector steht unter Windows-NT über TCP/IP als OPC-Client zur Verfügung.

5 Werkzeugunterstützung

Die E/A-Emulation wird in drei Schritten an die Prozeßperipherie einer neuen virtuellen Maschine angepaßt. Im *ersten Schritt* wird je nach Signalfrequenz und Feldgeräte-Schnittstelle die Kopplungsart ausgewählt und parametriert. Im *zweiten Schritt* erfolgt die logische Zuordnung der Sensoren/Aktoren der virtuellen Maschine zu Emulatorklemmen sowie deren Zuordnung zu Steuerungsklemmen. Im *letzten Schritt* wird die Hardware des Emulatorrechners aufgebaut und mit den Maschinensteuerungen physikalisch verbunden.

Zur Unterstützung dieser Arbeit wurde der *HW-Konfigurator* als grafisches Werkzeug unter Windows-NT implementiert. Der Benutzer wird bei seiner Arbeit von Assistenten geleitet. SPS-Symbol-Listen können importiert, Verdahtungslisten erzeugt und fertige Konfigurationen in einem offenen Datenformat abgelegt werden.

6 Beispielanwendungen

Am *itm* wurden Maschinen der Projektpartner unter Echtzeitbedingungen emuliert. Beispielhaft seien hier zwei der emulierten Maschinen kurz vorgestellt.

6.1 Transferhonmaschine, Firma Nagel

Die Transferhonmaschine der Firma Nagel hont Zylinderbohrungen in Motorblöcken. Die Motorblöcke durchlaufen verschiedene Honstationen der Transfermaschine. Die Steuerung besteht aus einer S5-115U mit 184 binären Eingängen und 136 binären Ausgängen mit Signalfrequenzen zwischen 1 Hz und 1 kHz. Alle Signale wurden über den *RtIoChannel* und über die *Register-Emulation* (CAN) an die S5-115U angeschlossen.

6.2 Folienreckmaschine, Firma Brückner

Die Folienreckmaschine der Firma Brückner erzeugt aus Kunststoffgranulat Folien definierter Dicke und Breite. Eine Teilmaschine, der Folienwickler, hat die Aufgabe, die erzeugte Folie aufzuwickeln und volle Rollen wärend des Betriebs gegen leere Rollen auszutauschen. Der Folienwickler wird von einer S5-135U gesteuert, die 7 Profibus-DP-Slave-Antriebe ansteuert und deren binäre (128 E, 80 A, 28 Relay-A) und analoge (12 E, 10 A) Signale über Profibus-DP-Slaves dezentral verteilt sind. Die E/A und die dezentralen Antriebe wurden über den *Profibus-DP-Slave-Emulator* und über den *RtIoChannel* an den virtuellen Folienwickler angeschlossen.

7 Zusammenfassung und Ausblick

Ziel im SEMI-Projekt ist es, der mittelständischen Maschinenbauindustrie die Vorteile der Hardware-in-the-Loop-Emulation zum Steuerungstest zu erschließen. Der Wert der Maschinen-Emulation steigt und fällt mit der Praxistauglichkeit (leichte Handhabbarkeit, Kosten) der Lösungen. Eine zentrale Rolle kommt dabei der Anbindung der Maschinenemulation an die reale und facettenreiche Steuerungswelt zu. Im vorliegenden Artikel wurde die E/A-Emulation als komplexes Kommunikationssystem zur Lösung dieser Aufgabe vorgestellt.

Details der Koppelarten, der Aspekt der Echtzeitfähigkeit des E/A-Emulators und das entwickelte Konfigurationswerkzeug hätten leider den Rahmen dieses Beitrages gesprengt und wurden von daher nicht oder nur am Rande behandelt.

Einige der entwickelten Komponenten der E/A-Emulation können auch in ganz anderen Anwendungsgebieten eingesetzt werden. So wird der Profibus-DP-Slave-Emulator für die Zertifizierung von Profibus-DP-Mastern eingesetzt. Der Device-Manager implementiert eine Schnittstelle zur neutralen Integration unterschiedlichster E/A-Systeme wie Feldbusse, I/O-Boards oder Ethernet-Anschaltungen. Er kann überall dort eingesetzt werden, wo es gilt, verschiedene Feldbusse neutral zu integrieren (s. a. Noah-Projekt).

Die Maschinenemulation wird künftig auch für die Schulung an Maschine und Leitstand sowie zur Unterstützung der Maschineninbetriebnahme eingesetzt. Vielfältige Anwendungen wie die Unterstützung des Verkaufs, im Rapid Prototyping, in der Fernwartung sowie bei der Online-Überwachung von Maschinen sind denkbar.

Literatur

1. Schaich, C.; Bender, K.: Simultaneous Engineering bei der Entwicklung intelligenter Maschinen. iwb-Seminarberichte (1996) 8.

2. Eversheim, W. (Hrsg.): Inbetriebnahme komplexer Maschinen und Anlagen. Düsseldorf: VDI-Verlag, 1990.

3. Simultaneous Engineering durch Maschinenemulation. cim 8, 1995.

4. Selic, B.; Gullekson, G.; Ward, T. W.: Real-Time Object-Oriented Modeling. New York: Wiley Professional Computing, 1994.

5. Bender, K.: Automatisierungstechnik. *itm* -Vorlesungsskript, 1994.

6. Effelsberg, W.; Fleischmann, A.: Das ISO-Referenzmodell für offene Systeme und seine sieben Schichten. Informatik-Spektrum, Springer, Jahrgang 9, Band 5.

Determinismus und Echtzeit bei Systemen mit INTERBUS-S

F. Raschbacher, H. Schweinzer

Institut für Elektrische Meßtechnik, Technische Universität Wien,
Gusshausstrasse 25-29, A - 1040 Wien

Abstract. INTERBUS-S ist vom Prinzip her deterministisch und für Echtzeitanwendungen besonders geeignet. In dieser Arbeit wird untersucht, bis zu welchen Größenordnungen der Zeitauflösung man bei INTERBUS-S von Determinismus sprechen kann. Weiters wird vorgestellt, wie INTERBUS-S in Echtzeitanwendungen und dezentralen busgekoppelten Regelkreisen mit hohen Taktraten eingesetzt werden kann.

Abstract. Due to its functional principle INTERBUS-S is deterministic and best suitable for real time applications. This work examines the timing resolution where INTERBUS-S can be considered to be deterministic. Further more it is shown how INTERBUS-S can be used with real time systems and distributed bus coupled closed loop systems with high clock rates.

1 Motivation

Determinismus und Echtzeitverhalten wirken sich stark auf die Anwendbarkeit eines Bussystems aus. Verbesserungen auf diesen Gebieten erschließen dem betreffenden Bussystem neue Märkte (z. B. Regelungstechnik in großräumigen Anlagen, sicherheitsrelevante Systeme im Bereich der Energieerzeugung und -verteilung, Medizin,...) und bieten damit universellere Einsetzbarkeit.

2 Grundlagen der Datenübertragung mit INTERBUS-S

2.1 Datenübertragung

INTERBUS-S ist ein Single-Master System und bildet ein Ringstruktur. Die einzelnen Geräte haben eine feste Datenbreite; die Datenbreite hängt von der Art des Gerätes ab und beträgt typisch 16-Bit oder 32-Bit (1 oder 2 Worte). Die gesamten Daten werden in einzelnen 8-Bit Datenblöcken übertragen. Die Datenübertragung kann man sich mit räumlich verteilten Schieberegisterstrukturen veranschaulichen. Jedes Feldgerät wird als Schieberegister mit entsprechender Datenbreite betrachtet. Zur Vereinfachung denken wir an 8-Bit Ein/Ausgabemodule.

Jedes Gerät liest zu Beginn seine Sensordaten (EDx) ein und legt diese in seinem Schieberegister ab (Abb.1). Der Busmaster schiebt nun das LoopBackWort (LBW, ein

Kennwort) und im Anschluß die Ausgangsdaten (ADx) für die Geräte, entsprechend ihrer physikalischen Lage im Ring, hinaus.

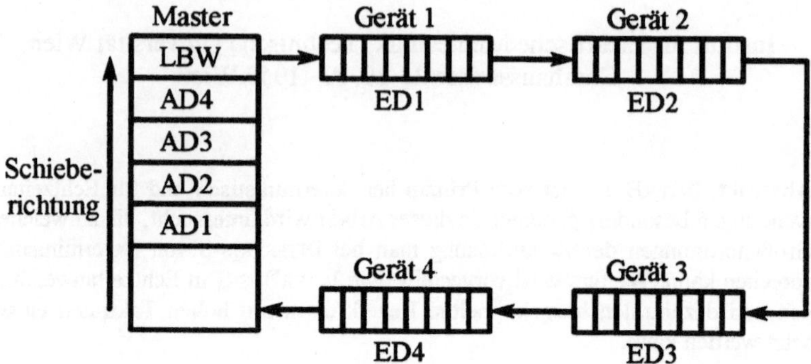

Abb. 1. Anordnung der Daten vor einer Datensequenz

Dadurch werden aber die Eingangsdaten, die bereits in den Schieberegistern der Geräte abgelegt sind, in den Busmaster zurückgeschoben.

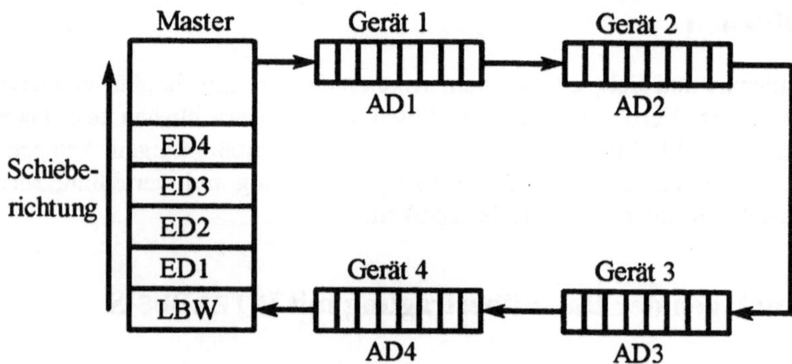

Abb. 2. Anordnung der Daten nach einer Datensequenz

Abb. 2 zeigt die Anordnung der Datenworte nach einer Datensequenz.

2.2 Timing

Die Übertragung der Daten wird nach folgendem Verfahren ausgeführt (siehe Abb.3). Der Sender (Busmaster, Anschaltbaugruppe) beginnt mit einer Datensequenz und wartet auf deren Beendigung. Ist das ausgesendete LoopBackWord wieder in den Busmaster zurückgekehrt, so ist die Datensequenz abgeschlossen und der Busmaster leitet die CRC-Phase ein. Erst nach abgeschlossener CRC-Phase beginnt der nächste Buszyklus.

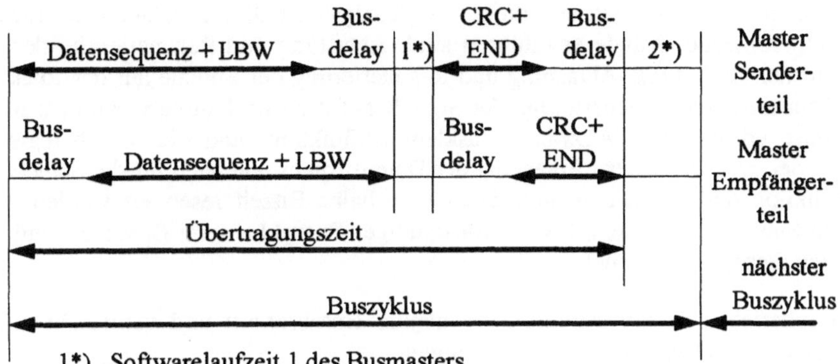

1*)...Softwarelaufzeit 1 des Busmasters
2*)...Softwarelaufzeit 2 des Busmasters

Abb. 3. Timing

Die Übertragungszeit besteht somit aus der Dauer einer Datensequenz, einer CRC-Phase, der Softwarelaufzeit 1 des Masters t_{sw} und der Dauer zweier Busdelays. Diese Busdelays ergeben sich aus dem Abwarten des Durchlaufs der Daten durch den Ring und setzen sich aus Laufzeiten der Signale auf den Leitungen t_{Ph}, Zeiten für die Abtastung der Signale (in jedem Teilnehmer) und Reaktionszeiten der Teilnehmer auf die ankommenden Daten zusammen.

Die Übertragungszeit $t_ü$ in einem InterBus-S System kann mit Hilfe der nachstehenden Formel im worst case-Fall abgeschätzt werden [1]:

$$t_ü = [13 \cdot (6 + n) + 4 \cdot m] \cdot t_{Bit} + t_{SW} + 2 \cdot t_{Ph}$$

$t_ü$ ....Übertragungszeit in Millisekunden
n ....Anzahl der Nutzdatenbyte
m ....Anzahl der installierten Fernbusteilnehmer
t_{Bit} ....Bitdauer $:= 2\mu s$ bei 500 kBit/s
t_{SW} ....Softwarelaufzeit 1 des Busmasters $:= 200\mu s$ (worst case)
t_{Ph} ....Laufzeit auf dem Übertragungsmedium, bei Cu: $16\mu s$/km Fernbuskabel (Hin- und Rückleitung)

Der Faktor 13 entsteht dadurch, daß jedes Byte mit 5 Bit Zusatzinformation (1 Startbit, 3 Controlbits und 1 Stopbit) übertragen wird. Zusätzlich zu den Nutzdatenbytes werden 2 Byte LoopBackWord, 2 Byte CR-Inhalt, und 2 Byte Endkennung übertragen.

InterBus-S verwendet Übertragungen mit Start/Stop-Synchronismus, d. h. jeder Teilnehmer hat einen eigenen Taktgenerator und diese Taktgeneratoren sind nicht synchronisiert. Mit der ersten Flanke des ersten übertragenen Bits synchronisiert sich der Taktgenerator im Teilnehmer mit dem Datenstrom und tastet die ankommenden Bits in der Bitmitte ab.

Damit treten Verzögerungen auf, die im ungünstigsten Fall den Faktor 4·m ergeben und wofür folgende Gründe anzuführen sind: Abtastung und Regenerierung der Signale auf der Hinleitung, Abtastung und Regenerierung der Signale auf der Rückleitung, Abtastung und Regenerierung der Signale auf der Rückleitung von einem möglicherweise vorhandenen Abzweig (Busklemmenfunktion) und die Schieberegisterfunktion der E/A-Daten. Pro Abtastung des Datenstromes und auch für die Schieberegisterfunktion der E/A-Daten muß daher eine halbe Bitzeit reserviert werden. Das ergibt in Summe 2 Bitzeiten. Da ein vollständiger Buszyklus zwei Busdelays enthält, wirkt sich dieser Wert doppelt aus.

Ein Peripheriebus ist ein synchron arbeitender Busabschnitt und verursacht daher bezüglich des Busdelays keine Verzögerungen.

2.3 Datenübernahme

Die Endkennung in Folge der CRC-Phase muß in diesem Zusammenhang noch näher betrachtet werden, denn hier wird die Datenübernahme eingeleitet.

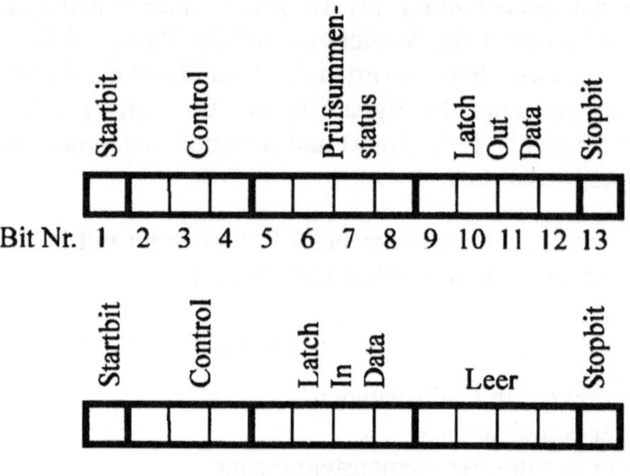

Abb. 4. Endkennung

Abb. 4 zeigt die Bedeutung der einzelnen Bits der Endkennung. Verursacht durch die Verzögerung auf Grund der mehrfachen Abtastung des Signals durch die Empfänger und die Laufzeit der Signale auf den Busleitungen, ist für jeden einzelnen Teilnehmer die LatchOutData-Phase je nach Anordnung des Teilnehmers im Ring früher oder später im Vergleich zu anderen Teilnehmern gegeben. Daher übernehmen die einzelnen Ausgänge die errechneten Zustände nicht gleichzeitig. Bei einer Standardübertragungsrate von 500 kBit/s und einer maximalen Anzahl von 512 INTERBUS-S Teilnehmern erhält man als Busdelay $2 \cdot m \cdot t_{Bit} + t_{Ph} = 2,048$ ms $+ t_{Ph}$ und als Zeitdifferenz $t_{dü}$ der Datenübernahme zweier benachbarter Teilnehmer $1,5 \cdot t_{Bit} + t_{Ph} = 3$ µs $+ t_{Ph}$, wenn diese Teilnehmer die Daten gleichzeitig in Bezug auf den in Abb. 4 angegebenen Datenstrom übernehmen, z. B. nach Empfang eines positiven Prüfsummenstatus mit der Flanke des ersten Bits der LatchOutData Phase (Abb. 4, Bit 9). Da definitionsgemäß die Datenübernahme zu einem beliebigen Zeitpunkt innerhalb der LatchOutPhase erfolgen kann, vergrößert sich $t_{dü}$ auf maximal 5.5 t_{Bit} +t_{Ph} (= 11 µs +t_{Ph}). Zusätzlich zu den oben beschriebenen Verzögerungen kommt es zu zeitlichen Schwankungen bei der Datenübernahme im maximalen Ausmaß der Taktperiodendauer des Teilnehmers (Jitter der Bitabtastung).

2.4 Busstörungen

Wir beschränken uns hier auf temporäre Busfehler, weil ·permanente Busfehler zu einem Stillstand des Busses führen und jedenfalls Fehlzustände darstellen. Bei temporären Kommunikationsfehlern werden die Ergebnisse eines fehlerhafter Buszyklus verworfen und ein neuer Zyklus durchgeführt. Dadurch kommt es zu unregelmäßigen Verzögerungen. Diese Verzögerungen, die jedem Determinismusgedanken widersprechen, können bis zu mehreren Buszykluszeiten betragen.

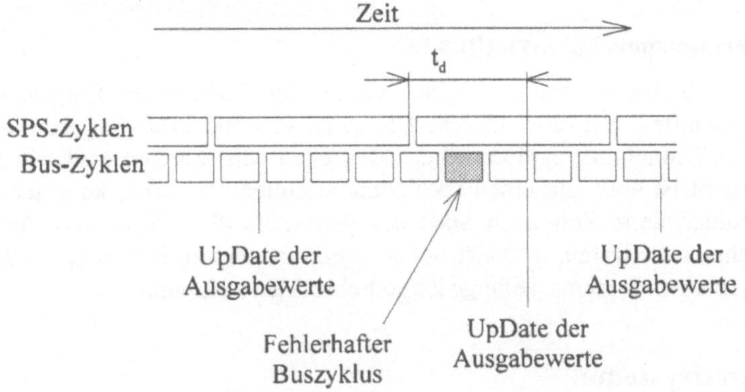

Abb. 5. Ausgabeverzögerungen durch Busstörungen

Das Auftreten von fehlerhaften Buszyklen verlängert t_d und ist nur in einem bestimmten Maß tolerierbar.

3 Typische Implementierung von Busanschaltungen

In typischen Implementierungen wird INTERBUS-S als Kommunikationssystem zwischen SPS und Sensoren/Aktoren geschaltet. Im allgemeinen sind die Buszyklen zeitlich nicht an die SPS-Zyklen gebunden. Ein freilaufender Betrieb des Busses sichert, daß die SPS immer aktuellste Eingabedaten zur Berechnung zur Verfügung hat. Diese Implementierungen sind nicht nur einfach zu realisieren, sondern sind darüber hinaus noch robust gegenüber gestörten Buszyklen. Auf Grund der nicht an die SPS-Zyklen gekoppelten Buszyklen ergeben sich aber unterschiedliche Update-Raten der Ausgangszustände. Dadurch erhält man unterschiedliche Verzögerungen t_d, die im fehlerfreien Fall zwischen ein und zwei Buszykluszeiten liegen (siehe Abb. 5).

4 Determinismus

4.1 Definition, Motivation

Determinismus bietet eine zeitlich festgelegte, beschreibbare Funktionsweise und damit die prinzipielle Möglichkeit, die Zeitpunkte von Ein-/Ausgabeoperationen vorhersagen zu können. Erfolgen die Ein-/Ausgabeoperationen darüber hinaus zeitlich periodisch, so sind damit auch die Bedingungen für Real-Time- und Abtastsysteme erfüllt (siehe Kap. 5). Bei räumlich ausgedehnten Steuerungssystemen mit mehreren hochdynamischen Vor-Ort-Reglern (z. B. bei Walzstraßen, Papierherstellung) müssen, um Zwischenzustände zu vermeiden, die Sollwerte von den einzelnen Reglern möglichst zu gleichen Zeitpunkten übernommen werden, da sonst ein Qualitätsverlust des Produktionsgutes befürchtet werden muß. Um die Synchronität der Ausgabe zu gewährleisten, ist Determinismus des Gesamtsystems Voraussetzung.

4.2 Determinismus bei INTERBUS-S

INTERBUS-S bietet, abhängig von der Anzahl der Teilnehmer, festgelegte Ein/Ausgabeeigenschaften. Bei einer üblichen Taktrate von 500 kBit/s ergeben sich, je nach Anzahl der Teilnehmer, typische Buszykluszeiten von wenigen Millisekunden. Diese Buszykluszeit ist aber, auf einen speziellen Busaufbau bezogen, konstant (siehe Kap. 2.2). Wohldefinierte Zeit nach Start des Buszyklus übernimmt bzw. übergibt jeder Busteilnehmer die Daten. INTERBUS-S ist damit eine deterministische Komponente für den Aufbau eines Systems, solange keine Fehlzyklen auftreten.

5 Echtzeitsysteme

5.1 Eignung von INTERBUS-S für den Aufbau von Echtzeitsystemen

Man unterscheidet grundsätzlich zwischen einem Hard-Real-Time System und einem Soft-Real-Time System [3]. Ein System ist dann ein Hard-Real-Time System, wenn das System zeitliche Anforderungen, im Sinn einer festgelegten Reaktionszeit (Deadline), unter allen spezifizierten Lastzuständen mit absoluter Sicherheit erfüllt. Bei Soft-Real-Time Systemen toleriert man, meist aus wirtschaftlichen Gründen, selten auftretende Verletzungen der zeitlichen Bedingungen. In Tab. 1 sind die wesent-

Tabelle 1. Anforderungen an ein Real-Time Übertragungssystem

Bedingung	von InterBus-S erfüllt
kleine maximale Übertragungszeit (vom Sender zum Empfänger)	Ja
minimaler Jitter	Ja
Abkopplung des Übertragungssystems von der Steuerung während einer Datenübertragung	Ja
Übertragung definierter Datenmengen	Ja
Übertragungsfehler müssen erkannt und sollten korrigiert werden, ohne den Jitter oder die Übertragungszeit zu beeinflussen	Erkannt, Daten verworfen

lichen Bedingungen angegeben, die ein Real-Time Übertragungssystem erfüllen soll [3]. InterBus-S kann auf Grund seines grundsätzlich deterministischen Verhaltens und seines in Kap. 2.2 gezeigten wohldefinierten worst case-Zeitverhaltens in Hard-Real-Time Systemen eingesetzt werden.

5.2 System „ohne Datenverlust"

An ein System SPS - INTERBUS-S „ohne Datenverlust" werden folgende Echtzeitanforderungen gestellt: im ungünstigsten Fall müssen Daten spätestens zum Zeitpunkt des Vorliegens neu berechneter Daten ausgegeben werden. Typische Implementierungen (siehe Kap. 3) erfüllen diese Anforderung. Aufeinanderfolgende Busfehler sind, abhängig vom Verhältnis SPS-Zyklusdauer t_{SPS} zu Buszyklusdauer t_{BUS}, und damit von der Anzahl der INTERBUS-S Teilnehmer, in einem bestimmten Ausmaß tolerierbar (Abb. 6). Die Modellvorstellung, daß die SPS streng zyklisch neue Ausgabedaten berechnet und zur Verfügung stellt, liegt diesem Gedankengang zugrunde.

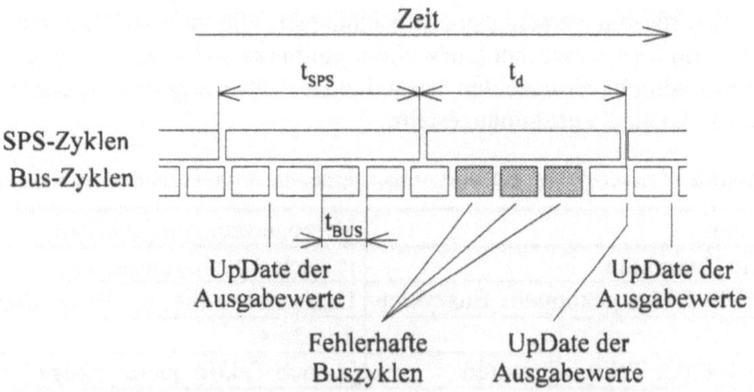

Abb. 6. Echtzeitanforderungen an ein System „ohne Datenverlust"

6 Abtastsystem

Will man über INTERBUS-S gekoppelte Module als Teile einer digitalen Regelung einsetzen, muß die Kombination SPS - INTERBUS-S deterministische Anforderungen erfüllen. Die zeitliche Kopplung der Buszyklen an die SPS-Zyklen ist eine wesentliche Voraussetzung, um eine konstante Zykluszeit der Gesamtanordnung, die für Abtastsysteme von entscheidender Bedeutung ist, zu gewährleisten [2]. Dazu ist nur eine definierte Auslösbarkeit der Buszyklen durch die Steuerung notwendig. Weiters kann das Auftreten von Busfehlern (siehe 2.4) in diesem Fall nicht akzeptiert werden. Für den Regler wirkt die Kombination INTERBUS-S - Regelstrecke als Strecke mit vergrößerter Totzeit (siehe Abb. 7), die durch verschiedene Regelstrategien berücksichtigt werden kann [2]. Für den Aufbau von Regelkreisen, bei denen die Komponenten über den Bus gekoppelt werden, kann die „Gleichzeitigkeit" von Ein- bzw. Ausgabe bei unterschiedlichen Busteilnehmern nötig sein. Die in Kapitel 2.2 angeführte genauere Betrachtung zeigt definierte kurze Verzögerungszeiten t_{du} zwischen den einzelnen Teilnehmern, die auf Grund der Signalabtastung und der Leitungslaufzeit der übertragenen Signale auf den Busleitungen zustande kommen. Diese Verzögerungszeiten zwischen je zwei Teilnehmern sind aber im wesentlichen konstant. Damit ergibt sich prinzipiell die Möglichkeit, diese Verzögerungen zu kompensieren und eine annähernd synchrone Datenausgabe zu gewährleisten.

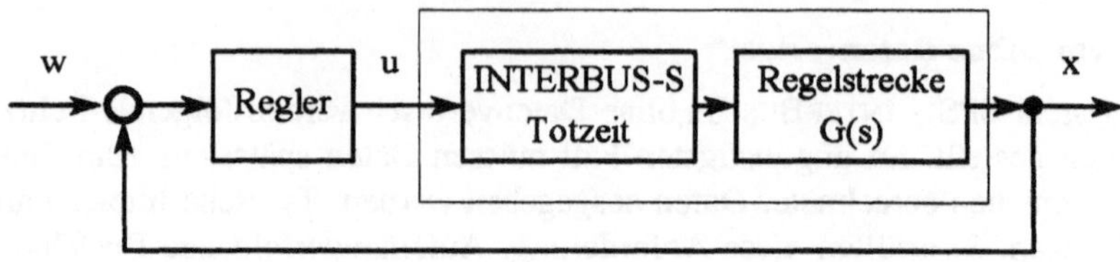

Abb. 7. Regelkreis mit INTERBUS-S

7 Ergebnisse, Zusammenfassung

In Kapitel 2 wurde auf verschiedene Mechanismen, die sich auf die Datenein- und Ausgabezeit zum Teil wesentlich auswirken, aufmerksam gemacht. Diese Mechanismen und die dadurch verursachten zusätzlichen Zeitverzögerungen oder -unsicherheiten sind in Tabelle 2 zusammengestellt.

Tabelle 2. Zeitverzögerungen oder -unsicherheiten verursachende Mechanismen

Mechanismus	Verzögerung, Unsicherheit
Fehlerhafte Buszyklen	je nach Anzahl, einige 10 ms
nicht an SPS-Zyklus gekoppelte Buszyklen	Unsicherheit bis zu 1 Buszyklus (~ ms)
Busdelay	2μs-2ms
Asynchronität der Taktgeneratoren	je nach Taktfrequenz, einige 10 ns

Auf Grund seines deterministischen Verhaltens ist INTERBUS-S prinzipiell für Echtzeitanwendungen gut geeignet. Determinismus im Gesamtsystem ist bei Auslösung der Buszyklen durch die Steuerung erreichbar. Damit sind sogar anspruchsvolle Hard-Real-Time Systeme und hochdynamische Abtastregelkreise realisierbar. Einschränkungen ergeben sich unter Umständen durch ev. auftretende Fehlzyklen. Für höchste Anforderungen kann die statische Zeitverzögerung zwischen den Teilnehmern korrigiert werden, während eine Ausgabeunsicherheit auf Grund der Asynchronität dagegen prinzipiell nicht korrigiert werden kann.

8 Literatur

1. Alfredo Baginski, Martin Müller: InterBus-S: Grundlagen und Praxis, Hüthig Verlag Heidelberg, 1994

2. Alexander Weinmann: Regelungen, Analyse und technischer Entwurf, Band 2, Springer Verlag Wien New York, 1984

3. Hermann Kopetz: Real-Time Systems, Kluwer Academic Publishers, Boston/Dordrecht/London, 1997

Test von Feldgeräten der Prozeßautomatisierung[i] (PROFIBUS-PA)

T. Bangemann

ifak - Institut für Automation und Kommunikation e. V. Magdeburg
Bereich Kommunikationssysteme
Steinfeldstr. 3, D-39179 Barleben

Abstract. Before a new product can be put on the market its functionalities have to be tested. Only by doing this costs caused by failures that occur during the whole lifecycle can be reduced. For Fieldbus devices this requires testing of single devices as well as testing the intercommunication between devices. The paper gives a brief overview about the different tests of devices designed to be used in process industry, an introduction into PROFIBUS-PA specification and presents a concept how to realize the tests.

1 Einleitung

Die Feldbustechnik hat sich in den letzten Jahren zu einer Schlüsseltechnologie in der Automatisierungstechnik entwickelt. Während in der Fertigungstechnik bereits der Feldbuseinsatz fest etabliert und ein Machtkampf zwischen den verschiedenen Feldbusanbietern und Nutzerorganisationen entbrannt ist, hat der Einzug dieser innovativen Technologie in der Verfahrenstechnik gerade begonnen.

Die durch die Fertigungstechnik aufgestellten Forderungen nach extrem hohen Übertragungsgeschwindigkeiten werden durch für die Verfahrenstechnik typische Forderungen nach Eigensicherheit, Stromversorgung über den Bus, Parametrierbarkeit der Geräte über das Medium, einheitliche Bedientools usw. ergänzt. Von entscheidender Bedeutung ist die Verfügbarkeit der Geräte.

Die Zuverlässigkeit eines Gerätes wird in allen Lebensphasen, beginnend mit der Spezifikation und Implementierung beeinflußt. In der Feldbustechnik ist nicht nur die Zuverlässigkeit eines Gerätes ausschlaggebend, vielmehr müssen alle Geräte einer Anlage fähig sein, eine Aufgabe gemeinsam zu erfüllen. Das heißt, die über das Medium ausgetauschten Informationen müssen für alle Geräte verständlich sein. Voraussetzung für den plausiblen Datenaustausch zwischen Geräten unterschiedlicher Hersteller ist eine normkonforme, interoperable Implementierung eines standardisierten Protokolls oder sogar von standardisierten Grundelementen der Anwendung.

---

[i]Teile der vorgestellten Konzeption wurden in einem durch das Kultusministerium des Landes Sachsen- Anhalt geförderten Projekt mit dem Titel „Erarbeitung einer Methodik zum Interoperabilitätstest" (FKZ: 1455A/0083) erarbeitet.

2 Tests von Feldgeräten

Die Durchführung verschiedener Tests im Vorfeld des praktischen Einsatzes eines Gerätes ist Voraussetzung für die Erfüllung der oben aufgestellten Forderungen. Einige Tests werden durch den Gesetzgeber (z.B. EMV- Tests) bzw. durch entsprechende Richtlinien (z.B. Tests von Geräten, die für den Einsatz in explosionsgefährdeten Bereichen bestimmt sind - u.a. Eigensicherheit) vorgeschrieben. Darüber hinaus sind Funktionstests der Geräte erforderlich.

Zusammenfassend kommen folgende Tests für Geräte der Verfahrenstechnik in Frage:

- EMV Tests
  Gegenstand des EMV- Tests sind Störungen auf der Versorgungsleitung sowie der Signalleitung (nach IEC 801-4) und Störfestigkeit gegenüber statischen Entladungen (nach IEC 801-2).

- Test der Eigensicherheit eines Gerätes
  Es erfolgt die Überprüfung der Einhaltung der in EN 50014 und EN 50020 getroffenen Festlegungen. Basis für diese Test ist z.B. das von der PTB Braunschweig entwickelte FISCO- Modell [1].

- Test der physikalischen Eigenschaften
  Überprüfung der schaltungstechnischen Umsetzung der Feldbusanschaltung sowie der elektrischen Eigenschaften des Gerätes am Medium

- Konformitätstest
  Test der Übereinstimmung (Konformität) einer Implementation Under Test (Protokollimplementierung) gegenüber der Spezifikation

- Interoperabilitätstest
  Überprüfung, ob ein Gerät bzgl. seiner Kommunikationsfähigkeit über ein gemeinsames Medium in der Lage ist, mit anderen Geräten eine Automatisierungsaufgabe zu lösen

- Interchangeability Test
  Test der Austauschbarkeit eines Gerätes durch Geräte anderer Hersteller

- Integrationstest
  Test der Integrationsfähigkeit eines Gerätes in eine bestehende Anlage

- Anwendungstest
  Überprüfung der Realisierbarkeit komplexer Automatisierungsaufgaben unter Verwendung der in den verschiedenen Geräten implementierten Applikationen (Function Block Algorithmen, andere Applikationen)

In der Praxis ist es nicht möglich, eine verbale Aussage wie: "Das Gerät funktioniert" oder "Das Gerät ist interoperabel" zu treffen. In Bezug auf Tests gibt es keinen

Absolutismus! Ein Test kann lediglich Aussagen bzgl. der korrekten Realisierung eines bestimmten Funktionsumfanges unter definierten Bedingungen treffen. Der Aussagegehalt eines Tests hängt wesentlich vom Testabdeckungsgrad sowie der Relevanz der durchgeführten Tests ab. Darüber hinaus wird die Akzeptanz eines Testes nicht unwesentlich von den für den Testkunden entstehenden Kosten beeinflußt.

Es ist notwendig, eine möglichst optimale Kombination aus den oben aufgeführten Tests für den Tests eines Feldgerätes festzulegen. Für eigensichere Feldgeräte der Verfahrenstechnik sind sowohl EMV- Tests als auch Test der Eigensicherheit vorgeschrieben. Der Funktionstest mit anschließender Zertifizierung ist optional, jedoch im Sinne der Marktakzeptanz eines Produktes sowie des Nachweises der Produktqualität dringend zu empfehlen.

3 Testgegenstand bei Geräten mit PROFIBUS-PA

Mit PROFIBUS-PA existiert eine Feldbusspezifikation für den Einsatz in der Verfahrenstechnik. Sie besteht aus den Teilen:

- DIN EN 61158-2 [2],
- DIN 19245 Teil 1, 3 (EN50170) [3],
- DIN 19245 Teil 4 [4],
- PROFIBUS-PA Profile for Process Control Devices Version 2.0. [5],
- Draft Specification PROFIBUS DP Enhanced [6].

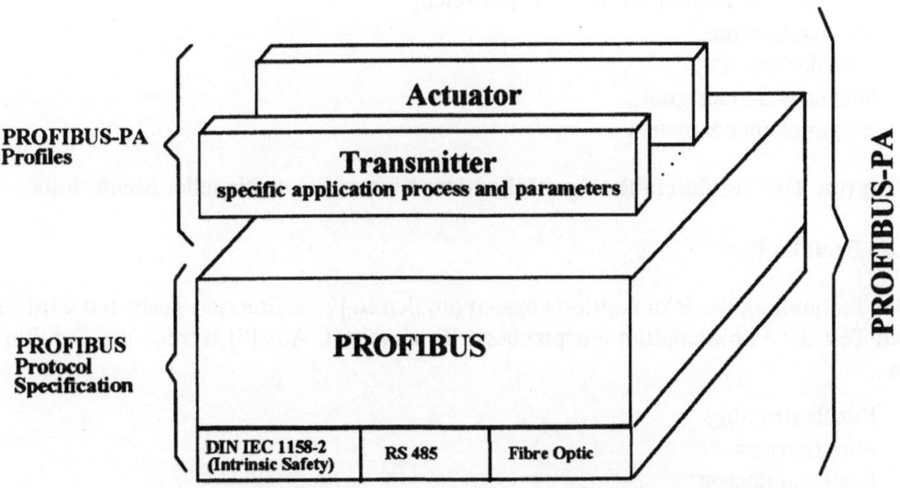

Fig. 1. PROFIBUS-PA Struktur der Spezifikationsdokumente [5]

Testgegenstand der nachfolgend beschriebenen Tests ist der komplette Spezifikationsumfang für PROFIBUS-PA. Der Umfang der zu realisierenden Tests orientiert sich in erster Linie an der Praxisrelevanz und ist eine Kombination aus den einleitend aufgeführten Funktionstests.

4 Testumfang

Basis für die Festlegung des Testumfanges sind die einleitend aufgeführten Spezifikationsdokumente. Die in den folgenden Abschnitten charakterisierten Testabschnitte setzen sich aus den eingangs definierten Testverfahren zusammen. Aus Effizienzgründen werden die Tests mit Ausnahme des EMV- Tests sowie des Tests der Eigensicherheit als ein Gesamtkomplex angeboten. EMV- Test sowie Test der Eigensicherheit werden als erfolgreich bestanden vorausgesetzt.

4.1 Statischer Check

Hierbei wird anhand der durch den Hersteller vorgelegten Dokumentationen überprüft, ob der Prüfling die Voraussetzungen für die Durchführung der Test (z.B. EVM Erklärung) erfüllt. Darüber hinaus wird die Korrektheit der zum Prüfling mitgelieferten Gerätestammdatendatei (GSD) Datei überprüft.

4.2 Test der Physik

Der Test der Physik basiert auf den in [2] getroffenen Festlegungen (siehe auch [10]). Es werden Eigenschaften wie:

- Funktion im definierten Spannungsbereich,
- Stromaufnahme,
- Signalerkennung,
- Form des Sendesignals,
- Eingangsimpedanz usw.

überprüft. Die Testdurchführung erfolgt unter Einsatz konventioneller Meßtechnik.

4.3 Protokolltest

Der Testumfang des Protokolltests basiert auf den in [7] definierten Tests und wird für den Test der Funktionalität entsprechend [6] erweitert. Aus [7] werden die Testgruppen:

- Busübertragung
- Funktionstest
- Konformitätstest

übernommen.

4.4 Profiltest [8]

Das "Profile for Process Control Devices" beschreibt Eigenschaften von Geräten der Profilklassen A und B. Die Gesamtheit der Variablen und Parameter eines Gerätes wird hier hinsichtlich ihrer Zuordnung zu Komponenten und Teilfunktionen des Gerätes in einzelne Blöcke zusammengefaßt. Das heißt, es wird festgelegt, welche Bedeutung ein Parameter für die zu realisierenden Blöcke hat, welche Parameterkombinationen gültig sind und wie das Mapping der Function Block Application auf den Protokollstack erfolgt. Einigen Parametern ist ein ausgeprägtes Objektverhalten implizit, das charakteristisch für einzelne Blocktypen ist. Anhand dieser Charakterisierung lassen sich folgende Testgruppen ableiten:

Test des System Under Test gegen das Testsystem ohne weiteren Busverkehr

Beim Test des System Under Test (SUT) gegen das Testsystem nach Testkonfiguration A soll überprüft werden, ob sich das SUT am Bus entsprechend der statischen Festlegungen verhält. Es wird davon ausgegangen, daß:

- der Datenverkehr ausschließlich zwischen dem Testsystem und dem System Under Test erfolgt,
- auf dem SUT keine Upper Tester entsprechend [9] realisierbar sind und ein Gerät in seinem standardmäßigen Auslieferungszustand zu testen ist,
- ausschließlich "Remote" Tests (Die Stimulierung der Tests erfolgt ausschließlich von einem entfernten Teilnehmer, dem Testsystem, aus über das Medium.) realisierbar sind.

Folgende Funktionalitäten sind Gegenstand der Tests:

- Mapping zwischen Profil und Protokoll,
- Existenz und Datentyp der Parameter/Objekte entsprechend Profil,
- Access Rights für Parameter/Objekte entsprechend Profil,
- definiertes Parameterverhalten (Alarm-, Warn-, Hystereseverhalten),
- Statusgenerierung,
- Zustandsmaschinen der Funktionsblöcke,
- Einhalten der Adressierungsregeln für die Parameter der Blöcke,
- Verbindungstypen (MSCY- Master/Slave zyklisch, MSAC_C1- Master/Slave azyklisch an die zyklische Verbindung gekoppelt, MSAC_C2- Master/Slave azyklisch von der zyklischen Verbindung entkoppelt) mit dem Testsystem.

Die Testkonfiguration besteht lediglich aus dem SUT, dem Testsystem sowie einem Busmonitor.

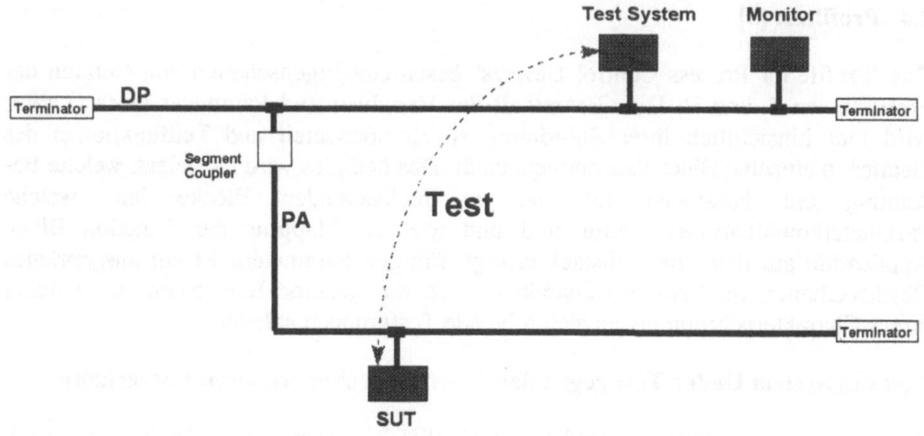

Fig. 2. Testkonfiguration A

Test des System Under Test (SUT) gegen das Testsystem sowie andere Master in der Multi- Vendor- Anlage (MVA)

Der Test des SUT nach Testkonfiguration B unterscheidet sich vom Test entsprechend Testkonfiguration A durch die Realisierung des Kommunikationsgeschehens zwischen dem SUT und einem Master- Gerät sowie dem Testsystem (in der MVA) und nicht ausschließlich gegen das Testsystem. Folgende Schwerpunkte können für diesen Test angegeben werden:

- Datenaustausch zwischen Master und SUT (Kommunikationsgeschehen entsprechend dem Datentransfer zwischen dem Master und anderen Multi- Vendor- Geräten),
- Entfernen und Hinzufügen von Geräten, mit denen das SUT kommuniziert,
- Zyklischer Datenverkehr (MSCY mit Master- Gerät der MVA),
- MSAC_C2 Datenverkehr mit Testsystem,
- Dauertest,
- Überprüfung, ob das SUT in eine bestehende Konfiguration (MVA) bei laufendem Betrieb eingefügt werden kann, ohne daß dieses zu Störungen des Systemverhaltens führt,
- die bereits beim Profiltest ohne weiteren Busverkehr überprüfte Funktionalität auch im gemeinsamen Betrieb mit anderen Geräten gewährleistet ist (Testumfang jedoch kleiner als im ersten Testschritt),
- ob der Wiederanlauf definiert erfolgt (nach Powerdown bzw. Kommunikationsausfall).

Zur Erreichung des Testzieles ist es notwendig, das SUT und das Testsystem gemeinsam mit anderen Geräten an einem gemeinsamen Medium zu betreiben. Dieses wird über zwei freie Plätze (1x DP- Segment, 1x PA- Segment) in einer MVA gewährleistet (siehe Testkonfiguration B).

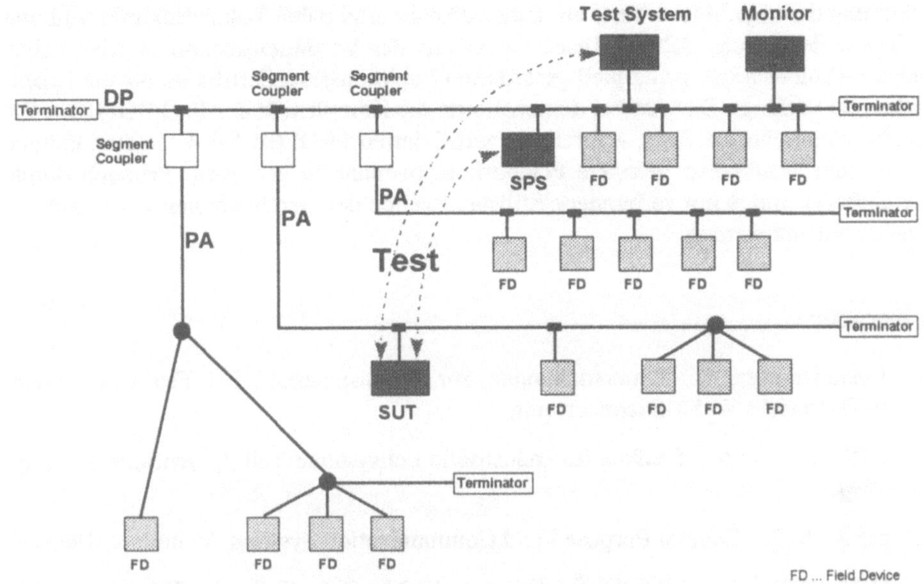

Fig. 3. Testkonfiguration B

5 Wie kommt man zu einem Zertifikat?

Der Zertifizierungsprozeß für ein PROFIBUS-PA Gerät läuft in folgender Reihenfolge ab:

1. Geräteentwicklung
2. EMV- Test (Ergebnis: Zertifikat)
3. Test der Eigensicherheit (optional, Ergebnis: Zertifikat)
4. Erstellung einer Gerätestammdatendatei (GSD)
5. Beantragung einer Ident- Nummer bei der PROFIBUS- Nutzerorganisation (Ergebnis: Zuweisung einer Nummer)
6. Beantragung eines Zertifizierungstests bei einem akkreditierten Testlabor (Ergebnis: Bestätigung und Terminzusage)
7. Testdurchführung (Ergebnis: Testbericht)
8. Beantragung eines Zertifikates bei der PNO auf der Basis des positiven Testberichtes (Ergebnis: Zertifikat)

6 Zusammenfassung

Die Qualität eines Produktes hängt nicht zuletzt von der Güte der im Vorfeld der Vermarktung durchgeführten Tests ab. Gütemaßstäbe sind dabei Testabdeckungsgrad und Relevanz der durchgeführten Tests. Für Geräte der Verfahrenstechnik werden neben den für Feldbusgeräte prinzipiell geforderten Funktionstests Zertifikate für die Eigensicherheit verlangt. Der Artikel demonstrierte die Komplexität der für PROFIBUS-PA Geräte vorgesehenen Tests. Käufer von zertifizierten PROFIBUS-PA Geräten können sicher sein, umfassend getestete Feldbuskomponenten zu erwerben. Prinzipbedingte Gerätefehler und damit verbundene Stillstandszeiten der Geräte können weitestgehend ausgeschlossen werden.

Literatur

1. Johannsmeyer, U.: Untersuchungen zur Eigensicherheit bei Feldbussystemen. PTB- Bericht W-53, Braunschweig, 1993.

2. DIN EN 61158-2: Feldbus für industrielle Leitsysteme Teil 2, Deutsche Fassung, 1994.

3. prEN 50170: General Purpose Field Communication Systems, Volume 2, 1995.

4. DIN 19245-4: PROFIBUS Process Field Bus Teil 4, Physical Layer (Bitübertragungsschicht) und Data Link Layer (Datensicherungsschicht) für die Prozeßautomatisierung, 1996.

5. PNO-Richtlinie PROFIBUS-PA, Profile for Process Control Devices Version 2.0, PNO Karlsruhe, 1996.

6. Draft Specification PROFIBUS- DP Enhanced, Version 1.14, PNO Karlsruhe, 1997.

7. Test Specification for PROFIBUS- DP Slaves, PNO Karlsruhe, 1996.

8. Bangemann, T.: Prüfrichtlinie für Feldgeräte entsprechend „PROFIBUS-PA Profile for Process Control Devices", Entwurf Version 0.5, ifak, Barleben, 1997.

9. ISO IS 9646: Information Processing Systems - OSI Conformance Testing Methodology and Framework, 1991.

10. Hähniche, J.: Eigensichere Übertragungstechnik für PROFIBUS-PA, In: Elektrotechnik und Informationstechnik, ÖVE- Verbandszeitschrift, S.: 249- 253, 5/1997

DP-Erweiterungen - neue Entwicklungen für den Profibus

Pöschmann, A.; Hähniche, J.

Institut für Automation und Kommunikation (ifak)
Steinfeldstr.3 (IGZ)
39179 Barleben
email: {apoe|haeh}@ifak.fhg.de
Telefon: ++49 39203 81072
Fax: ++49 39203 81100

Abstract. The paper presents an overview on some selected facilities of the extended PROFIBUS-DP-Protocol. This protocol is worked out by the DP-working group of the PROFIBUS User Organisation and published as a PNO Guideline. We introduce the main principles of the new asynchron communication facilities, the device model and principles of addressing, the multi-network facilities and the alarm model. These new capabilities open new application fields for the protocol (e.g. in process automation).

1 Einleitung

Der vorliegende Beitrag beschäftigt sich mit den derzeit in der PROFIBUS Nutzerorganisation diskutierten Erweiterungen für PROFIBUS - DP. Seit der Herausgabe der DIN 19245 Teil 1 und 2 gab es eine kontinuierliche Weiterentwicklung des PROFI-BUS, um die verschiedenen Einsatzbereiche in der Automatisierungstechnik abdecken zu können. Mittlerweile ist der PROFIBUS auch europäischer Standard geworden. Stillstand bedeutet Rückschritt - diesen Grundsatz hat auch die starke Anwendergemeinschaft des europäischen Feldbusstandards aufgegriffen, um sich den zukünftigen Anforderungen der industriellen Kommunikation zu stellen. Natürlich geschieht das alles, ohne die stabile, in der EN50170 [1] genormte Basis zu verlassen. Kompatibilität und damit der Investitionsschutz besitzen einen hohen Stellenwert bei allen Aktivitäten zur Pflege und Weiterentwicklung des Standards.

Es werden nachfolgend die Neuerungen vorgestellt, die das Einsatzspektrum von PROFIBUS - DP für intelligente Feldgeräte erweitern und besonders für die Prozeßautomatisierung benötigt werden. Gerade die Prozeßindustrie steht an der Schwelle, die geeignete Feldbustechnologie als Rationalisierungsmaßnahme einzusetzen. Mit PROFIBUS - PA auf der Basis des erweiterten DP-Protokolls wird dem Anwender in der Verfahrenstechnik die maßgeschneiderte Technologie zur Verfügung gestellt. Die in Fig. 1 dargestellte Anwendung soll beispielhaft ein Szenario in der Verfahrenstechnik aufzeigen. Zum einen ist auch in der Verfahrenstechnik ein zyklischer Datenaustausch zwischen den dezentralen Feldbuskomponenten und einer zentralen Steuerung gefordert. Zum anderen besteht gerade in der Verfahrenstechnik zunehmend die Notwendigkeit, alle Geräte von einer zentralen Konsole aus zu konfigurieren und auch zentral alle Parameter und Daten der Feldbuskomponenten zu verwalten.

Hierdurch ist zu jedem Zeitpunkt sichergestellt, daß die aktuellsten Konfigurationsdaten bereitgestellt werden können. Weiterhin erfordert der Feldbuseinsatz in der
Verfahrenstechnik, daß mit Hilfe von Bedien- und Beobachtungskonsolen einzelne
Prozeßwerte der dezentralen Feldbuskomponenten gelesen bzw. geschrieben werden
können (gleiches gilt auch für Hand-Held-Terminals).

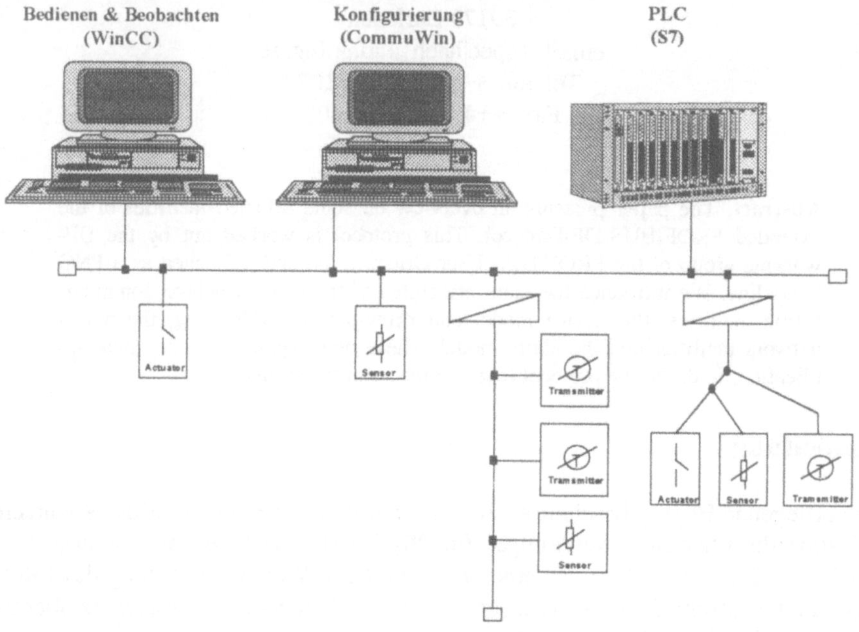

Fig. 1. Beispielhafte Konfiguration in der Prozeßautomation

Der in der Fertigungsautomatisierung im Hinblick auf Übertragungseffizienz und
unter Kostengesichtspunkten im I/O-Bereich etablierte Feldbus wird auf Grund der
zusätzlichen Forderungen um azyklische Kommunikationsfähigkeiten ergänzt. Diese
bilden die Grundlage für die Einführung bestätigter Alarmmeldungen, einer umfas-
senden Geräteparametrierung mit dem Laden von Daten- und Parametersätzen jetzt
auch in intelligente Slavegeräte und eröffnet neue Perspektiven für den Übergang in
andere Netzwelten. Bei der Spezifikation dieser neuen „Quality of Service" wird
besonderer Wert auf kleine und effiziente Modelle gelegt, die mit wenig Programmcode implementiert werden können. Es bleibt also spannend um den PROFI-
BUS, was für die Attraktivität dieser Technologie sicher nicht von Nachteil ist.

2 Azyklische Kommunikation - der Transportmechanismus

PROFIBUS-DP besitzt folgende Kommunikationsmöglichkeiten :

- Master - Slave - Kommunikation zwischen DP-Master Klasse 1 (z.B. Steuerungen) und DP-Slaves für den zyklischen, gepufferten Produktivdatenaustausch, einfachste Parametrierung und Konfigurierung (vornehmlich in der Anlaufphase), die Bereitstellung gepufferter Diagnosemeldungen durch den Slave und das Absetzen von Steuerkommandos (Sync, Freeze) durch den Master (auch als Multicast / Broadcast)

- Master - Slave - Kommunikation zwischen DP-Master Klasse 2 (z.B. Diagnosegerät) und DP-Slaves für das Auslesen der aktuellen Ein- und Ausgänge, der Diagnose und der Konfigurationswerte. (Auslesen der Werte auf eigene Initiative zu beliebigen Zeitpunkten)

- Master - Master - Kommunikation für die Parametrierung (Laden, Lesen und Aktivieren der Parametersätze) des Masters Klasse 1 durch den Master Klasse 2 sowie das Auslesen der im Master Klasse 1 gespeicherten Diagnosemeldungen

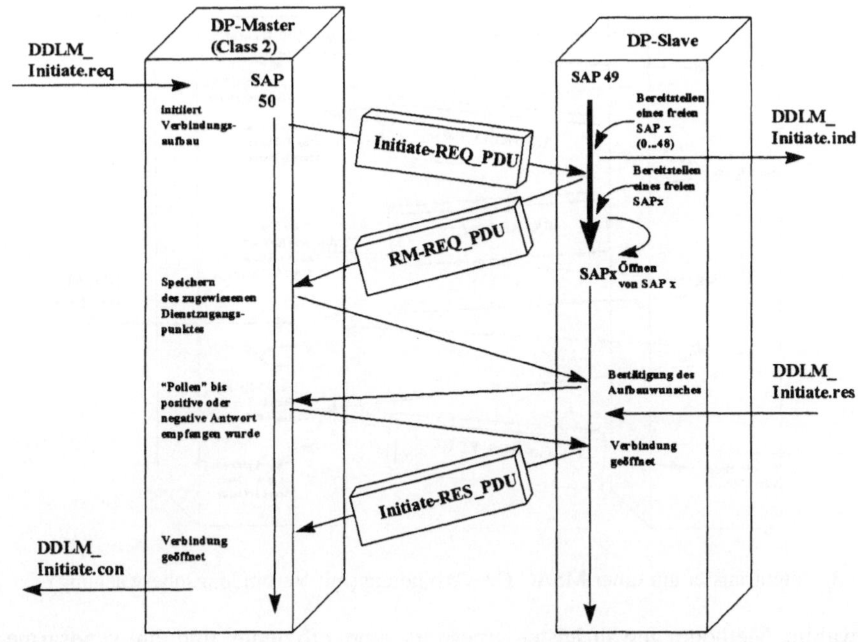

Fig. 2. Verbindungsaufbau aus einer DP-Master (Class 2) DP-Slave Kommunikationsbeziehung

All diese Eigenschaften sind für einen sicheren und effizienten Datenaustausch im Bereich der dezentralen Peripherie zugeschnitten. Benötigt die Applikation jedoch azyklischen, nachrichtenorientierten Transport von Daten, muß dieser mit entsprechendem Overhead als Container innerhalb der zyklischen Daten emuliert werden [3]. Die Flußkontrolle muß dann beispielsweise die Applikation übernehmen. Die PROFI-

BUS-DP Erweiterungen definieren zwei Methoden für eine direkte azyklische Kommunikation, die in einfacher und effizienter Weise das bisherige PROFIBUS-DP Protokoll optimal ergänzen. Als Transportmechanismus werden zwei azyklische Kommunikationstypen unterschieden. Der Typ MSAC_C1 charakterisiert eine Verbindung zwischen DP-Master Klasse 1 und einem zugehörigen DP-Slave. Auf dieser Verbindung werden die Dienste DDLM_Read, DDLM_Write und DDLM_Alarm_Ack definiert. Diese Kommunikationsbeziehung ist von der Existenz der zyklischen Kommunikation der I/O-Daten abhängig, d.h. die Verbindung wird mit dem Eintritt in den zyklischen Datenaustausch implizit geöffnet, ohne daß ein Verbindungsaufbauprotokoll über den Bus gefahren werden muß. Die zyklische Verbindung ist gleichzeitig die Verbindungsüberwachung für den azyklischen Kanal.

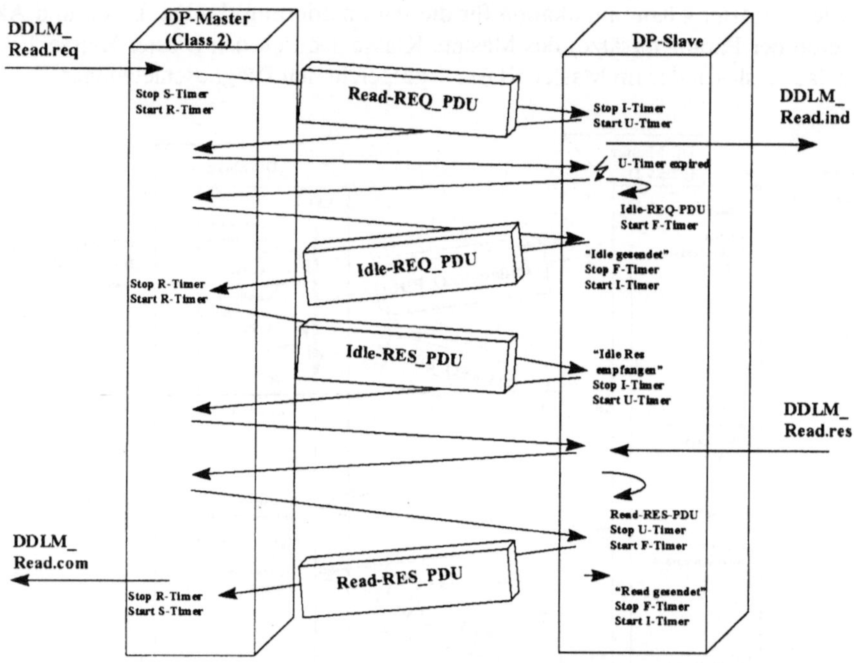

Fig. 3. Datentransfer auf einer MSAC-C2 Verbindung (mit Verbindungsüberwachung)

Die gewählte Methode gewährleistet einerseits eine effiziente und aufwandsarme Implementierung, andererseits wird eine Sicherheitsfunktion erfüllt, indem nur der steuernde Master Klasse 1 den MSAC_C1 Kanal benutzen darf. Jedes Slave-Gerät kann nur einen solchen Kanal besitzen. Der Verbindungstyp MSAC_C2 (siehe Fig. 2) wird zwischen Master Klasse 2 - und Slave-Geräten benutzt. Es ist der typische Kanal zwischen Projektierungstool und Feldgerät. Die Dienste DDLM_Initiate und DDLM_Abort werden zum Einrichten und Beenden der Verbindung eingesetzt. Unterstützend wurde hier ein Ressourcen-Manager eingeführt, der die Dienstzugangspunkte des Data Link Layer effizient und transparent verwaltet. Jeder Verbindungsaufbau des Masters richtet sich zunächst an den Ressourcen-Manager des Slaves, der in einer immediate Response den nächsten freien Service Access Point (SAP) zurückspiegelt. Die DDLM_Initiate-Response-PDU wird dann bereits an diesem SAP abgeholt.

Als Produktivdatendienste sind auf dieser Verbindung die Dienste DDLM_Read, DDLM_Write und DDLM_Data_Transport definiert. Außerdem wird die Verbindung durch projektierbare Idle - Zyklen überwacht (siehe Fig. 3). Feldgeräte können gleichzeitig mehrere MSAC_C2 Verbindungen auch zu unterschiedlichen Master Klasse 2 Geräten besitzen. Auf einer azyklischen Verbindung ist ein ausstehender Service zugelassen. Der DP-Master Klasse 2 pollt den entsprechenden Slave bis die Antwort empfangen und bevor ein erneuter Service-Request zugelassen wird. Die Flußkontrolle wird vom Protokoll realisiert. Es existieren 3 verschiedene Timer im Server (DP-Slave) und 2 verschiedene Timer im Client auf einer MSAC_C2 Verbindung, wobei jeweils nur 1 Timer zu einem Zeitpunkt aktiv ist. Der Client überwacht die Aktivitäten der Applikation auf einer bestehenden Verbindung mit dem S(end)-Timer und der Server mit dem U(ser)-Timer. Laufen die Timer ab, werden jeweils Idle-Zyklen eingeschoben. Der R(receive)-Timer des Masters und der I(ndication)-Timer des Slaves überwachen die Verbindung, wobei der Ablauf dieser Timer den Verbindungsabbruch bewirken. Der Slave überwacht mit dem F(etch)-Timer das Abholen (Pollzyklus) einer bereitgestellten Antwort durch den Master.

3 Gerätemodell und Architektur

Das Gerätemodell bei PROFIBUS-DP ging von modularen bzw. kompakten Slave-Geräten aus. Bei einem kompakten Slave-Gerät bestand die Möglichkeit, virtuelle Module zu definieren. Je nach Anwendungsfall enthalten die Module ihre modulspezifischen Konfigurations- und Diagnoseinformationen. Es erfolgte keine Adressierung der einzelnen Module, da alle Ein- bzw. Ausgänge in einem Block übertragen wurden. Eine Zuordnung zu den Modulen ist bei PROFIBUS-DP Aufgabe der Anwendung.

Mit der Erweiterung der PROFIBUS-DP-Funktionalität wurde das vorhandene Gerätemodell weiterentwickelt (siehe Fig. 4). Es wird jetzt davon ausgegangen, daß bei einem DP-Slave-Gerät mehrere Anwendungsprozesse integriert sein können, die durch entsprechende Anwendungsprozeßinstanzen (API) repräsentiert werden. Die in den APIs enthaltenen virtuellen oder physischen Datenobjekte sind in sogenannten Slots enthalten. Adressierung der Daten in den Produktivdiensten erfolgt über Slot_Number und Index. Die Benutzung von Slot_Number und Index ist herstellerspezifisch, wobei jedoch für modulare und kompakte DP-Slaves eine einheitliche Verfahrensweise em-pfohlen wird. Für modulare Slaves sollte Slot_Number ein Modul (auch Steckplatz) beginnend mit eins in aufsteigender Reihenfolge repräsentieren. Die Slot_Number 0 ist dem Gerät selber vorbehalten. Für Kompaktgeräte wird die gleiche Vorgehensweise mit virtuellen Modulen empfohlen. Ein Master Klasse 2, kann zu den verschiedenen APIs eine Verbindung realisieren (MSAC_C2-Verbindung). Die Adressierung der jeweiligen APIs erfolgt während des Verbindungsaufbaus. Mit verschiedenen APIs können verschiedene Adreßräume aufgespannt werden, da

Slot_Number und Index immer relativ zum API liegen. Einfache Geräte besitzen jedoch nur ein API, wodurch eine flache Adressierung erreicht wird.

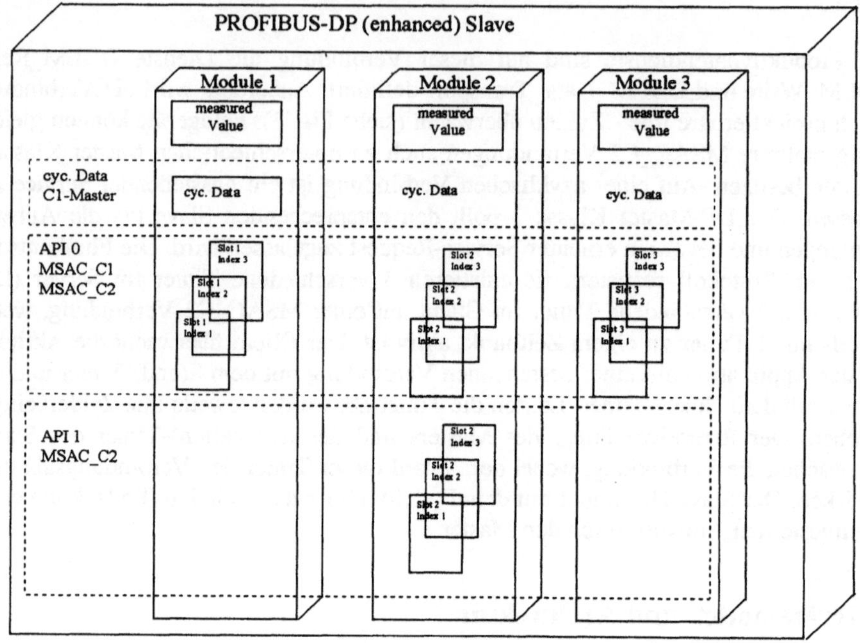

Fig. 4. Gerätemodell

API 0 besitzt per Definition einen besonderen Status. Über API 0 adressiert der Master Klasse 2 die gleichen Objekte wie der Master Klasse 1 über die MSAC_C1 - Verbindung.

4 Netzübergreifende Adressierung

Die netzübergreifende Adressierung bezieht sich einerseits auf die Möglichkeit, Subnetze innerhalb eines DP-Netzes aufzubauen und zu adressieren und andererseits Verbindungen eines Teilnehmers aus einem beliebigen Netzwerk (z.B. Internet) zu einem DP-Slave zu unterhalten. Die im Abschnitt 2 beschriebene Timerfunktionalität unterstützt diese Fähigkeit durch eine zeitliche Entkopplung der Anwendungsfunktionen. Der Dienst zum Verbindungsaufbau einer MSAC_C2 Verbindung unterstützt diese Eigenschaft durch zusätzliche Adreßparameter. Fig. 5 zeigt das eingeführte Modell für die über ein DP-Netzwerk hinausgehende Adressierung. Die dafür notwendigen Parameter werden nur im Verbindungsaufbau eingesetzt, d.h. die Produktivdatendienste kommen weiterhin mit Slot_Number und Index als Adreßparameter aus. Mit dem Verbindungsaufbau wird der gesamte Kanal bis zum Endgerät hergestellt. Zur Realisierung wurde neben der Zieladresse (Rem_Add) der normalen DP-Station bzw. des Links eine zusätzliche Adreßerweiterung eingeführt. Die Erweiterung unterscheidet wiederum zwei Typen.

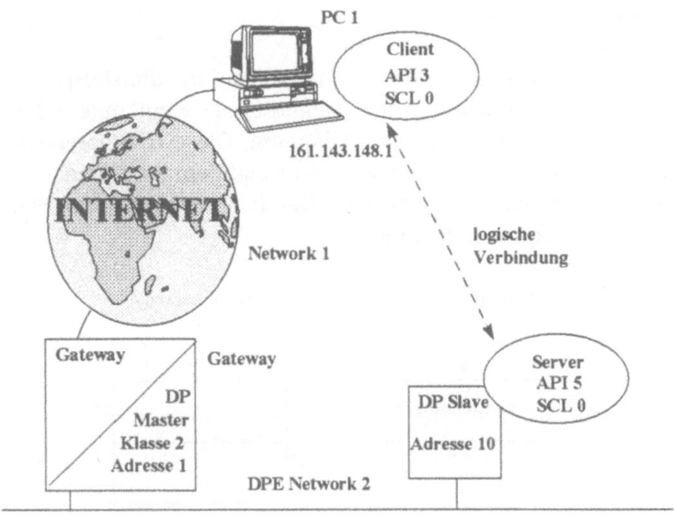

Fig. 5. Netzübergreifende Adressierung

Der Typ 0 beinhaltet die Adressierung der API ohne netzübergreifende Adreßerweiterungen. Diese werden im Typ 1 mit der Network_Address des Netzwerkes und mit der MAC_Address des Knotens definiert. Für das in Fig. 5 dargestellte Beispiel einer Verbindung eines Clients aus dem Internet (Network 1) zu einer Server-Applikation auf einem DP-Slave (Network 2) ergeben sich folgende Adreßparameter in der „Initiate_REQ_PDU" vom Master Klasse 1 zum DP-Slave:

Tab. 1. Aufbau der Initiate-REQ-PDU (Adreßparameter)

Initiate-REQ-PDU 1 → 10	
Remote Address	10
Source Type	1
Source Length	14
Destination Type	0
Destination Length	2
Source	
Application Process Instance	3
Security Level	0
Network	1
MAC	161.143.148.1
Destination	
Application Process Instance	5
Security Level (SCL)	0

5 Alarmmodell

PROFIBUS-DP enthält bisher auf Anwendungsebene ein dreistufiges Diagnosekonzept, bestehend aus der gerätebezogenen Diagnose, der Kennungs- oder modulbezogenen Diagnose und der kanalbezogenen Diagnose. Diese Diagnosetypen haben eine Gemeinsamkeit, indem sie einen Momentanzustand repräsentieren und somit über einen Puffer arbeiten. Es ist nicht gesichert, daß die Anwendung im Master Klasse 1 jede Änderung in der Diagnose empfängt.

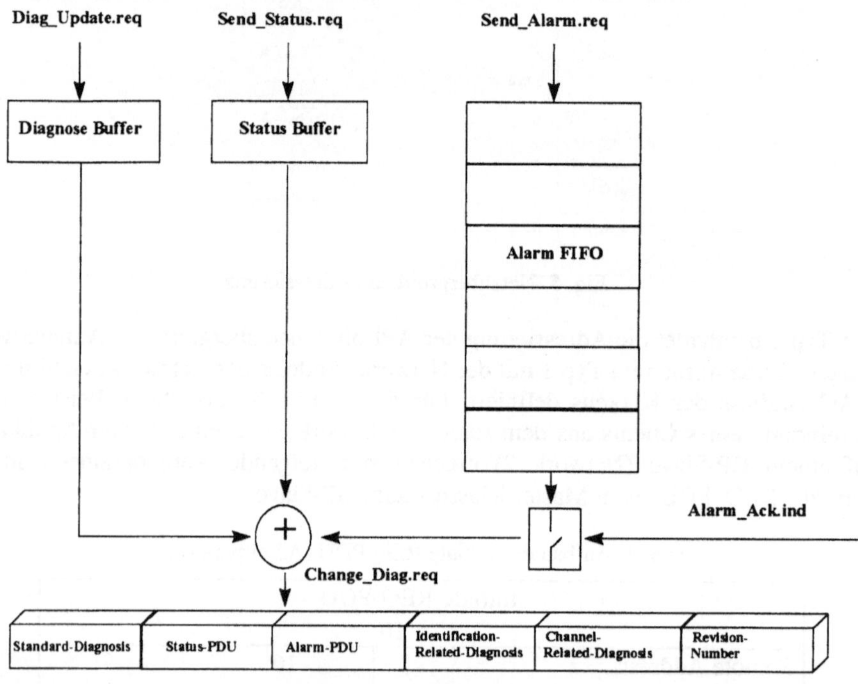

Fig. 6. Alarmmodell

Für bestimmte Anwendungen ist es jedoch unerläßlich, Alarmmeldungen in der Reihenfolge ihres Auftretens entgegenzunehmen und zu quittieren. Ein solches Alarmmodell ist im Rahmen der erweiterten PROFIBUS-DP Funktionen definiert. Aufbauend auf der Diagnose ist ein Warteschlangenmodell mit Flußkontrolle installiert. Das Prinzip ist vereinfacht in Fig. 6 dargestellt. Eine Diagnosenachricht kann in verschiedene Abschnitte eingeteilt werden, wobei jedoch nicht zwingend alle Teile vorhanden sein müssen. Somit kann die PDU neben den Standard-Diagnose-Parametern (nach EN50170 [1]) Statusmeldungen, Alarmmeldungen sowie kanal- und kennungsspezifische Diagnose enthalten. Die gerätespezifische Diagnose ist nur dann erlaubt, wenn Alarm- und Statusmeldungen als Spezialisierung der gerätespezifischen Diagnose nicht eingeschaltet sind. Es sind folgende sechs Alarmtypen vordefiniert:

- Diagnose-Alarm (z.B. Kurzschluß oder Übertemperaturen)
- Prozeß-Alarm (Ereignis im angeschlossenen Prozeß)
- Ziehen-Alarm (Baugruppe entfernt)
- Stecken-Alarm (Baugruppe hinzugefügt)
- Status-Alarm (Zustandswechsel, z.B. Stop)
- Update-Alarm (Veränderung von Parametern durch Fremdeingriffe)

Eine Statusmeldung ist eine nicht zu quittierende, vom Anwender definierte Nachricht, die im Rahmen der gerätespezifischen Diagnose kodiert wird. Die Verknüpfung der Modelle bietet ein sehr leistungsfähiges Konzept, wobei je nach Applikation die geeigneten Eigenschaften ausgewählt werden können. Alarmmeldungen werden basierend auf dem Diagnosemodell nur zum DP-Master Klasse 1 übertragen und auch von diesem über die MSAC_C1-Verbindung bestätigt.

6 Zusammenfassung

In diesem Beitrag wurden die in dem DP-Arbeitskreis der PNO erarbeiteten funktionalen Erweiterungen von PROFIBUS-DP vorgestellt. Basierend auf dem in breitem Einsatz befindlichen PROFIBUS-DP Protokoll wurden neue Transportmechanismen (azyklischer Datenaustausch), ein neues Adressierungsmodell und ein neues Alarmmodell definiert. Diese funktionalen Erweiterungen ermöglichen den Einsatz von PROFIBUS-DP neben der Fertigungsautomatisierung auch in der Prozeßautomatisierung. Zur Gewährleistung der Interoperabilität zwischen den Feldbusgeräten werden basierend auf der Festlegung minimaler Funktionalitäten sogenannte Konformitätsklassen festgelegt. Derzeit sind für den Master Klasse 1 die Conformance Class A und für den Master Klasse 2 die Conformance Class B definiert.

Literatur

1. prEN 50170 Final Draft European Standard General Purpose Field Communication System, CENELEC, March 1996

2. PROFIBUS - DP/V1, Draft Specification, Karlsruhe, März 1997

3. Jünger, B.: *PROFIBUS-DP erschließt sich neue Einsatzbereiche*, Elektronik 1997, Heft 7, S.32 ff

Mehrkanaliger skalierbarer Master für AS-Interface

V. Douridanova[1], K. Bender[1,2], P. Wenzel[1]

[1]Forschungszentrum Informatik, Karlsruhe (FZI)

[2]Lehrstuhl für Informationstechnik im Maschinenwesen (itm)
Technische Universität München, München

Abstract. The AS-Interface was developed with the purpose to replace the cable tree of the automation field. The 31 slaves AS-Interface provides are not sufficient for many applications. The master tool, presented below, gives the possibility to generate an AS-Interface master with a wished number of channels, which can address any number of sensors and actuators.

1 Das Aktuator/Sensor-Interface (AS-Interface)

1.1 AS-Interface als der Ersatz für den Kabelbaum

Zwei Jahre nach der Markteinführung hat das Aktuator-Sensor-Interface bereits eine beachtliche Verbreitung gefunden und seine wirtschaftlichen und technischen Vorteile bewiesen. Die Gründe dafür liegen hauptsächlich in der bewußten ziel- und anforderungsorientierten Konzeption dieses Systems.

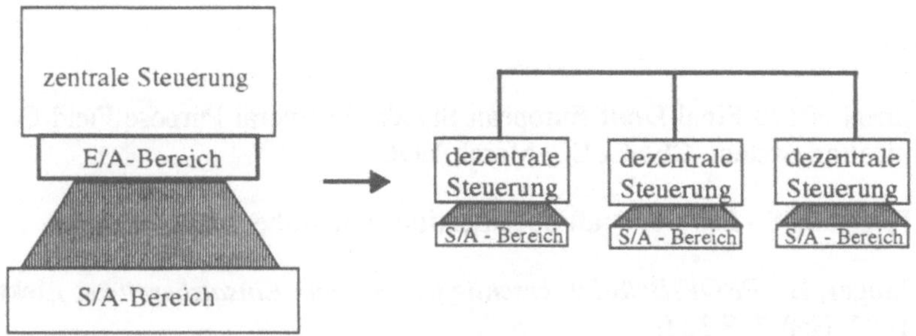

Abb. 1. Verteilte Steuerungssysteme

Das Einsatzgebiet des AS-Interface ist die unterste Ebene der Automatiesierungshierarchie, der Bereich der Aktuatoren und Sensoren. Auf diesen Bereich zielen auch andere Feldbusse ab (Interbus-S, CAN, Profibus-DP u. s. w.), die in manchen Literaturquellen unter dem Sammelbegriff Sensor/Aktuator Busse zusammengefaßt werden. Sie sind jedoch unter einem ganz anderen Aspekt zu betrachten, da sie konzipiert worden sind, um eine Dezentralisierung und Verteilung der Automatisierungsaufgabe zu erreichen (Abb. 1). Anstelle einer

großen zentralen Steuerung werden mehrere kleinere Steuerungen im lokalen Bereich eingesetzt, die mit der Regelung des technischen Prozesses „vor Ort" beauftragt sind.

Die Vorteile einer solchen dezentralen Steuerung drücken sich hauptsächlich in einer einfachen und übersichtlichen Struktur und einer Aufsplittung und Verringerung des Program-
mieraufwands aus, was sowohl die Projektierung als auch Erweiterungen und Änderungen
im Automatisierungssystem vereinfacht. Der Informationsaustausch untereinander muß eine
Reihe von globalen Anforderungen im Gesamtsystem erfüllen. Charakteristisch für die
Datenübertragung sind dabei die Größen:

- Datenmenge pro Nachricht, z.B. einfache Sensor-/Aktuatordaten: ein bis zwei Byte,
  größere Strukturen: einige bis einige hundert Byte, Steuerungsprogramme: einige
  tausend Byte.

- Multi-Master-Betrieb, bedingt durch die vielfältigen Kommunikationsverbindun-
  gen zwischen den lokalen Steuerungen

- mittlere Buslaufzeiten (um die 100ms), da die Steuerung „vor Ort" erfolgt.

Das Konzept einer dezentralen Steuerung kann für viele bereits existierende konventionelle
Steuerungssysteme beträchtliche Probleme mit sich bringen. Die Steuerungsaufgabe muß
neu verteilt werden, was auf Grund der erforderlichen längeren Stillegungs- und Ausfallzeiten der Anlage meistens technisch nicht möglich ist. Ein Ausweg ist die Beibehaltung der
Struktur des konventionellen Steuerungssystems und das Ersetzen der Verbindungen zu den
Peripherieelementen (d.h. des unübersichtlichen Kabelbaums) durch eines der obengenann-
ten Bussysteme. Diese Bussysteme erscheinen, bezogen auf die einfachen und kostengünstigen Peripherieelemente mit ihrem aufwendigen Kommunikations-Overhead zu kompliziert
und zu teuer.

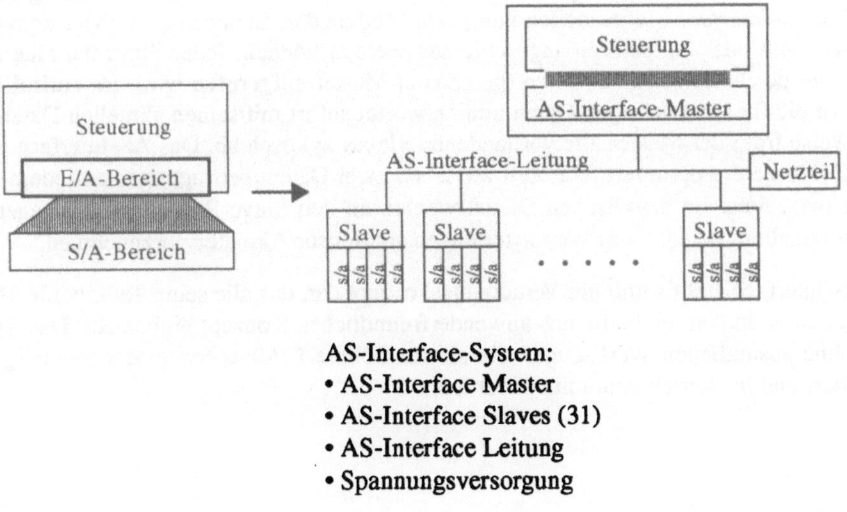

Abb. 2. AS-Interface-System als Ersatz des Kabelbaums

Mit der Entwicklung vom AS-Interface sollte genau diesem Problem durch eine kosteneffektive Lösung begegnet werden. Die Schnittstelle zwischen Steuerung und Peripherie sollte in ihrer Funktion und Handhabung nicht wesentlich von dem konventionellen Anschluß der binären Sensoren und Aktuatoren abweichen. Der Anwender kann ohne spezielle Anpassungen seine handelsüblichen Sensoren/Aktuatoren an einen AS-Interface-Slave anschließen. Auch andere nicht weniger wichtige Anforderungen wurden berücksichtigt: Echtzeitfähigkeit, hohe Zuverlässigkeit, hohe Fehlertoleranz, EMV, flexible Topologie, einfache Installations- und Montagetechnik. Um die Einfachheit und Kostengünstigkeit des Systems zu unterstreichen, wird bei der Systemcharakterisierung die Betonung auf „Interface" (Schnittstelle) gelegt.

1.2 Systemeigenschaften von AS-Interface

Abb. 2. zeigt das AS-Interface und seine Bestandteile. Der *AS-Interface Master* ersetzt die bisher verwendeten Ein- und Ausgabemodule einer Steuerung (SPS), ohne dabei die Schnittstelle wesentlich zu verändern. Er wird als eine Einschubkarte realisiert und kann in jede Steuerung problemlos integriert werden. Alle Ein- und Ausgabedaten werden wie üblich parallel abgebildet. Der Master übernimmt die Umwandlung des parallelen Datenstroms in einen seriellen. Mit dem Übergang der parallelen Datenübertragung zur seriellen Kommunikation kommt man zwangsläufig zum Busprinzip, auch wenn es explizit nicht unterstrichen wird.

Das AS-Interface ist ein Bussystem mit stark ausgeprägter Single-Master-Multi-Slave-Struktur. Die Kommunikationsverwaltung ist im Master konzentriert. Er bietet jedoch bedeutend mehr als die E/A Karten, die er ersetzt. Dem Anwender stehen mehrere Diagnose- und Hilfsfunktionen zur Verfügung, die die Installation, den Betrieb und die Wartung des Bussystems wesentlich erleichtern. Als *Übertragungsmedium* dient ein ungeschirmtes Zweidrahtkabel, das gleichzeitig zur Spannungsversorgung und Datenübertragung benutzt wird. Die *AS-Interface Slaves* stellen kompakte Module dar, an denen bis zu vier konventionelle Sensoren oder Aktuatoren angeschlossen werden können. Jeder Slave hat eine nichtflüchtig gespeicherte Adresse, durch die er vom Master aufgerufen wird. Im Aufruf erhält der Slave die für ihn wichtigen Daten und antwortet sofort mit seinen aktuellen Daten. Auf diese Weise fragt der Master alle vorhandenen Slaves zyklisch ab. Das AS-Interface ist auf kurze Zykluszeiten optimiert. Die Zeit zwischen zwei Datenübertragungen an jedem Slave beträgt max. 5ms. Im azyklischen Dienst werden an den Slave Parameter übertragen. Sie werden vor allem genutzt, um Voreinstellungen am Sensor/Aktuator vorzunehmen.

Das AS-Interface stellt somit ein Vernetzungskonzept dar, das alle seine Teilmodule *Master, Leitung, Slave* in sein einheitliches anwenderfreundliches Konzept einbezieht. Das System kann ohne zusätzlichen Werkzeuge, wie sie für höhere Feldbus-Systeme notwendig sind, projektiert und in Betrieb genommen werden.

2 Mehrkanaliger Master für das AS-Interface

2.1 Das Architekturkonzept

Bei der Entwicklung des AS-Interface ist man bewußt Kompromisse eingegangen. Zum Beispiel ermöglichte die Einschränkung des Datenaustausches auf vier Bit je Teilnehmer minimale Abmessungen des Slave-Schaltkreises (ASIC). Somit konnte das Modul auch in kleinen Sensoren und Aktuatoren integriert werden. Weiterhin sichert eine Zykluszeit von maximal 5 ms die Echtzeitfähigkeit in den meisten Anwendungsfällen. Für die so gewählte Zykluszeit ist die Übertragungssicherheit für bis maximal 31 Teilnehmer (100 m Buslänge) gewährleistet. Dies erlaubt die Realisierung des Masters durch einen einfachen 8-Bit-Mikrocontroller. Die Begrenzung der Anzahl der Slaves auf 31 erweist sich jedoch als unzureichend, besonders bei großen Anlagen mit hunderten von Anschlüssen auf kleinem Raum. Bei Automatisierungsaufgaben mit mehr als 124 binären Peripherieelementen müssen in die jeweilige Steuerung weitere Master integriert werden, was den Platzbedarf, aber auch die Kosten erhöht.

Das vom Wirtschaftsministerium des Landes Baden Württemberg geförderte Projekt ASIMA hat zum Ziel, einen AS-Interface Master zu entwickeln, mit dem der Industrie eine wirtschaftlichere Lösungsvariante für Automatisierungssysteme mit mehr als 124 Peripherieelementen zur Verfügung gestellt wird. Als wichtig hat sich dabei die Eigenschaft der Skalierbarkeit erwiesen, da zum Beispiel von der Anzahl der Kanäle die erforderliche Rechenleistung des Masters abhängt. Ein Master mit vielen Kanälen würde sich für kleinere Steuerungsaufgaben als überdimensioniert und somit als kostenungünstig erweisen. Dies führte zur Entwicklung eines Mastertools, das dem Anwender (dem Hersteller von Masteranschaltungen) erlaubt, die Leistung des Masters nach branchen-, profil-, und applikationsspezifischen Anforderungen und nach Kostenkriterien selbst zu bestimmen. Folgende Aspekte wurden bei der Entwicklung des Mastertools berücksichtigt:

- Einhaltung der Master-Spezifikation: Die Master-Spezifikation sollte für den mehrkanaligen Betrieb weder verändert noch erweitert werden. Jeder Kanal mußte alle Prüfvorschriften bezüglich des Standards problemlos erfüllen können.

- Kompaktheit: Die Hardwarerealisierung des mehrkanaligen Masters durfte die geometrischen Maße eines einkanaligen Masters nicht übersteigen (Platine in Europaformat oder kleiner).

- Skalierbarkeit der Kanalanzahl: Die Anpassung an die gewünschte Anzahl von Kanälen sollte vollautomatisch durch Veränderung von wenigen Parametern bzw. Switches in der Software, aber auch bei der Generierung der Hardwaremodule erfolgen.

Abb. 3. zeigt die Architektur des mehrkanaligen Masters. Eine schon bewährte Partitionierung für einfache AS-Interface Master in die drei Funktionsebenen Übertragungs-, Ablaufkontrol- und Masterebene wurde beibehalten. Diese Struktur erweist sich deshalb als vorteilhaft, da sie die erforderliche Gesamtleistung in Teilleistungen zerlegt, die sich in ihrer Größenordnung um Potenzen unterscheiden:

- Die Übertragungsebene bestimmt das Verhalten und die Überwachung der seriellen Kommunikation auf der AS-Interface Leitung. Sie ist verantwortlich für die Einhaltung der Zeitverhältnisse während eines Telegrammzyklus, d.h. vom Masteraufruf bis zur Slaveantwort. Nach der AS-Interface Spezifikation benötigt dieser Zyklus zwischen 150–162 µs.

- Die Ablaufkontrollebene ist für die Steuerung der Betriebsphasen und der Telegrammsequenzen zuständig. Sie muß in der Lage sein, in einem Zeitfenster von 5ms für die n Kanäle die maximale Anzahl von je 33 Telegrammen zu generieren, wie es die AS-Interface Spezifikation in der Phase des Normalbetriebs erfordert.

- Die Masterebene ist mit zeitunkritischen Funktionen beauftragt, die sie an der Schnittstelle zum Host als Hilfsdienste anbietet: Auslesen von Slavekonfigurationsdaten und Slaveparametern, Informationen über den aktuellen Zustand im Netz u. s. w..

Die härtesten Leistungsanforderungen werden an die Übertragungsebene gestellt. Da die Zeitverhältnisse im µs-Bereich liegen und dadurch extrem kritisch sind, wurde für diese Ebene eine gleichzeitige (parallele) Bearbeitung auf n Verarbeitungseinheiten vorgesehen. Jeder einzelne Kanal bekommt ein Rechenmodul (z.B. einen RISC-Prozessor) zugewiesen, das für die Kommunikation auf dem jeweiligen AS-Interface Bus zuständig ist.

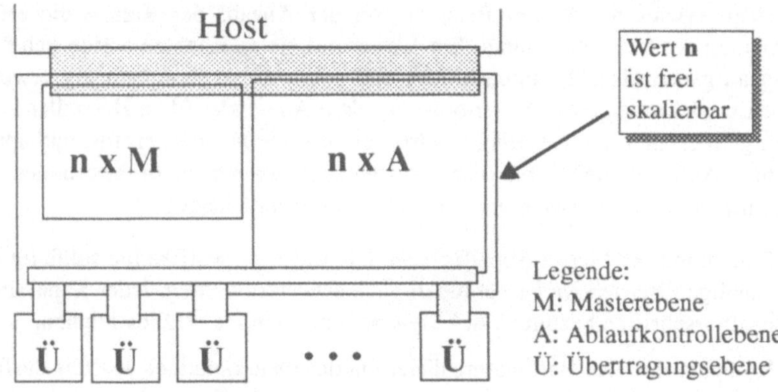

- n Ü   getrennte Verarbeitung auf **n** Recheneinheiten
- n x M,
- n x A: parallele Verarbeitung auf **einer** Recheneinheit

Abb. 3. Architektur des n-kanaligen Masters

Die Zeitverhältnisse auf der Ablaufkontrollebene erlauben die Beauftragung einer Recheneinheit mit der Bearbeitung der Dienste dieser Ebene für alle n Kanäle. Der Rechenkern muß in dem fest vorgegebenen Zeitfenster von 5 ms die notwendigen Telegrammsequenzen für den Normalbetrieb für alle Kanäle generieren. Da das Zeitfenster unabhängig von der Anzahl der Kanäle ist, hängt die Leistungsanforderung an den Rechenkern ungefähr proportional von der Kanalanzahl des Masters ab. Eine feinere Anpassung der Leistung und Senkung der Kosten für den Rechenkern kann bei Anwendungen erzielt werden, bei denen bereits vor der Implementierung feststeht, daß die einzelnen Kanäle unsymmetrische Last aufweisen, d.h. wenn die Anzahl der zu bedienenden Slaves in einigen Kanälen viel kleiner ist als die maximal mögliche Anzahl von 31 Slaves (z.B. ein zweikanaliger Master, bei dem ein Kanal mit 31 Slaves, der andere jedoch nur mit wenigen Slaves besetzt ist). Durch Verleihung einer höheren Priorität für die Bearbeitung des vollbesetzten Kanals kann man praktisch erreichen, daß der Datenaustauschzyklus bei jedem Slave von dem unterbesetzten Strang auch bei schwächerer Leistung des Rechenkerns kleiner ist als die kritische Marke von 5 ms. Läßt man diese Verfeinerung zu, dann hängt die geforderte Leistung für das Rechenmodul proportional von der Anzahl der vorhandenen Slaves im Gesamtsystem ab.

Die Funktionsdienste der Masterebene müssen nach der ASI-Spezifikation keine zeitkritischen Eckdaten einhalten. Ihre Ausführung wurde deshalb dem gleichen Rechenmodul zugewiesen, das die Ablaufkontrollebene realisiert. Die zusätzliche Leistung, die man für den Rechenkern planen sollte, ist unabhängig von der Anzahl der Kanäle, da die Dienste dieser Ebene azyklisch und nicht zeitgleich in mehreren Strängen abgearbeitet werden. Selbst im Extremfall ist es nicht notwendig, zusätzliche Leistung für diese Ebene einzuplanen, da ihre Dienste nur in Ausnahmesituationen (Inbetriebnahme, Wartungsarbeiten) und während der Betriebsphasen des Masters ausgeführt werden, welche keine zeitkritischen Zyklen enthalten (z. B. Offline-Phase).

Die Verarbeitungseinheit kann neben der Realisierung der Ablaufkontroll- und Masterebene weiterhin folgende Funktionen bearbeiten:

- Funktionen der Hostebene (Bearbeitung der Slavedaten, Datenaustausch zwischen den Kanälen) und

- Funktionen der Anwendung (Bedienung einer seriellen Schnittstelle).

Diese Dienste treten in den meisten Fällen zyklisch auf und erfordern die Einplanung zusätzlicher Rechenleistung. Das Hauptziel, das bei der Ausarbeitung des Architekturkonzeptes verfolgt wurde, war eine modulare Struktur mit voneinander unabhängigen Funktionsebenen und klar definierten Schnittstellen zu entwerfen, die gestattet:

- Funktionsebenen bzw. Module zu ersetzen, ohne die gesamte Struktur zu beeinflussen

- hohe Skalierbarkeit bezüglich der Anzahl der Kanäle zu erreichen

- Software- und Hardware weitgehend unabhängig voneinander zu entwickeln

2.2 Realisierung

Die Realisierung umfaßt sowohl die Hardware als auch die Software des Masters.

Die Softwarerealisierung des n-strangigen Masters konzentriert sich auf die Entwicklung eines Programms, das die Ablaufkontroll- und Masterebene für n Kanäle (n frei wählbar) auf einem Rechenmodul durchführt.

112

Die Schnittstellen zu den benachbarten Ebenen (Übertragungs- und Hostebene) wurden nach den folgenden Prinzipien gestaltet:

- Zu den Kommunikationsprozessoren der Übertragungsebene sind entweder einfache Register oder zweistufige Pipelines für die Übergabe des Auftrags und Übernahme der Antwort vorgesehen.

- Die Schnittstelle zur Hostebene ist ein Speicherbereich, der groß genug ist, um die Daten von allen Strängen aufzunehmen (minimal n x 64 Byte).

Für den Rechenkern ergab sich die typische Situation des Mehrprozeßbetriebs (Multitasking). Der Einsatz eines Multitaskingbetriebssystems steigert jedoch die Leistungsanforderungen an die Prozessoreinheit und erschwert die Portierbarkeit der Software auf verschiedene Hardwareplattformen. Der Verarbeitungsprozeß kann als typischer Vertreter eines SPMD-Prozesses (Same Programm, Multiple Data) klassifiziert werden. Die einzelnen Tasks beanspruchen außer Prozessorzeit und Code keine weiteren gemeinsamen Ressourcen. Die Dienste eines Multitaskingsystems werden also nur in beschränktem Umfang gebraucht. Aus diesen Gründen wurde das Steuerwerk für die Regelung der Prozessorzeit eigens dafür realisiert.

Die Software wurde in der Programmiersprache ANSI-C entwickelt. Die Anzahl der Kanäle „n" ist eine vor dem Compilierungsvorgang einzugebende Konstante. Wenige Parameter gestatten die Anpassung an die konkrete Hardware. Eine Reihe von Funktionen und Variablen dienen der Abschätzung der Leistung des ausgewählten Prozessors und können bei Bedarf in die Compilierung aufgenommen werden, um die Portierung zu erleichtern (z.B. Messen der Datenaustauschzykluszeit im vorgegebenen Strang, Anzeige der Datenüberschreibungen an der Schnittstelle zur Übertragungsebene, u.s.w.). Für die Realisierung der Übertragungsebene jedes einzelnen Kanals wurde ein Kommunikationsprozessor des Typs PIC 16C56 vorgesehen. Bei Bedarf kann diese Ebene neu gestaltet werden: Es können beispielsweise leistungsstärkere Recheneinheiten eingesetzt werden, die die Übertragung auf mehr als einem Kanal bewältigen können, außerdem ist der Einsatz von ASIC-Bausteinen, die die analogen und digitalen Teile der Übertragungsebene vereinen, möglich.

Abb. 4 stellt die Hardwarearchitektur eines zweikanaligen Masters dar. Er besteht aus den drei Hauptmodulen (in Abb. 4 grau dargestellt):

- Rechenkern 80x51 Prozessor, der die Leistung für die Realisierung der Master- und Ablaufkontrollebene für die zwei Kanäle erbringt. (Untersuchungen ergaben, daß der 8-Bit Prozessor Dallas 80C320 für zwei Kanäle vollkommen ausreicht. Bei höheren Taktfrequenzen kann der Prozessor sogar noch zusätzliche Aufgaben übernehmen.)

- Zwei RISC Prozessoren PIC 16C56 für die Übertragungsebene

- FPGA Baustein, der die Logik für die Schnittstellen zwischen der Übertragungs- und Ablaufkontrollebene bzw. Master- und Hostebene beinhaltet. (Die Anpassung der Logik an eine variable Anzahl von Kanälen erfolgt automatisch, womit ein hoher Grad an Skalierbarkeit nicht nur bei Software, sondern auch bei Hardware erreicht wurde.)

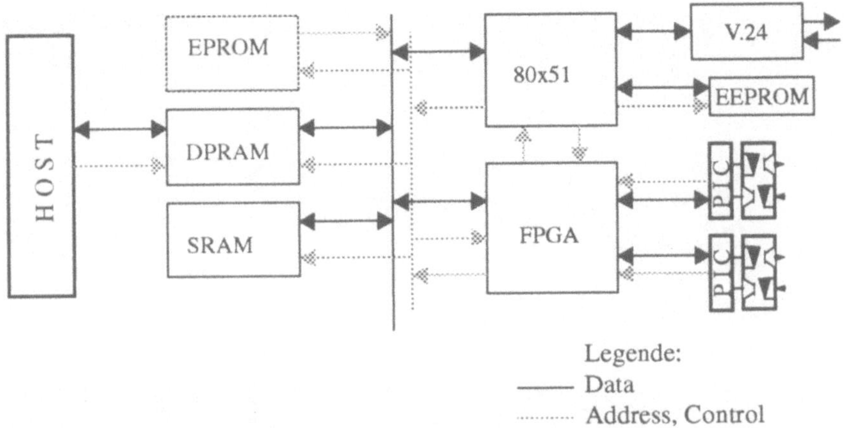

Abb. 4. Dual-Master Controlle

Da der dargestellte Dual-Master-Controller auch Entwicklungszwecken dient, enthält er einige zusätzliche Module (32 kByte SRAM, 32 kByte EPROM, eine V.24 Schnittstelle). Die durchgängige Skalierbarkeit und Modularität des Mastertools sowohl bezüglich Hardware als auch der Software versetzt den Entwickler in die Lage, in kurzer Zeit mit geringem Entwicklungsaufwand, einen auf seine spezielle Anwendung angepaßten Master zu generieren. Er muß dazu anhand einer Kriterienliste, die ihm zur Verfügung gestellten Module zusammenstellen. Die Anpassung der Module an die geforderte Anzahl von ASI-Kanäle erfolgt dabei weitgehend automatisch.

3 Zusammenfassung

Das AS-Interface konnte in seinem zentralen Einsatzgebiet, dem Feldbereich, der durch Kabelbäume und Parallelverdrahtung gekennzeichnet ist, beachtliches Echo erfahren. Es bietet sich zugleich als ideales Subsystem für höhere Bussysteme an (Profibus, Interbus-S, u.s.w.), wo es die Schnittstelle zum technischen Prozeß standardmäßig übernimmt. Der Beitrag zeigt eine Möglichkeit zur Erweiterung des AS-Interface-Systems zur Erfassung einer unbegrenzten Anzahl von Peripherieelementen auf. Das System kann somit ohne Einschränkung in jede lokale Steuerung eines dezentralen Steuerungssystems eingebaut werden. Damit kann eine klare Trennung zwischen Datenerfassung, Steuerung vor Ort und Kommunikation mit höheren Systemen erzielt werden, was den Weg für eine Vereinheitlichung und Vereinfachung von verteilten Systemen eröffnet.

Literatur

1. ASI Association (Hrsg.): Actuator Sensor Interface, Complete Specification, Version 2.0, Germany, 1995.

2. Kriesel, W.; Madelung, O. W. (Hrsg.): Das Aktuator-Sensor-Interface für die Automation, Carl Hanser Verlag, München, 1994.

2 Gebäudeautomation

Building automation

Zentrales Thema ist das Netzwerk-Management: Die hohe Anzahl der Knoten, die daraus resultierenden komplexen Konfigurationswerkzeuge, die noch immer viele Wünsche offen lassen, sowie die Wartungsproblematik der Netze im allgemeinen schaffen eine völlig neue Situation in der Netzwerktechnik.

Zusätzliche Aspekte, die mehr und mehr Interesse wecken, sind die Text- und Sprachübertragung über Feldbussysteme und die Anbindung von Internet an Feldbussysteme.

The focal topic is network management: the large number of nodes, the resulting complex configuration tools still leaving much to be desired, and the maintenance problem in general create an entirely new situation in network technology.

Additional aspects provoking more and more interest are the transmission of text and speach data over fieldbus systems as well as the connection of the Internet and fieldbus systems.

Ein Feldbus-Profil auf dem Weg zur europäischen Norm

Peter Fischer

Fachhochschule Dortmund
Informations- und Kommunikationstechnik
Sonnenstr. 171
D-44137 Dortmund
e-mail: fischer@fh-dortmund.de

Abstract. Das im Rahmen einer deutschen VDMA-Arbeitsgruppe entwickelte universelle Profil für Feldbussysteme in der Gebäudeautomation wird in einer europäischen CEN/CENELEC Task Group zu einer europäischen Vornorm weiterentwickelt. Diese Vornorm ermöglicht die einfachere Anbindung verschiedener Feldbussysteme untereinander und zu Kommunikationnssystemen auf Management- oder Automationsebene. In diesem Aufsatz werden die Grundlagen des Profils sowie die Objekte und Dienste vorgestellt, wobei insbesondere die Behandlung von Alarm- und Ereignismeldungen bei einfachen Geräten erläutert wird.

Abstract. The universal Profile for Building Automation Systems on Fieldnet Level developed by a German VDMA working group is becoming a European prestandard under the development of an European CEN/CENELEC task group. With this prestandard the link to fieldbus systems amongst each other and to communication systems on management or automation level is possible with less effort. In this paper the basics and the objects and services of the profile will be presented. Especially the alarm and COV (change of value) handling with simple datapoints will be shown.

1 Feldbus-Normung für die Gebäudeautomation

Im März 1997 hat die Working Group WG 4 im Technical Committee TC 247 des CEN vier Kommunikationssysteme für die Feldebene im Bereich der Gebäudeautomation als europäische Vornorm vorgeschlagen. Diese vier Systeme sind

- Batibus,
- Europäischer Installationsbus EIB,
- European Home System Bus EHS und
- LonTalk Protocol.

Batibus ist eine französische Norm, die für die Kommunikation auf Feldebene in Gebäudeautomationssystemen entwickelt wurde und in Frankreich weit verbreitet ist.

Der EIB ist eine deutsche Vornorm (DIN V VDE 0829 "Elektrische Systemtechnik für Heim und Gebäude (ESHG)"), die auf einem Busstandard aus dem Bereich der Installationstechnik basiert, aber auch für komplexe Anwendungen aus dem Bereich der Gebäudeautomation eingesetzt wird, wie z. B. Einzelraumregelung. Die meisten EIB-Installationen finden sich im deutschsprachigen Raum Europas.

Der EHS wurde im Rahmen mehrerer ESPRIT-Projekte hauptsächlich für den Einsatz im Heimbereich entwickelt und unter dem Titel "EHS - An Application and Communication Protocol for Electronic Home and Building Systems and Networks" veröffentlicht. Es gibt zur Zeit (fast) keine Installationen mit dem EHS.

Das LonTalk Protocol ist das einzige der aufgeführten Kommunikationssysteme, das für den Einsatz in allen möglichen industriellen Automationsanwendungen konzipiert wurde und das weltweit eingesetzt wird.

Bei dem Vergleich der Applikationsschichten der erwähnten Kommunikationssysteme zeigt sich sehr schnell, daß ein kleinstes, gemeinsames Vielfaches an Objekten und Services der Schicht 7 nicht vorhanden ist. Somit müßte jeder Hersteller, der alle Protokolle anbieten will, entsprechende Gateways entwickeln und einsetzen mit allen damit verbundenen Nachteilen. Innerhalb eines VDMA-Arbeitskreises wurde daher ein Papier mit dem Titel "Profile Building Automation - Fieldnet Level (PBA-FLN)" (siehe auch [1]) erarbeitet und in die WG 4 von CEN TC 247 eingebracht. Ausgehend von diesem Papier wurde vom CEN ein Work Item an eine gemeinsame Task Group von WG 4 und CENELEC TC 205 WG 3 vergeben mit der Maßgabe, eine Objekt- und Dienstebeschreibung für die Feldebene in der Gebäudeautomation zu spezifizieren, damit diese als europäische (Vor-)Norm veröffentlicht werden kann.

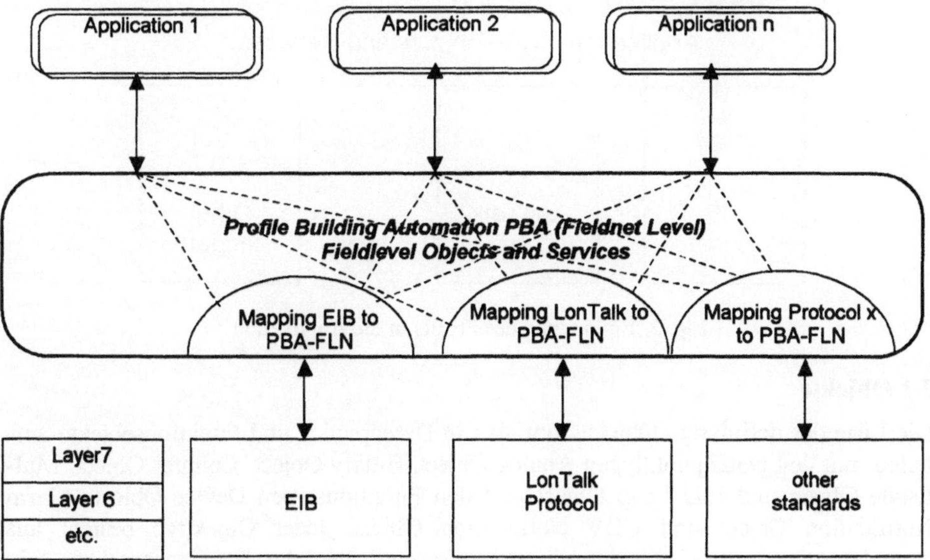

Fig. 1. Universelles Profil für Feldbussysteme in der Gebäudeautomation

2 Profile Building Automation - Fieldnet Level

Grundlage für dieses Profil waren die Funktionen, die in verschiedenen Papieren, wie z. B. VDI 3814 [2], CEN TC 247 WG3 [3] und EN 50090 [4], spezifiziert sind. Dabei wurde auch die datenpunkt-orientierte Sichtweise übernommen, die sich in dieser Form in modernen Protokollen der Anwendungsschicht, wie z. B. beim objektorientierten LonTalk Protocol, nicht direkt wiederfindet. Der Vorteil dieser Sichtweise liegt jedoch in der großen Flexibilität zur Realisierung von Diensten und Funktionen.

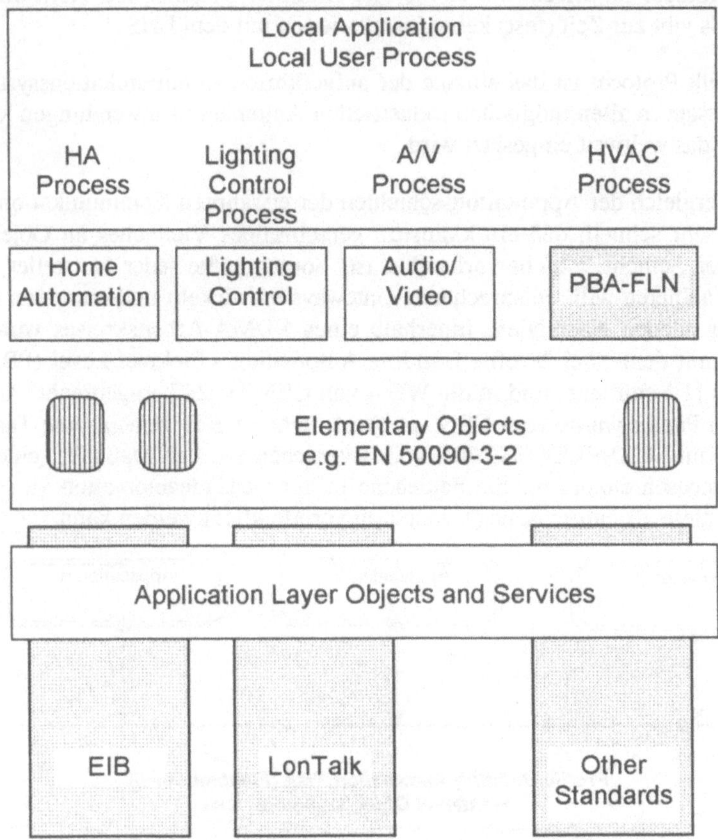

Fig. 2. Einbindung des Profils in die Applikation

2.1 Objekte

Die Liste der definierten Objekte läßt sich in Datenpunkt- und Funktionsobjekte aufteilen, mit den Datenpunkttypen Analog Object, Binary Object, Counter Object, Multistate Object und PID Loop Object und den Funktionstypen Device Object, Alarm Notification Object und COV Notification Object. Jeder Objekttyp besteht aus "mandatory properties", die immer vorhanden und über das Netz ansprechbar sein müssen, "optional properties", wie z. B. Alarmgrenzen, die vorhanden sein können

und, wenn sie vorhanden sind, ansprechbar sein müssen, und "engineerable properties", die im Objekt vorhanden sein können, aber nicht über das Netz ansprechbar
sind, wie z. B. voreingestellte Werte.

Als Beispiel ist in Fig. 3 die Tabelle des "Analog Object" abgebildet. Die engineerable
properties für diesen Objekttyp können z. B. sein der Default-Wert für den
Present_Value, das Update-Intervall und das Deadband.

Analog Object FLN		
Object_Id_Number	m	R
Object_Type	m	R
Object_Name	m	R
Value_Presentation	m	R
Present_Value	m	W
Status	m	R
Units	m	R
Priority	o	W
Description	o	R
Commanded_Value	o	R
High_Alarm_Limit	o	W
High_Warning_Limit	o	W
Low_Warning_Limit	o	W
Low_Alarm_Limit	o	W
Maintenance/Out_Of_Service_Flag	o	W
Event_Disabled	o	W
Warning/Alarm_Disabled	o	W
Acknowledged_States	o	W
Alarm_Acknowledge	o	R

m: mandatory
o: optional
R: read only
W: read and write

Fig. 3. Analog Object FLN

2.2 Dienste

Die Kommunikation auf der Feldebene gliedert sich in

- prozeßorientierte Kommunikation für Interaktionen zwischen verteilten
  Prozessen auf Feldebene,

- managementorientierte Kommunikation für Interaktionen zwischen Prozessen auf Feldebene und Anwendungen auf höheren Ebenen und

- "engineering"-orientierte Kommunikation für Interaktionen zwischen Prozessen auf Feldebene und installationsbezogenen Prozessen in Engineering oder Service Tools.

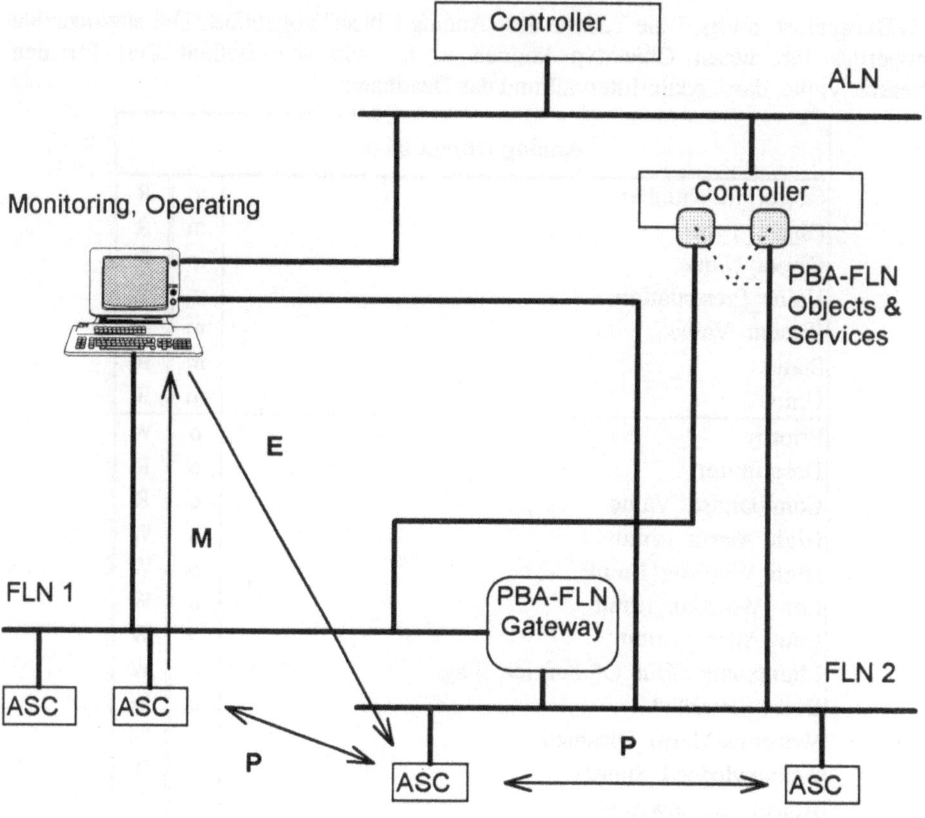

Legende:

ALN   Automation Level Network
ASC   Application Specific Controller
FLN   Field Level Network
E   Engineering Communication
M   Management Communication
P   Process Communication

Fig. 4. Kommunikation zwischen Feld- und Automationsnetzen

Die Prozeßkommunikation zwischen verschiedenen Feldbussystemen innerhalb eines Gebäudeautomationssystem wird eher die Ausnahme als die Regel sein und müßte mit Hilfe eines Gateways realisiert werden. Auch die Kommunikation während der Engineering-Phase spielt keine bedeutende Rolle. Am wichtigsten ist die Kommunikation zwischen einem Feldbussystem und einer Bedien- und Beobachtungsstation (Monitoring, Operating) oder einem Kommunikationssystem auf Automations- oder Managementebene (Controller).

Unter diesen Gesichtspunkten wurden die drei abstrakten Dienste ReadPropertyService, WritePropertyService und NotificationService als hinreichend zur Beschreibung der Kommunikation angesehen. Der Lese- und Schreibzugriff geschieht immer nur auf ein Attribut (Property) eines Objektes. Mit Hilfe des NotificationService werden Änderungsmeldungen aus dem Feld (Wertänderungen (COV), Zustandsänderungen (COS) oder Alarmmeldungen) an andere Einrichtungen weitergeleitet. Da auf Feldebene die Kapazität der Geräte sehr begrenzt ist, gibt es einen besonderer Mechanismus, der die verschiedenen Adreßinformationen für die verschiedenen Meldungen auch einfachen Geräten zur Verfügung stellt.

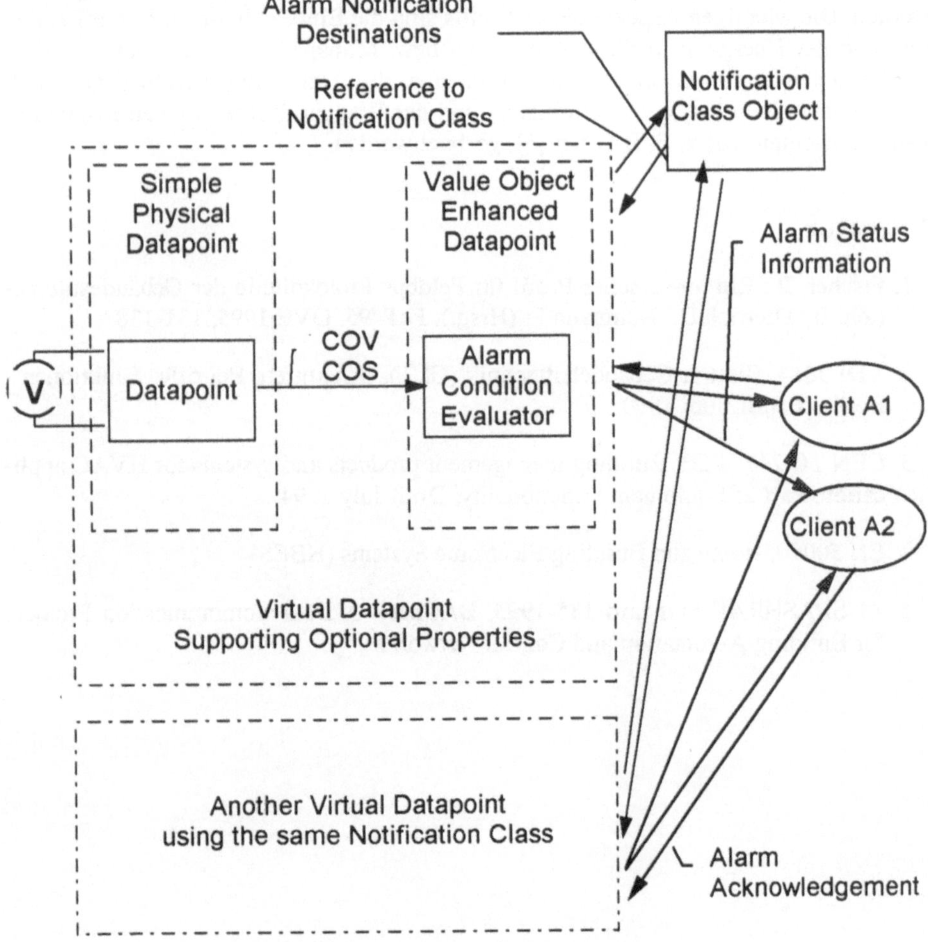

Fig. 5. Erzeugung der Alarminformation bei einfachen Feldgeräten

Wenn eine Wert- oder Zustandsänderung in einem einfachen Feldgerät auftritt und die Auswerteeinheit des dazugehörenden erweiterten Datenpunktes diese Änderung erkennt, dann wird eine Anforderung an das Notification Class Object gesendet. Die Antwort enthält die Adressen der Clients, an die die Wert- oder Zustandsänderung gemeldet werden soll. Eine Bestätigung der Meldung erfolgt, falls notwendig, von einem der Clients. Es können mehrere Notification Class Objects in einem System vorhanden sein, aber ein Datenpunkt kennt immer nur ein Notification Class Object.

3 Ausblick

Das Profile Building Automation - Field Net Level wird wahrscheinlich im Herbst 1997 als Vorschlag für eine europäische Vornorm unter dem Titel "Data Communication for HVAC Application - Field Level Objects" bei CEN TC 247 eingereicht werden. Die wichtigen Aspekte dieses Profils sind die Einbeziehung einfacher Feldgeräte und der Übergang in die Automations- bzw. Managementebene. Damit soll für die Hersteller von Gebäudeautomationssystemen die Anbindung verschiedener Feldbus-Kommunikationssysteme vereinfacht und der Weg zu übergeordneten Kommunikationssystemen, wie z. B. BACnet [5], geebnet werden.

Literatur

1. Fischer, P.: Ein universelles Profil für Feldbus-Protokolle in der Gebäudeautomation. In: Dietrich D., Neumann P. (Hrsg.): FeT '95. ÖVE 1995, 131-138.

2. VDI 3814, Blatt 1, Gebäudeleittechnik (GLT), Strukturen, Begriffe, Funktionen - Berlin: Beuth, Juni 1990

3. CEN TC 247 WG3, Building management products and systems for HVAC application, Part 2: Equipment Functionality, Draft July 1994

4. EN 50090, Home and Building Electronic Systems (HBES)

5. ANSI/ASHRAE Standard 135-1995, BACnet - A Data Communication Protocol for Building Automation and Control Networks

Beispiel einer Vernetzung für die Heimautomation mit LonWorks RTR-10 Router

Christian Wirth, Ratko Posta, Maximilian Ochensthaler
Gerhard Helge Schildt

Institut für Automation / Institut für Computertechnik,
Technische Universität Wien

Abtract. Heutzutage bewerkstelligen Haushaltsgeräte „nicht-zusammenhängende" Aufgaben. Daher ist es die derzeitige Praxis getrennte Steueranwendung - sog. "Automations-Inseln" - zu planen und zu entwickeln. Dieser Beitrag beschreibt eine mögliche Lösung der Kopplung weitverzweigter Netzsegmente in vernetzen Haushalten mittels Router. Es werden Anforderungen an die Netzinfrastruktur des Steuernetzwerkes, sowie die Ergebnisse eines durchgeführten Projektes erläutert.

Nowadays, home appliances manage incoherent tasks. Hence, it is a current practice to plan and to develop separate control applications, so called "islands" of technology. This paper describes a solution for the variety of wiring topologies and network segment interconnections in smart homes by means of routers. Requirements on the network infrastructure of home control buses are shown and results of a recent project are explained.

1   Einführung

Ziel der Haushaltsautomation ist die Vernetzung aller in einem Haushalt befindlichen Geräte (*siehe Abb.1*). Damit sind nicht nur die Elemente des Versorgungsstromkreises wie Schalter, Taster, Lampen, Temperaturfühler, Jalousiemotoren etc. gemeint, sondern auch Geräte der „braunen" und „weißen" Ware, welche in ein gemeinsames Haushaltsnetz integriert werden sollen.

Dieses Ziel stellt technisch sehr unterschiedliche Anforderungen an das zugrunde liegende Netzwerk [2]. Der dabei wichtigste Parameter ist die notwendige Übertragungsrate der einzelnen Geräte. Die Bandbreiten dieser Geräte liegen zwischen einigen Bits pro Minute bis zu einigen hundert MByte pro Sekunde [21] und legen damit eindeutig die Anforderungen an die Physik des HomeNets fest: Zweidraht-, Funk-,

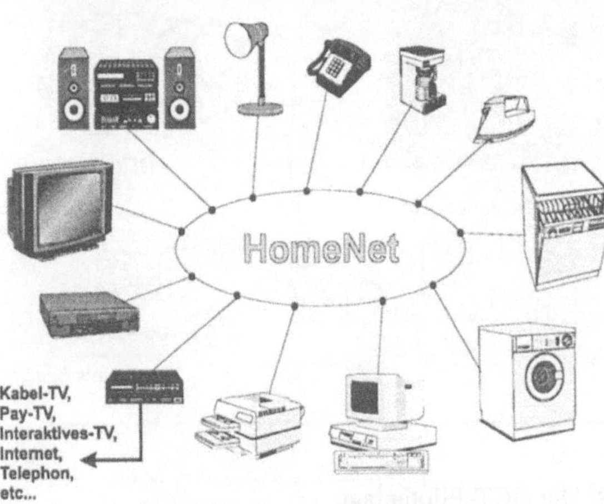

Abb. 1: Der vernetze Haushalt

124

Infrarot- und PowerLine-Übertragung mit gängigen Übertragungsprotokollen für niedrige Übertragungsraten - Glasfaser und spezielle Verfahren für die Datenkomprimierung für hohe Übertragungsraten.

Technisch stellen diese Anforderungen in der heutigen Zeit keine Probleme mehr dar, da der Erfindungsprozeß der zugrunde liegenden Basistechnologie – der Feldbussyteme [2], [19] im großen und ganzen als bereits abgeschlossen gilt. Das derzeitige Hauptproblem, welches einen breiten Durchbruch der Haushaltsautomation verhindert, liegt in der Uneinigkeit der Hersteller einen einheitlichen HomeNet-Bus-Standard zu schaffen. Weltweit streiten ca. acht Bussysteme um die Vorherrschaft im HomeNet-Sektor [3]. Erste Bestrebungen sich zu einigen sind (*leider nur europaweit*) bereits im Gange [4], [5]. Bis zur endgültigen Protkoll-Verabschiedung werden aber noch mindestens zwei Jahre verstreichen.

2 Das HCT-Projekt

Genauso wichtig wie ein einheitliches, leistungsfähiges Bussystem ist eine geeignete Softwareumgebung, mit deren Hilfe vernetzte Haushalte projektiert, gewartet und bedient werden können. Unabhängig von der bestehenden Busproblematik wird z.Z. am Institut für Automation in Zusammenarbeit mit dem Institut für Computertechnik, eine verteilte PC gestützte Softwareplattform zur Erstellung herstellerunabhängiger Applikationen für vernetzte Haushalte entwickelt [1], [13], [14].

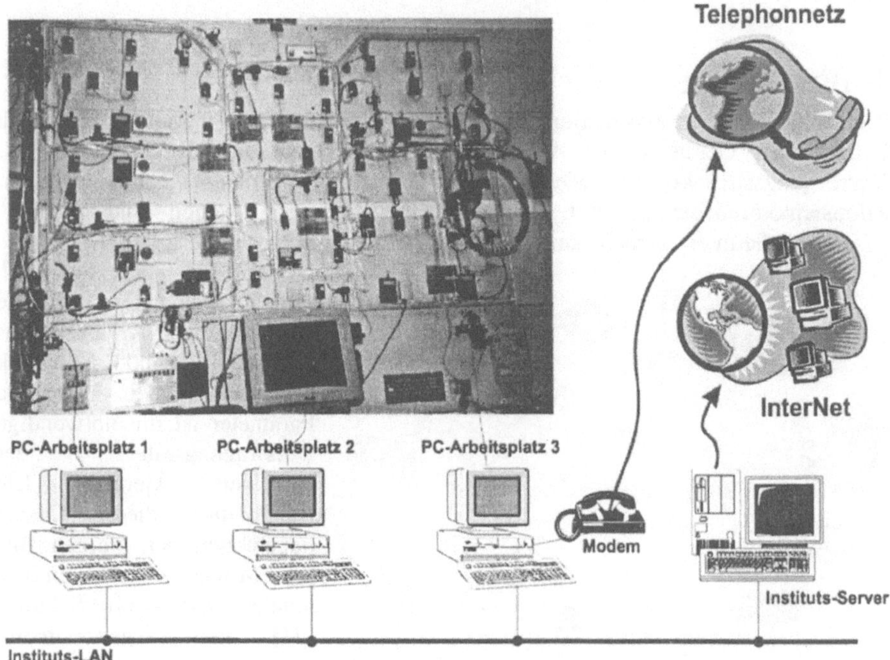

Abb.2: Die HCT-Pilotanlage

Um solche Applikationen in einer geeigneten Umgebung testen zu können wurde ein Modell eines vernetzen Haushaltes entwickelt (*siehe Abb.2*). In diesem "Modell-Haus" befinden sich - voll verkabelt - die gängigsten elektrischen Verbraucher, sowie eine vollständige Wasserver- bzw. entsorgung und ein Heizungssystem [12], [15], [16], [17]. Bis jetzt wurden ausschließlich Komponenten der LonWorks-Technologie[1] verwendet [18]. Alle Haushaltsgeräte werden von einem separaten LON-Knoten simuliert. Auf diese Weise können jederzeit Daten wie Wasser-, Lufttemperatur, Durchfluß, Wasserleitfähigkeit, Strom, Spannung, cos φ, Wetter, Luftzusammensetzung usw. im Haus-Modell dezentral erfaßt, ausgewertet und an die HCT-Softwareplattform[2] weitergeben werden. Die Softwareplattform selbst ist verteilt auf drei PCs installiert [15], [22].

Das dem HCT-Modellhaus zugrunde liegende Netzwerk (*HomeNet*) selbst besteht aus Gründen der Ausfallsicherheit und der notwendigen Übertragungsraten der einzelnen Geräte, aus verschiedenen Subnetzen, unter der Verwendung unterschiedlicher Übertragungsmedien (*Zweidrahtleitung, Starkstromleitung, Funk, Infrarot, Glasfaser*). Die Integration der Subnetze erfolgt über Router.

3 Die HCT-Subnetze

Das LON Netzwerkmodell [7] bildet die Grundlage zur Planung und Realisierung einer Anwendung in LONWorks-Technologie. Dieses Modell definiert *wie* die Netzwerkknoten logisch miteinander verbunden werden. Die Vernetzung der Knoten erfolgte unter Berücksichtigung folgender Faktoren:

- **Erhöhung der Ausfallssicherheit**: Wichtige Knoten können in separate Netzwerke gehängt werden.
- **Datenübertragungsrate**: Geräte mit hohen Datenübertragungsraten können in separaten Netzen hängen, um die Netzwerkauslastung möglichst gering zu halten.
- **Übertragungsmedien**: Zusammenfassung von Knoten mit den gleichen Datenübertragungsmedien.
- **Router**: Der Netzwerkplan hilft bei der Konfiguration und bei der Einsatzplanung von Routern

Der HCT-Modellhaushalt besteht aus insgesamt 5 Subnetzen. (*siehe Abb. 3*). Ein Backbone (*im konkreten Fall eine 1,25 Mbit/s Zweidrahtleitung*) verbindet drei PCs miteinander. Von jedem PC aus kann auf jeden Netzwerkknoten zugegriffen werden. Diese Netzwerkstruktur gewährleistet somit eine örtliche Unabhängigkeit der Anwendungen. Daher ist es für den Anwender unerheblich von welchem PC aus er seine Softwareapplikation startet. Das Haushaltsnetz besitzt damit für den Anwender maximale Transparenz. Durch gezielte Wahl der Übertragungsmedien und Übertragungsraten kann ein sog. „Flaschenhals" zwischen zwei Anwendungen oder Knoten vermieden werden. Router helfen diese Konzepte praktisch zu realisieren.

---

[1] LON = Local Operating Network
[2] HCT = HomeNet Configuration Tool

4 Das Routing

Das Routing in LonWorks-Netzen baut auf eine hierarchische Adressanordnung (*Domain-, Subnet- und Node- Adressen*), sowie auf diverse Kommunikationsdienste der Netzwerkschicht des LonTalk-Protokolls [8], [9] auf. Die Addressierungsarten in LonTalk unterscheiden sechs verschiedene Mechanismen: Die Nachrichten können entweder an alle Knoten innerhalb einer Domain (*Broadcast*), oder an alle bzw. eine Gruppe von Knoten die sich innerhalb verschiedener Subnetze befinden, gesendet werden. Einzeln können die Knoten über ihre eindeutige 48bit lange Neuron-Identifizierungsnummer, oder über die Knotennummer innerhalb der Gruppe bzw. des Subnetzes adressiert werden. Zur Gestaltung von weitverzweigten Lonworks-Netzwerken stehen verschiedene Netzwerkkomponenten wie Repeater, Bridges oder Router zur Verfügung:

- **Repeater** verbinden Nachrichtenkanäle und leiten den gesamten Nachrichtenfluß von einem Nachrichtenkanal zum Anderen weiter.
- Eine **Bridge** wird hauptsächlich zum Verbinden und Weiterleiten von Nachrichtenflüssen in Domänen verwendet.
- Ein **Learning-Router** lernt im laufendem Betrieb die bestehende Netzwerk Topologie; d.h. also welche Knoten sich auf den Seiten seiner Ports befinden. Dies geschieht durch Auswerten der Source-Subnet-Adressen in der NPDU[3] unter Ausnützung der Tatsache, daß sich über einen Router niemals *ein* Subnet erstrecken darf.
- Ein **Configured-Router** leitet nur dann Nachrichten in eine andere Domäne weiter, wenn diese bestimmten Regeln entsprechen. Diese Router müssen von einem Netzwerk-Management-Tool entsprechend konfiguriert werden. Dabei müssen alle "Forwardingtabellen" (*sowohl für die Gruppen- als auch für die Subnet/Node- Adressen*) vollständig ausgefüllt werden.

Um die im HCT-Hausmodell erstellten Konzepte auch praktisch realisieren zu können werden Router [10] benötigt, da diese:

- die Grenzen eines einzelnen Kanals, die Kapazität und die Länge der Netzwerkstränge erweitern.
- Interfaces zwischen Subnetzen unterschiedlicher Kommunikationsraten und – medien bilden. Sie bilden z.B. die Schnittstelle zwischen einem 1,25 Mbps und einem 625 kbps Zweidrahtkanals.
- die Sicherheit in einem LonWorks-Netzwerk erhöhen. Wenn z.B. Kanäle durch Router physikalisch isoliert werden, hat ein Fehler im Kanal A keine Auswirkung auf die Knoten im Kanal B.
- je nach Konfiguration den Datenverkehr in den Subsystemen isolieren und dadurch die Netzwerkleistung verbessern. Intelligente Router leiten nur dann Nachrichten in das andere Subsystem, wenn sie auch für dieses bestimmt sind.

---

[3] NPDU = Network-Protokoll Data Unit

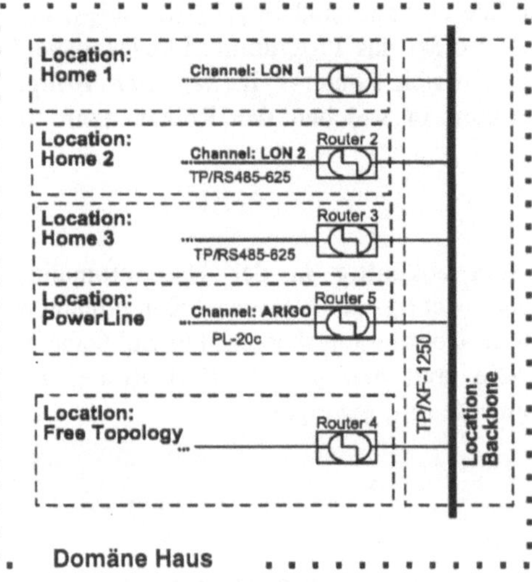

Abb. 3: Die Organisation des HCT-Netzwerkes

Der im HCT-Haushaltsnetz verwendete LonWorks-Router ist das RTR-10-Router-Core-Module [10], [22]. Die fünf Router welche im HCT-Modellhaushalt verwendet werden, verbinden drei unterschiedliche Kommunikationsmedien: 220V~ Wechselstrom, Zweidrahtleitung mit einer Übertragungsrate von 625 kbps und 1,25 Mbps und eine 78 kbps Free-Topology-Leitung. Im Modellhaushalt befinden sich sechs Subnetze (*Locations*). Jedes Subnetz versorgt einen bestimmten Teil des Modellhauses. Der Router 5 (*Abb.3*) verbindet den Backbone mit 220V~Wechselstrom. Für den Backbone, welcher die drei Projekt-PCs verbindet, wurde ein besonders leistungsfähiges Übertragungsmedium gewählt, da dort die höchsten Datenübertragungsraten zu erwarten sind.

5 Der RTR-10-Router

Die LonWorks-Router bestehen aus 2 Seiten. (*siehe Abb.4*). Jede Seite, auch Port genannt besteht aus einem NEURON-CHIP und ist, von der Netzwerkseite aus betrachtet, wie ein normaler Netzwerknoten zu erreichen. Als Grundregel für die Verwendung von Router gilt: *Ein Subnetz darf sich niemals über einen Router erstrecken*. Der Router erlaubt also nicht dasselbe Subnetz (*innerhalb der gleichen Domäne*) auf beiden Routerseiten [20].

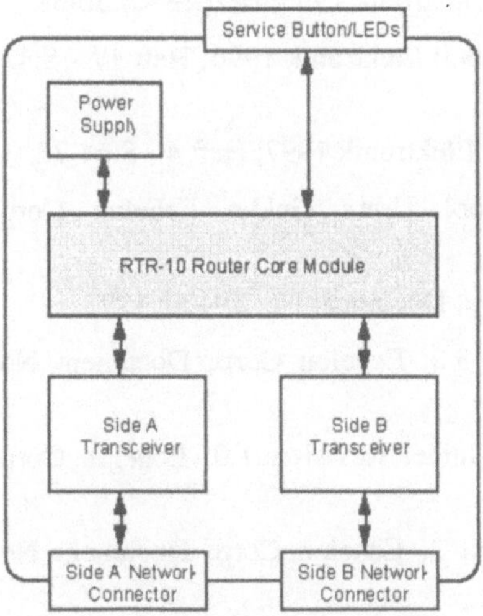

Die Router selbst bestehen aus dem Router-Core-Module, einer 5V-Stromversorgung, 2 Transceivern sowie einem Service und einem Reset-Pin [11], (*siehe Abb.4*). Es existieren auf jeder Seite Anschlüsse für die eindeutige Typenerkennung der SMX[4]-kompatiblen Router. Die Platine wurde so konstruiert, daß sie selbständig erkennt, welche Transceivertypen an den jeweiligen Seiten des Routers angeschlossen sind.

Abb. 4 :Blockschaltbild des RTR-10-Routers

---

[4] SMX . . . LonWorks Standard Modular Transceiver

Die Transceivertypen die das RTR-10-Module erkennt, sind in [10] zu finden. Für die Installation der Router wurde im HCT-Projekt das Programm "LONManager" verwendet [6], [7]. Mit dieser Software lassen sich Router-Typ (*Repeater, Bridge oder Router*) sowie die Kanäle konfigurieren in welchen der Router dann im Netzwerk hängen wird.

6 Schlußfolgerung

Unabhängig davon, welcher Standard sich zukünftig in der Hausautomation etablieren wird, Routing bleibt einer der wichtigsten Netzwerk-Komponenten. Dieser Beitrag soll zeigen, daß verschiedene Übertragungskanäle und Subnetze in vernetzten Haushalten unbedingt notwendig sind und sich RTR-10-Router zur Kopplung solcher weitverzweigter Netzsegmente sehr gut eigen.

Literatur

1. Dietrich D., Leeb G., Ochensthaler M., Posta R., Schildt G. H.; A Configuration Tool for HomeNet; IEEE Transactions on Consumer Electronics - Volume 42 - Number 3 - S.387 - S.393; Ausgabe August 1996.

2. Dietrich D., Reiter H., Schweinzer H.J.; Neue Herausforderungen der Feldbustechnologie; S.85 - S.94 des Tagungsbands zur internationalen Konferenz: Technologie und Lebensraum - Intelligente Gebäudetechnik & Elektronik für Lebensqualität; Congress Center Villach - Österreich; 25.-26. November 1996.

3. Happacher M.; Das vernetzte Eigenheim; Elektronik 1995; Heft 18 - S.50-58.

4. Happacher M.; Die Aufwärmphase ist vorbei; Elektronik 1996; Heft 17 - S.42-S.47.

5. Happacher M.; Eine gemeinsame Sprache; Elektronik 1997; Heft 4 - S.64-72.

6. LonManager LonMaker Installation Tool Users Guide; Echelon Corp; Document No. 39500; 1993.

7. LonManager Profiler User's Guide; Echelon; Document No. 39400; 1993.

8. LonTalk® Protokol Specification; Vers. 3.0; Echelon Corp; Document No. 19550; 1994.

9. LonWorks PC Lontalk Adapter User's Guide; Revision 1.0; Echelon Corp; Document No. 79310; 1994.

10. LonWorks Router User's Guide; Revision 3; Echelon Corp; Document No. 79100; 1992.

11. LonWorks Transceiver Installation Instructions; Echelon; Part No. 078-0145-01C; 1995.

12. Neubauer P.; Entwicklung einer LON-basierten Wetterstation mit WWW Anbindung; Diplomarbeit; Institut für Computertechnik TU Wien; 1997.

13. Ochensthaler M., Reiter H., Redlein A., Schildt G.H., Dietrich D.; HomeNet Configuration Tool – ein Projektbericht; Proceeding zum FACILITY-KONGRESS der Akademie für technische Gebäudeausrüstung (ATGA); 23.–25. April 1997.

14. Ochensthaler M.; Redlein A.; Reiter H.; Rohrhofer R.; IUCCIMNEWS: Zeitschrift des InterUniversitären Centrums für CIM; HomeNet - das Intelligente Haus der Zukunft; 1996.

15. Palensky P.; Demand-Side-Management in privaten Haushalten mit LonWorks®; Diplomarbeit; Institut für Computertechnik TU-Wien; 1997.

16. Preining B.; Aufbau einer LON basierten Ökostation für private Haushalte; Diplomarbeit; Institut für Computertechnik TU-Wien; 1997.

17. Reiter H., Posta R., Ochensthaler M., Schildt G.H., Dietrich D.; Energieoptimierung im vernetzten Eigenheim durch verteilte Wetterdienste im Internet; E&I – Elektrotechnik und Industrie - ÖVE-Verbandszeitschrift; Ausgabe 5/1997.

18. Schluckebier W.; Anwendung der LonWorks Technologie in der Gebäudetechnik - der intelligente Raum; S.153 - S.158 des Tagungsbands zur internationalen Konferenz: Technologie und Lebensraum - Intelligente Gebäudetechnik & Elektronik für Lebensqualität; Congress Center Villach - Österreich; 25.-26. November 1996.

19. Schnell G.; Bussysteme in der Automatisierungstechnik; Vieweg; 1994

20. Schweinzer H-J.; Multi Media Router Untersuchung bei LonWorks; Institut für Computertechnik TU Wien; 1997.

21. VESA Open Set Top; *Home Network - First Call for Proposals - Version 1.0*; 1 August 1995.

22. Wirth C.; Entwicklung und Aufbau einer Netzinfrastruktur für ein Haushaltssteuernetzwerk; Diplomarbeit; Institut für Automation TU Wien; 1997.

Demand Side Management in private homes by using LonWorks®

Peter Palensky, Ratko Posta

Institut für Computertechnik, TU-Wien
Gusshausstrasse 27-29/384, 1040 Wien

Abstract - This paper presents a system to distribute power consumption in private homes uniformly over time as well as to reduce total power consumption to meet the interests of the energy producers. We proclaim autonomous and intelligent domestic appliances that communicate with each other. Every electrical device is supported by a built-in LonWorks-Node that incorporates distributed algorithms. The main objectives are the to provide logical relations between the devices, and to coordinate the energy consumption policy. All this happens in an autonomous and distributed way. The paper elaborates four different algorithms and indicates strategies to save energy in global dimensions

1 Introduction

Demand Side Management (DSM) is a method to coordinate the activities of energy consumers and energy providers and is very much in the interest of energy producers. DSM seeks to avoid peaks of energy consumption [1] in order to achieve an approximately constant energy consumption that meets the characteristics of power stations. The typical pattern of energy consumption in a household stands in contrast to this; a lot of energy is consumed in the morning and at noon and almost none at night.

There are already simple DSM-solutions with programmable timers for washing machines etc., but they cannot be used for more complex problems and they must be programmed by the user. Other solutions (used at mountain huts where energy consumer and producer are the same person) offer a lock for certain devices when another device is running to avoid power peaks. These solutions are, however, difficult to control and often quite expensive when used for a larger number of appliances.

The objective is to smooth energy consumption by using intelligent domestic appliances that communicate with each other. The most important feature of an interconnected home is its multifunctionality [2]. A home network can be used to increase comfort as well as safety and security. DSM is a further application and reason for using home networks. Alternative energy sources, reservoirs, and sinks like solar panels, heatpumps, hot water tanks, transparent insulation [3], intelligent lighting systems [4] and air conditioners can also be coordinated by the DSM system. This DSM System is not limited to private households. It can be also used in large business facilities, machine halls and wherever electrical appliances are in use.

The chosen home-network for the control system was LonWorks [7]. This system offers an architecture of distributed intelligence and all kinds of communication-media like twisted-pair, fiberoptics, power-line, radio, etc.

2 Software Structure of a DSM-House

A domestic appliance for an energy saving system has to have certain properties. This includes the possibility of an economy-mode, a sleep-mode to finish its work later and other modes to fit the device into the actual electrical state of the system. Furthermore it is necessary to control the device from another point in the home net and to get information about it such as current and future power consumption and general future behavior. Also new strategies for switching on and shutting down a device in an energy-saving manner are required. All these features have to be controlled and therefore every electrical device in the household (shaver, washing machine, TV,...) is supported by a built-in LonWorks-Node that includes 3 software layers (Fig. 1):

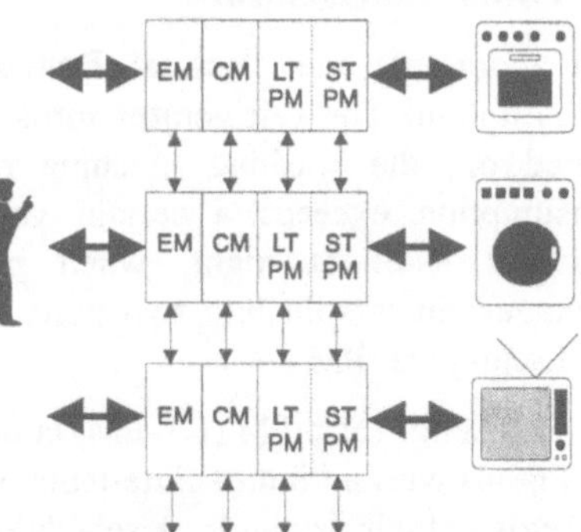

Fig. 1: Software Layers

- Energy-Management (EM), the user side layer
- Case Management (CM), the middle layer
- Power Management (Short Term- and Long Term-PM), the device side layer

2.1 Energy Management

EM tries to save energy by reducing and correcting excessive demands by the user like 27°C room temperature and hints for optimizing the lighting (Fig. 2). This supposes an easy-to-use user interface.

EM analyzes the circumstances why someone wants to turn on the light owing to architectural reasons, personal preferences, the weather and other things. This requires a large number of sensors and

Fig. 2: EM - Energy Management

a distributed database system. EM represents an expert system for energy-tuning. It suggests to change the light bulbs in the often frequented rooms to neon tubes which save energy, produce less heat and thus relieve the air conditioner [8].

Furthermore the best position for a new floor lamp can be calculated considering personal preferences.

2.2 Case Management

CM deals with logical relations and connections between the devices (Fig. 3). For example it makes no sense to project slides while the light is on and presence-sensors control the lighting. These connections depend on the person who is living in and with this system.

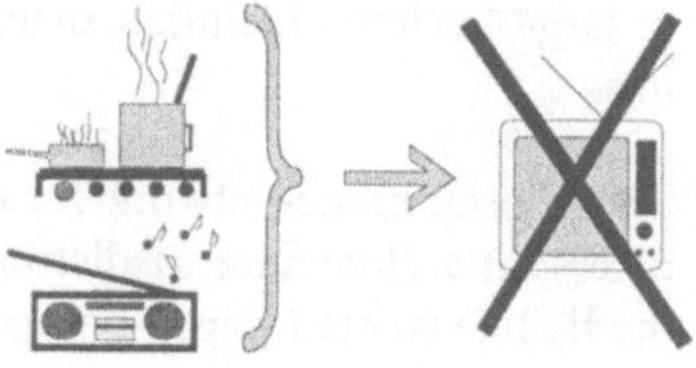

Fig. 3: CM - Case Management

It might make sense for someone to inhibit the TV Set during cooking, but would be not acceptable for someone else. The system has to identify the person's lifestyle and to adapt itself to singles, grandmothers, double-income-no-kids and to other types. Even more so, the system has to be a mixture between a system analyzer and a sociologist. A simple adjustment of the system by a symbolic programming language is also conceivable.

2.3 Power Management

PM represents the classical Demand Side Management; it smoothes power consumption. The refrigerator turns itself off for 20 minutes when the stove is turned on, the washing machine runs in economy-mode if the total power consumption exceeds a certain value and the dishwasher activates itself at night, when power is cheap. Power Management is split into two parts, the Short-Term PM and the Long-Term PM.

- The Long-Term PM (LT-PM) coordinates the devices (
- Fig. 4) with a kind of time-table, where the appliances can register their demands. A scheduler tells the device when it may do its work depending on the price of power and other entries in the table. A table entry may include a current profile and a point of time when the device wants to have its work finished. A washer, for instance, predicts its power consumption for heating the water, washing, and spin drying and tells that it wants to finish by 8:30 am.

Fig. 4: Long-Term-Power-Management

- The Short-Term PM (ST-PM) does not care about time, it decides for the moment (Fig. 5). Appliances that have the possibility to change their source of energy or their energy consumption might do so if it is necessary. A simple example is that the dishwasher changes into economy-mode when the overall current consumption exceeds a preset value of amperes. The difference to ordinary DSM-Systems is that the future behavior needs not to be known because the system reacts on actual changes that are predicted by the devices (the automatic curtain predicts, for instance, 1 ampere current

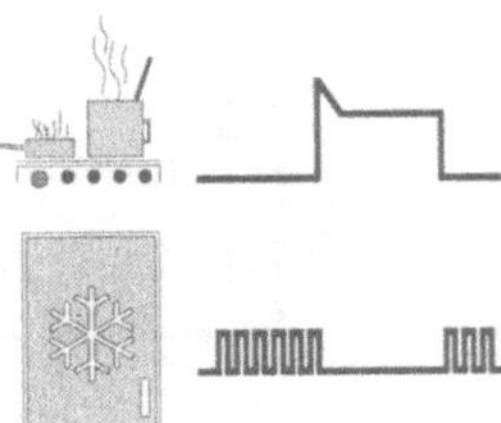

Fig. 5: Short-Term-Power-Management

consumption before it gets into action due to the sun). Real current peaks are not only detected, they are avoided. The predicted energy consumption may exceed the desired maximum current, but after the DSM-System has reached its equilibrium the devices put their now balanced forecasts into action.

PM does not save energy because economy-modes may need less power, but they take longer and in terms of energy it makes no difference if the washer runs at night or not. The difference is that energy costs less at night and the consumption profile is smoothened which will also be a financial advantage in future

Let us focus on one of the layers:

3 Short-Term Powermanagement (ST-PM)

If, in case of a too high (predicted) peak load, some appliances have the possibility to switch to another electric circuit, to switch off or to enter another mode, it must be guaranteed that not all devices economize their operation but only as many as necessary. Therefore, the appliances must act one after another while communicating with each other. Let us have a look on three different strategies and compare them:

a.) Centralized ST-PM

One ST-PM node (P) collects the predicted data (pred) and organizes all devices (D) by sending out control data (ctrl) (Fig. 6). The required algorithm is quite complex and powerful. The optimum can always be found.

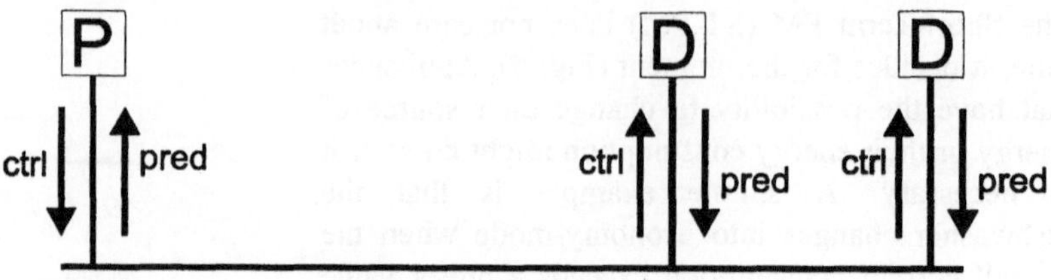

Fig. 6: ST-PM Method a

It is obvious that remote control of the devices is a potential security gap and the functionality collapses when the PM node breaks.

b.) Decentralized ST-PM

Every device has a ST-PM node of its own (Fig. 7) that collects all predictions of all appliances and joins them to an overall prediction. In case of a too high energy consumption prediction the PM nodes will influence their particular devices to decrease consumption.

Fig. 7: ST-PM Method b

Every device can offer different possibilities for a lower consumption and the PM node chooses the best one. A PM node can only control one device and cannot influence any other device which increases safety but decreases the flexibility of the system. If a device decides to change its mode it adds this wish into a queue and waits until it is its turn. All nodes have identical queues that are permanently checked for consistence.

c.) Decentralized and redundant ST-PM

This mechanism is a mixture between approaches a) and b). Every device has a PM node that can control its device and every other device in the system (Fig. 8). The proper function of every node is monitored by the rest of the nodes.

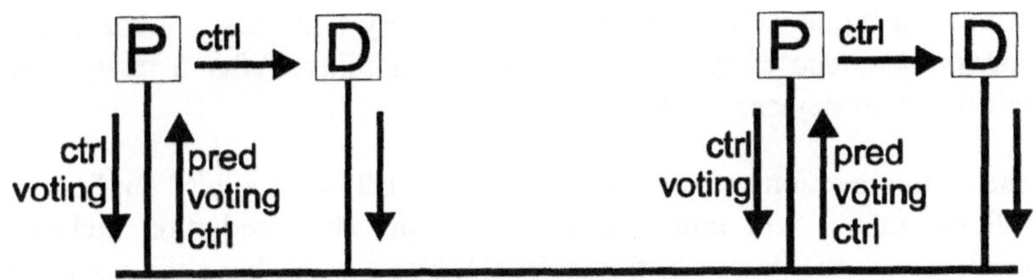

Fig. 8: ST-PM Method c

Every node calculates an optimum configuration for all the devices and marks its solution with a system-wide valid key. This key assesses the future current and the amount of the changes. This mark is broadcast over the net and received by every other node. Every node calculates a different solution and the best will be taken. After the winner node is found it sends its changes to the devices which adopt accordingly. The algorithms on the nodes are ideally different ones and include random mechanisms. This provides fault tolerance because at least one correct node can control the whole system if all other nodes cannot solve the problem.

It is no problem if a node marks itself wrong, because all other nodes check the mark after the broadcast control data.

4 Reached Objectives and Aspects for the Future

The HomeNetConfigurationTool-Project[1] [9] at the University of Technology in Vienna includes a model of a private home interconnected with LonWorks. This model was extended by adding simulation nodes which simulate the behavior of a TV Set, a washing machine and so on. These simulation nodes can be controlled remotely as well as manually (open the door of the refrigerator, turn on the microwave,...etc.). Their interface to the rest of the Home Net consists of a network variable[2] that represents the remote control functions as an input and another one that offers the needed prediction as an output. Some of the devices are supported by a ST-PM node, others are not, in order to test how the system deals with non-intelligent appliances which do not send out a prognosis of their power consumption.

The actual realization corresponds to a self-ordering strategy b.) (decentralized ST-PM) because of safety reasons. The queues are administered by network variables and are of stochastic order to ensure a statistical deviation: The devices should take

---

[1] granted by FWF Austria, P-10699 ÖMA

[2] Network Variables are LonWorks communication objects that are based on layer-7-services

their chances to influence the system in turns. Every node has its own queue and every incoming queue entry is checked for plausibility. After a new broadcast prognosis the other devices react on it.

Fig. 9 shows the reactions of two devices (D1 and D2) after a third appliance (D3) has predicted that it will immediately switch on. This prediction includes the current, needed by the device in future (at t1). The devices have the possibility to change into sleep-mode or save-mode, where the consume less power. Device D2 realizes that the common power consumption is too high and changes into sleep-mode (t2). After this, the overall power consumption allows device D1 to switch from save-mode to on-mode (t3).

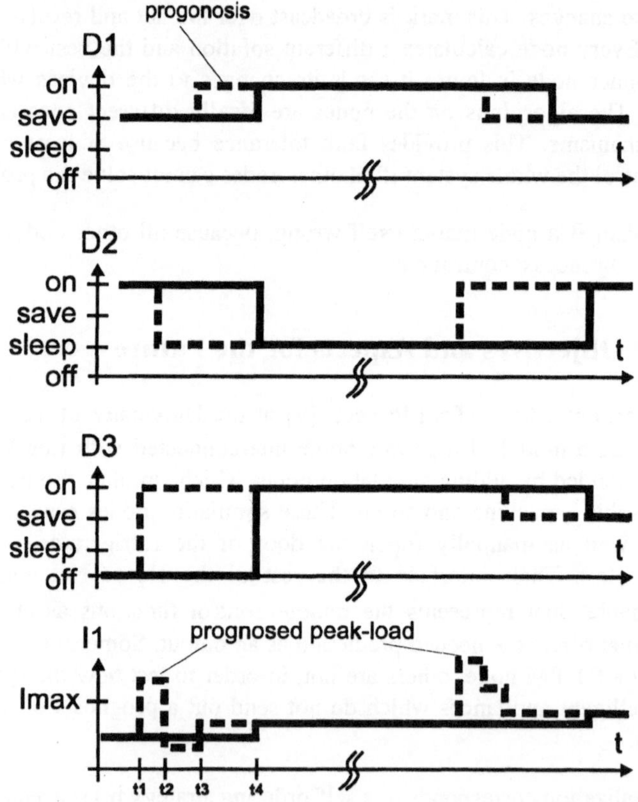

Fig. 9: reactions of the power management system after a new prognosis

The reactions are random-ordered, and the situation would look different, if device D1 had been faster than D2. As seen at the bottom of the figure, the current-peaks occur only as an information on the network. The real, physical current stays smooth and changes after the system has reached its equilibrium (t4). After a certain time device D3 leaves sleep-mode (for instance D3 is a refrigerator and its internal temperature is rising) and again the other members of the net have to react.

The current objective is to add the behavior of typical users to the simulation and to increase the number of devices and sensors. The development of new strategies and algorithms will be done on simulation environments where multiple households consisting of more than 500 devices and several users can be simulated. The results of these simulations will help to find new strategies to minimize overall energy consumption. In consequence, new power plants can be prevented and existing ones optimized.

Another future objective is to implement all four layers of the model to offer full functionality and an easy-to-operate user-interface. This work finds its continuation in a new project together with STEWEAG, an Austrian utility company, where the results of this work will be used build up a network that supports the customers of STEWEAG to save energy (Layer 1 - Energy Management). This network offers a lot of possibilities for a new kind of service offered by the energy providers.

References

1. R. Rollet, Diploma work; Institut für Energiewirtschaft TU-Wien, Vienna, Austria, 1993

2. K. Scherer, V.Grinewitschus, Integrated Home Systems for Resource Conserving Living, Tagungsband LonWorks-Tagung; Villach, Austria, 1996

3. E. Rummich, Alternative Energiequellen, Alternative Energiespeicher, VO-Skripten, TU-Wien, Vienna, Austria, 1995

4. Energiesparende Beleuchtungsanlagen, Energiesparinformation, Hessisches Umweltministerium, Referat für Öffentlichkeitsarbeit, Wiesbaden, Germany, 1994

5. Demand Side Management with LonWorks; LonWorks Engineering Bulletin, Echelon Corporation, Palo Alto, CA 94304, USA, 1996

6. Lee Goldberg, „green engineering: designing for a brighter future", Electronic Design Jan. 1996; p.108ff, USA, 1996

7. LonWorks Technology Device Data, Motorola Inc., USA 1995

8. Ashare Journal Dec. 1996, page 35ff, USA 1996

9. Leeb, G.; Posta, R.; Ochensthaler, M.; Schildt, G.-H.; Dietrich, D., „A Configuration Tool for HomeNets", International Conference on Consumer Electronics 1996; Chicago, Illinois, USA, 1996

Konfiguration und Simulation des intelligenten Hauses

Werner Dilger
Technische Universität Chemnitz, Fakultät für Informatik
D-09107 Chemnitz

Abstract

With the progress in building automation technology the realization of the *intelligent home* comes into reach. The design of such a technical system is a typical configuration task because the whole system is composed of given parts in such a way that the requests of customers are fulfilled, certain constraints are satisfied, and an optimal solution according to different criteria is produced. This paper presents a solution of the configuration problem for the intelligent home.

Die fortschreitende Entwicklung der Gebäudesystemtechnik macht die Idee des *intelligenten Hauses* realisierbar. Der Entwurf eines solchen technischen Systems ist eine typische Konfigurationsaufgabe, da das gesamte System aus vorgegebenen Teilen zusammengesetzt wird, wobei die Kundenanforderungen erfüllt, bestimmte Randbedingungen eingehalten und eine nach verschiedenen Kriterien möglichst optimale Lösung hergestellt werden müssen. Im vorliegenden Papier wird eine für das intelligente Haus spezifische Lösung des Konfigurationsproblems vorgestellt.

1 Einleitung

Unter den Stichwörtern *smart house* oder *intelligentes Haus* dringt die Informationstechnik seit einigen Jahren in ein neues Gebiet vor. Technische Geräte bestimmen die Funktionen eines Gebäudes, auch eines Wohnhauses, in immer stärkerem Mass. Bisher werden diese Geräte auf herkömmliche elektromechanische Weise bedient und sie sind lauter Einzelsysteme, d.h. sie arbeiten völlig unabhängig voneinander.

Die Entwicklungen der Informationstechnik in den letzten Jahren zeigen, wie man durch Integration von Einzelsystemen eine höhere Funktionalität erreichen kann. Solche Systeme werden aber auch zunehmend komplexer und ihre an den Anforderungen des Nutzers orientierte Realisierung immer aufwendiger. In diesem Papier wird dargestellt, dass die Entwicklung eines Automationssystems für das intelligente Haus (kurz: IH-System) eine Konfigurationsaufgabe ist, wie diese gelöst werden kann und wie der Entwurf durch Simulation getestet werden kann.

In Abschnitt 2 wird begründet, warum IH-Systeme eine Konfigurierungsaufgabe darstellen und ihre besonderen Merkmale werden angegeben. In den Abschnitten 4 bis 6 werden die drei Hauptschritte der Konfigurierung von GA-Systemen ausführlich beschrieben. Es wird ausgeführt, dass sich an die Konfiguration eines GA-Systems in naheliegender Weise eine simulative Verifikation des Entwurfs anschliessen läßt. Dafür wird in Abschnitt 7 ein Algorithmus angegeben. Abschnitt 8 stellt den vorgelegten Ansatz in den Kontext der Konfigurationssysteme.

2 Das intelligente Haus als Konfigurationsaufgabe

Zum Entwurf von Gebäudeautomationssystemen für grössere Gebäude werden üblicherweise existierende Planungssysteme verwendet. Der Entwerfer muss dabei aber schon die Planungsvorgaben gedanklich in eine Konfiguration umgesetzt haben, die er dann mit Unterstützung der Systeme realisiert. Bei solchen Projekten ist das Auftragsvolumen genügend gross, daß es sich lohnt, ausreichend viel Zeit für einen möglichst optimalen Entwurf aufzuwenden. Für heutzutage verwendete Techniken bei der Realisierung und für die Planung von Gebäudeautomationssystemen vgl. [3, 7, 8].

Bei der Realisierung eines Automationssystems für das intelligente Haus liegen andere Randbedingungen vor als bei der Automation grösserer Gebäude. Sie lassen sich in den folgenden Punkten zusammenfassen:

1. Jedes Haus hat seine Eigenheiten und jede Bauherrschaft ihre besonderen Wünsche.
2. Es gibt bei einem Haus in der Regel spezielle Anforderungen an das Energiemanagement, an die Kommunikation und an den Gebäudeschutz.
3. Das Auftragsvolumen erlaubt aus Gründen der Wirtschaftlichkeit keinen zeitaufwendigen, bis ins Letzte ausgefeilten Entwurf.

Die prinzipiell mögliche vollständige Vernetzung erlaubt aber ganz neue und komplexe Möglichkeiten der Steuerung und Regelung. Um diese unter den gegebenen Randbedingungen realisieren zu können, muss der Entwerfer durch ein Assistenzsystem unterstützt werden.

Der Entwurf eines IH-Systems ist eine typische *Konfigurationsaufgabe*. Auf der einen Seite sind die Hardware und die Software fest vorgegeben oder leicht aus vordefinierten Elementen zu erzeugen, sie stellen also eine Art Baukasten dar, aus dessen Teilen das System aufgebaut wird. Auf der anderen Seite liegt eine Beschreibung vor, welche Funktionen das zu installierende Netz nach den Wünschen des Nutzers erfüllen soll. Die Aufgabe ist nun, mit den vorhandenen Mitteln ein System zu entwerfen, das die spezifizierten Anforderungen erfüllt.

Die Voraussetzungen für die Entwicklung des IH-Systems liegen im Wesentlichen in Form von Anforderungen seitens der Nutzer, als Baupläne und als umgebungsbedingte Besonderheiten vor, also teils in graphischer und teils in verbaler Form. Bei der graphischen Information wird vorausgesetzt, dass sie bereits formalisiert ist, z.B. als CAD-Zeichnung. Die verbal vorliegenden Information muss noch in formale Notation umgesetzt werden, damit sie maschinell verarbeitet werden kann.

Allgemein beschreiben die Nutzeranforderungen Funktionen, die in bestimmten Situationen ausgeführt werden sollen (z.B. Licht bei Anwesenheit in einem Raum), und Reaktionen auf bestimmte Ereignisse (z.B. Öffnen einer Tür bei Annäherung). Sie enthalten also einerseits Angaben über Situationen und Ereignisse, andererseits über erwartete Reaktionen des IH-Systems. Aus der ersten Art von Angaben lassen sich Situationsbeschreibungen und Ereignisfolgen (genannt *Umgebungen*) ableiten, aus der zweiten Art Spezifizierungen der erforderlichen Geräte und ihrer Vernetzung, d.h. praktisch der Entwurf des IH-Systems. Mit Hilfe der Umgebungen lässt sich der Entwurf simulativ testen. Es sind also die folgenden Schritte durchzuführen:

1. Entwurf des IH-Systems
2. Ableitung von Umgebungen
3. Test durch Simulation

3 Systementwurf

Beim Entwurf des IH-Systems sind zwei Aufgaben zu lösen:

1. Spezifizierung der Funktionsweise der einzelnen Gerätetypen.
2. Festlegung der Wirkungszusammenhänge zwischen den Geräten durch Definition von Gruppen und Nachrichten.

Die erste Aufgabe dient dazu, die Anforderungen an die Gerätetypen zu definieren. Diese Gerätetypen können über die spezifizierten Funktionen hinaus noch weitere Funktionen besitzen; hier kommt es nur darauf an, die Funktionen zu beschreiben, die sie auf jeden Fall haben müssen. Die zweite Aufgabe dient der Festlegung der Funktionen, die sich durch das Zusammenwirken der Geräte ergeben. Die Gerätemodelle werden gemäss folgender Syntax beschrieben:

Gerät	⟨Name des Gerätetyps⟩
Typ	⟨Typ der Oberklasse von Gerät⟩
Parameter	⟨Liste der Parameter, die von Gerät manipuliert werden können⟩
Variablen	⟨Liste lokaler Variablen⟩

$Zustand_1, ..., Zustand_n$

Die Beschreibung des Verhaltens erfolgt mittels Zuständen, deren wesentliche Bestandteile Regeln sind. Ein Zustand hat die Form:

status = ⟨Wert des Zustands⟩
Liste von Regeln

Eine Regel hat die Form

<u>wenn</u> $Bedingung_1$ <u>und</u> ... <u>und</u> $Bedingung_m$ <u>dann</u> $Aktion_1, ..., Aktion_k$

wobei $m \geq 0$ und $k \geq 1$. Eine Bedingung ist entweder eine Nachricht oder ein Boolescher Ausdruck über den Parametern des Geräts. Eine Aktion ist entweder eine Wertzuweisung an einen Parameter oder das Senden einer Nachricht. Eine Sende-

Aktion wird mit sende ⟨Nachricht⟩ bezeichnet. Eine Nachricht ist ein Tripel der Form (Absender, Inhalt, Empfänger)

Einzelne Geräte können verschiedenen Zwecken dienen, d.h. für mehrere verschiedene Teilsysteme verwendet werden. Aus diesem Grund müssen die für die einzelnen Teilsysteme erstellten Entwürfe gegeneinander abgeglichen werden, um die Installierung des gesamten Systems mit möglichst geringer Zahl von Geräten durchzuführen. Die Teilsysteme können, ausgehend von der verbalen Information, unabhängig voneinander und deshalb parallel zueinander entworfen werden. In den einzelnen Phasen des Entwurfs sind folgende Schritte durchzuführen:

Entwurf von Teilsystem i
1. Definition von Gerätemodellen.
2. Festlegung von Gerätegruppen.
3. Bestimmung der auszutauschenden Nachrichten.

Abgleich der Entwürfe der Teilsysteme
1. Bestimmung lokaler Überlappungen von Teilsystemen.
2. Identifizierung gleichartiger Geräte in Überlappungsbereichen.
3. Anpassung der Lokalisierung der Geräte.

Vervollständigung des Entwurfs
1. Vereinigung der Funktionen identifizierter Geräte.
2. Revision der Gerätegruppen.
3. Endgültige Festlegung der Nachrichten.

Soweit spielt sich der Entwurf nur auf der Modellebene ab. Auf dieser Ebene wird er auch getestet. Erst danach können nach den Vorgaben des Modells die physikalisch existierenden Geräte bestimmt werden. Sollten sich für ein oder mehrere Gerätemodelle keine geeigneten Geräte finden lassen, muss der Entwurf revidiert werden.

4 Ableitung von Umgebungen

Eine Umgebung ist durch eine Folge von Situationen und Ereignissen in der Umwelt gegeben, die für die Funktionsweise des zu konfigurierenden IH-Teilsystems relevant sind. Als Parameter kommen hier die für Sensoren definierten in Frage. Eine Situation ist dann durch die Menge dieser Parameter zusammen mit ihren aktuellen Werten gegeben und ein Ereignis durch den Wertewechsel mindestens eines dieser Parameter. Situationen und Ereignisse können auch unvollständig durch Teilmengen der relevanten Parameter beschrieben sein. Ist S eine Menge von Situationen und E eine Menge von Ereignissen, dann ist die Menge der Umgebungen über S und E folgendermassen definiert:

$$U = \{(s_1, ..., s_m, e_1, ..., e_n) \mid s_i \in S, e_j \in E, m, n \in \mathbb{N}\}$$

142

Die Ableitung der Umgebungen ist in [1] mit Rückgriff auf die Umsetzung der Anforderungen in eine formale Notation detailliert beschrieben. Die Ereignisse in einer Umgebung sind in ihrer Reihenfolge zeitlich geordnet, d.h. es gilt (e_1, t_1), ..., (e_n, t_n) mit $t_1 < ... < t_n$.

5 Verifikation

Bei der Verifikation geht es um die Überprüfung des Verhaltens des entworfenen IH-Systems anhand der Umgebungen. Ausgehend von einer Anfangssituation soll bei Vorgabe einer Umgebung von dem IH-System eine Folge von Reaktionen erzeugt werden, die das erwünschte Verhalten des intelligenten Hauses beschreiben. Die Reaktionen sind die in den Regeln beschriebenen Aktionen. Bei Ausführung eines Sende-Befehls wird die Nachricht als Reaktion notiert. Bei Ausführung eines Zuweisungsbefehls wird eine Reaktion erzeugt. Die Bedingungsteile der Regeln sind entweder Nachrichten oder Boolesche Ausdrücke. Im ersten Fall ist die Bedingung erfüllt, wenn eine Nachricht des beschriebenen Typs vorliegt, im zweiten Fall, wenn der Ausdruck wahr ist. Ist dies aufgrund eines Ereignisses in der Umgebung der Fall, dann muss gelten, daß die aktuelle Zeit grösser oder gleich dem Zeitparameter des Ereignisses ist. Der folgende Algorithmus führt die Verifikation durch.

Verifikation

Eingabe: Eine Liste von Geräten, eine Umgebung.
Ausgabe: Eine Folge von Reaktionen durch das intelligente Haus.

1. $t \leftarrow$ Anfangszeit
2. $U \leftarrow$ aktuelle Umgebung.
3. $G \leftarrow$ Liste der Geräte des Szenarios.
4. $S \leftarrow$ Menge der Anfangswerte der Geräteparameter.   ; aktuelle Situation
5. $S' \leftarrow S$.   ; dient zum Vergleich mit S, $S = S' \Rightarrow$ Stop
6. $A \leftarrow \varnothing$.   ; Liste der Reaktionen
7.   Führe für jedes Gerät g in G die folgenden Schritte durch:
     7.1.   Bestimme den aktuellen Zustand s von g aus S.
     7.2.   Arbeite die Regeln von s nach folgender Vorschrift ab:
            Ist die Bedingung erfüllt, dann führe die Aktionen entsprechend ihrer Form aus:
            7.2.1.   sende (g, Inhalt, g'): Führe die Aktion durch, füge die Nachricht an
                     das Ende von A an.
            7.2.2. $p \leftarrow$ Wert: Führe die Aktion durch, füge (p, alter_Wert, Wert) an
                     das Ende von A an.
                     Falls (g.p, alter_Wert) in S enthalten, ersetze den
                     Eintrag durch (g.p, Wert).
            Ist die Bedingung aufgrund des ersten Ereignisses $e \in U$ erfüllt, entferne e
            aus U.
     7.3.   Falls $S = S'$, Stop mit Ausgabe von A.
8. $t \leftarrow t+d$.
9.   Gehe zu Schritt 7.

6 Diskussion

Als die beiden wichtigsten Arten des Konfigurierens werden das strukturorientierte und das ressourcenorientierte Konfigurieren bezeichnet, vgl. [9, 4]. Der hier vorgestellte Ansatz zur IH-Konfigurierung ist am ehesten dem ressourcenorientierten Konfigurieren zuzuordnen, vgl. [6]. Für ein IH-System ist keine feste Struktur vorgegeben und verschiedene IH-Systeme können aus ganz unterschiedlichen Teilen zusammengesetzt sein, so daß ein strukturorientierter Ansatz nicht in Frage kommt. Die Konfigurierung geht von den zu erfüllenden Funktionen aus und die erforderlichen Teile werden nach dem jeweiligen Bedarf ausgewählt. Die Ressourcen, um die es dabei geht, sind auf der physikalischen Ebene verschiedene Materialien und Energie, auf der logischen Ebene Informationen.

In [5] sind mehrere Kriterien dafür genannt, wann der Einsatz der Simulation sinnvoll ist und analytische Modelle nicht in Frage kommen oder nicht möglich sind. Das erste Kriterium ist die zu große Komplexität der Struktur des Objekts. Dies trifft für ein IH-System zu. Hader nennt als einen der Zwecke der Simulation den Test von Domänenobjekten auf die Erfüllung vorgegebener Restriktionen. In diesem Sinn wird die Simulation hier benutzt. In Haders Vorschlag für eine Simulationskomponente in einem Konfigurationswerkzeug werden sog. *Simulationsmethoden* als Voraussetzung für das Simulieren genannt. Diesen entsprechen hier die Gerätemodelle.

7 Schluss

Die IH-Technik bietet die Möglichkeit einer besseren Nutzung der Ressourcen in einem Gebäude, insbesondere der Einsparung von Energie und der flexibleren Nutzung der installierten technischen Systeme. Diesen Vorteilen steht die erhöhte Komplexität eines IH-Systems gegenüber, die erhöhte Anforderungen an die Installateure technischer Systeme in einem Gebäude stellt. Traditionell sind dies Handwerker. Es ist zu erwarten, daß diese Berufsgruppe mehr und mehr mit den Anforderungen bei der Einrichtung von IH-Systemen überfordert sein wird. Aber auch Fachleuten mit spezieller Qualifikation sind kaum in der Lage, optimale Systeme aufzubauen, die alle von der Technik angebotenen Möglichkeiten nutzen. Hier können Konfigurationssysteme dem Entwerfer helfen, einen nach verschiedenen Kriterien möglichst günstigen Entwurf zu erstellen. Dies kann dazu beitragen, die Qualität der Lösungen und die Nutzung von IH-Systemen zu verbessern.

Die nächste Stufe der Entwicklung könnte die evolutionäre Genese von IH-Systemen sein. Danach erfolgt die Entwicklung eines technischen Systems auf der Modellebene durch schrittweise Änderungen des Entwurfs, ausgehend von einem ersten Vorentwurf, und Test durch geeignet zu definierende Fitnessfunktionen, vgl. [2].

Literatur

1. Dilger, W.: Das intelligente Haus als Konfigurationsaufgabe. In: Sauer/Günter/ Hertzberg (Hrsg.): Planen und Konfigurieren 96, infix-Verlag, Proceedings in AI, 1996, 160 - 171.

2. Dilger, W.: Decentralized Autonomous Organization of the Intelligent Home According to the Principle of the Immune System. In: Proceedings of the 1997 IEEE Intl. Conf. on Systems, Man, and Cybernetics, Orlando, FL, 1997.

3. EIBA: EIB Handbook Issue 2.1. Brüssel 1993.

4. Gülden, O., Günter, A.: Konzept für die Realisierung einer ressourcenorientierten Vorgehensweise in KONWERK. In: Günter (Hrsg.): Wissensbasiertes Konfigurieren. Ergebnisse aus dem Projekt PROKON. Infix, St. Augustin, 1995.

5. Hader, S.: Einsatz von Simulationsverfahren bei der Konfigurierung. In: Günter (Hrsg.): Wissensbasiertes Konfigurieren. Ergebnisse aus dem Projekt PROKON. Infix, St. Augustin, 1995.

6. Heinrich, M.: Ressourcenorientiertes Konfigurieren. Künstliche Intelligenz 1 (1993) 11 - 15.

7. Müller, R.: LON - das universelle Netzwerk. Teil 1 und 2. Elektronik 22/23, 1991.

8. Scherg, R.: EIB planen und installieren. Vogel Buchverlag, Würzburg, 1995.

9. Tank, W.: Wissensbasiertes Konfigurieren: Ein Überblick. Künstliche Intelligenz 1 (1993) 7 - 10.

A Systems- and Network Management Framework for Open Home Systems[*]

Ratko Posta

Institute of Computer Technology, University of Technology Vienna

Abstract. Contemporary proposals for Home Automation Systems standards in Europe [3, 5, 6] include several data communication protocols and define common objects for control, monitoring and management of home networks. Accordingly, the author proposes two goals for future Home Systems software tools. First, a generic integration framework should be designed that allow network management to converge towards single and heterogeneous management architecture. Second, to ease the use and to lower the requirements on persons involved in development, installation and maintenance of home networks, operational and management models should be the same. This paper proposes to manage various management systems from a generic management framework which is based entirely on CORBA [2]. Benefits of this scheme are a uniform programming model and capability to dynamically manipulate agents at runtime.

1 Introduction

Home Automation Systems aim to enhance functionality of a household by adding distributed control and information technology. With the decentralization of computing resources into a variety of possibly heterogeneous networks and with the advent of distributed computing, management of home networks has become more important and complex. The increasing complexity and heterogeneity is pushing consumer industry and research to look for a user-centred and consistent way of managing networks. The purpose of the *systems- and network framework for open home systems (SNMF)* is to identify software concepts and to design an infrastructure for the further development from today's heterogeneous and often isolated management services towards a universal and open Home Automation management [1]. With the advent of open object-oriented distributed computing models such as CORBA, it is feasible to transparently manage systems developed in one management model from a distributed object-oriented processing model, i.e. to unify management services under a common network network management based on CORBA architecture. The structure of the paper is as follows: first, we will give a short overview of home networks. Then,

---

[*] This research work is granted by the Austrian Science Fund (FWF P10699-ÖMA)

integration models for network management will be presented and compared with a dynamic approach to use CORBA for network management, with the focus being on the dynamic and scaleable management platform. Finally, some conclusion will be drawn.

2 Home Networks

The services in a residential network might include high bandwidth access to the Internet and to private networks, cable television and voice telephony services. Other home network services might be home control, file transfer, network security, electronic mail, electronic games, along with many others yet unimagined applications. Fig. 1 shows an example of a Home Automation System with significant sub-networks and their interconnections.

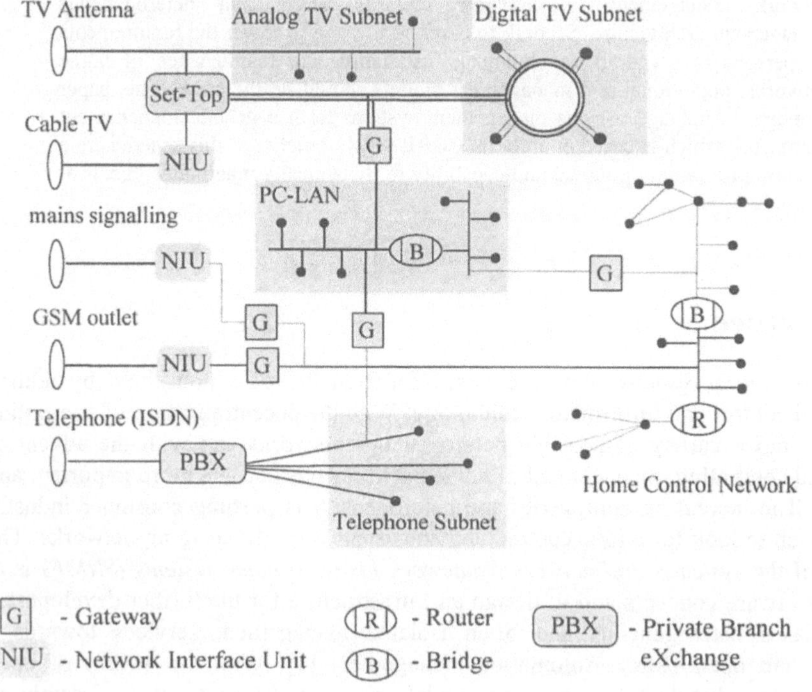

Fig. 1. Home Automation System - Home appliances are part of a distributed control system.

To handle interaction with a wide selection of equipment and different network interfaces, *home networks* must comprise a variety of network technologies and mechanisms for multiservice interconnections. This technology diversity arises from the fact that dozens of incompatible systems are currently present on the market [3]. *Residential gateways* [4] provide functions of intelligent cross-connect devices transferring residential services and external services being delivered over access networks. The primary goal in a *home network* is to adapt the communication services to the needs of the consumer and to hide complexity of the system environment.

3 Integrated Systems- and Network Management

Large and technically complex distributed systems cannot be put together and managed by human effort alone. To configure, secure, optimize and maintain appliances in a home network, efficient and user friendly tools for systems and network management must be provided. Residential networks comprise various specific management systems employing heterogeneous network- and systems management models. There are several network management systems which are in the focus of consideration for Home Automation Systems in Europe. Home automation buses, such as European Installation Bus (EIB) [5], Process Field Bus (PROFIBUS) [6] Profile Building Automation, European Home System (EHS) [7], Batibus, Building Automation and Control Network (BACnet) [8], and Local Operating Network (LON) [5] have been accepted nationally and internationally yet. In network management of PC-LAN and telecommunication systems, Simple Network Management Protocol [9] and Common Management Information Protocol [10] are predominantly used today. An integrated systems- and network management platform for residential networks need to manage both, fieldbus and LAN management models.

Given the example that for the management of the home lighting system the European Installation Bus (EIB) technology [11] is used and that a heterogeneous residential network maintains most of the management data on personal computers in a LAN. Building manager may need to have some of the customer's LAN management information to exchange control messages between the lighting system and the PC.

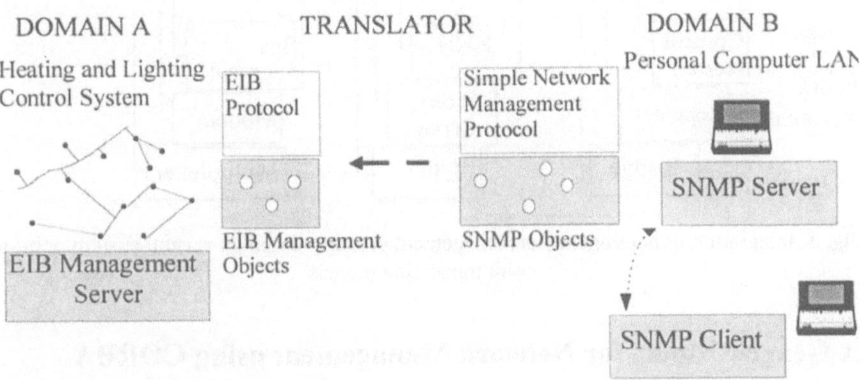

Fig. 2. Conversion of management information between different management models

Converting information from one management model to the other is a complex task and requires implementation knowledge from both management models used. *Management-gateways* represent one possible solution for management information and management protocol translation. In Fig. 2 a translation based on algorithms for the static and operational translation of object models between a SNMP management system and a EIB management system is shown. In most cases it is performed by a compiler which accepts a specification of domain A and outputs a specification for domain B. Such mapping is usually unidirectional and irreversible. The reason for this is that transforming a specification from one domain to another, in most cases, results in loss of information since all domains are syntactically and semantically different. The interchange of management information becomes even more difficult when multiple network managers and several different network management models are involved. In the worst case, each of the managers involved has to know the union of all management models used to perform operations in other management domains. Figure 3 shows a typical custom integration scheme where several legacy network management systems are designed separately and presented to end user as an integrated network management interface. Due to the complexity and poor structure, point-to-point integration method usually results in a variety of individual and incompatible custom solutions which are unstable, error-prone and difficult to extend. Therefore, a strong focus on software architecture and use of standard object classes and interfaces must be considered.

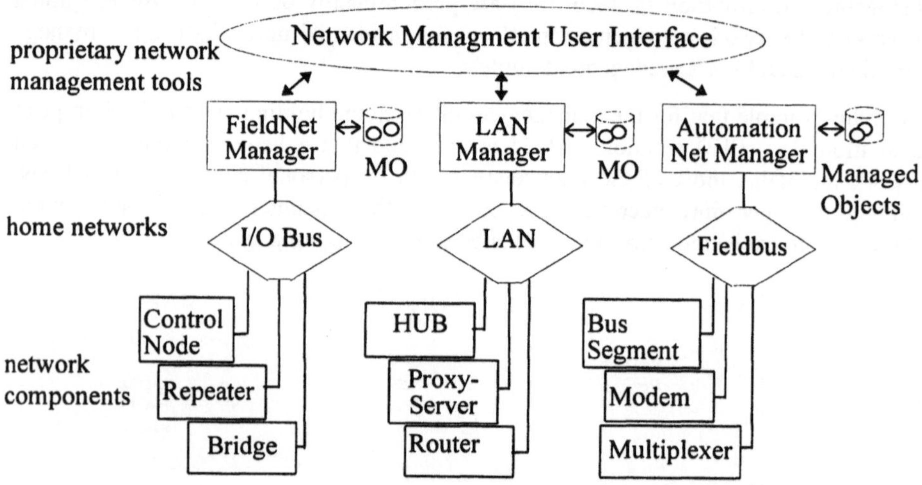

Fig. 3. Integration of heterogeneous management systems based on several custom point-to-point translation models

4 A Generic Model for Network Management using CORBA

A novel approach for multi-domain management framework is based on a uniform generic object model that can be used to transparently manipulate instances of various specific object models. Object-orientation and distributed object technology allow system management to converge towards one single, homogeneous and global management architecture which is capable of representing any target network management object.

4.1 CORBA Overview

Distributed object computing represents the synergy of two major areas of software technology: distributed computing systems and object-oriented design. A widely used distributed object computing model is the Common Object Request Broker Architecture (CORBA), which is being standardized by over 700 companies members of Object Management Group. The main features of CORBA are the transparency in facilitating client-object communication; an object's interface that specifies operations and types that the object supports; the CORBA Interface Repository that allows the object interface system to be accessed and written at runtime; language mappings between OMG-Interface Definition Language and implementation languages (e.g. C++, C, Ada95, Java and Smalltalk); the Static Invocation Interface that allows an invoking object to access a remote object using syntax natural to implementation language; dynamic invocation over the Dynamic Invocation Interface and Dynamic Skeleton interface that allows an object invocation to be constructed manually, without static knowledge about object being invoked; and Inter Object Request Broker (ORB)-Protocols that provide a general ORB interoperability architecture. CORBA provides a stable model for distributed object-oriented systems that helps developers cope with heterogeneity and complexity of distributed systems.

4.2 Using CORBA for Network Management

Enabling distributed applications written in many languages and using a variety of networking protocols is a challenging problem for an integrated systems- and network management framework. CORBA offers the basic infrastructure to abstract the communication layers. The dynamic invocation, which allows explicitly building messages and argument lists at runtime, assists developers in building gateways between CORBA and non-CORBA systems. Figure 4 shows a generic framework for integration of network management models which are managed from CORBA object model. The framework consists of an object manager capable of representing multiple management models and adapters constituting bridges between the specific target management systems. Additionally, an object metadata repository maintains information about the classes, types, interfaces and methods used in specific object models. The metadata repository maintains information about the structure of data of several target models. It is needed to type check arguments to requests at runtime and to convert values between generic object manager and target models. Metadata repository make it possible to integrate and make use of existing and tested legacy management applications without the need of new implementation. The generic object manager wraps instances of target systems in a generic representation and offers them as proxy instances to client applications. Accordingly, a uniform and transparent interface (GOM) allows management tools to disregard the specific underlying object models used.

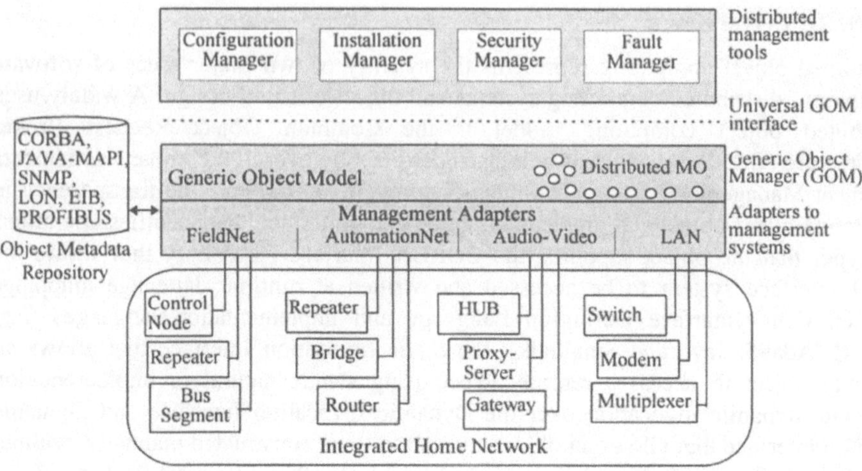

Fig. 4. Integration of heterogeneous management systems in residential networks based on CORBA and a distributed generic object oriented framework

For management of Home Automation Systems we chose configuration language designed for Home-Net Configuration Tool [13]. The uniform programming model for management of residential networks defines a homogeneous and global management and simplify the design of home network management tools.

5 Conclusion

The goal of this paper was not to demonstrate that one approach is better than another but to understand the benefits of the distributed object computing and then to identify a possible solution of an integrated management model for residential networks. The article shows a feasible solution for management of heterogeneous home network resources through CORBA. The growing impact of CORBA in the telecommunications sector [14] and several developments which allow to manage and operate different network- and systems management models using CORBA [12] indicate that CORBA will become important in the network- and systems management. Two main directions of our research in heterogeneous home network management are the generic management framework and the uniform management programming model.

References

1. Posta, R.: Project Proposal Extended Management Platform for Home-Network Configuration Tool, Institut für Automation, University of Technology Vienna, July 1997

2. Object Management Group, The Common Object Request Broker: Architecture and Specification, Revision 2.0, July 1995.

3. Quirighetti, R.: Report on CENELEC TC205 and it's Working group activities to CEN TC247, October 1996, Landis&Gyr AG

4. Holliday, C.: The residential gateway, IEEE Spectrum, Vol. 34, No. 5, May 1997

5. Building management products and systems for HVAC applications, Draft of European Prestandard, Ref. No. prEN###, CEN/TC247, 1996

6. General purpose field communication system, European Standard EN50170, European Committee for Electrotechnical Standardization (CENELEC), Dec. 1996, Ref. No. EN 50170:1996 E

7. Home Systems Specification 1.1, European Home Systems Association, Excelsioralaan 11-Bus 1, B-1930 Zaventem, Belgium, 1992

8. Newman, M.: Integrating Building Automation and Control Products using the BACnet, ASHRAE Journal Nov. 1996, ASHRAE

9. J. Case, M. Fedor, M. Schoffstall and C. Davin, The Simple Network Management Protocol (SNMP), RFC 1157, May 1990.

10. ISO/IEC, CCITT, Information Technology-OSI, Common Management Information Protocol (CMIP)-Part 1: Specification ISO/IEC 9596-1, CCITT Recommendation X.711, 1991.

11. Rose, M.: Gebäudesystemtechnik in Wohn- und Zweckbau mit dem EIB, Hüthig 1993, ISBN 3-7785-2239-6

12. Ban, B., Deri, L.: Static vs. Dynamic CMIP/SNMP Network Management Using CORBA, Proceedings IS&N'97, May 27-29, 1997, Como, Italy.

13. Leeb, G., Posta R., Ochensthaler, M., Schildt, G.-H., Dietrich, D.: A Configuration Tool for Homenet, IEEE Transactions on Consumer Electronics, Vol. 42, No. 3, Aug. 1996, pp. 387-395

14. Schmidt, D.: Distributed Object Computing, IEEE Communications Magazine, Feb. 1997, pp. 42-44

Object-based Distributed Application Design

M. Goossens

The EIB Association (EIBA),
Tinklaan 5, B-1160 Brussels, Belgium

Abstract. A series of examples introduce the concept of EIB Distributed Objects: they illustrate the advanced object-oriented features defined by the EIB system for Home & Building Automation. This approach is then investigated in more detail. It is shown how a set of dedicated services allow efficient (connectionless) point-to-point access to individual properties across the EIB network. A call-back interface abstracts the underlying Application Layer from the precise way the Objects are managed by the application.
Finally, an outline methodology for the design of Distributed Applications is presented, which allows a set of appropriate interface objects to be determined.

1. Preface

As Home & Building Electronic Systems based on peer-to-peer automation networks increasingly establish themselves as the successor to centralised master-slave buses, new approaches are called for in the design of distributed solutions. From the point of view of Interworking Standards, this is reflected in an evolution from merely standardising interfaces via (shared) variables towards a more powerful description in terms of Distributed Objects.

The power of Distributed Objects lies largely in the abstraction level they provide: they hide the underlying implementation details from management clients (such as configuration managers, tools for diagnostics, supervision and control, etc.). The EIB system [1] provides the necessary models, for which standard implementations are available.

2. Introducing EIB Distributed Objects (EdObject)

2.1 General Structure and Purpose of EIB Distributed Objects

The entities we here refer to as 'Objects' bundle a certain number of related properties - usually bound to one particular logical functionality or "channel", which is implemented in a device or node. This is depicted schematically in fig. 1, which also shows that a device may support several EdObjects. Each object consists of a number of properties. The (value of) an individual property is actually a one-dimensional indexed array of elements; each element is itself a record which may combine one or more of fields.

EdObjects are 'Distributed' in the sense that they may be accessed across the network, rather like components in Microsoft's Distributed Component Object Model (DCOM) [2]. In fact, this is the EdObjects main purpose: to allow structured access to the process image via the network, both for supervision as well as for control purposes. Status or output information may be obtained from a device by reading certain properties. By changing (writing) property values, the device's behaviour may be influenced.

Still from fig. 1, it is clear what we mean by *structured* access. The fact that this structure information is *explicitly* registered in the device (see section 3 for details), isolates the outside world from implementation details such as the precise memory location of the data. As a result, access via EdObjects is far superior to access via DMA for supervision and control purposes.

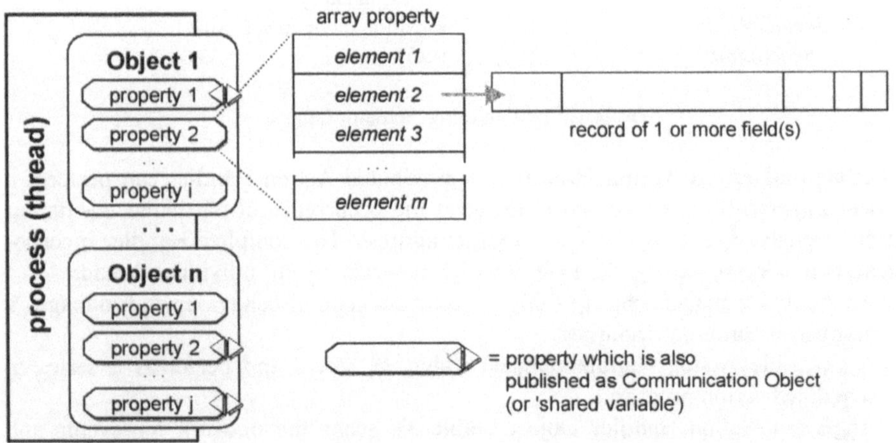

Fig. 1. The interface to a process consists of objects built up of properties, which are an array of records.

2.2 Publishing Properties as Shared Variables

In addition to replacing DMA in many cases, the EdObject concept also complements shared variables quite nicely. Typical runtime communication among EIB devices relies heavily on multicast (or 'group') communication between local instances of a shared variable. For EIB, shared variables are usually known as (Group) Communication Objects. This mechanism is extremely efficient [3].

Considerable strength may be derived from combining the benefits from *both* access mechanisms. The developer achieves this by simply publishing a property (or even a concatenation of properties from more than EdObject object!) as a shared variable with the flick of the wrist. The property and the variable are said to *coincide*.

2.3 Some Building Automation Objects

Exactly how powerful the EIB Object approach is, may be glanced from the following examples. In the tables, the names of properties coinciding with a shared variable are preceded by a + sign. The final EIB Interworking specification may differ slightly from the examples presented here.

Dimming Actuator. One popular example from lighting control is the Dimming Actuator Object. As may be seen from table 1, it allows full control over the behaviour of a single actuator channel. Although it forms an essential part of the object's specification, we omit the state machine from this description (but see [4]).

Property	Size (bit)	Comment
+Switch	1	switch on/off
+DimmingControl	4	increment/decrement
+SetValue	8	self-explanatory
CurrentValue	8	self-explanatory
JumpOrGlide	1	transition mode to set value
DimmingRate	8	(sec)

Table 1. The Dimming Actuator Object

Schedule and Event Action Handler. A distributed Action Handler can perform a (series of) action(s) at scheduled times or at the occurrence of particular events; an action consists of writing a value to a target address. The complete Handler incorporates three objects (see Fig. 2), each of which is served by an individual thread:

1. An Action Manager Object (Table 4), controls general behaviour and manages a number of common resources.
2. A Schedule Action Handler Object (Table 2), stores and performs a series of scheduled action programs.
3. An Event Action Handler Object (Table 3), scans the network for events and stores and performs corresponding action programs.

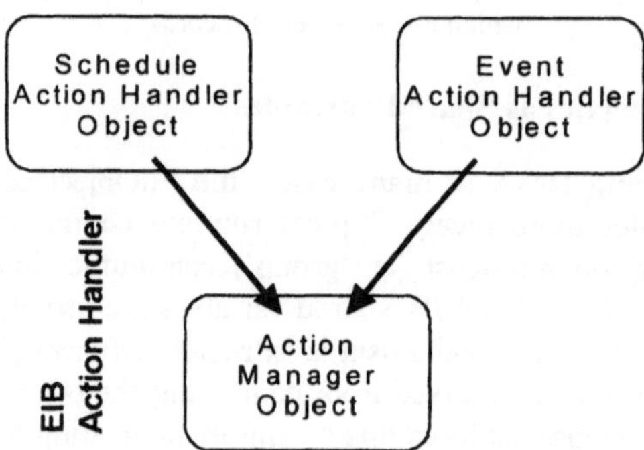

Fig. 2. Three EIB Objects cooperate to build an Action Handler.

The full specification exceeds the scope of this contribution; we can only present the major characteristics and many interesting implementation details (such as the *structured array* mechanism, which allows access to individual fields within a property record) will be skipped.

Property	Size (byte)	Comment
Schedule Action	n x 12	array of Schedule Action entries
Day Cycle	m x 47	array of Day Cycles
Calendar	q x 12	array of Calendar entries

Table 2. The Schedule Action Handler Object

Scheduling. A Schedule Action entry has the following formal record structure:

⟨*ScheduleActionEntry*⟩ = ⟨*DayProgramID*⟩⟨*Instant*⟩⟨*Action*⟩

where Instant indicates *when* something is to be done:

⟨*Instant*⟩ = ⟨*Hour*⟩⟨*Minute*⟩⟨*Second*⟩,

and Action *what* is to be done (always in the form of updating either a shared variable or a property value):

⟨*Action*⟩ = ⟨*Target*⟩⟨*Value*⟩,

with

⟨*Target*⟩ = ⟨*TargetType*⟩⟨*TargetReference*⟩.

According to the type, the Target Reference is either a local ID of a shared variable (known as Communication Object number), or an index in the Object Reference table of the local Action Manager Object (see below).

The ⟨DayProgramID⟩ logically groups single Schedule Actions with identical ID into a Day Program. Such a Day Program is actually a template, which may now be assigned to a particular day in either:

- a Day Cycle:

  ⟨*DayCycle*⟩ = ⟨*StartDate*⟩⟨*NumDays*⟩
      {⟨*DayProgramID*⟩(1)...⟨*DayProgramID*⟩(N)}

  where N is the value of ⟨NumDays⟩. In other words, a Day Cycle is a cycle which consists of N Day Programs. These will be carried out cyclically from the Start Date onwards.

- or a Calendar Entry:

  ⟨*CalendarEntry*⟩ = ⟨*Date*⟩⟨*ValidFrom*⟩⟨*ValidUntil*⟩⟨*DayProgramID*⟩

  where the Day Program will be carried out each year on the specified Date, within the period of validity.

Where useful, time and date fields support the use of wild cards; this allows expressions like '*each hour, at 17 minutes past the hour*', or '*every 1st day of the month*'.

Property	Size (byte)	Comment
Event Action	m x 14	array of Event Action entries
Event	n x 6	array of Events

Table 3. The Event Action Handler Object

Events. An Event Action entry ('*what* is to be done') exhibits the following structure:
$$\langle EventActionEntry\rangle = \langle EventProgramID\rangle\langle EventAction\rangle$$
with
$$\langle EventAction\rangle = \langle UpdateSource\rangle\langle Delay\rangle\langle Action\rangle.$$
The Action is identical to the one in the Schedule Action Entry. The sequence and relative timing of individual actions in an Event Program (grouping entries with the same ID) is controlled via the Delay field. The Manager Object can update the action's value from the Update Source address. Now turning to events as such:
$$\langle Event\rangle = \langle TriggerCondition\rangle\langle EventProgramID\rangle,$$
we see that the Trigger Condition ('*when* it's to be done') tells the Event Action Handler to 'listen' on the Trigger Source group address:
$$\langle TriggerCondition\rangle = \langle TriggerSource\rangle\langle TriggerOperator\rangle\langle ThresholdValue\rangle$$
The received value is compared to the Threshold Value according to the Trigger Operator. The Handler can scan for the occurrence of a particular value, as well as transitions of the threshold. Type information is required to allow correct interpretation of the received value:
$$\langle ThresholdValue\rangle = \langle ValueType\rangle\langle Value\rangle.$$
When an event fulfilling the condition takes place, the Trigger's Event Program will be executed. Typically, this is a 'scenario' which combines a number of lighting, temperature control and security settings, which set in when the event occurs.

Property	Size (byte)	Comment
Enable	1	enable or disable complete handler
LastChangedOn	6	time stamp (for multi-client usage)
ProcessTime	11	last minute processed before power failure
Catching-up	3	when to catch up after power failure
Object Reference	n x 7	object/property reference for non-group actions
Text	m x 33	array of text strings

Table 4. The Action Manager Object

Manager Object. A local Action Manager instance must be available as a 'server' for Handler Objects to work, and manages a number of resources for them. Its Object Reference property contains an indexed array of 'target' Object References:
$$\langle ObjectReference\rangle = \langle PhysicalAddress\rangle\langle ObjectNumber\rangle\langle PropertyID\rangle\langle ArrayIndex\rangle.$$
It is used as a lookup table by actions with Target Type *object_reference*, in which case the Target Reference represents the index in this array. Although this was omitted above, all actions also have a reference to a text string. These string resources are managed by the Manager Object as well, via its Text property.
Also not shown for simplicity is a 2-bit field which allows each action, cycle, program or calendar entry to be temporarily disabled or globally invalidated. (For Schedule Action entries, this field controls the Catching-up behaviour.) To this end, the Manager's Command property $\langle Command\rangle = \langle CommandOperation\rangle\langle OperandID\rangle$ supports a list of Command Operations like *start_event_program, update_event_program, disable_schedule_action_entry* etc. on the record defined by Operand ID.

3. Implementing Distributed Objects for EIB

3.1 A Property's Abstract Record Structure

As an Object's building blocks, properties have the abstract record structure shown in Fig. 3. "Abstract", because the call-back interface (see section 3.3) completely encapsulates the actual implementation by the application developer.

property description
property_id (unsigned8)
type (unsigned8)
max_no_of_elem (unsigned 16)
access (unsigned8)
property value
array[0] = no. of elements
array[1... max_no_of_elem] = value

Fig. 3. The abstract record structure of an EdObject.

Every property has some description attributes in addition to an array of values. The property_id uniquely identifies the property within the object it belongs to. The communication partner can also check specifically for the property type. The access property indicates the access level required to read from or write to the property value. The other attributes are self-explanatory.

3.2 Constructing Objects from Properties

Now where are the Objects themselves in all of this? Each object instance in an EIB device has a unique identifier, the object_id, which serves as a handle for accessing its properties: each EIB Distributed Object consists of at least one property. The property with ID PID_OBJECT_TYPE (= 1) is the 'header' description of the object itself. This property is mandatory for every object. (For simplicity, the Object Type property has been omitted from the examples in section 2 above.)

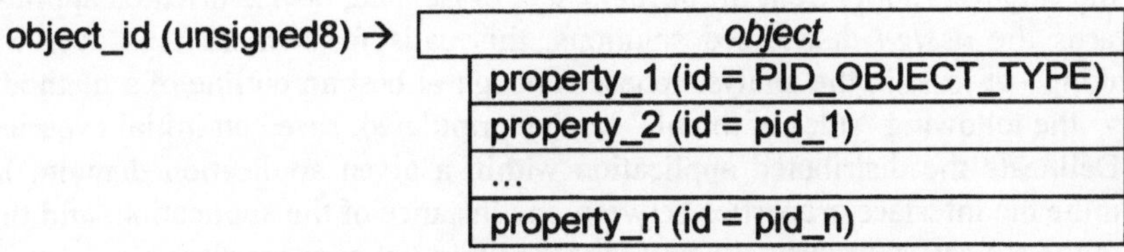

Fig. 4. Constructing objects from properties.

In this way, a communication partner can scan a device's objects and for each object it finds, determine the object type.

3.3 The EdObject Call-back Interface

EIB Distributed Objects are managed entirely by the application. A standardised call-back interface completely isolates the access to the property via the EIB Application Layer from the implementation chosen by the application developer. For example, this allows different objects to (locally) share a common property, or the underlying memory mapping to be optimised for concatenation of properties as a shared variable, or the use of external memory resources. Of course, the developer each time has to implement the appropriate call-back function. Existing firmware implementations of the EIB Operating System offer a default call-back.

3.4 Services for Accessing Properties

In contrast with systems where objects are supported via a general data-block or "file" transfer mechanism, the EIB operating system actually "knows" both the EdObjects and their internal structure. This awareness is embodied in the following set of Application Layer services dealing with properties: A_Read_Property_Description, A_Read_Property_Value, A_Write_Property_Value. When zooming in on the request primitive of the read service by way of example, this becomes even clearer: in *A_Read_Property_Value.req (cr_id, class, object_id, property_id, no_of_elem, start_index)*, the argument *cr_id* identifies the communication relationship (for connection-oriented communication), and *class* its priority. With the remaining arguments, it is possible to select a subset [*start_index, no_of_elem*] from the value array of a particular property from the specified object.

For these services, no connection is required between the communication partners. In this case, the relationship is referred to as one-to-one connectionless communication. If the property's access attribute requires access authorisation, an authorised connection has to be built up [4].

4. Towards a Methodology for Distributed Application Design

Designing Objects as we have encountered them in this paper goes beyond determining communication interfaces. Interfaces also have to be assigned to threads or processes distributed on the network, and they have to guarantee a certain behaviour. As the emphasis shifts from the *development* of isolated, device-oriented applications towards the *design* distributed solutions, this calls for a more top-down way of thinking - at least in the analysis phase. Though at best an outline of a methodology only, the following "rules of thumb" can be formulated, based on initial experience:

1. Delineate the distributed application within a given application domain. Determine the interface properties between one instance of the application, and the rest of the system; appropriately group the resulting list of properties.
2. Maximally divide the application into processes / threads, describing the precise functionality of each one (state machine, ...).
3. Determine the (runtime) communication interfaces between the processes. Resulting properties should probably be published as shared variables.

4. (Standardisation.) Determine the *minimal* interface, which fixes the mandatory properties. Beyond the *optional* properties, the Interworking Standards for EIB Distributed Objects leave open the possibility to add proprietary ones. Luckily, no amount of standardisation will ever be able to quell the creativity of the developer.
5. Determine additional properties needed for configuration and supervision. Make sure each status variable has a Current Value property.
6. Appropriately group the resulting list of properties into objects (different channels should have different objects). Distinguish clearly between client (resp. producer) and server (resp. consumer) side.
7. To the extent possible, assign the properties of step 1 to the objects of step 6, defining additional objects if necessary.
8. Consider the impact of various ways the threads can be grouped locally or distributed across the network.

Some additional guidelines or meta-rules may be imposed by management models for configuration, supervision and control, such as the EIB Home Management model [5]. Other boundary conditions are imposed by the impact of resource requirements (EEPROM, RAM, multi-threading, ...) on cost - particularly in the home domain. (With EIB's focus on group communication, runtime communication bottlenecks on slower media may be largely ignored.) For the time being, cost vs. resource considerations also make direct application of formal methodologies from structured or object-oriented analysis and design unrealistic for building automation.

Admittedly, no methodology can replace the combination of domain expertise and developer's intuition. But at least, the steps suggested above may serve to gain a foothold when starting the analysis, and as a guideline to assess the outcome.

Acknowledgements

The author wishes to thank P. Ferstl (Bosch-Siemens Hausgeräte GmbH) and K. Adler, M. Hartmann, Dr. W. Kristen, N. Stroick and P. Tomiç (Siemens AG) for the useful input and the examples (and for coming up with objects to begin with).

References

1. Goossens, M.: A Survey of the EIB System. In: EIBA Proceedings, EIBA (1997).
2. Rogerson, D.: Inside COM, 11. Microsoft Press (1997).
3. Goossens, M.: Communication and Addressing on EIB. In: EIBA Proceedings, EIBA (1997).
4. The EIB Handbook Issue 3 - Vol.3 (preliminary edition). EIBA (1997).
5. Goossens, M.: Easy Installation and Home Management. In: FeT Proceedings 1997, Springer (1997).

Eine Lösung für die Power-Line-Kommunikation im CENELEC C-Band

Gerrit Telkamp

ATICON Home Automation GmbH
Rebenring 33, D-38106 Braunschweig, Germany
☎ (+49) 531 3804 340, FAX (+49) 531 3804 342
E-Mail: g.telkamp@tu-bs.de

Überblick. *Power-Line-Kommunikation* ist die Mitbenutzung des Stromversorgungsnetzes zur Datenübertragung. Als physikalisches Medium für Feldbussysteme eingesetzt ermöglicht das eine erhebliche Reduzierung des Installationsaufwandes und der damit verbundenen Einrichtungskosten. Interessante Anwendungen ergeben sich im Bereich der Automatisierungstechnik, vor allem aber bei der Haus- und Gebäudeleittechnik (*Home- and Building-Automation*). Dieser Artikel gibt grundlegende Informationen über die Power-Line-Kommunikation und beschreibt die kostenoptimierte Realisierung eines Kommunikationsknotens, der speziell für Anwendungen in der Haus- und Gebäudeleittechnik entwickelt wurde. Mit einem Gateway können diese Kommunikationsknoten auch als Erweiterung für den EIB (*European Installation Bus*) eingesetzt werden.

Abstract. *Power-Line-Communication* is data transmission via mains. Used as a physical layer for field bus systems, it makes the installation easier and reduces the expense. Useful applications arise in the field of automation, particularly in the area of home- and building automation. This paper gives basic information about power-line-communiation. Furthermore, it describes the cost-optimized realization of a communication node, which was designed especially for applications of home- and building automation. With a gateway these communication nodes can easily be used as extension for the EIB (*European Installation Bus*).

1 Einleitung

Die Mitbenutzung des Stromversorgungsnetzes zur Datenübertragung *(Power-Line-Kommunikation)* ist keine völlig neue Entwicklung. Bereits im Jahre 1922 wurden in Europa die ersten Trägerfrequenzsysteme für den Frequenzbereich von 15-500 kHz zur Übertragung von Schaltinformationen über Hochspannungsnetze entwickelt. Bis heute sind die meisten Anwendungen der Power-Line-Kommunikation bei den Energieversorgungsunternehmen und Stadtwerken zu finden, die damit z.B. Straßenbeleuchtungen ein- und ausschalten oder Zähler fernablesen. Im privaten Wohnbereich sind allenfalls sogenannte „Baby-Phones" verbreitet, bei denen analoge Sprachsignale in relativ schlechter Qualität mit Hilfe des 230V-Stromversorgungsnetzes übertragen werden. Die digitale Datenübertragung über das Stromversorgungsnetz bietet dagegen ein sehr elegantes Medium für Systeme der Haus- und Gebäudeleittechnik, da hier flexible und vor allem kostengünstige Kommunikationskonzepte gefragt sind. Das Verlegen spezieller Steuerleitungen in Altbauten ist in den meisten Fällen zu aufwendig und daher nicht praktikabel. Die Vernetzung intelligenter Haushaltsgeräte mit Power-Line-

Kommunikation (z.B. Waschmaschine und Kühlschrank für das Energiemanagement) ist ohne zusätzlichen Aufwand möglich, weil diese Geräte ohnehin an das Stromversorgungsnetz angeschlossen werden müssen.

Bei der Power-Line-Kommunikation können heute mit geeigneter Technik und vertretbarem Aufwand Datenraten von 2400 Bauds und mehr erreicht werden. Für die meisten Anwendungen in der Haus- und Gebäudeleittechnik ist das mehr als ausreichend.

2 Standardisierung des Power-Line-Mediums

Das Europäische Komitee für Elektrotechnische Normung in Brüssel (CENELEC) hat mit seiner Norm EN 50065-1, *„Signalübertragung auf elektrischen Niederspannungsnetzen im Frequenzbereich 3 kHz bis 148 kHz"* [1] den Rahmen für die Power-Line-Kommunikation festgelegt. Sie wurde von der Deutschen Elektrotechnischen Kommission im DIN und VDE als DIN-EN 50065-1, Klassifikation VDE 0808 übernommen und sieht vier unterschiedliche Frequenzbänder vor:

- das *A-Band* (3 kHz - 95 kHz), das für Energieversorgungsunternehmen reserviert ist,

- das *B-Band* (95 kHz - 125 kHz), welches von allen Anwendungen ohne Zugriffsprotokoll genutzt werden kann,

- das *C-Band* (125 kHz - 140 kHz), das für Home-Automation-Produkte vorgesehen ist. Ein vorgeschriebenes Zugriffsprotokoll (CSMA/CA = Carrier Sense Multiple Access/Collision Avoidance) ermöglicht die Koexistenz verschiedener, inkompatibler Systeme in diesem Frequenzband; und schließlich

- das *D-Band* (140 kHz - 148.5 kHz), das Alarm- und Sicherheitssystemen ohne Zugriffsprotokoll vorbehalten ist.

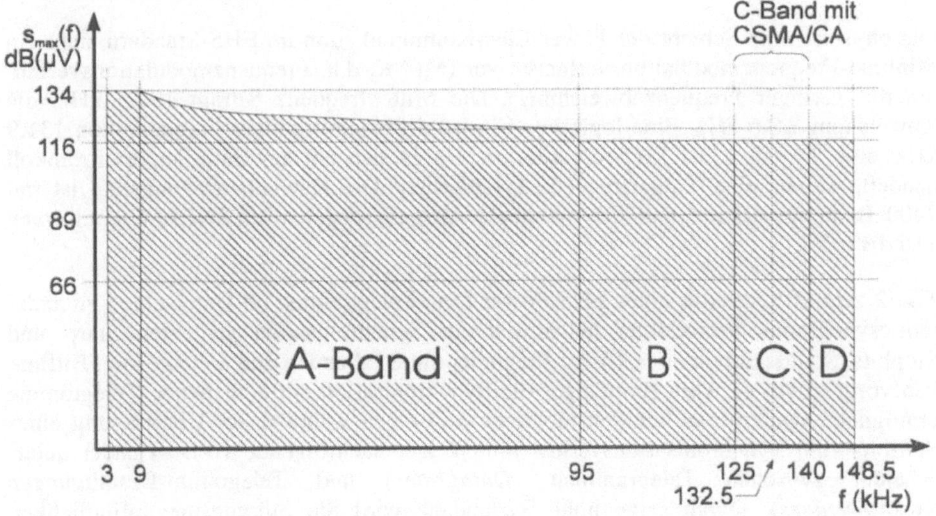

Fig. 1. Maximale Ausgangspegel von 3 kHz bis 148.5 kHz in dB (µV) nach EN 50065-1

In allen vier Frequenzbändern sind maximale Ausgangspegel für die Signalübertragung über das Stromversorgungsnetz festgelegt (*Fig. 1*). Für Anwendungen in der Hausleit-technik ist das C-Band mit höchstens 116 dB (µV) vorgesehen. Der maximale Aus-gangspegel muß mit einem Spitzenwertdetektor und vorgeschriebenem Meßempfänger über einen Zeitraum von 1 Minute festgestellt werden.

Deneben legt die EN 50065-1 für alle elektrischen Geräte am Stromversorgungsnetz Störleistungsgrenzwerte in den entsprechenden Frequenzbändern fest. So werden pa-rasitäre Beeinflussungen der Power-Line-Kommunikation verringert. Das ebenfalls vom CENELEC-Standard geforderte Zugriffsprotokoll im C-Band beschreibt, welche Fre-quenz ein Sender testen muß, bevor er ein Power-Line-Telegramm verschickt und die maximale Dauer, die ein Sender das Power-Line-Medium belegen darf. Eine physikali-sche Schicht wird hingegen nicht definiert.

3 Ein Übertragungsprotokoll für die Power-Line-Kommunikation

Modulations- und Codierungsverfahren haben beträchtlichen Einfluß auf die Sicherheit der Datenübertragung. Insbesondere bei der Kommunikation über das Stromversor-gungsnetz, ein nicht speziell für die Datenkommunikation geschaffenes Medium, kann die Störungsunempfindlichkeit durch geeignete Modulationsverfahren, Fehlerkorrektur-codes und Prüfsummen entscheidend verbessert werden. Nach dem OSI-Referenzmodell sind die physikalische Schicht (*physical Layer*) und die Medienzu-griffssteuerung (*Medium Access Control, MAC*) eines Übertragungsprotokolls für alle medienabhängigen Dienste vorgesehen. Vom European Home Systems-Standard (*EHS*) [2,3,4] werden diese Dienste für die Power-Line-Kommunikation im Rahmen der CENELEC-Bestimmungen definiert. Dank des offenen und frei verfügbaren EHS-Standards können einzelne Komponenten ohne die technologische Bindung an einen bestimmten Hersteller entwickelt werden, wodurch sich die Herstellungskosten der Kommunikationsknoten weiter reduzieren.

Die physikalische Schicht der Power-Line-Kommunikation im EHS-Standard sieht ein Minimal-Frequenzmodulationsverfahren vor (MFSK, d.h. Frequenzmodulationsverfah-ren mit geringer Frequenzabweichung). Die Mittenfrequenz beträgt 132,5 kHz, die Abweichung ±0,6 kHz. Eine logische '1' wird demnach mit einer Frequenz von 131,9 kHz, eine '0' mit 133,1 kHz repräsentiert. Da es sich um ein Halb-Duplex-Protokoll handelt, ist nur eine Trägerfrequenz erforderlich. Die Datenübertragungsrate ist mit 2400 Baud spezifiziert und der maximale Ausgangspegel mit CENELEC-konformen 116 dB (µV).

Fig. 2 zeigt die Struktur eines EHS Power-Line-Telegramms. Im Unterschied zu ande-ren asynchronen Protokollen fehlen die zur Synchronisation genutzten Start- und Stopbits. Stattdessen ist eine 16-Bit-Preambel mit 8 fallenden und 8 steigenden Bitflan-ken vorgesehen, die dem Empfänger die Synchronisation auf eingehende Telegramme ermöglicht. Andernfalls könnten mögliche Störungen während der Übertragung eines Startbits zum Synchronisationsverlust führen. Der nachfolgende 16-Bit-Header unter-scheidet zwischen Telegrammen (*Datagrams*) und Telegramm-Bestätigungen (*Acknowledges*). Durch seine hohe Redundanz wird die Störungsunempfindlichkeit zusätzlich verbessert. Alle Bytes des nachfolgenden Datenbereichs werden um einen 6-

Bit-FEC-Code (Forward Error Correction) auf 14 Bits erweitert. Das Generatorpolynom der FEC ($x^6+x^5+x^4+x^3+1$) erlaubt die Korrektur von bis zu 3 zusammenhängenden Fehlern in einem 14-Bit-Feld. Damit können auch z.B. durch Lichtdimmer oder Schaltnetzteile verursachte phasensynchronen Störungen mit einer Dauer bis zu 1 ms korrigiert werden.

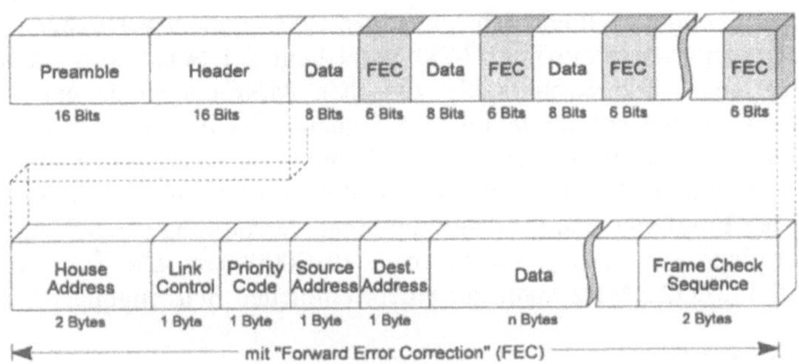

Fig. 2. Das EHS-Power-Line-Telegramm

Mit dem ersten Feld des Datenbereichs wird die 16-Bit-Hausadresse (*House Address*) übertragen. Das ermöglicht die logische Unterteilung des gesamten Übertragungsmediums in separate Unternetzwerke, wodurch die gegenseitige Beeinfussung verschiedener Hausleitsysteme am gleichen Stromversorgungnetz verhindert wird. Die Hausadresse wird vom ersten Gerät eines Hausleitsystems nach einem vorgegebenen Pseudo-Zufallszahlenalgorithmus zufällig gewählt und bei der Installation folgender Geräte auf Knopfdruck weitergereicht. Alle Telegramme enthalten außerdem einen Übertragungszähler (*Link Control*), der sicherstellt, daß kein Kommando aufgrund einer verlorenen Telegramm-Bestätigung doppelt ausgeführt wird. Weil sich priorisierte Telegramme im Power-Line-Netzwerk kaum realisieren lassen, enthält das folgende Feld *Priority Code* lediglich Informationen zur Unterscheidung von Gruppen- und Individualadressen. Der Name wurde aus Kompatibilitätsgründen zu anderen Medien beibehalten. Die nächsten beiden Felder definieren die Quell- und Zieladresse (*Source Address* und *Destination Address*). Der anschließende Datenbereich überträgt die Nutzinformationen. Das letzte Feld der EHS-Power-Line-Telegramme enthält eine 16-Bit-Prüfsumme (*FCS*, d.h. *Frame Check Sequence*) und stellt die Konsistenz des gesamten Telegramms sicher. Die FCS wird nach korrektem Telegrammempfang als Bestätigung an den Sender zurückübertragen.

4 Kostenoptimierte Power-Line-Kommunikationsknoten

Der hier vorgestellte kostenoptimierte Power-Line-Kommunikationsknoten der Firma ATICON Home Automation GmbH integriert Power-Line-Modem, Power-Line-Interface, Stromversorgung, Mikrocontroller mit externem ROM sowie eine optogalvanisch getrennte Ein-/Ausgabeschnittstelle auf einer kleinen Platine von nur 61 x 113 mm. Mit einem maskenprogrammierten Mikrocontroller und verbessertem Schaltnetzteil sind weitere Platzoptimierungen in Vorbereitung. *Fig. 3* zeigt das Blockschaltbild des Power-Line-Kommunikationsknotens.

164

4.1 Power-Line-Modem

Die Funktionseinheit *Power-Line-Modem* übernimmt die Modulation und Demodulation der Trägerfrequenz, überwacht das Medienzugriffsprotokoll und stellt dem Mikrocontroller Takt, Reset und einen Watchdog zur Verfügung. Es wurde um den EHS-Power-Line-kompatiblen Baustein ST 7537 von SGS Thomson aufgebaut [5], der in großen Stückzahlen und in einem sehr günstigen Preissegment verfügbar ist. Mit einer Datenübertragungsrate von maximal 2400 Baud liegt der Schaltkreis im Rahmen der EHS-Spezifikationen. Die Schnittstelle zwischen Mikrocontroller und ST 7537 ist denkbar einfach und besteht aus lediglich 4 Signalen: *Tx* für zu sendende Signale, *Rx* für empfangene Signale, *Rx_Tx* für die Umschaltung zwischen Sende- und Empfangsmodus, und schließlich _CS als Indikator für die Trägerfrequenz (*Carrier Sense*). In einem speziellen Empfangsmodus beträgt die Signalverstärkung des ST 7537 ca. 70 dB, das entspricht einer Verstärkung um das mehr als 3000fache. Die Taktversorgung des Bausteins von 11.05922 MHz kann mit einem einfachen Schwingquarz realisiert werden.

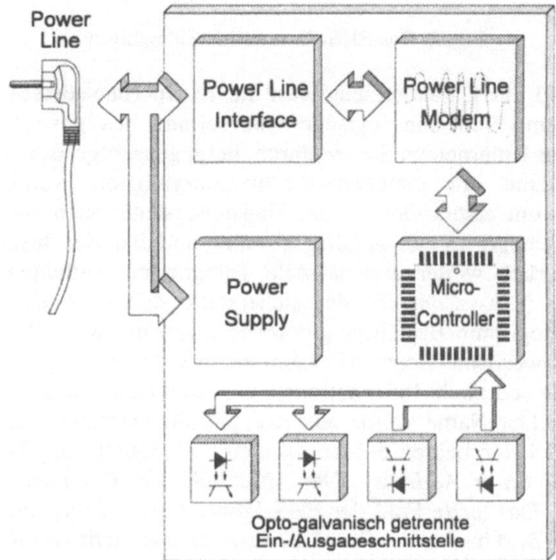

Fig. 3. Blockschaltbild der Power-Line-Kommunikationsknotens

Untersuchungen von Malack und Engstrom (IBM Electromagnetic Compability Laboratory) an 86 Stromversorgungsnetzen in 6 europäischen Ländern haben gezeigt, daß die Impedanz der stromführenden Leitungen im Bereich um 100 kHz durchaus kleiner als 1,5 Ω werden kann. Dieser Wert wird hauptsächlich von den angeschlossenen Verbrauchern und dem Innenwiderstand des Verteilungstransformators bestimmt. Damit derart niedrige Impedanzen nicht zu Störungen bei der Datenübertragung führen, muß die Leistungsendstufe des Power-Line-Modems entsprechend dimensioniert werden. Die Endstufe des hier vorgestellten Kommunikationsknotens ist nicht im ST 7537 integriert und kann Leitungen mit einer Impedanz bis unter 1 Ω treiben.

4.2 Power-Line-Interface

Die Signale aus dem Power-Line-Modem werden über ein *Power-Line-Interface* kapazitiv in das Stromversorgungsnetz eingespeist. In umgekehrter Richtung gelangen die aus dem Netz gefilterten Signale über das Power-Line-Interface zum Power-Line-Modem. Um Bauteilkosten zu optimieren, wurde die Schaltung ohne induktive Signalübertrager aufgebaut, weshalb diese Funktionsgruppe nicht galvanisch vom Netz isoliert ist. Ein Varistor schützt die Schaltung vor Spannungsspitzen auf dem Stromversorgungsnetz, die durch induktive Lasten wie z.B. elektrischen Heizungen oder Motoren verursacht werden können.

4.3 Stromversorgung

Für die interne Stromversorgung des Kommunikationsknotens wurde ein abgewandelter Tiefsetzsteller entwickelt, der die Eingangsspannung von bis zu 250V AC in 10V Gleichspannung (bis 200 mA) umwandelt. Ein nachgeschalteter Linearregler versorgt den Digitalteil mit 5V Gleichspannung. Eine speziellen Boot-Schaltung sorgt für die geringe Ruhestromaufnahme von unter 1 Watt. Durch den Verzicht auf induktive Übertrager in der Stromversorgung konnten die Herstellungskosten der Schaltung weiter gesenkt werden. Da auch das Power-Line-Interface von der Stromversorgung nicht isoliert ist, führt das zu keinen zusätzlichen Nachteilen.

Die Stromversorgung wurde derart dimensioniert, daß dem Power-Line-Modem auch bei geringen Netzimpedanzen von bis zu 1 Ω noch genügend Sendeenergie zur Verfügung steht. Außerdem wurde die Schaltfrequenz des Tiefpaßstellers so gewählt, daß die fundamentalen Harmonischen die Mittenfrequenz des Power-Line-Modems nicht überschneiden. Dadurch sind mögliche Beeinflussungen der Power-Line-Datenübertragung durch die Stromversorgung auf ein Minimum reduziert worden.

4.4 Mikrocontroller

Der Mikrocontroller kommuniziert über die Ein-/Ausgabeschnittstelle mit der Umgebung des Kommunikationsknotens und sorgt für die Power-Line-gerechte Umsetzung der Informationen. Zur Kostensenkung wurde ein Intel-8052-Derivat mit 256 Byte internem RAM gewählt. Die aktuelle Version ist mit 32 kByte externem ROM ausgestattet. Zusätzlich ist ein externes EEPROM mit 256 Byte vorgesehen, in dem z.B. Netzwerkvariablen abgelegt werden können. Wie im Abschnitt über das Power-Line-Modem bereits erläutert wurde, wird die Taktversorgung und Resetsignalgenerierung von dieser Funktionseinheit übernommen, was den Bedarf an zusätzlichen Bauteilen vermindert. Der Systemtakt des Mikrocontrollers beträgt daher 11.05922 MHz.

4.5 Ein-/Ausgabeschnittstelle

Eine galvanisch entkoppelte Ein-/Ausgabeschnittstelle stellt die Verbindung des Kommunikationsknotens zur Außenwelt her. Es stehen zwei Ein- und zwei Ausgänge mit einer Isolationsspannung von bis zu 5300 $V_{AC\ RMS}$ zur Verfügung. Die Isolation ist unbedingt notwendig, da die übrige Schaltung von der Stromversorgung nicht galvanisch getrennt ist. Die Ein- und Ausgänge sind TTL- und CMOS-kompatibel und für Signalübertragungen bis zu 19200 Baud geeignet.

5 Anbindung an den EIB

Über die Ein-/Ausgabeschnittstelle und geeignete RS-232-Wandler können die Kommunikationsknoten an eine EIB-Datenschnittstelle angeschlossen werden. Bei entsprechender Programmierung lassen sich so andere Geräte mit Power-Line-Anbindung von einem EIB-System aus fernsteuern. Das serielle EEPROM des Kommunikationsknotens arbeitet in diesem Zusammenhang als Konfigurationsspeicher zur Umsetzung der EIB-Steueradressen in Power-Line-Geräteadressen. Ein derartiges EIB-Power-Line-System wurde auf der Hannover-Messe Industrie von der Firma ATICON bereits vorgestellt. Die Schaltung findet zusammen mit der RS-232-Schnittstelle in einem 105 mm-Hutschienengehäuse Platz.

6 Zusammenfassung

Die Power-Line-Kommunikation ist gerade wegen des außerordentlich geringen Installationsaufwandes ein willkommenes Medium für die Haus- und Gebäudeleittechnik. Viele Anwendungen würden vom Verbraucher kaum in Erwägung gezogen, wenn eigens dafür Steuerleitungen verlegt werden müßten. Beispielsweise wäre eine Waschmaschinenfernabfrage vom Fernsehgerät aus sicherlich ein verkaufsförderndes Argument, aber bei hohem Installations- und Kostenaufwand kaum vermarktungsfähig. Aus gleichem Grunde erfordern die Kommunikationsknoten ein weitgehend „intelligentes" Protokoll, das einfachste Installationen unterstützt, damit Komponenten auch ohne ausgebildete Fachleute in Betrieb genommen werden können. In Verbindung mit dem EIB könnte die Installation von Power-Line-Komponenten mit einem speziellen Installationstool, das Zugang zur ETS-Datenbank hat, größtenteils automatisch realisiert werden.

Für Anwendungen in der Automatisierungstechnik werden Anfang 1998 Power-Line-Kommunikationsknoten im Hutschienengehäuse nach IEC 1131 verfügbar sein.

Literatur

1. EN50065-1; *„Signalübertragung auf elektrischen Niederspannungsnetzen im Frequenzbereich 3 kHz bis 148 kHz"*; herausgegeben von der CENELEC; Genf, Juli 1993

2. EUREKA Projekt *„Integrated Home Systems"*; ESPRIT Projekt 2431 *„Home Systems"*; ESPRIT Projekt 5448 *„Integrated Interactive Home"*

3. *„Home Systems Specification 1.3"*; herausgegeben von der European Home Systems Association; Brüssel, 1996

4. KUNG, JEAN-BART und andere; *„The European Home System Network"*, TRIALOG; 1. Auflage 1996

5. *„Power-Line Modems & Applications"*; SGS-Thomson Microelectronics; Mai 1994

LonVoice - Transmitting Control Data and Voice by using the LonTalk Protocol

Hans-Jörg Schweinzer

Vienna University of Technology
Institute for Computer Technology
Gußhausstraße 27
A-1040 Vienna
schweinzer@ict.tuwien.ac.at

Abstract. In the field of building and home automation, the need of transmitting voice using the installed fieldbus system increases. The presented LonVoice system is based on the LonWorks technology to transmit voice between LonVoice nodes in real time. LonVoice extends typical LonWorks applications in order to use the same transport media for transmitting control data and voice data. Several approaches how to interface voice data to a Neuron Chip as well as the demands on the LonWorks communication channel are discussed. Based on these results a low cost implementation is presented in this document.

1 Introduction

LonWorks technology is especially designed to implement control networks. These networks consist of intelligent nodes (LonWorks nodes) that interact with their environment and communicate with other nodes using a variety of communication media. The communication protocol implemented on every LonWorks node is called the LonTalk protocol. LonTalk is a message based protocol designed for the use in Field Area Networks (FANs) that supports peer to peer networking. The message size is limited to fulfill the requirement of a short reaction time.

The presented LonVoice system is based on the LonWorks technology to transmit voice between LonVoice Nodes in real time. In order to use the LonWorks technology in this manner several problems need to be managed. First of all, it is necessary to find a proper technology to digitize the voice signal and to reduce the voice data rate to a minimum. Then methods have to be evaluated to bring the voice data stream into the Neuron Chip and to transmit the data using the LonTalk protocol. The receiving node then needs to output the voice data to a device, that reconverts the digital voice data into an analog signal (Fig. 1).

2 Typical Applications for LonVoice

LonVoice extends typical LonWorks applications in order to use the same transport media for transmitting control data and voice data. Especially in the field of building and home automation as well as in the field of industrial applications, installation costs can be minimized by using only one transport media.

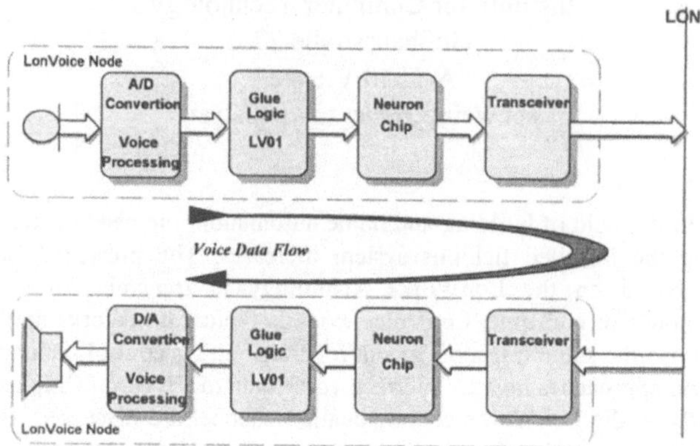

Fig. 1. Voice Data Flow using LonWorks Technology

Let's take a look at the entrance of a typical European apartment complex. For every apartment there is a bell, a phone and a common ground wire. Assuming the apartment complex consists of 40 apartments that makes 120 wires to be installed. Using the LonVoice system, only a LonWorks network connection needs to be established.

LonWorks Technology is used by some companies for elevator control. In every elevator cabin there is an emergency phone that needs to be wired in addition to the control system. Using LonVoice, the expensive wiring between the control station and the cage can be cut in cost.

In modern building automation applications, fire alarm systems play a central role. An important part of fire alarm systems is to guide people to save exits in case of an alert. In addition to the well known EXIT-signs, fire alarm systems using LonVoice can guide people by announcements that are automatically controlled by the fire alarm system. Also the fire brigade can use the system for important announcements.

For system integrators, LonVoice can be a helpful and convenient tool during network installation and maintenance. The already installed communication media - the LonWorks network - can be used as a phone line. Thus, workers can talk to one another by just connecting the LonVoice node to the communication media where ever they like. This can save a lot of time and money.

3 Digitizing the Voice Signal - The ADPCM CODEC

Human voice is a non periodical analog signal that needs to be processed to be transmitted over a digital communication media. The telephony, as a typical application that transmits digitized human voice in real time, has produced a lot of standards (CCITT, ITU-T) how voice signals have to be handled. Important parameters are the channel bandwidth and the resolution of the quantized signal. In case of the LonVoice system the bandwidth utilization on the communication channel should be as small as possible. Another constraint is that all the voice processing needs to be located outside of the Neuron Chip in low cost parts. Thus, a PCM (Pulse Code Multiplex) - CODEC (Coder / Decoder) has been selected as a proper voice signal processing unit.

In a PCM-system, the voice signal is sampled at 8 kHz and each sample is quantized into 12 bits. A special companding filter is used to convert the digital signal with a resolution of 12 bit into one with 8 bit. This digital signal now requires a bit rate of 8kHz * 8 bit = 64kbps for a simplex (one way) voice transmission. For a fieldbus system like LonWorks, a net data rate of 64kbps for a simplex voice communication is unacceptable high. A much better solution is to use ADPCM (Adaptive Differential PCM) instead of PCM. A typical ADPCM circuit is based on a Digital Signal Processor (DSP), that has a special algorithm implemented to convert the 64kbps PCM-signal into a 16kbps ADPCM-signal and vice versa. 16kbps appears suitable for most LonWorks applications. After evaluating several products that seem to be appropriate for the discussed ADPCM-application a Motorola part, the MC145540 ADPCM CODEC, was the first choice (Fig. 2). The MC145540 is a single channel μ- or A-Law companding PCM-CODEC-filter. This ADPCM CODEC is a complete solution for digitizing and reconstructing voice in compliance with several international standards. For the LonVoice system it is operated in the 16kbps mode in compliance with CCITT G.726. The used companding scheme is μ-Law.

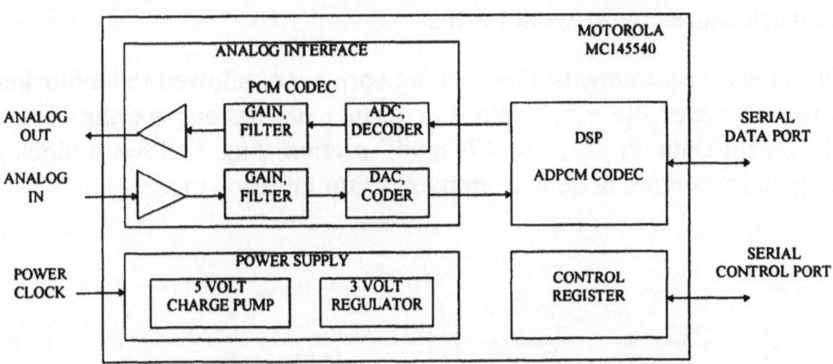

Fig. 2: Simplified Block Diagram of the MC145540

4 The Neuron Chip

The Neuron Chip, produced by Motorola and Toshiba, is the heart of every LonWorks node. The whole LonTalk communication protocol as well as the application program are implemented in one chip. Neuron Chips are available in two different base types - the 3150 and the 3120 family. The most important difference between the two types is that the 3150 has an externally accessible memory bus and needs a non volatile external memory (EPROM, Flash, etc.) that holds at least the firmware, whereas the 3120 can get along with its internal memory. Thus, peripheral components can be connected to the 3150 in two different ways:

- Memory mapped by using the memory bus interface or
- through the I/O-pins

The 3120 family only supports peripherials connected via the I/O-pins. All the different Neuron Chips are programmed in Neuron C, a modified ANSI C, that is extended to support special communication and I/O hardware resources supported by the Neuron Chip.

5 Voice Data vs. LonTalk Messages

The digitized voice signal, i. e. the ADPCM data signal, is represented by a continuous bit stream whereas LonTalk is a message based protocol. In order to transmit a continuous bit stream using LonTalk messages, the sending node has to convert the voice data stream into LonTalk messages that are sent to the receiving node. The receiving node then needs to reconvert the LonTalk messages into the ADPCM data stream. In order to transmit a continuous bit stream using LonTalk at least the following requirements must be met:

- The communication channel throughput must be higher than the data rate of the continuous bit stream.

- The buffers on the receiving and the transmitting side have to be large enough to store the ADPCM data in case of a transmission delay on the LonWorks network caused by network traffic.

In addition to these requirements a control network is not allowed to be blocked by the transmission process of the voice data. For further discussions see the 'Transmitting Voice and Control Data on the same Channel ' section. Fig. 3 shows a block diagram for a sending and receiving node in a simplex communication mode.

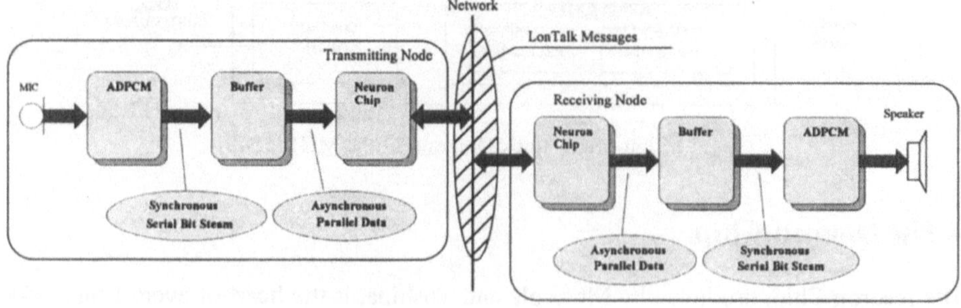

Fig. 3. Transmitting and Receiving Node in a Simplex Communication Mode

6 How to Interface the ADPCM Data to the Neuron Chip

As mentioned in the Neuron Chip section, Neuron Chips can interact with peripherals using their I/O pins, or in case of the 3150 family additionally by using the external memory bus. In case of the LonVoice system, the 16 kbps continuous data stream needs to be input by the Neuron Chip when sending voice data and has to be output when receiving voice data from the network. Evaluating the abilities of the Neuron Chips, no possibility was found to overcome this requirement without external buffers. One of the most important problems is that the application processor communicates with the network processor by using the application buffer [4, page 3-2]. During this communication process, the application processor cannot handle any software driven I/O objects. For a further discussion take a look at the following example:

Suppose explicit messages are used to transmit voice data, the following timing relations occur for a Neuron Chip at 10 MHz..

To perform a *memcpy()*: $\qquad t_{memcpy()} = 123\mu s + n \cdot 18\mu s \qquad$ n ... byte count

To perform a *msg_send()*: $\qquad t_{msg\_send} \approx 1ms.$

To evaluate a *msg_arrives*-event: $\quad t_{msg\_arrives()} \approx 550\mu s$

If a LonWorks node wants to send, for instance, 16 data bytes, it has to copy these 16 bytes into the *msg_out* object and send it by calling the function *msg_send()*. This takes $123 + (18*16) + 1000 = 1410\mu s$. If 16 bytes are to be received, the *msg_arrives* event must be evaluated and then the data has to be copied out of the *msg_in* object. Overall the whole process takes $123 + (18*16) + 550 = 960\mu s$. During this time period the application processor cannot handle any software driven I/O object. On the other hand ADPCM words need to be clocked every 125μs. Thus, an external buffer has to be used to give the Neuron Chip application processor enough time to handle the transmission process to the application buffer. This external buffer is implemented as a FIFO (First In First Out). For incoming and outgoing voice data, there must be a separate FIFO. If the Neuron Chip receives voice data from the network, this data is to be copied into the WRITE FIFO, where it can be read by the ADPCM-part. In the reverse direction, when the ADPCMs data stream is to be sent by the Neuron Chip, the Neuron Chip reads the voice data out of the READ FIFO and sends it to the network.

In addition to the storing task, the FIFO has to report its storage status by using flags. These flags are necessary for the Neuron Chips data flow control. As mentioned above the ADPCM data is represented as a synchronous bit serial data stream, whereas the Neuron Chip needs to read/write the voice data in a bit parallel/byte serial manner. Therefore another task of the FIFO circuitry is to perform the necessary serial/parallel conversion.

6.1 Using the Neuron Chip I/O pins

Using the Neuron Chip I/O pins to interface the external FIFO buffers, 8 I/O pins are used to perform a Byte I/O. The remaining 3 I/O pins are used to control the FIFO buffers data flow into and out of the Neuron Chip (Fig. 4).

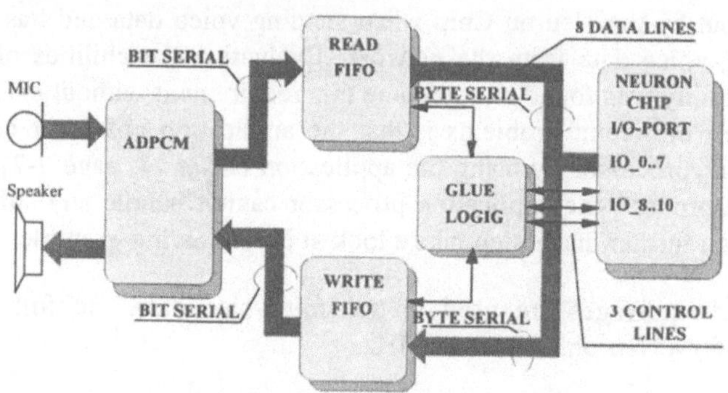

Fig. 4. Interfacing the Neuron Chip to the LonVoice peripherals using the I/O port

In this configuration, only a half duplex or simplex voice communication can be set up. For a full duplex voice communication, the Neuron Chip application processor is too slow. Performed measurements showed (see Table 1) that the performance of the involved software objects like Byte I/O (data) and Bit I/O (control) is not capable of achieving the data throughput on the I/O port needed for a full duplex communication.

Table 1. Neuron Chip I/O Port Performance Interfacing the external FIFOs

Action	Performance of the Neuron Chip I/O port	
	1 Byte	2000 Bytes
Neuron® Chip reads from FIFO	265µs	530ms
Neuron® Chip writes into FIFO	315µs	630ms

Looking at Table 1 it is evident that the bottleneck is located in the receiving node that needs to write the voice data into the FIFO. With the values of Table 1 as a basis and the knowledge that the Neuron Chip application processor needs for example approximately 1.83ms to perform a critical section and to copy 45 data bytes, the maximum achievable data throughput of the receiving node is approximately 2800kbps. The minimum FIFO size has to be set accordingly (Fig. 5).

The recommended FIFO size for the above application is 16bytes for each direction (READ an WRITE FIFO).

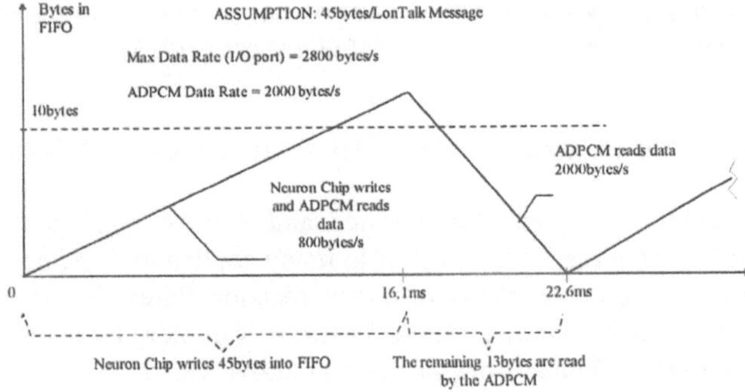

Fig. 5. FIFO Memory Usage in Case of 45 Data Bytes per LonTalk Message

6.2 Using the Neuron Chip Memory Bus

A more powerful way to interface the Neuron Chip to the voice data buffers is to use the memory bus (Fig. 6). However, this interface is only available for the 3150 family. Using the memory interface offers the possibility to set up a full duplex communication between communicating LonVoice nodes. Towards this benefit, the use of a 3150 increases the costs due to the requirement of external memory and due to a more complex board design.

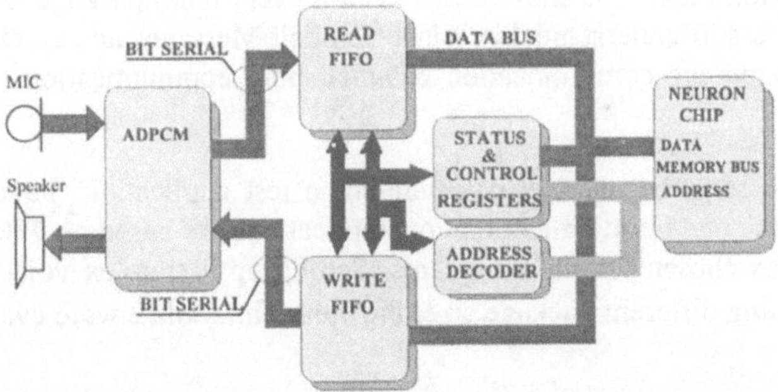

Fig. 6. Interfacing the Neuron Chip to the LonVoice Peripherals using the Memory Bus Interface

Another benefit of using the memory bus is that the data throughput between the buffers and the Neuron Chip is much better than when using the I/O pins. Thus, the Neuron Chip application processor can process some other non time consuming tasks and use the available I/O port. During the evaluation process of the memory mapped voice data interface, a data throughput of approximately 4.2kbytes/s has been determinedfor a full duplex transmission.

For the recommended FIFO buffer size the same considerations apply as when using the I/O port interface with the modification that the FIFO is written or read in a shorter time. Assuming the buffer size is again 16 bytes, the Neuron Chip is only allowed to write smaller data packages into the WRITE FIFO and to read data out of the READ

FIFO until it is empty. The needed flow control can be realized by memory mapped registers that reflect the actual status of the FIFO buffers (Fig. 6).

7 Transmitting Voice and Control Data on the same Channel

Using the opportunity of transmitting voice and control messages on the same LonWorks channel, some restrictions apply. In many applications the network data rate is a given parameter. Also the allowed system reaction times should be considered. Transmitting voice in only one direction (simplex, half duplex) by using the LonVoice system means to transmit data at a net data rate of 2kbytes/s.

LonTalk Protocol Data Units (PDUs) embed the raw data into a framework of protocol overhead data. The percentage protocol overhead decreases with the message size. Thus, to achieve a high efficiency on the channel the message size should be large as possible (221 bytes). On the other hand, a large LonTalk package size increases the system reaction time because the network is busy during the transmission process. Therefore, the appropriate message size is a compromise between channel efficiency and system reaction time.

LonTalk supports multiple message services [4], for instance acknowledged and unacknowledged services. LonVoice messages should use the unacknowledged service to save bandwidth. Field tests showed that even if every tenth package is discarded, the voice message is still understandable. Since Explicit Messages and Network Variables behave equally on the communication channel, both communication objects can be used.

To evaluate the required channel data rate some test applications have been set up. Since many home and building automation applications are based on 78kbps networks, this channel was chosen for measurements. Setting up a simplex voice channel (one direction) by using different package sizes the values in Table 2 were evaluated[1].

Table 2. Transmission Parameters as a Function of the Message Data Size on a 78kbps Channel

Channel Parameters (78kbps)	Message Data Size			
	31 bytes[2]	36 bytes	45 bytes	200 bytes
Bandwidth Utilization	39%	37%	33%	27%
Packets/sec	63,8	55,6	44,4	10
Packet Length	4,3ms	4,88ms	5,9ms	22,2ms
Packet Gap	11,3ms	13,1ms	16,3ms	77,8ms

Fig. 7 shows the channel condition during the voice transmission.

---

[1] All measurements have been made by using the LonTalk unacknowledged service

[2] Network Variable - all other Explicit Messages

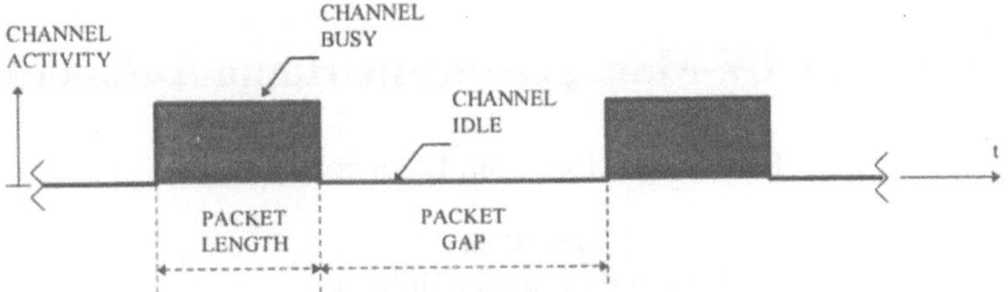

Fig. 7. Channel Properties

During a package gap, any LonTalk package(s) can be transmitted. Discussing the values printed in Table 2, a package size of 45 data bytes seems to be a good compromise on a 78kbps channel for most applications. Setting up another voice channel (for instance: full duplex voice communication) the bandwidth utilization and the packets/s need to be doubled. On channels with a higher data rate, the same boundary conditions apply.

8 Conclusion

Typical LonWorks applications can be extended by the feature of voice communication if the boundary conditions discussed in the above section are respected. Thus, the LonVoice system is a convenient and easy to use extension for LonWorks applications.

Voice data can be handled on a LonWorks network like common LonTalk messages. Thus, Network Variables and Explicit Messages can be used to transport voice data. During the binding process, multiple LonVoice nodes can be bound with the common binding tools. Due to the LonTalk capabilities not only one node can be used as a communication terminal equipment but also groups of nodes are allowed. This is especially useful for announcement applications. When using Explicit Messages with explicit addressing, LonVoice Nodes can be equipped with a keypad to dial the number of the wanted communication terminal equipment. In this manner, a wide field of applications can be covered.

References

1. Motorola: MC145540 Advanced Information , USA, 1995

2. Echelon: LonTalk Protocol Specification , Version 3.0

3. Motorola: LonWorks - Technology Device Data, USA, 1995

4. Motorola: LonWorks - Technology Device Data, USA, 1996

5. Echelon: Neuron C Prgrammers Guide, USA, 1995

LonText - Ein TV-Manager für die Home Automation

Hans-Jörg Schweinzer, Dietmar Loy

TU Wien
Institut für Computertechnik
Gußhausstraße 27
A-1040 Wien
{schweinzer,loy}@ict.tuwien.ac.at

Abstract. Providing an intuitive user interface for controlling intelligent devices in home automation applications will ease the success of intelligent homes. LonText picks up this idea by realizing a TV-based human-machine-interface using teletext-like screens to intuitively navigate the user through the information he requires or inputs. In order to provide interoperability the software structure presented complies with the LonMark interoperability guidelines. Finally the target hardware is addressed.

1 Motivation

Im Bereich der Building und Home Automation werden Feldbussysteme in einem immer größerem Maß eingesetzt. Das gesamte Tooling zur Inbetriebnahme und Wartung als auch zur Visualisierung und Steuerung basiert auf PCs bzw. Hand Helds. Der typische Anwender im Home-Bereich verfügt aber oftmals nicht über die Kenntnisse einen PC zu bedienen oder er scheut die für die Anschaffung anfallende Investition. Damit wird klar, daß im kostensensiblen Bereich der Home Automation die Notwendigkeit eines Tools gegeben ist, das von den üblichen Konventionen abweicht, indem auf für die User bereits verfügbare Anzeigeeinrichtungen (Mensch-Maschine-Interface) zurückgegriffen wird.

Genau dieser Ansatz führte zur Entwicklung des hier vorgestellten LonText-TV-Manager. Für Netzwerke, basierend auf der LonWorks-Technologie, wird als Ein-/Ausgabemedium für die Bedienung ein Fernsehapparat mit Infrarot Fernbedienung verwendet. Die von LonText verwendeten Ein-/Ausgabemasken erinnern den Benutzer sehr stark an den Teletext (Videotext). Damit ist die Basis für ein leichtes Bedienen der angeschlossenen Feldbusgeräte geschaffen.

2 Das User Interface - die Ergonomie

Mit LonText ist es dem Anwender intuitiv möglich, Funktionen des LonWorks-Netzwerkes im Haus bequem mit einer IR-Fernbedienung über ein TV-Gerät zu steuern und zu überwachen. Das User-Interface, basierend auf wohlbekannten Videotext- (Teletext-) Seiten, unterstützt eine Internet-Browser-ähnliche-Benutzerführung, die den Anwender intuitiv zu jenen Informationen und Steuerblöcken leitet, die er benötigt. Abb. 1 zeigt, wie beispielsweise die von einem Temperatursensor ermittelte Raumtemperatur, aber auch Statusinformationen des Systems dem Anwender bereitgestellt werden. Sollen Parameter wie die Temperatur verändert werden, so bewegt man mit Hilfe der IR-Fernsteuerung den Cursor auf das jeweilige Eingabefeld und gibt über die Fernsteuerung den gewünschten Wert ein (Abb. 2).

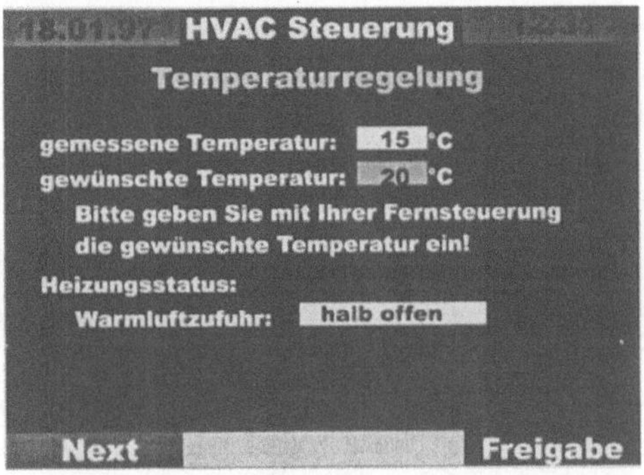

Abb. 1: Screen Shot einer LonText-Seite

Nach der Eingabebestätigung (Datenfreigabetaste auf der IR-Fernbedienung) führt LonText automatisch die notwendigen Updates im Netzwerk durch. Über die Betätigung der Home-Page-Taste kann ein Inhaltsverzeichnis der verfügbaren Anzeigemasken (Templates) aufgerufen werden. Werden Alarmmeldungen aus dem LON empfangen, so verfügt LonText über Methoden, diese Alarmmeldungen automatisch auf dem TV-Gerät anzuzeigen oder auch auf einem Video-Recorder aufzuzeichnen.

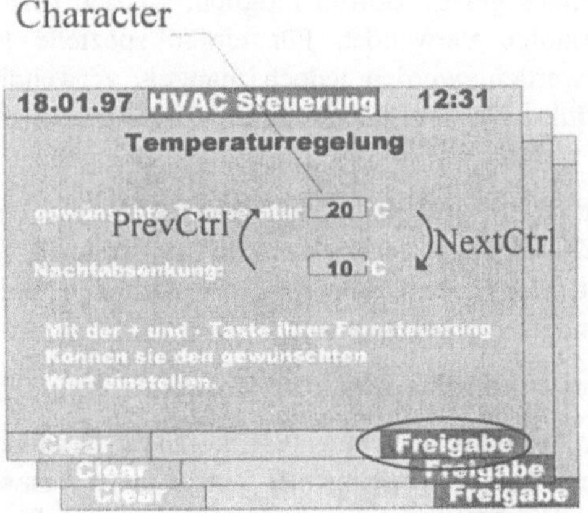

Abb. 2: Funktionen des User Interfaces

3 LonMark basierende Software-Modellierung

Interoperable Feldbuslösungen werden speziell in der Home und Building Automation die heute vielerort installierten proprietären Lösungen ablösen. Bei LonWorks-Netzwerken wird eine Basisinteroperabilität durch Einhalten der von der LonMark Interoperability Association herausgegebenen Interoperability Guidelines sicher-gestellt. Die nunmehr in der Version 3.0 vorliegenden Richtlinien [1] basieren auf funktionellen Objekten (Sensor, Actuator, und Controller Object) sowie auf einem Knotenobjekt, das die auf einem physikalischen Knoten implementierten funktionellen Objekte verwaltet. Für die Kommunikationsobjekte in Form von Netzwerkvariablen werden vordefinierte Variablentypen (Standard Network Variable Types - SNVTs) verwendet. Abb. 3 zeigt beispielhaft die Darstellung für ein allgemeines Sensor-Objekt.

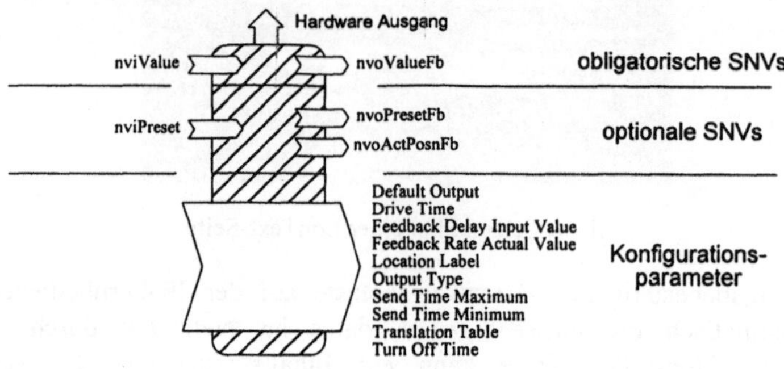

Abb. 3: Sensor-Objekt

3.1 Objektdefinitionen

Beim Software Design von LonText wurde besonderer Wert auf die Einhaltung der Interoperabiliy Guidelines gelegt. Soweit möglich, wurden bereits definierte SNVTs für die Netzwerkvariablen verwendet. Für einige spezielle Funktionen, die von LonText unterstützt werden, wurden jedoch mangels verwendbarer SNVTs eigene Variablentypen eingeführt. Die Aufgaben von LonText lassen sich folgenden Objekten zuordnen:

- IR-Sensor- und IR-Aktor-Objekt: bearbeiten von Fernsteuerbefehlen,
- Tele-Controller-Objekt: Steuerung des LonText-Systems,
- Template-Controller-Objekt: Steuerung der Aktionen und Anzeigen, die vom jeweiligen Template vorgegeben sind.

3.1.1 Infrarot-Sensor- und Infrarot-Aktor-Objekt

Der IR-Sensor (Abb. 4a) wird über eine IR-Empfänger-Hardware angesteuert. Wird ein neues Codewort empfangen, generiert der Sensor einen Netzwerkvariablen-Update über `nvoCommand`. In `nvoCommand` wird der Name der gedrückten Taste als String `SNVT_str_asc` gespeichert, also zum Beispiel „+". Die Übersetzung von IR-Code-wort auf Tastenname wird im Sensor vorgenommen. Der Name selbst besitzt noch keine semantische Bedeutung zur Ablaufsteuerung in den Eingabemasken. Diese

Zuordnung übernimmt der Tele-Controller. Der IR-Aktor (Abb. 4b) übernimmt Netzwerkvariable-Updates in nviCommand, wobei derselbe Datentyp SNVT_str_asc wie beim Sensor verwendet wird. Die Tastennamen werden in entsprechende IR-Kommandosequenzen übersetzt und über einen IR-Transmitter ausgesendet.

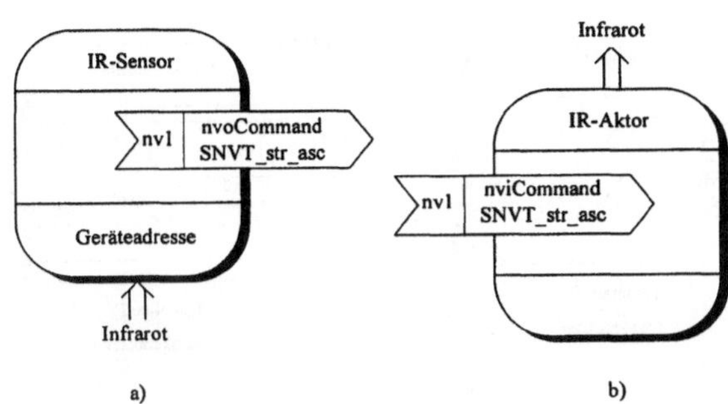

Abb. 4: IR-Sensor und Aktor Objekt

3.1.2 Template-Controller-Objekt

Im Template-Controller-Objekt (Abb. 5a) sind alle templatespezifischen Eigenschaften implementiert. Dazu zählen statische Texte am Template, Anzeigeelemente und Eingabefelder. Das Template-Controller-Objekt reagiert auf Befehle vom Tele-Controller und sendet Feed-Back-Meldungen an ihn zurück. Kurze Ausgaben auf den Teletext-Bildschirm werden ebenfalls über diese Feed-Back-Meldungen implementiert.

Das Neuzeichnen beim Anzeigen eines Templates wird vom Tele-Controller-Objekt initiiert. Die Übertragung der graphischen Repräsentation des Templates wird dann mit dem LON-File-Transfer durchgeführt. Durch diese Konstruktion kann ein Tele-Controller einen Teletext-Baustein mit mehreren Seiten nützen und ein einfaches Template-Caching realisieren. Alle Netzwerkvariablen die zur Kommunikation mit funktionellen Objekten notwendig sind, um sie über das Template-Controller-Objekt zu bearbeiten (das sind Netzwerkvariablen deren Inhalt angezeigt oder verändert werden soll), erfolgt über Definitionen in der User-Defined-Section des Objektes. Damit wird klar, das Template-Controller-Objekte an die angeschlossenen funktionellen Objekte angepaßt werden müssen. Idealerweise werden die Template-Controller-Objekte gleich vom Hersteller der funktionalen Objekte erstellt und am gleichen physikalischen Knoten untergebracht auf dem sich auch ein oder mehrere funktionale Objekte befinden.

3.1.3 Tele-Controller-Objekt

Das Tele-Controller-Objekt (Abb. 5b) besitzt Ein- und Ausgangsnetzwerkvariablen für das IR-Sensor-Objekt und das IR-Aktor-Objekt (nviIRcmd und nvoIRcmd). Zur Kommunikation mit den Template-Controller-Objekten dienen die Netzwerkvariablen nvoCmd und nviCmdFb. Über nvoCmd sendet das Tele-Controller-Objekt Kommandos an die Template-Controller-Objekte. Diese Kommandos sind vom Typ `tmpl_cmd_t`, in dem das Kommando, die Template-ID und Daten angegeben werden. Über `nviCmdFb` erhält das Tele-Controller-Objekt Rückmeldungen von den Templates. Das können Alarmnachrichten, Teletextschreibanforderungen oder auch Heartbeat-Signale sein.

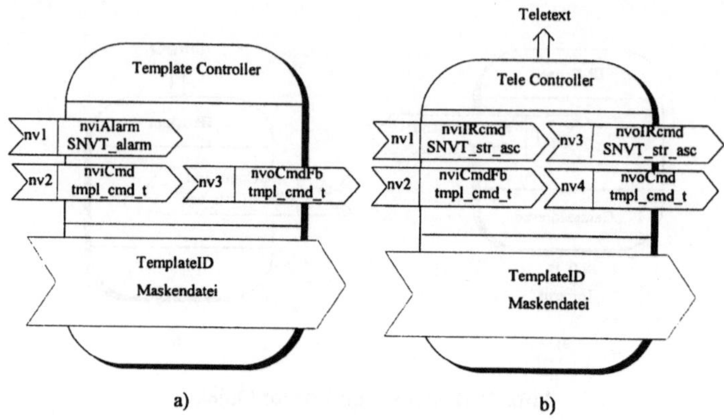

Abb. 5: Template- und Tele-Controller-Objekt

4 LonText Hardware

Die LonText Hardware stellt die physikalische Verbindung zwischen den im Home-Bereich wohl etablierten Mensch-Maschine-Schnittstellen wie Fernsehgerät und IR-Fernbedienung und dem LonWorks Netzwerk her. Entscheidend ist ein Kosten senkender Ansatz, der auf Standardkomponenten der Konsumelektronik zurückgreifen muß.

Die LonMark basierenden Objekte IR-Sensor, IR-Aktor und Tele-Controller sind auf einem Neuron-3150-Chip mit externem RAM sowie externem FLASH implementiert. Da in der Gebäudeautomatisierung unterschiedlichste Netzwerkverkabelungen Anwendung finden, muß auch die LonText-Hardware an die unterschiedlichsten Kommunikationsmedien anpaßbar sein. Es werden sowohl zweidrahtgebundene Interfaces wie RS-485 und Free-Topology (FTT-10) als auch Power-Line-Kommunikation und Funkanbindung unterstützt. Der Kommunikationsmultiplexer MUX in Abb. 6 stellt die entsprechende Verbindung zwischen Neuron-Chip und Netzwerktransceiver her.

Die Darstellung von Informationen aus dem Netzwerk erfolgt mittels Teletextbaustein, der einerseits via I2C (Inter IC) Bus mit dem Neuron-Chip verbunden ist, andererseits durch ein Scart-Kabel mit dem Fernsehgerät oder einem Videorecorder verbunden wird. Alarmmeldungen oder Statusinformationen aus dem Netzwerk können somit auf einem Videorecorder mitprotokolliert werden.

Der Informationsfluß kann auch in umgekehrter Richtung erfolgen. So können aktuelle Informationen wie z. B. Uhrzeit, Datum, Wettervorhersage etc. dem Teletext entnommen und dem LonWorks Netzwerk zugeführt werden. Alarmmeldungen werden durch ein akustisches Signal (Signalton, Melodie etc.) im Fernsehgerät angekündigt.

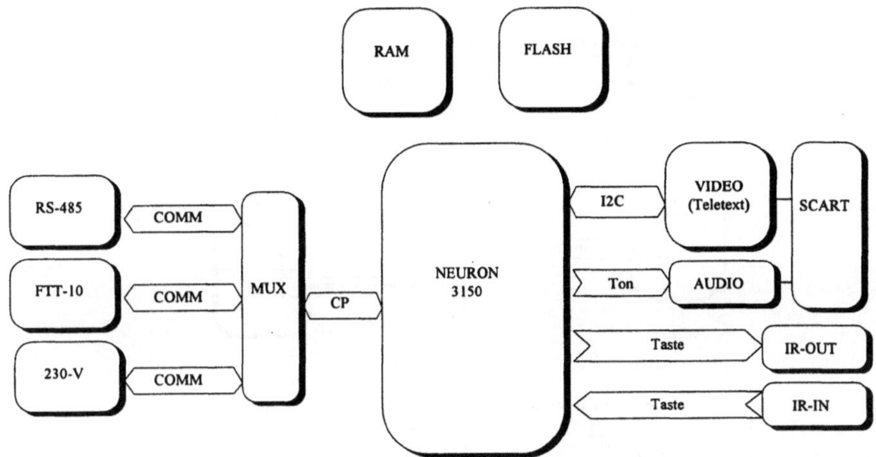

Abb. 6: Blockschaltbild der LonText Hardware

Mittels Infrarot-Ausgang werden externe Geräte ferngesteuert. Fernsehgeräte können ein- und ausgeschalten, das Programm gewechselt und die Lautstärke verändert werden. Videorecorder können in den Aufnahmemodus geschalten werden oder ereignisgesteuert eine Bandsequenz wiedergeben. Dazu muß LonText die Kommandos der Originalfernbedienung erlernen und im FLASH-Speicher ablegen.

Der Infrarot-Eingang dient einerseits zur Benutzerführung in den Eingabemasken der Teletextseiten mit Hilfe der mitgelieferten IR-Fernbedienung und andererseits zum Erlernen der Tastencodes fremder Geräte.

5 Applikationsbeispiel

Als Applikationsbeispiel soll nun eine typische Anwendung aus dem Bereich Heizung-Lüftung-Klima skizziert werden. Ein VAV-Controller steuert über einen Aktor die Warmluftzufuhr eines Raumes und verwendet zur Bestimmung der Raumtemperatur einen Temperatursensor. Über LonText kann nun der Anwender die aktuelle Raumtemperatur abfragen und dem VAV-Controller die gewünschte Temperatur mitteilen.

Folgende Objekte sind damit zur Implementierung des Anwendungsbeispiels notwendig (Abb. 7):

- VAV Controller (eigentliche Raumklimasteuerungseinheit)
- Damper Actuator (Regler für Warmluftzufuhr)
- Temperatursensor
- Infrarotsensor (Eingabeinterface für den Anwender)

- Infrarotaktor
- Tele-Controller
- Template-Controller

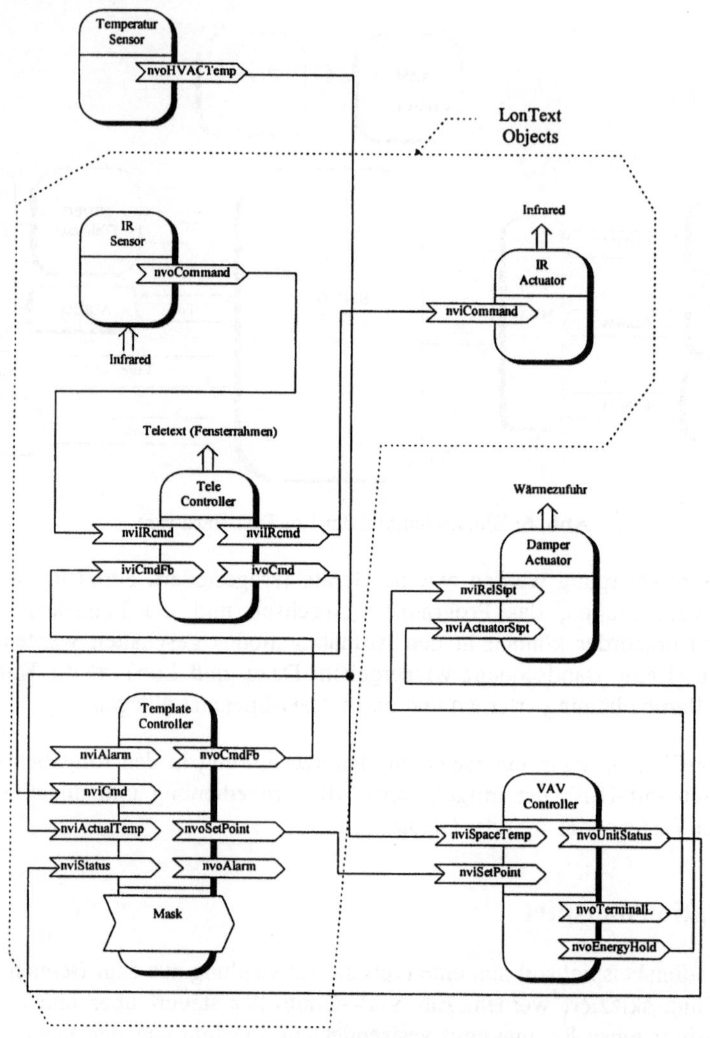

Abb. 7: Beispielapplikation mit LonText-Objekten

Der Tele-Controller übernimmt die Anweisungen vom Eingabeinterface, baut das auszugebende Teletextfenster auf und kann das Fernsehgerät über den Infrarot Aktor aktivieren. Die Definition über die Gestaltung des Teletextbildschirms hat der Tele-Controller vom zuvor aktivierten Template-Controller übernommen. Teile des Teletextfensters sind statische Texte, aber auch Ein- und Ausgabefelder. Über diese Felder kann Information, gesteuert von der IR-Fernsteuerung, eingelesen und Information, die vom System bereitgestellt wird, ausgegeben werden.

Abb. 1 zeigt das mögliche Aussehen des Teletextbildschirms. Ausgabefelder sind gelb hinterlegt, Eingabefelder grün. Die Datenfreigabefunktion teilt dem System mit, daß eine Dateneingabe abgeschlossen ist, und die eingegebene Informationen zu übernehmen sind. Im Beispiel bedeutet das also, daß eine gültige Wunschtemperatur mit einer Fernsteuerung eingegeben wurde. Die neu einzuregelnde Wunschtemperatur wird nach der Datenfreigabe an den VAV Controller übergeben. Dieser ermittelt durch einen Vergleich mit der vorherrschenden Raumtemperatur die notwendigen Parameter für den Damper-Actuator und übergibt diese über eine Netzwerkvariable. Der VAV-Controller ist ebenfalls mit dem Template-Controller verbunden, und kann zum Beispiel beim Auftreten von Unregelmäßigkeiten im System, den Template-Controller veranlassen, eine Anforderung zur Ausgabe einer Alarmmeldung am Teletextbildschirm an den Tele-Controller zu richten.

6 Literatur

1. LonMark Interoperability Association: LonMark Layer 1-6 Interoperability Guidelines, Version 3.0, Palo Alto, USA, 1996

2. LonMark Interoperability Association: LonMark Application Layer Interoperability Guidelines, Version 3.0, Palo Alto, USA, 1996

3. LonMark Interoperability Association: HVAC - LonMark Functional Profiles, Version 1.0, Palo Alto, USA, 1996

4. Echelon Corporation: File Transfer, LonWorks Engineering Bulletin, Echelon, Palo Alto, USA, 1995

5. Echelon Corporation: The SCPT Master List, Echelon, Palo Alto, USA, 1996

6. Echelon Corporation: The SNVT Master List and Programmer's Guide, Echelon, Palo Alto, USA, 1996

7. SGS-Thomson Data Book, Teletext Decoder, 1995

8. Motorola: LonWorks - Technology Device Data, Austin, USA, 1996

9. Echelon: Neuron 3150 Chip External Memory Interface, LonWorks Engineering Bulletin, Echelon, Palo Alto, USA, 1995

LonWorks basierende Standbild- und Sprachübertragung

Norbert Stampfl, Ratko Posta
Institut für Computertechnik, Technische Universität Wien

Abstract. Dieser Beitrag beschäftigt sich mit zwei atypischen Anwendung des Feldbussystems LON. Es wird untersucht, in wieweit LonWorks den Anforderungen der Standbild- und Sprachübertragung gerecht werden kann. Zwar wird durch entsprechende Kompressionsverfahren eine Reduktion der Daten erreicht, dennoch ist aber ein Kompromiß zwischen Übertragungszeiten und -qualität, sowie der Beibehaltung einer maximalen Busbelastung zu finden, um andere Anwendungen nicht zu blockieren. Die Vorteile und das Anwendungspotential der Standbild- und Sprachübertragung über einen Feldbus rechtfertigen die Bemühung um diese Thematik.

This paper deals with two atypical applications using LonWorks technology. The main objectives are to develop a system to transmit grey-scaled still images and voice data and examine demands on LonWorks for these applications. Although compression algorithms are used, a compromise between update-time and quality must be found, to avoid detraction of other LonWorks applications in the same network.

1 Einleitung

Gab man sich noch vor einigen Jahren damit zufrieden, einfache Sensordaten und Stellgrößen über Feldbusse zu übertragen, so ist man in zunehmenden Maße bemüht, dieses Medium auch für die Übertragung großer Datenmengen, wie sie bei der Bild- und Sprachverarbeitung entstehen, zu nutzen. Die Vorteile dieser Strategie liegen auf der Hand. Einerseits wird die Ressource Feldbus universeller genützt, andererseits führt ein breites Anwendungsspektrum zu vermehrten Einsatzmöglichkeiten des Feldbusses. Einmal in einem Gebäude bzw. in einer Fertigungsanlage installiert, kann durch einfaches Zuschalten entsprechender Knoten (Abb.1.), eine Bild- bzw. Sprachübertragung, aufgebaut werden. Es muß auf keine andere Technologie zurückgegriffen werden, was sich in weit geringeren Installations- und Wartungskosten manifestiert.

Abb.1. Knoten für die Standbild- und Sprachübertragung

Die Anwendungen für Standbild- und Sprachübertragung in den Einsatzgebieten der Feldbusse sind vielfältig. So geht diese Arbeit ursprünglich aus der Forderung hervor, in einem mit LonWorks voll automatisierten Haushalt, auch Bilder und Sprache mit dieser Technologie zu übertragen. Dies erlaubt den Aufbau einer Gegensprechanlage am Hauseingang, ein Bildtelefon im Gebäude sowie die optische und akustische Überwachung und Kontrolle von Räumen oder Kleinkindern. Aber auch in der Feldebene von Industrieanlagen, wo Feldbusse vorwiegend eingesetzt werden, besteht Bedarf an akustischer und optischer Information. Die Überwachung von Fertigungsstraßen oder Kesselanlagen kann über dasselbe Medium erreicht werden, mit welchem auch die zum Betrieb notwendigen Stell- und Sensordaten ausgetauscht werden.

2 Übertragung von Daten über LonWorks

LonTalk, das Übertragungsprotokoll von LonWorks unterstützt alle 7 Schichten des OSI-Modells und stellt dem Anwendungsprogramm grundsätzlich 2 Kommunikationsobjekte zur Verfügung. Es sind dies Netzwerkvariablen und Explicit Messages. Erstere haben die Eigenschaft, daß sie im Anwendungsprogramm wie herkömmliche Variablen behandelt werden können, jedoch wird jedes Update automatisch von der Firmware an die verbundenen Knoten gesendet. Bei Explicit Messages muß sich der Programmierer selbst um diesen Vorgang kümmern, allerdings können mit einer Explicit Message bis zu 238 Byte Nutzdaten übertragen werden, mit einer Netzwerkvariable nur 32 Byte. Da sich die beiden Kommunikationsobjekte im Header nur um ein Bit unterscheiden, ist klar einzusehen, daß der Protokolloverhead bei der Übertragung mit Hilfe von Explicit Messages viel geringer ist, als mit Netzwerkvariablen. Besonders bei großen Datenmengen, wie sie bei der Bild- und Sprachübertragung auftreten ist dies der entscheidende Faktor für die Verwendung von Explicit Messages zur Datenübertragung.

Um eine effiziente Übertragung zu erreichen, und die Busbelastung möglichst gering zu halten wird das Datenaufkommen bei beiden Anwendungen mit, in der weiteren Folge noch näher erläuterten Komprimierungsverfahren reduziert. Als Busmedium wird RS485 eingesetzt, da dieses bis zur maximalen Busdatenrate von 1,25MBit/s betrieben werden kann. Im folgenden wird jeweils auf die Sprach- und Standbildübertragung näher eingegangen.

3 Sprachübertragung

Die Sprachübertragung wurde im Projekt LonVoice [1] bereits erfolgreich realisiert, und wird deshalb an dieser Stelle nur kurz erläutert.

3.1 Anforderungen

Zur Sprachübertragung werden 2 gleichwertige Knoten verwendet, die sowohl als Sender als auch als Empfänger wirken. Das Sprachsignal soll dabei mit 8KHz abgetastet werden, was bei der verwendeten PCM-Kodierung zu einer Datenrate von

64kbit/s führt. Um die Busbelastung jedoch so gering wie möglich zu halten, muß für eine weitere Datenreduktion gesorgt werden.

3.2 Realisierung

Es wurden 2 Prototypen entwickelt welche die angeführten Anforderungen erfüllen. Zur Datenreduktion wird ein ADPCM-Codec eingesetzt, der den Datenstrom von 64kBit/s um den Faktor 4 auf 16kBit/s verringert. Um den Bus in Sprechpausen nicht unnötigerweise zu belasten, wird ein Verfahren zur Sprechpausenerkennung eingesetzt, das die Übertragung in diesem Zeitraum unterbindet.

Die Übertragung der komprimierten Sprachdaten erfolgt über Explicit Messages, die kontinuierlich übertragen werden. Bereits bei einer Busdatenrate von 78kBit/s kann mit diesem System annähernd Telefonqualität erreicht werden.

4 Standbildübertragung

4.1 Anforderungen

Auch bei der Übertragung von Standbildern ist ein Kompromiß zwischen Bildqualität, Busbelastung und Übertragungsdauer zu finden. Um dies teilweise dem Benutzer selbst zu überlassen, können Bilder mit Auflösungen 80x60, 160x120 und 320x240 Bildpunkten bei 16 Graustufen übertragen werden. Dazu werden ein Kameraknoten, der mit Hilfe einer CCD-Kamera die Bilder aufnimmt, komprimiert und in den Bus speist, sowie ein Anzeigeknoten aufgebaut. Eine ausführliche Beschreibung ist [2] zu entnehmen.

4.2 Realisierung

Anzeige- und Kameraknoten stehen in einer Client-Server Beziehung, wobei allein der Client einen Datenaustausch initiieren kann. Als Anzeigeknoten wird ein PC eingesetzt, um eine komfortable Oberfläche zu gewährleisten, und eine Weiterverarbeitung der Bilder, etwa die Einspeisung ins Internet zu erreichen.

Um Bildparameter vom Anzeigeknoten aus zu verändern, werden Helligkeit, Kontrast und Auflösung über Netzwerkvariablen zum Kameraknoten übertragen. Über eine spezielle Netzwerkvariable wird der Aufnahmevorgang am Kameraknoten gestartet, worauf Bilddaten mit Hilfe von Explicit Messages so lange gesendet werden, bis das vollständige Bild übertragen wurde.

Um den Protokolloverhead so gering wie möglich zu halten, werden die Pakete im Unacknowledged-Mode übertragen. Übertragungsfehler werden aber vom Anzeigeknoten erkannt, der in diesem Fall ein Update des Bildes unterbindet.

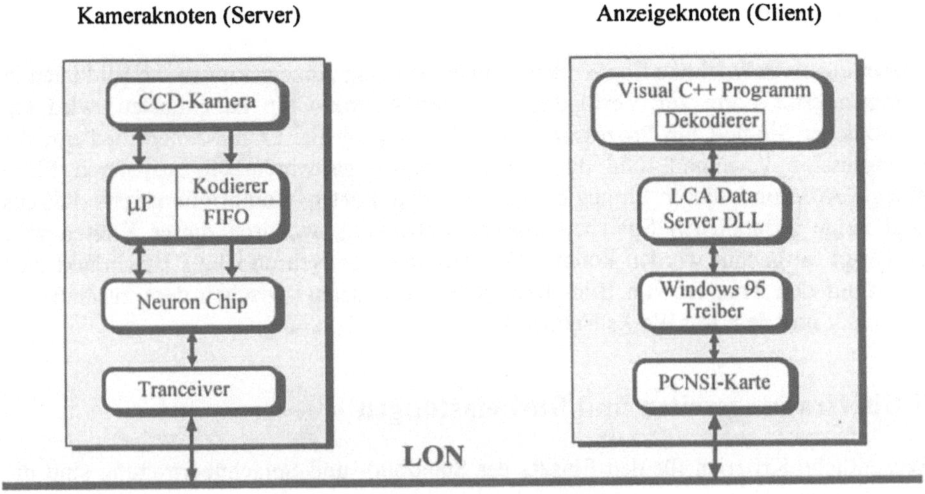

Abb.2. Blockschaltbild von LonVision

Als Komprimierungsverfahren wird eine statische Entropiekodierung der Bildpunkt-differenzen verwendet. Der Code wird dabei so gewählt, daß er einerseits leicht zu berechnen ist, also nicht in einer Tabelle abgelegt werden muß, und andererseits für möglichst viele Bilder eine hohe Datenreduktion erreicht. Der Vorteil einer statischen Codetabelle ist außerdem, daß sie dem Sender und Empfänger bekannt ist und somit nicht mitübertragen werden muß. Mit diesem verlustlosen Verfahren wird je nach Bildinhalt und Auflösung eine Datenreduktion auf rund 50% erreicht.

Anzeigeknoten

Die Bilder werden auf einem PC unter Windows 95 dargestellt. Um Anwendungs-programmen den Zugriff auf andere LonWorks-Knoten zu ermöglichen, stehen eine Vielzahl von Hard- und Softwarekomponenten für den PC zur Verfügung. Ein für diese Anwendungen zufriedenstellender Datendurchsatz läßt sich aber nur mit Hilfe des Data-Servers erreichen. Dieses Softwaretool ist Teil des Object-Servers, einem Softwarepaket, auf dessen Funktionen ein Anwendungsprogramm über OLE zugreifen kann, und dadurch Netzwerkmanagement und Monitoring-Aufgaben erfüllen kann.

Als Hardwarekomponente wird eine PCNSI-Interfacekarte verwendet, da nur diese den Data-Server unterstützt und einen Zugriff auf das verwendete RS485-Busmedium erlaubt.

Das Anwendungsprogramm ist in Visual C++ geschrieben, um eine rasche Dekodierung und Anzeige der Bilder zu erreichen. Eine komfortable Oberfläche erlaubt neben der Anzeige der Bilder auch die Einstellung von Aufnahmeparametern und die Steuerung des Bildzykluses.

Kameraknoten

Dieser eigens entwickelte LonWorks Knoten stellt dem Anzeigeknoten die Bilddaten in komprimierter Form zur Verfügung. Um den Neuron-Chip zu entlasten, wird ein zusätzlicher Single-Chip Prozessor verwendet, welcher die CCD-Kamera bedient, die Kompression vornimmt, und die Daten in einem softwaremäßig realisieren FIFO ablegt. Aufgrund dieser Strategie und des verwendeten Komprimierungsverfahrens sind keine zusätzlichen Speicherbausteine notwendig, wodurch dieser Knoten sehr preiswert aufgebaut werden konnte. Die Aufgabe des Neuron-Chips beschränkt sich somit auf den Transfer von Bild- bzw. Kommandodaten zwischen dem zusätzlichen Prozessor und dem LonWorks Netzwerk.

5 Übertragungszeiten und Busbelastungen

Wesentliche Kriterien für den Einsatz der Standbild- und Sprachübertragung sind die erreichbaren Übertragungszeiten und die dabei auftretenden Busbelastungen. Aus Sicht des Anwenders wird eine hohe Bild-Update-Rate und eine zufriedenstellende Sprachqualität gefordert. Allerdings dürfen der Bus, und daraus resultierend, andere zeitkritische Anwendungen nicht blockiert werden. Um dies zu erreichen bietet LonWorks die Priorisierung von Datenpaketen an. Da verzögerte Pakete bei der Sprach- und Bildübertragung nur zu einem Qualitätsverlust, nicht aber zu kritischen Situationen führen können, findet der Datenaustausch bei diesen Anwendungen unter Verwendung von niederprioren Nachrichten statt. Dadurch wirken sich auch die hohen, bei der Sprach- und Standbildübertragung auftretenden Busbelastungen nicht negativ auf andere kritische Anwendungen aus.

Zur Ermittlung repräsentativer Meßwerte wird ein Netzwerk mit 2 Sprachknoten, einem Kamera- und einem Anzeigeknoten, sowie einem Protokollanalyser-Knoten aufgebaut. Im Einzelbetrieb führen beide Anwendungen bei einer Busdatenrate von 78kBit/s zu einer Busbelastung von 70% und bei 625kBit/s zu rund 30% Belastung. Ein gemeinsamer Betrieb ist somit erst ab einer Busdatenrate von 625 kBit/s sinnvoll, da darunter die Sprache nur unverständlich wiedergegeben werden kann und bei der Bildübertragung mit einer Verdopplung der Übertragungszeiten gerechnet werden muß. In Tab.1. sind Übertragungszeiten und Kompressionsfaktoren eines, mit LonVision übertragenen Bildes bei unterschiedlichen Auflösungen und bei einer Busdatenrate von 625 kBit/s angeführt.

Tab.1. Einzelbildübertragung bei 625kBit/s

Auflösung	Übertragungszeit/Bild in [s]	Kompressionsfaktor
80 x 60	0,22	1,5
160 x 120	0,44	1,8
320 x 240	1,37	2,1

6 Zusammenfassung

Durch diese Arbeit konnte das Potential von LonWorks in Hinblick auf die Standbild- und Sprachübertragung aufgezeigt werden. Auch wenn noch nicht dieselbe Qualität wie bei anderen Technologien erreicht wird, ist es trotzdem sinnvoll sich um diese Thematik zu bemühen. Da die eingangs erwähnten Vorteile bestehen bleiben und mit einer weiteren Zunahme der Leistungsfähigkeit von LonWorks zu rechnen ist, sowie die Komprimierungsverfahren noch ausgebaut werden können, ist mit dem kommerziellen Einsatz dieser Anwendungen unter LonWorks schon bald zu rechnen.

Literatur

1. H.J. Schweinzer, ICT, LonVoice - Transmitting Voice by using LonTalk

2. N. Stampfl, Diplomarbeit am ICT der TU-Wien, Standbildübertragung über LON

LON-basierte dezentrale Wetterbeobachtung

Peter Neubauer, Ratko Posta

Technische Universität Wien
Institut für Computertechnik

Abstract. Der vorliegenden Beitrag beschreibt eine lokale Wetterstation, die in das Modell eines automatisierten Haushalts integriert ist. Die Daten, die die Wetterstation zu Verfügung stellt, können von „intelligenten„ Haushaltsgeräten weiterverwendet werden. Bei der Entwicklung der Wetterstation wurde das Konzept der „verteilten Intelligenz„ verfolgt, sodaß bei Ausfall eines Knotens das übrige System weiter funktioniert. Statistische Daten, wie beispielsweise Temperaturmittelwerte, werden in einer Datenbank auf dem PC gespeichert. Außerdem ist das System mit dem World Wide Web (WWW) verbunden, wobei die PCLTA-Karte von Echelon zur Verbindung zwischen dem LonWorks-Netz und dem PC Verwendung findet. Der Zugriff auf die Daten via WWW erfolgt mit Hilfe des Common Gateway Interface (CGI).

Abstract. This article deals with an automatic weather station which is integrated into the model of an automated household. Data supplied by this weather station can be used be „intelligent" components of the household model. In order to ensure the function of all other devices in case a node fails the concept of „distributed intelligence" applies to the system. Statistical data as temperature averages is stored in a database on a PC. Furthermore the system is connected to the World Wide Web (WWW). Echelon's PCLTA is used for the connection between the LonWorks network and the PC. The data access via WWW is implemented with CGI.

1 Einleitung

Sehr wichtige Anwendungen in vernetzten Haushalten und Gebäuden sind die Erfassung und Auswertung von Wetterdaten, weil damit die Heizungssteuerung beeinflußt werden kann. An eine Wetterstation, die sich auf einem Gebäude befindet, werden die Anforderungen gestellt, daß sie flexibel und redundant ausgelegt und einfach zu installieren und kalibrieren ist. Im Rahmen des Projekts Home-Net Configuration Tool (FWF P10699-OEMA) entwickelten wir eine auf LonWorks [1] basierende Wetterstation, die in das Modell eines Privathaushalts integriert ist und an dieses Daten liefert [2]. Dieser Beitrag beschreibt das Konzept für eine effiziente Integration der Wetterdaten in ein Heimnetz und ihre wichtigsten Anwendungen im Zusammenhang mit automatisierten Haushalten. Dazu soll zunächst einmal die Wetterstation selbst erläutert werden, dann die Erfassung von statistischen Daten in der Datenbank, die

Verbindung der Wetterstation mit dem Internet und schließlich noch eine Kamera, die via Internet zur Wetterbeobachtung geeignet ist. Abbildung 1 gibt einen Überblick über das hier beschriebene System.

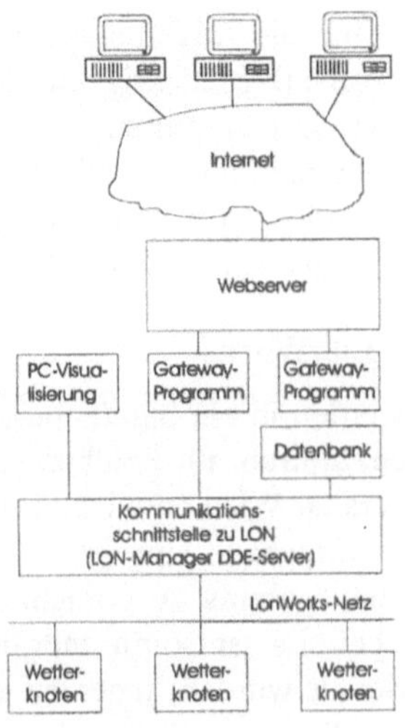

Abb. 1: Struktur des Systems

2 Die Wetterstation

Zuerst wird grundsätzlich erörtert, welche Werte überhaupt erfaßt werden können, und zu welchen Zwecken diese Daten – außer zur reinen Information des Benutzers – noch verwendet werden können. Anschließend wird der Aufbau der Wetterstation mit Hilfe von LonWorks beschrieben.

2.1 Verwendete Daten

Da es das Ziel ist, die Meßwerte weiterzuverarbeiten, sollen diejenigen Daten erfaßt werden, die zur Gebäudeautomation geeignet sind. Da im Rahmen dieses Projekts nicht die Wetterbeobachtung, sondern vielmehr die Gebäudeautomatisierung im Vordergrund steht, werden sowohl Daten, die im Gebäudeinneren gemessen werden, als auch solche, die im Freien anfallen, erfaßt.

Werte im Gebäudeinneren werden zur Gebäudeautomation herangezogen. Zu diesen Werten zählen nach [3] die Lufttemperatur, die Strahlungswärme, die Luftgeschwindigkeit und die relative Luftfeuchtigkeit. Nicht nur diese Werte haben Einfluß auf das subjektive Empfinden des Raumklimas, zwei weitere wichtige Parameter sind von Mensch zu Mensch verschieden: die Kleidung und die Aktivität. In [3] werden mehrere Modelle vorgestellt, mit deren Hilfe die Zufriedenheit mit dem Raumklima ermittelt werden kann. Man wird es allerdings nie schaffen, daß alle Benutzer eines

Gebäudes mit dem herrschenden Klima zufrieden sind. Man wählt daher entweder einen Mittelwert, oder man teilt den Raum in kleinere Klimazonen ein. Erfaßt werden müssen auf jeden Fall die Raumtemperatur und die relative Luftfeuchtigkeit, um die Klimaanlage entsprechend regeln zu können.

Die interessanten Daten im Freien sind die Außentemperatur, die zur Heizungssteuerung verwendet werden kann, die Außenluftfeuchtigkeit, der Luftdruck, die Windgeschwindigkeit und –richtung, die Niederschlagsmenge und die Außenhelligkeit. Mit Hilfe dieser Werte können auch Wetterberichte erstellt werden. So wäre es denkbar, eine Vielzahl solcher lokaler Wetterstation zu installieren, um damit von einer zentralen Stelle genaue Wetterprognosen erstellen zu lassen.

2.2 Aufbau der Wetterstation mit LonWorks

Die oben erwähnten Wetterdaten werden in ein bereits bestehendes Modell eines vernetzten Haushalts, das im Kompetenzzentrum für Feldbussysteme am Institut für Computertechnik der Technischen Universität Wien entwickelt worden ist, eingespeist. Das Modell ist mit Hilfe von LonWorks aufgebaut. Um die Verwendung von Geräten verschiedener Hersteller innerhalb eines Systems zu erleichtern, wurden standardisierte Kommunikationsobjekte definiert [1]. Es ist somit möglich, die Sensoren, die im Rahmen dieser Arbeit selbst entwickelt wurden, jederzeit gegen kommerziell erhältliche auszutauschen. Die Sensoren erfassen zunächst einmal die aktuellen Werte der oben angeführten Meßgrößen – mit Ausnahme der Niederschlagsmenge, hier ist die Angabe eines Bezugszeitraums notwendig. Außerdem werden die Mittelwerte der letzten halben Stunde direkt von den Knoten ermittelt. Auch Minima und Maxima werden von den Knoten berechnet. Sollten die vorhin erwähnten selbst gebauten Knoten durch fertige ersetzt werden, geht die Fähigkeit zur Erfassung von Mittelwerten verloren, weil Standardsensoren diese Möglichkeit normalerweise nicht bieten.

3 Die Verbindung von LonWorks und PC

3.1 Übertragung von Daten zwischen LonWorks und PC

Die Verbindung des LonWorks-Netzes mit einem PC erfolgt mit Hilfe der PCLTA-Karte von Echelon [4]. Diese Karte kommuniziert mit dem PC über den ISA-Bus und mit dem LonWorks-Netz über einen Transceiver, dessen genaue Type vom verwendeten Übertragungsmedium bestimmt wird. Der weitere Datenaustausch wird über den LonManager DDE-Server abgewickelt [5]. Über die standardisierte DDE-Schnittstelle von Microsoft Windows können somit Anwendungsprogramme auf dem PC Daten mit den Knoten im LonWorks-Netz austauschen. Hier kam Microsoft Visual Basic zur Anwendung, es ist jedoch auch die Verwendung jeder anderen Programmiersprache denkbar. In diesem Zusammenhang besonders interessant ist die Möglichkeit, Daten aus dem Haushalt direkt in ein CAD-Programm zu übertragen, womit beispielsweise eine Fehlererkennung automatisch durchgeführt werden kann. Dies wird gerade im Rahmen einer Diplomarbeit untersucht.

3.2 Wetterstatistiken

Nachdem die hier vorgestellte Arbeit Teil eines größeren Projekts ist, sollen die Daten auch weiteren Benutzern zu Verfügung gestellt werden, die nicht unbedingt mit dem LonWorks-Netz verbunden sind. Daher werden die von den LON-Knoten ermittelten Halbstundenmittelwerte, Tagesminima und Tagesmaxima in einer Datenbank erfaßt, außerdem werden die Halbstundenmittelwerte jeweils um Mitternacht zu Tagesmittelwerten zusammengefaßt. Um die gespeicherte Datenmenge in Grenzen zu halten, werden jeweils nur die Halbstundenmittelwerte der letzten 24 Stunden gespeichert. Als Datenbank zur Speicherung der statistischen Daten wurde Microsoft Access benutzt.

4 Die Verbindung von LonWorks und Internet

Hier waren Überlegungen anzustellen, wie man Daten zwischen zwei so unterschiedlichen Medien wie LonWorks und dem Internet austauscht. Es bietet sich natürlich an, die Verbindung über den ohnehin im System bereits vorhandenen PC herzustellen. Eine andere Möglichkeit bestünde darin, einen Microcontroller zu benutzen, in dem sowohl das LonTalk-Protokoll als auch das im Internet verwendete TCP/IP-Protokoll und das darauf aufbauende HTTP-Protokoll implementiert sind. Diese Variante bietet den Vorteil, ein LonWorks-Netz über Internet abfragen und kontrollieren zu können, ohne daß im LonWorks-Netz ein PC eingebunden sein muß.

4.1 CGI und Java

Dokumente, die im World Wide Web (WWW) angeboten werden, sind in Hyper Text Markup Language (HTML) verfaßt [6]. Der Austausch von Daten zwischen dem Client – hier Browser genannt – und dem Server erfolgt mittels dem HTTP-Protokoll. Der Server kann jedoch nur HTML-Dokumente zur Verfügung stellen. Sollen Daten, die der Benutzer an den Server sendet, verarbeitet werden, so erledigt dies ein anderes Programm, ein sogenanntes Gateway-Programm. Die Kommunikation zwischen diesem Gateway-Programm und dem Server läuft über das Common Gateway Interface (CGI).

JAVA stellt eine alternative Form des Datenaustauschs zwischen Browser und Server dar. Es handelt sich hierbei um eine objektorientierte Programmiersprache. Die JAVA-Programme - *Applets* genannt - werden als Quelltext zum Client geschickt, dieser compiliert sie dann und führt sie aus. Dadurch werden viel komplexere Anwendungen möglich als mit reinen HTML-Files. So ist beispielsweise die Verarbeitung von Benutzereingaben möglich, ohne daß deswegen eine neue Verbindung zum Server aufgebaut werden muß. Serverseitig gehört zu jedem Applet ein *Java-Servlet*. Ein Applet, das auf einem Client läuft, tauscht Daten nicht mehr mit dem Webserver aus, sondern kommuniziert direkt mit dem Servlet, wodurch die Verbindung schneller wird als über CGI.

4.2 Aktuelle Werte und Statistiken über Internet

Wenn der Benutzer die Internet-Adresse anwählt, wo er aktuelle Wetterwerte angeboten bekommt (Abbildung 2), wird zunächst vom Browser eine Anfrage an den HTTP-Server gesendet. Dieser ruft das Gateway-Programm auf, das das entsprechende HTML-Dokument erstellt. Dieses Programm kommuniziert über die DDE-Schnittstelle mit dem LonManager DDE-Server, der wiederum über die PCLTA-Karte die Verbindung zum LonWorks-Netz herstellt. Das Gateway-Programm erstellt dann das HTML-File aus Abbildung 2. Somit können die aktuellen Wetterwerte ins Internet übertragen werden. Werden von Benutzer statistische Daten angefordert, liest ein entsprechendes Gateway-Programm direkt das Datenbankfile aus. Details zur Datenübertragung sind in [6] und [2] beschrieben.

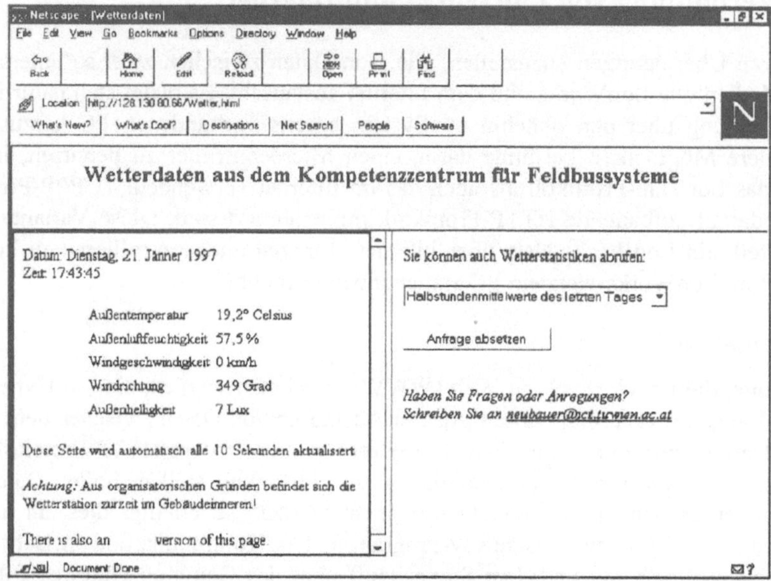

Abb. 2: Aktuelle Wetterwerte

5 Die Wetterkamera

Eine interessante Erweiterung der Wetterstation besteht im Einsatz einer Kamera, die Bilddaten über das LonWorks-Netz überträgt [7]. Diese kann dazu dienen, Wetterbilder zu übertragen. So kann beispielsweise Bewölkung nicht nur mit einem Helligkeitssensor, sondern auch optisch erkannt werden kann. Die automatische Erkennung der Art des Niederschlags - Regen oder Schnee - sowie von Nebel wäre möglich. Weiters ist eine Fernüberwachung des Haushalts via Internet dadurch machbar. Eine Fernsteuerung der Kamera wurde - unter Berücksichtigung von Sicherheitsaspekten - realisiert, um die genannten Anwendungsarten zu demonstrieren. Dem Benutzer wird dabei die Möglichkeit geboten, die Kamera über die Bedienoberfläche seines WWW-Browsers zu positionieren. Dazu dienen vier Buttons, die rund um das aktuelle Kamera-Bild auf der Browser-Oberfläche gezeichnet sind (siehe Abb. 3).

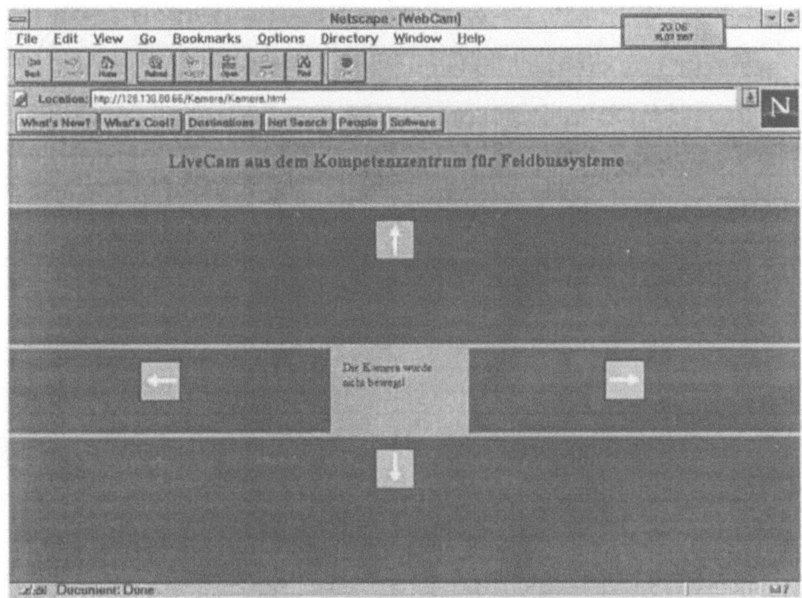

Abb. 3: Übertragung der Wetterbilder mit Hilfe des World Wide Web

Literatur

1. LonWorks Technology Device Data, Motorola, Inc., 1995

2. Neubauer, P.: Entwicklung einer LON-basierten Wetterstation mit WWW-Anbindung, Diplomarbeit, Institut für Computertechnik, TU Wien, 1997

3. A Thermal Comfort Production Tool, ASHRAE Journal, September 1996, S. 39ff

4. OnWorks PC LonTalk Adapter User's Guide, Revision 1.0, Echelon Corporation, Document No. 79310

5. LonManager DDE Server User's Guide, Version 1.5, Echelon Corporation, Document No. 39300

6. Graham, I.: The HTML Sourcebook Wiley Computer Publishing, 1996

7. Stampfl, N.: Standbildübertragung über LON, Diplomarbeit, Institut für Computertechnik, TU Wien, 1997

Drei Aufgaben, eine Software:

Konfigurieren, Programmieren und Visualisieren mit **PC WORX** für **INTERBUS**

PC WORX: die Automatisierungssoftware der Zukunft für PC-basierende Steuerungen mit INTERBUS. Offen und herstellerunabhängig durch weltweite Standards wie INTERBUS, Windows NT und IEC 1131. Für die durchgängige Kommunikation auf einer Datenbasis. Mit SYSTEM WORX zur Systemkonfiguration, PROGRAM WORX zur Programmierung nach IEC 1131 und GRAPH WORX für die Visualisierung von Steuersignalen.

Phoenix Contact, Postfach 13 41, 32819 Blomberg, Tel. 0 52 35/34 02 22, Fax 0 52 35/34 20 51

Der Bus zur Zukunft

Die Anforderungen an Gebäudemanagementsysteme steigen kontinuierlich. EIB setzt heute den Standard von morgen.

Über EIB, den zukunftssicheren Bus, können alle Komponenten einer Gebäudeausrüstung gesteuert werden. Ein Vorteil, der dem Gebäudebetreiber deutliche Einsparungen ermöglicht und eine hohe Flexibilität bei Nutzungsänderungen garantiert.

Gern informieren wir Sie näher, mit unserer Infobroschüre. Bitte anfordern:
European Installation Bus Association,
Avenue de la Tanche 5
B-1160 Brussels
Fax: ++32 2/6 75 50 28
http://www.eiba.com

Europäischer Installations Bus

ACORN 1479 - Design Tools for Fieldbus

I Brownlie, J Hoerger (Siemens), F Iwanitz (ifak), P Barbosa (EDP), R Dionisi
(ENEL)

Eutech Engineering Solutions Ltd
PO Box 99, Belasis Hall Technology Park,
Billingham, Cleveland, TS23 4YS
e-mail iain.brownlie@gbeut02.eutech.com

Abstract. Future Fieldbus based process control systems will be characterised
by a large number of interconnected smart field devices. The hitherto applied
centralised network design (in which the individual field devices are directly
linked to the automation system) will be extended by distributed structures. As a
result tomorrow's engineering tools will need to satisfy a range of new demands.

This will also generate new customer demands as they require significantly
enhanced engineering processes which can best be characterised by keywords
like "device independent plant engineering", "criteria driven device selection",
"allocation of functions to real devices as late as possible" (late distribution).

Within the framework of the EU-sponsored Brite/EuRam Project ACORN 1479
and in close collaboration with major users, prototype tools supporting the
pertinent engineering methods to be used at the field level are currently being
developed and tested.

1 Introduction

The ACORN 1479 project receives financial support from the Commission of the
European Communities under the Brite EuRam III Workprogram for collaborative
research and development in the area of Industrial and Materials Technologies.

The objective of the project is to develop a software toolkit to support the design and
implementation of the emerging new generation of open, highly distributed control
systems, based on fieldbus technologies and to validate these tools in the context of
real process applications.

The end user partners in the project include ICI Eutech (UK) representing the
Chemical sector, and ENEL (IT) and EDP (PT) representing the Power sector. The
toolkit developers include Siemens (DE), CISE (IT), ifak (DE), and Fraunhofer IITB
(DE). Intrasoft (GR) are responsible for integrating the individual tools. The ACORN
consortium represents a unique group of end users, system integration and research
organisations with considerable knowledge and experience of fieldbus technologies.

3 Bereichsübergreifende und weitere Beiträge

Interdisciplinary and other contributions

Die im Folgenden dargestellten Aufsätze behandeln unter anderem die Anbindung von LonWorks an Internet, womit die überraschend schnell um sich greifende Sprache Java ins Spiel kommt. Ein weiteres interessantes Thema dieses Kapitels ist die Smart-Card-Technik im Zusammenhang mit Home Systems. Es wird ein Security-Prinzip angeboten, das die Fragenkomplexe innerhalb der Gebäudetechnik und Home Systems lösen könnte. Aber auch Fragen der EMV und der Feldbuswerkzeuge werden hier angesprochen.

The following papers discuss among other topics the interconnection of LonWorks and the Internet, an application calling for the surprisingly fast spreading programming language Java. Another interesting subject of this chapter is smart card technology in combination with home systems. A security principle is proposed that could help solve the problems of building automation and home systems. But also questions of EMC and fieldbus tools are addressed.

The ACORN 1479 project started on 1st February 1996 and is due to complete at the end of July 1988. Some 40 man years of research and development effort will be spent by the consortium over the 30 month duration of the project. The consortium would like to thank the European Commission for their generous financial support towards the 5.4 MECU budget.

2 User Requirements for Design Tools

The Piping and Instrumentation Drawing (P&I D) has long been used as the primary reference for communication between the disciplines of Process, Control and Instrumentation engineering. Even with the advances of modern distributed control systems these drawings are still used to define the basic control strategy, even if supplementary drawings and datasheets are needed to capture all the details.

While the price/performance ratio of modern hardware continues to improve, the level of expertise required to successfully deploy systems is also increasing. This means that the cost of system engineering is growing as a percentage of total plant cost. End users therefore require design methodologies and tools which will improve the efficiency of the design process.

The introduction of fieldbus will increase the amount of choices available to the designer. There will be decisions whether to use simple or smart instruments, whether to execute control in the centre or distribute it to the field. There will be many options for the detailed design of the network topology and allocation of instruments to bus segments. These decisions will have a direct impact on overall system integrity, reliability and performance. There is a strong need for intelligent tools which can understand the design context, and can manage ever increasing application complexity.

With fieldbus, it is no longer possible to consider the different requirements for process, control and instrumentation engineering in isolation. A design philosophy for the whole system is required. Many of the first generation fieldbus tools are expected to support configuration of single instruments, or possibly a fieldbus segment. End users expect to distribute control functions across a range of equipment. Traditional centralised control platforms will continue to be used, but will be combined with more radical distribution of control into field devices. It is clear that an integrated approach to design is required.

The design of the process control systems implemented using a proprietary control architecture is heavily influenced by the available features of the target system. As the time for design, construction and commissioning is increasingly constrained by market forces, there are strong incentives to start the design process earlier, to improve the quality of design to minimise commissioning delays, and to use component based design methods which offer design portability and options for re-use.

Recent trends towards open systems and function block standards means that a common process control design method supported by tools is becoming feasible.

3 Toolkit Overview

The ACORN 1479 project has responded to these requirements by defining the set of tools illustrated in Fig. 1 below.

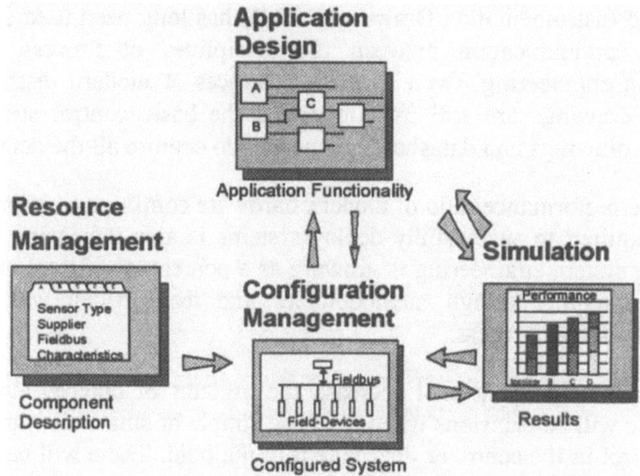

Fig. 1 ACORN 1479 Toolkit Framework

The Application Design Tool (ADT) is used to define the required application functionality independent from the implementation technology. The ADT captures information about both the instrumentation requirements and control algorithms.

The Resource Management Tool (RMT) maintains a library describing the capabilities of the hardware devices used to construct real systems. These components include instrumentation, dedicated controllers, fieldbus segments and network devices.

The Configuration Management Tool (CMT) defines the distributed control architecture by connecting hardware components from the RMT to form a network, and allocating the functions defined in the ADT to devices which offer that capability.

The Simulation Tool (ST) allows the user to execute the control strategy for a given configuration to assess dynamic performance. This will allow early testing of the control philosophy defined within the ADT, and identify network loading and other performance metrics based on the CMT configuration.

An Administration Tool is also provided to maintain user and project information.

4 Case Study

Project results gained up to now have been evaluated in the context of an existing multi-vendor application example based on the installation illustrated in Fig. 2 below.

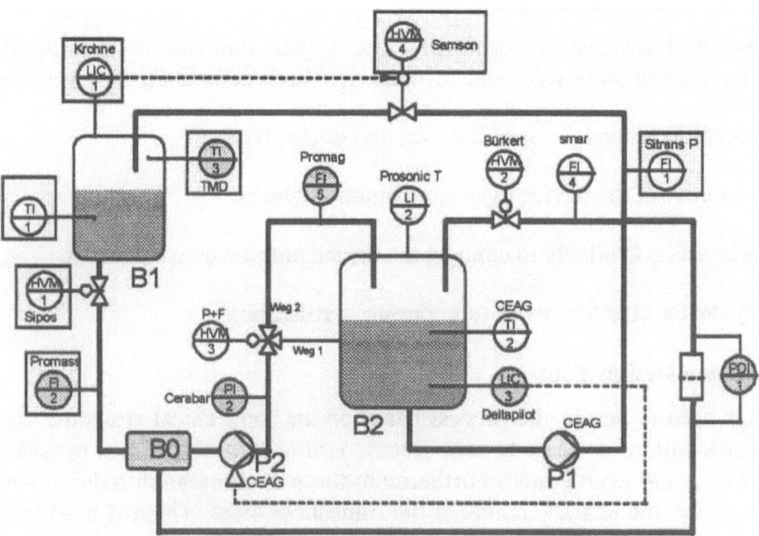

Fig. 2 Case Study Application Example

The liquid is drained off with positioner HVM 1 from Tank B1. The mass flow is measured with the mass flowmeter FI 2. The level in tank B2 can be varied using three way valve HVM 3.

The regulation of the flow to tank B1 made with a cascade control (LIC 1 and HMV 4) and keeps the level in the tank B1 stable at the set value.

By opening the valve HVM 2, the flow rate of pump P1 can be influenced. This is used as the disturbance for the control loop. The diverted flow over HVM 2 is measured with the flowmeter FI 4.

Temperatures are measured in tank B1 and tank B2 with TI 1 and TI 2. The level in tank B2 is measured with an ultrasonic system LI 2 and a hydrostatic measurement device LI 3.

The functionality of the regulation can be tested with adjusting valve HVM 1 or opening and closing valve HVM 2.

The application example is based around the level control scheme for the Vessel B1, including the flows in and out of the vessel and their respective control valves.

We will find this structure as we follow the workflow using the various ACORN tools.

4.1 Administration Tool

The Administration Tool is first used to define projects and users. Then the design process can start.

4.2 Resource Management Tool

Device types and bus segment types are first written into the resource library. The attributes for the chosen devices and fieldbus have been defined for the case study

- identification of model, manufacturer, and device type

- the capability of the device to execute function blocks

- the interface(s) available to connect the device onto fieldbus networks

- the key device characteristics (e.g. power, certification)

4.3 Application Design Tool

The ADT is used to design the process based on an hierarchical structure. Using the function block editor, the cascade control loop used to control the level in tank B1 and shown in Fig. 3 has been edited. Furthermore the instrumentation requirements have been defined, i.e. the characteristics of the transducer block which is used to connect the function block to the process.

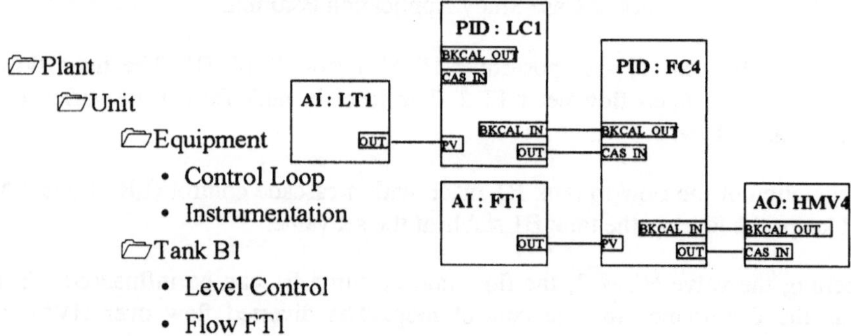

Fig. 3 Case Study Application Design

There are additional attributes which characterise the control loop as a whole which also need to be captured during design. These include its duty (trip, control, alarm or indication), performance requirements (e.g. frequency and timing constraints) and integrity requirements (e.g. redundancy and hazardous area certification).

An important objective of ACORN is to reduce risk and improve the quality of design by providing a range of consistency, completeness and design integrity checks. For example, the ADT can check whether all mandatory connections have been defined.

4.4 Configuration Management Tool

The CMT is used to design the fieldbus configuration. The user must select devices from the RMT library, connect them together to form networks, and allocate the function blocks from the ADT into appropriate devices as illustrated in Fig. 4 below.

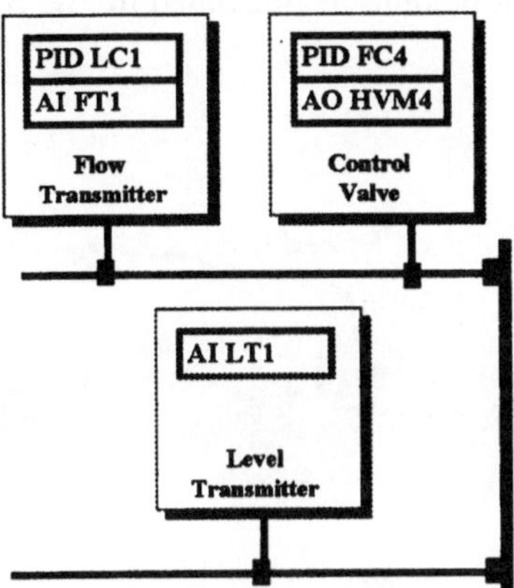

Fig. 4 Case Study Configuration

The functional instrumentation requirements may be augmented with additional constraints (e.g. fieldbus technology and preferred supplier), to identify appropriate devices. In addition to the instrumentation, the user may select other devices required to implement the control scheme (e.g. DCS or PLC controllers).

The user has many design options relating to topology and allocation strategies. As with the ADT tool, the CMT implements a number of design checks. The example in Fig. 4 would fail the "single loop integrity" check because all devices involved in the control loop are not connected to the same fieldbus segment.

4.5 Simulation Tool

SMT is the last tool in the chain. It can be used to simulate the application and the network. Three levels of mathematical models have been realised to do this.

- environment layer is a very simple model of the physical process

- application layer is a rigorous emulation of function block algorithms

- communication layer is a semi-rigorous stochastic model of network traffic

The user has a number of options to parameterise and interact with the running simulation which can be visualised using an animation tool. Results will normally be used to modify either the Application Design or the Configuration to improve dynamic performance, or to compare the efficiency of alternative implementation strategies.

5 Pilot Applications

5.1 Chemical Pilot

ICI are developing a Chemical Pilot Application which will implement a site monitoring application using fieldbus as illustrated in Fig. 5 below.

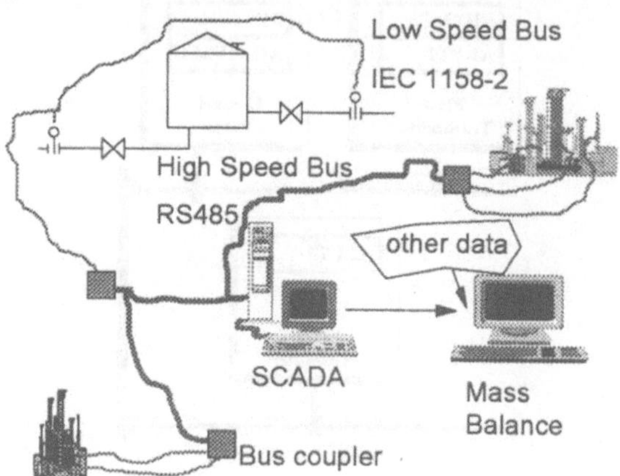

Fig. 5 Chemical Pilot

This application involves the collection of 59 flow, temperature and pressure measurements from remote areas across a large site. These measurements must be integrated into higher level management systems where a mass balance is calculated.

The principle business and technical objectives of the Chemical Pilot are:-

- demonstration of IEC1158-2 [1] low speed fieldbus segments running from the plant control rooms into hazardous areas. The architecture also includes a high speed fieldbus running between the plant control rooms to integrate measurements into the site wide management information systems.

- demonstration of device interoperability using devices sourced from several suppliers conforming to Profibus-PA device profiles [2]

- automation of manual readings with the associated improved accuracy of data

- real time integration of fieldbus instrumentation with business planning functions

- immediate alarms to a central location should a measurement be lost

- practical experience of design, installation, commissioning and operating fieldbus

This will be achieved through the design, configuration and simulation of a suitable fieldbus network using the ACORN 1479 toolkit and it's subsequent implementation and trial.

5.2 Power Pilot

The ACORN project will also implement a Power Pilot application. This will consider the impact of fieldbus on the control of the steam generation cycle of conventional power stations as illustrated in Fig. 6 below.

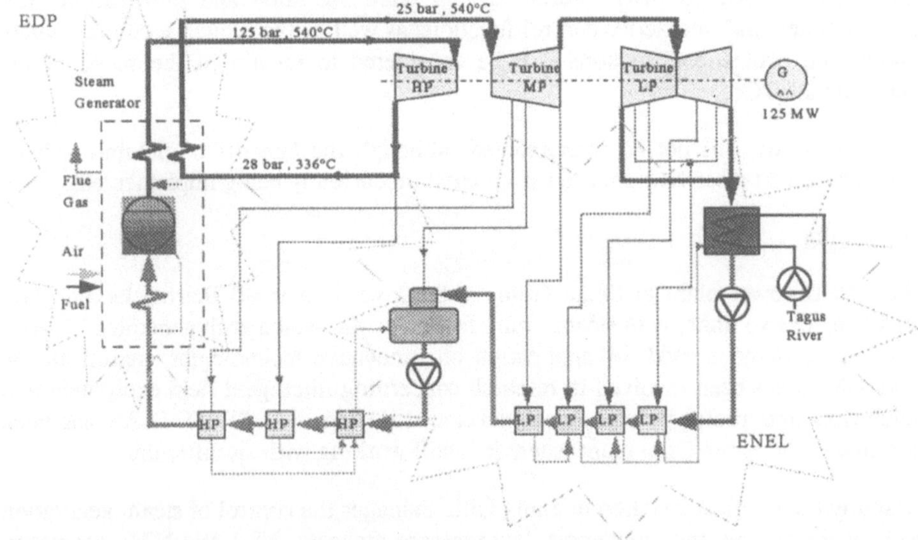

Fig. 6 Power Pilot Application

<u>EDP Pilot</u>

The objective of EDP Power Pilot Application is to apply and assess the toolkit developed in the ACORN Project. It will be used as an engineering support tool for fieldbus technology requirements during the up-front design of a fieldbus based control system in the context of ongoing refurbishment of the Carregado Power Station.

The fieldbus design study will be applied to a selected set of plant areas of the steam generator, which comprise:

- the boiler feedwater system

- the superheated steam attemperation

- the fuel oil heating and pressure control station

- the auxiliary steam delivering system

- the steam pre-heating of the combustion air

The selection of those plant areas was mainly based on the ability to define several clusters of field devices, closely located in the installation, and concerning to control loops with low functional interaction to each other and to the other plant areas.

Through this way, the advantages from the distributed control concept, by means of the fieldbus technology, may be better exploited and profited.

The fieldbus based control system will be conceived as integrated within the conventional DCS, to which it shall interface. The conventional DCS will provide an integrated platform for user interfacing and plant operation and information. The more complex and interactive control functions as well as the sequence control, safety related and protection functions will be considered to remain in the scope of the conventional DCS.

The EDP study will not be implemented, although the benefits of fieldbus will be assessed against a conventional reinstrumentation currently being implemented.

ENEL Pilot

The ENEL power plant at Santa Gilla, in Sardinia, is a small thermoelectric plant consisting of two units, with steam drum, fuelled by oil, which each generates 35 MW. This power plant is used for application on innovative technologies, mainly in the field. ENEL has been involved in research concerning intelligent field equipment and fieldbusses and prototype equipment developed during the ESPRIT DIAS has been installed at the Santa Gilla plant where it is still working without difficulty.

Presently the fieldbus installed in Santa Gilla manages the control of steam generation cycle of the second unit: condenser, low pressure preheater N° 1 and N° 2, deareator, feedwater pumps and high pressure heater N° 4. The intelligent field devices include level, temperature, pressure, flow transmitters and several control valves.

The fieldbus application at the moment used in the Santa Gilla control sub-system is based on one of the first releases of FIP protocol and consequently, from the technical and normative points of view, it does not comply with the CENELEC Standard EN 50170-3 (i.e. WorldFIP).

The field devices are still working well and consequently there are no compelling reasons to change them. The existing old FIP application will be maintained and a new one will be implemented with the following characteristics:

- compliance with the European Fieldbus Standard EN 50170 [3]. In particular, physical layer and communication rules of the EN 50170-3, i.e. WorldFIP

- installation a range of intelligent devices currently available on the market (multiplexor, motor control centre, modulating actuator, intelligent temperature transmitter, robot video camera are all being considered).

- One control loop of the existing control system will be implemented in the new ACORN pilot providing availability of suitable devices can be confirmed.

The tools developed in the ACORN project will be used for this application in Santa Gilla and are considered by ENEL as a reference for designing and configuring of future plants automation systems.

6 Validation

A validation exercise will be carried out to measure quantitative (objective) benefits and assess the qualitative (subjective) benefits of fieldbus technology in general, and the ACORN 1479 design toolkit in particular.

The benefits from fieldbus throughout the plant lifecycle will be considered based on the experience of implementing the pilot applications.

- Design Phase. The impact of fieldbus on the design process is expected to be significant, and the ACORN toolkit is expected to help by improving efficiency and design integrity. Smart instrumentation will also affect the design process.

- Installation and Commissioning Phase. Significant savings are expected from plant wiring costs and installation time. Commissioning of smart instruments should be simple with tool support.

- Operation. Improved reliability, accuracy and availability of measurements is expected. Simulation tool results will be compared against actual results.

- Maintenance. Advanced developments in sensor validation and device diagnostics will be assessed. Ability to maintain devices and fieldbus during operation will be considered.

- Human Factors. The project will review the skills, training required to deploy fieldbus successfully

The results of this validation exercise will be disseminated by the project in due course.

7 Conclusions

Fieldbus technologies are maturing rapidly. The technology will give benefits but there are associated risks, and End Users need to understand the impact that it will have on their design processes. A strategy for introducing fieldbus needs to consider *system* and *lifecycle* implications of the technology, and small scale pilot applications can be effective in building confidence and understanding.

We believe that the availability of tools like ACORN 1479 and the widespread dissemination of information relating to the experiences of practical implementation of the pilot applications will help facilitate early introduction of a very promising technology which has been a long time in gestation.

The ACORN 1479 project would welcome any comments on the information presented in this paper.

References

1. IEC 1158-2, Fieldbus Physical Layer Standard

2. Profibus-PA-Profile for Process Control Devices Class A and B (draft), PNO, 1996

3. CENELEC EN50170 General Purpose Fieldbus Communication Systems, EN50170-1 P-NET, EN50170-2 Profibus, EN50170-3 WorldFIP

Java und LonWorks - von der Fernsteuerung zur Visualisierung

Heinrich Reiter*, Christian Kral**, Marcus Carotta*

* Institut für Automation, Treitlstr. 1, A-1040 Wien
Tel.: +43-1-58801-3986, Fax: +43-1-5863260, Email: hreiter@auto.tuwien.ac.at

** HTC Kral, Malfattig. 1/2/1, A-1120 Wien
Tel.+Fax: +43-1-8100269, Email: Kral@aon.at

Kurzfassung. Ziel dieses Beitrags ist es, eine Realisierung eines Gateways zwischen Java und LonWorks zu beschreiben[1]. Der Zweck dieses Gateways liegt im Bereich komfortabler Fernsteuerungen, es lassen sich aber mit geeigneten Verfahren auch Anwendungen zur Visualisierung, Konfiguration oder gar Wartung von LonWorks-Kontrollnetzen entwickeln. Bei der Implementierung wurde viel Wert auf Einfachheit und Nachvollziehbarkeit gelegt; ebenso wurden Standardentwicklungsumgebungen und weit verbreitete Entwicklungsplattformen verwendet. Sicherheitsaspekte und ein Blick auf zukünftige Java-basierte Technologien zur Fernsteuerung von Kontrollnetzen runden die Arbeit ab.

Abstract. The objective of this paper is to describe the implementation of a Java-LonWorks gateway[1]. Such a gateway has a lot of applications, like comfortable remote control, visualisation, configuration or maintenance of LonWorks networks. We want to point out that the implementation is easy to understand; common platforms and development environments are used in order to decrease development costs. Security and Java-based future developments are discussed at the end of this paper.

1 Einleitung

Im Jahre 1996 war die Fernsteuerung von LonWorks-Netzwerken mit Java nur eine Vision[2], mittlerweile gibt es schon ausgereifte Produkte am Markt zu erstehen: Ein kalifornischer Hardware-Hersteller bietet einen LonWorks-fähigen Web-Server an, der

---

[1] Dieser Arbeit liegt das vom Fonds zur Förderung der wissenschaftl. Forschung geförderte Forschungsprojekt HomeNet Configuration Tool (Projektnr. P-10699 ÖMA) zugrunde. Dieses Projekt wird zusammen mit dem Institut für Computertechnik der TU Wien durchgeführt.
[2] Scott McNealy, CEO von Sun Microsystems, deutete in seiner Ansprache zur „Spring 1996 LonUsers Conference" an, daß die Java-basierte Fernsteuerung von Kontrollnetzwerken eine vielversprechende Zukunft hat.

mittels Java-Technologie einen komfortablen Fernzugriff auf LON-Knoten erlaubt. Ein ähnliches Produkt ist aus Schweden zu erwarten.

Für manche Kontrollnetzwerke ist eine reine Softwarelösung oft schon ausreichend. Ein gewöhnlicher PC, ausgestattet mit Windows 95 (oder NT) und einem (als Shareware frei erhältlichen) Web-Server, ist die Basis, um eine komfortable Fernsteuerung, Visualisierung oder Konfiguration von LonWorks-Netzwerken zu gewährleisten. Der Vorteil einer reinen Software-Lösung ist, abgesehen von den geringeren Kosten bei der Anschaffung, die leichte Wartbarkeit der jeweiligen Programmteile. Diese Arbeit will die grundlegenden Mechanismen der Interaktion zwischen Java und LonWorks erklären, wobei viel Wert auf leicht nachvollziehbare Methoden bei Entwurf und Implementierung gelegt wurde.

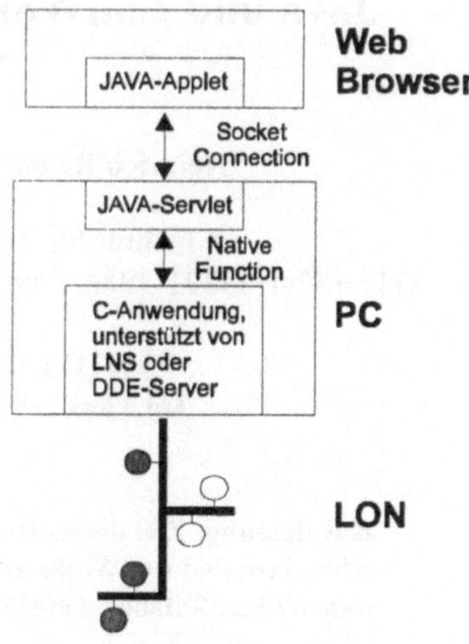

Abb. 1 Schema eines Java-LonWorks Gateways

2 Von Java zu LonWorks

Java ist eine interpretierte Programmiersprache, das heißt, sie ist hardware-unabhängig. Das einzige, was eine Java-Anwendung benötigt, ist ein Java-Interpreter, der mit Hilfe des jeweiligen Betriebssystems die Java-Befehle ausführt[3]. Mittlerweile gibt es für jede erdenkliche Plattform einen Java-Interpreter, sodaß das Prinzip „write once, run anywhere" voll ausgeschöpft werden kann.

Entgegen den ursprünglichen Erwartungen ihrer geistigen Väter hat sich die Programmiersprache Java nicht zu einer universellen Sprache für Haushaltsgeräte, sondern zum state-of-the-art für die Gestaltung von interaktiven Web-Seiten entwickelt.

Um Kontrollnetzwerke fernzusteuern, zu visualisieren und zu warten, ist Java besser geeignet als bloße HTML-Seiten, die zwar viele Möglichkeiten für komfortable Benutzerschnittstellen bieten, jedoch aufgrund umfangreicher *Common Gateway Interface* Programmierung den Entwicklungsaufwand in die Höhe treiben.

---

[3] Der Interpreter ist nichts anderes als die sogenannte *Java Virtual Machine*. Darunter versteht man den in Software realisierten Java-Prozessor (siehe [5]), der konzeptionell als *Stack Machine* aufgefaßt werden kann.

Abb. 1 zeigt den schematischen Aufbau des Java-LonWorks Gateways. Der Benutzer verwendet einen gewöhnlichen Web-Browser, um das Java-Applet[4] vom Web-Server, der am PC läuft, zu laden. Das Applet nimmt nun Kommandos vom Benutzer entgegen und schickt diese über eine *Socket Connection* zum Java Servlet. Das Servlet aktiviert letztendlich eine Windows-Anwendung, die auf den LonWorks-Treiber (das kann die LonWorks Component Architecture, der DDE-Server oder ähnliches sein) direkt zugreift.

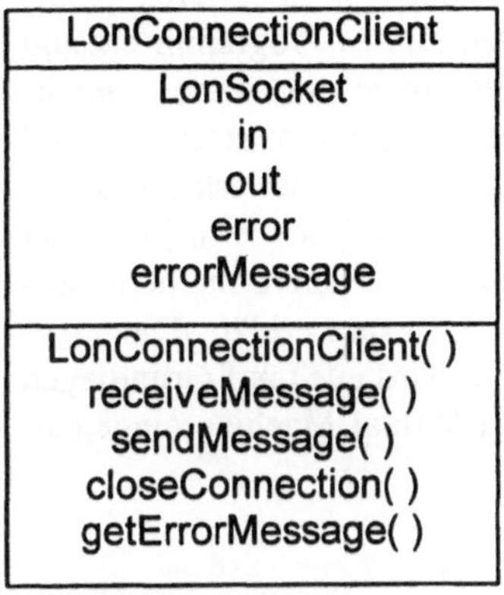

2.1 Applets, Servlets, Sockets: Internet-Kommunikation mit Java

Die Kommunikation zwischen Applet und Servlet basiert auf Sockets. In [2] findet man einen guten Überblick über die Sockets-Programmierung mit Java, in [1] kann man anhand einer konkreten Implementierung eines Java-LonWorks Gateways Teile des Sourcecodes nachlesen.

LonConnectionClient
LonSocket in out error errorMessage
LonConnectionClient( ) receiveMessage( ) sendMessage( ) closeConnection( ) getErrorMessage( )

Abb. 2 Eigenschaften und Methoden der Klasse LonConnectionClient

Auf der Client-Seite (im Web-Browser) gibt es eine Java-Klasse, die die entsprechenden Methoden zur Sockets-basierten Kommunikation mit der Server-Seite (PC) beinhaltet. Analog sieht es beim Servlet aus: Auch hier existiert eine eigene Klasse, die die Kommunikation mit dem Applet regelt. Abb. 2 und Abb. 3 zeigen die unterschiedlichen Schnittstellen der Kommunikationsklassen bei Servlet und Applet. Während auf der Client-Seite lediglich Methoden zum Senden und Empfangen vorhanden sind, findet man beim Servlet die Funktion *main( )*, die für eine außerhalb eines Web-Browsers lauffähige Java-Anwendung unbedingt notwendig ist. Das Servlet hat in unserem Fall nur die Aufgabe, auf einlangende Befehle zu warten und entsprechend zu reagieren.

Die Verbindung zwischen Applet und Servlet ist somit auf einfache Weise realisiert.

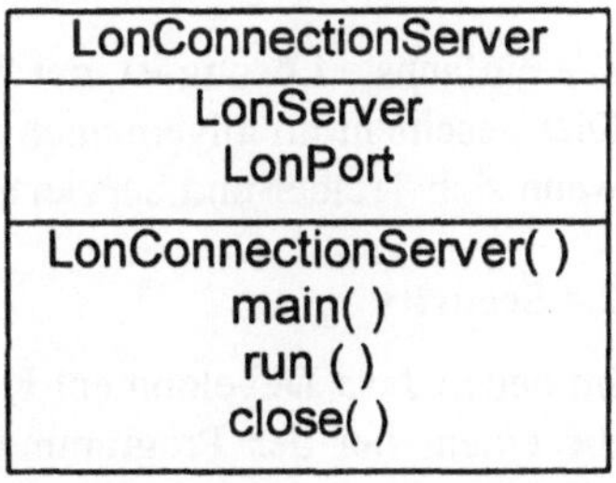

LonConnectionServer
LonServer LonPort
LonConnectionServer( ) main( ) run ( ) close( )

Abb. 3 Eigenschaften und Methoden der Klasse LonConnectionServer

---

[4] Das Wort *Applet* leitet sich von *Application* ab, ebenso wie sich sich *Servlet* von *Server Application* ableitet. Da *Applets* und *Servlets* im allgemeinen relativ kleine Anwendungen sind, hat man sich auf diese sprachliche Verniedlichung geeinigt.

2.2 Native Methods: Java und C arbeiten zusammen

Unter einer *native method* versteht man eine Methode, die zwar in Java deklariert, jedoch in einer anderen Programmiersprache implementiert ist. In Java kann man (in der Version 1.0.2) Funktionen der Sprache C in Java-Anwendungen integrieren. Dabei wird man von einem Werkzeug namens *javah* unterstützt. Dieses Programm erzeugt aus der Deklaration der native method ein C-Headerfile sowie eine Datei namens stubs.c. Danach ist es nur mehr erforderlich, die C-Funktion selbst zu schreiben und daraus mit einem geeigneten Compiler, unter Zuhilfenahme des generierten Header-files sowie stubs.c, eine Dynamic Link Library zu erzeugen. In [1] sowie in [6] findet man eine konkrete Implementierung einer Java-Anwendung, die native methods aufruft, in [2] sind sämtliche dabei verwendeten Mechanismen exakt erklärt. Die Java-interne Vorgehensweise wird in [5] exakt beschrieben, hier wird auch auf Grundlagen der Implementierung der native methods seitens der Java Virtual Machine eingegangen.

2.3 Windows und LonWorks

Hat man einmal von Java aus eine C-Funktion aufgerufen, so ist die Verbindung zu LonWorks nicht mehr weit[5]. Idealerweise greift die native method selbst auf den Lon-Works-Treiber zu (das wäre beispielsweise mit dem LonManager DDE Server denkbar). Falls die Treiberanwendung zu komplex ist (und man es dem Java-Laufzeitsystem nicht zumuten will, eine riesige C-Anwendung einzubinden), so kann mittels diverser Windows-Kommunikationsmechanismen eine parallel zum Java-Servlet laufende Anwendung gesteuert werden. Insbesondere die Möglichkeit, unter Windows NT mit Hilfe von *named pipes* das Servlet und den LON-Treiber auf zwei unterschiedlichen Rechnern laufen zu lassen, klingt für LANs sehr interessant.

Die einfachste Lösung ist, der Treiberanwendung eine Windows-Message zu schicken: Dies geschieht im allgemeinen schnell und zuverlässig, allerdings funktioniert dies nur, wenn sich Treiber und Servlet auf ein und demselben Rechner befinden.

2.4 Security

Im neuen Java Development Kit (Version 1.1.1) findet man eine Vielzahl an Klassen, die einem bei der Programmierung von Verschlüsselungs- und Signaturprozeduren behilflich sind. Diese bauen auf dem Secure Sockets Layer der Firma Netscape Communications Corp. auf, womit eine abhörsichere Übertragung von heiklen Daten möglich wird. In [1] findet man einen Überblick über derzeitige Entwicklungen auf dem Sektor der sicheren WWW-Protokolle, in [9] werden grundlegende Sicherheitsaspekte von Java behandelt. Man darf sich als Entwickler des Applets nicht nur auf Verschlüsselung und Authentifizierung verlassen; schon beim Laden des Applets muß sichergestellt werden, daß es das richtige (d. h. eigene) Applet ist. Ein „böses" Applet

---

[5] Verwendet man als Treiber die LonWorks Component Architecture (siehe [4]), so kann man auf die native method gänzlich verzichten, da es hier möglich ist, direkt von Java aus auf das LON zugreifen zu können. Siehe dazu [8].

könnte wichtige Information vom Benutzer abfragen, während dieser im Glauben lebt, seine Fernsteuerungsbefehle gelangen an die richtige Adresse, nämlich ins LON.

Es soll an dieser Stelle darauf hingewiesen werden, daß heikle Daten aus dem LON nie unverschlüsselt über das Internet geschickt werden sollten.

3 Ein Beispiel: Java-Applet schaltet Lampe ein

Gehen wir von dem Beispiel aus, daß der Benutzer des Java-Applets eine per LON ansteuerbare Lampe einschalten will. Zu diesem Zweck wählt er den entsprechenden Befehl im Applet aus (man denke sich einen *Button*, der den Schalter repräsentiert).

Danach geschieht folgendes:

1. Das Applet schickt den (verschlüsselten) Befehl „Lampe einschalten" an das Servlet.
2. Das Servlet empfängt den Befehl und invoziert eine Klasse, die sich fortan um die Aktivierung des LonWorks-Treibers kümmert.
3. Das Servlet beginnt wieder, auf einlangende Befehle zu warten. Durch Multi-Threading ist das technisch möglich.
4. Die Klasse, die für den Zugriff auf den LonWorks-Treiber verantwortlich zeichnet, ruft eine *native method* auf und wartet auf eine etwaige Fehlermeldung.
5. Die *native method* schickt der LON-Treiber-Anwendung eine Windows-Message mit dem Inhalt „Lampe einschalten".
6. Die Message wird von der Anwendung empfangen, und die Lampe wird eingeschaltet.
7. Eine Rückmeldung („Lampe ist eingeschaltet") seitens der Anwendung kann über eine (hier nicht behandelte) Windows-Message in die umgekehrte Richtung erfolgen. Die Dynamic Link Library muß zu diesem Zweck eine Methode für die Verwaltung von Messages besitzen.
8. Die *native method* liefert den Rückgabewert an die LON-verantwortliche Klasse zurück.
9. Das Servlet schickt diesen (nun verschlüsselten) Rückgabewert an das Applet, worauf jenes die Meldung „Lampe ist eingeschaltet" ausgibt.

4 Die Zukunft: VRML, Java-Prozessoren und Neuron Java

Dreidimensionale Visualisierung wird bei der Wartung und Konfiguration von Kontrollnetzwerken bald selbstverständlich sein. Die *Virtual Reality Modelling Language* (VRML) ist dabei jene Sprache, die schon jetzt grafische Modellierung in drei Dimensionen erlaubt - und das in den gängigsten Web-Browsern. VRML mit Java zu verbinden ist in [7] genau beschrieben. Es kann hier aus Platzgründen nicht näher auf VRML eingegangen werden, jedoch soll der Quantensprung in der Benutzerfreundlichkeit, der sich durch Einsatz von VRML ergibt, nicht unerwähnt bleiben. Gebäude lassen sich „virtuell" fernsteuern, und komplizierte Anordnungen im Kontrollnetzwerk eines Fertigungsbetriebes werden mit Hilfe der dritten Dimension leichter handhabbar.

Ein weiterer Punkt in der Reihe der zukünftigen Entwicklungen ist der Einsatz von Java-Prozessoren (siehe [10]) im LON. Diese könnten durchaus den PC als Gateway zum WWW verdrängen. Ein Java-Prozessor, der einen Neuron Chip als Kommunikationsmittel verwendet, würde neue Dimensionen in der Entwicklung von verteilten Anwendungen eröffnen.

Bis man eine Java Virtual Machine **im** Neuron Chip selbst vorfindet („Neuron Java"), wird aber noch eine lange Zeit vergehen - die Speicheranforderungen sowie die hohe erforderliche Rechenleistung lassen solch eine Vision (vorerst) nicht zu.

Literatur

1. Carotta, M.: HomeNet Control - Internetbasierende Fernsteuerung von Heimnetzen unter Berücksichtigung von Sicherheitsaspekten. Diplomarbeit, Institut für Automation, Technische Universität Wien, Österreich 1997.

2. Bartlett, N. et al.: JAVA Programming Explorer. Scottsdale, AZ, USA. The Coriolis Group, 1996.

3. Gamma, E. et al.: Design Patterns, Elements of Reusable Object-Oriented Software. Addison-Wesley, 1995.

4. Blomseth, R. et al.: The LonWorks Network Services (LNS) Architecture Technical Overview. Echelon Corp., Doc. Nr. 39310, Palo Alto, USA, 1996.

5. Lindholm, T. et al.: The Java Virtual Machine Specification. Addison-Wesley, 1997.

6. Reiter, H.: Remote Control of LonWorks Networks with Java. Proceedings of the Spring 1997 LonUsers International Conference, Santa Clara, CA, USA, 1997.

7. Lea, R. et al.: JAVA for 3D and VRML Worlds. New Riders Publishing, Indianapolis, USA, 1996.

8. Raji, R.: Creating LNS Applications with Java. Proceedings of the Spring 1997 LonUsers International Conference, Santa Clara, CA, USA, 1997.

9. McGraw, G. et al.: Java Security - Hostile Applets, Holes, and Antidotes. Wiley Computer Publishing, New York, 1996.

10. Wayne, P.:Java Chips. In: Byte 11/96, S. 79, McGraw-Hill, USA, 1996.

Kopplung von LONWORKS-Netzen mit verteilten Java-Applikationen

Jürgen Gausemeier, Gerrit Gehnen

Heinz Nixdorf Institut, Universität-GH Paderborn
Fürstenalle 11, D-33102 Paderborn, Deutschland

Abstract: In diesem Beitrag wird eine Kopplung von LONWORKS-Netzen auf der Basis von LONMARK-Objekten mit verteilten Java Anwendungen unter Verwendung von JavaBeans-Komponenten vorgestellt. Die LONMARK-Objekte werden dabei auf einzelne Beans abgebildet. Durch die Kopplung auf Objektbasis wird der Aufwand für die Erstellung der Schnittstelle im Gegensatz zu einer Kopplung auf Variablenbasis erheblich reduziert. Als Beispiel wird die Integration eines Materialflußsystems in eine verteilte Fertigungssteuerung präsentiert.

This paper discusses the integration of LONWORKS-networks and distributed Java-Applications using LONMARK-objects and JavaBeans components. The LONMARK-objects are mapped on JavaBeans. With integration on object level the effort for creation of the interface is reduced significant in comparison to integration on networkvariable level. Example for integration is a distributed material flow control system connected to a manufacturing control system.

1 Problematik

Mit den Möglichkeiten, die durch die rasche technologische Entwicklung im Bereich der Mikroelektronik eröffnet werden, steigen auch die Anforderungen an Systeme der Automatisierungstechnik. Während früher nur Steuerungsfunktionen in abgeschlossenen Systemen gefordert wurden, kommen mit der wachsenden Vernetzung immer mehr Anforderungen der Systemintegration in unternehmensweite Datenverarbeitungsnetze auf. Voraussetzung für eine Integration ist eine Vernetzung, die bis hinunter zu den einzelnen Sensoren und Aktoren reicht. SPS-Steuerungen bieten heute die Möglichkeit, in übergeordnete Netze integriert zu werden, einzelne Systeme integrieren sogar WWW-Server in der SPS, um den Prozeß zu überwachen und Eingriffmöglichkeiten bereitzustellen.
Mit den wachsenden Anforderungen an die Qualitätsdokumentation, die Realisierung flexibler Fertigungsprozesse mit Losgröße eins und die steigende Komplexität der Fertigungsprozesse werden SPS-Steuerungen jedoch immer mehr zum Flaschenhals. Das

---

LON, LONWORKS, LONTALK, LONMARK sind (eingetragene) Warenzeichen der Echelon Corp.; Java, JavaBeans sind (eingetragene) Warenzeichen der Sun Microsystems, Inc.

Konzept der zyklischen Verarbeitung, die SPS-Steuerungen eigen ist, kommt immer mehr an Grenzen. Auch schnelle Feldbussysteme können an dieser Stelle nur lindernd wirken, die grundsätzlichen Probleme beheben sie nicht. Mit dezentralen, intelligenten Automatisierungssystemen auf der Basis von CAN und LONWORKS wurden in den letzten Jahren erfolgreich sehr flexible und leistungsfähige Automatisierungssysteme geschaffen, die zeigen, daß der dezentrale intelligente Automatisierungsansatz für viele Anwendungen Vorteile bringt.

Nachteil der dezentralen intelligenten Ansätze ist, das eine Vielzahl von Daten verteilt in der Anlage erzeugt werden, die für Anwendungen außerhalb des Automatisierungsnetzes relevant sind. So muß z.B. eine Prozeßvisualisierung in einem LONWORKS-Netz Verknüpfungen zu fast allen Knoten aufnehmen, wenn sie den Anlagenzustand umfassend darstellen soll. Auch die Steuerung der Anlage nach vorgegebenen Anweisungslisten oder Rezepturen erfordert eine Verteilung von Daten z.B. von einem Datenbankserver in das Netz hinein. Hier sind intelligente Lösungen gefragt, um die hier entstehende Komplexität zu beherrschen.

Die Kommunikation von Knoten der dezentralen intelligenten Automatisierung erfolgt mit Hilfe eines Kommunikationsprotokolls. Dieses Kommunikationsprotokoll, bei LONWORKS z.B. das LONTALK-Protokoll, ist nach dem ISO/OSI 7-Schichten Standard spezifiziert. Damit ist eine reibungslose Kommunikation zwischen den einzelnen Teilnehmern gewährleistet. Da jedoch mit dem Protokoll die Strukturierung der Schnittstellen der Teilnehmer und die Semantik der Daten nicht definiert ist, müssen für den praktischen Betrieb weitere Vereinbarungen getroffen werden. Erst solche Vereinbarungen erlauben es, mit einfachen Werkzeugen Knoten zu Netzen zusammenzufügen. Fehlen solche Vereinbarungen, ist bei jeder Anlage erneut ein immenser Aufwand zur Konfiguration aufzubringen, der heute nicht mehr akzeptabel ist.

2 Lösungselemente

Für LONWORKS sind solche Vereinbarungen mit dem LONMARK-Standards getroffen. Diese Standards, die von der LONMARK Interoperability Association definiert werden, schaffen Objekte mit klar umrissenen Schnittstellen und Verhalten. [5]
Ein Knoten, bestehend aus Prozessor, Feldbusinterface, Prozeßkopplung und der auf dem Prozessor laufenden Software, wird in der Software als Set von einem oder mehreren Anwendungsobjekten und einem Knotenobjekt strukturiert. Das Knotenobjekt übernimmt die Verwaltung des Knotens. Die Anwendungsobjekte können Sensoren, Aktoren oder Controllerobjekte sein. Durch diese Strukturierung wird der Anwender in die Lage versetzt, Knoten ohne Programmieraufwand durch das Kombinieren von einzelnen Objekten zu gestalten. Dabei müssen die Objekte nur noch mit einem graphischen CASE-Tool aus einer Bibliothek ausgewählt und untereinander verbunden werden.
Zur Vernetzung von LONWORKS-Netzen mit übergeordneten Anwendungen gibt es eine Vielzahl von verschiedenen Möglichkeiten. Es bietet sich an, Verfahren anzuwenden, welche die Struktur von LONWORKS möglichst gut abbilden und gleichzeitig eine universelle Verwendbarkeit auf möglichst vielen Plattformen erlauben.

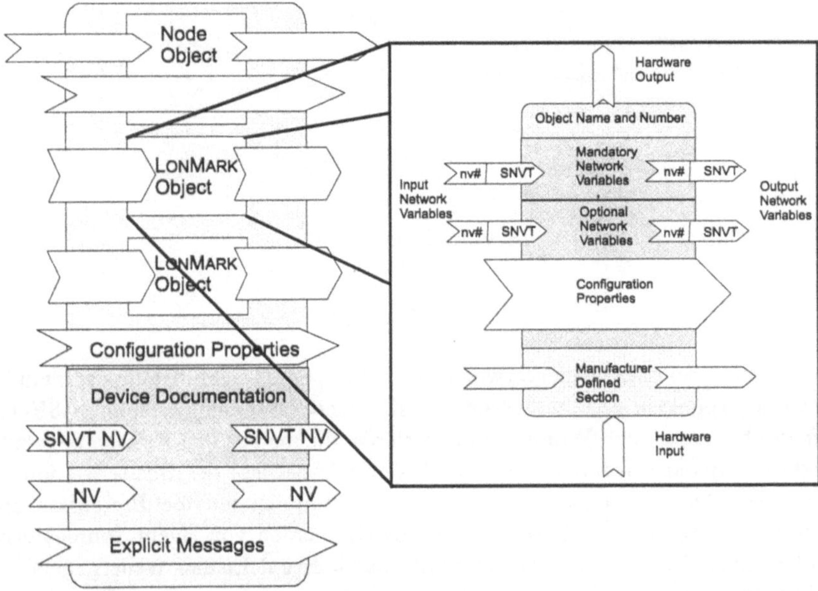

Fig. 1. LONMARK-Knoten: Softwaresicht auf den Gesamtknoten und LONMARK-Objekt

In den letzten Jahren wurde in vielen Bereichen die Programmiersprache Java propagiert. Java, ursprünglich für Embedded Systeme wie Set-Top-Boxen konzipiert, hat durch die Verbreitung des WWW und die Unterstützung von sogenannten Applets in WWW-Dokumenten starke Beachtung gefunden. Diese Applets, kleine Programme, die lokal im Web-Browser ablaufen, stellen aber nur einen Aspekt von Java dar. Weitgehend unbeachtet ist geblieben, daß mit Java und der RMI-Bibliothek (RMI=Remote Method Invocation) verteilte Anwendungen möglich sind, die über TCP/IP-Netze kommunizieren [7]. Dabei sind Visualisierungen in Web-Browsern, Zugriffe auf Datenbanken auf entsprechenden Servern und paralleles Rechnen in Netzwerken ohne großen Entwicklungsaufwand möglich. Die Kommunikation zwischen den verteilten Programmteilen erfolgt über Ereignisse, die über das TCP/IP-Netzwerk ausgetauscht werden

Die Objektorientierung von Java erlaubt eine weitgehende Kapselung von Funktionalitäten mit definierten Schnittstellen. Um den Entwicklungsprozeß weiter zu beschleunigen, wurde auf der Basis von Java mit JavaBeans eine Komponentenarchitektur entwickelt. Die JavaBeans-Architektur, definiert die Schnittstellen und das Verhalten von wiederverwendbaren Java-Komponenten [6]. Diese Beans lassen sich mit Hilfe von graphischen Entwicklungswerkzeugen mit nur wenig zusätzlicher Applikationslogik zu vollständigen Anwendungen zusammenführen [1].

218

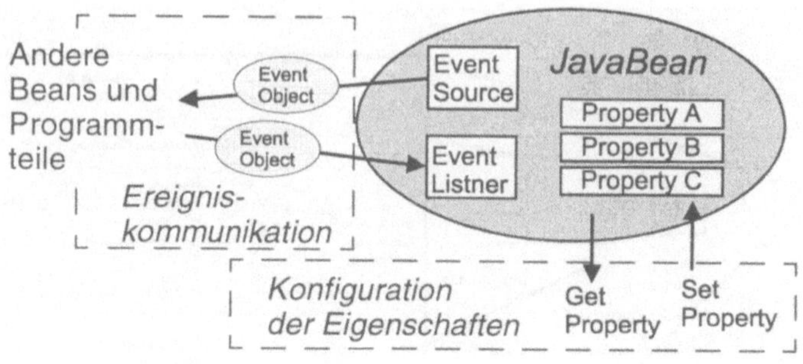

Fig. 2. JavaBean

Ein Bean ist dabei eine wiederverwendbare Klasse, deren Schnittstelle nach einheitlichen Gesichtspunkten gestaltet wurde. Die Klassenbeschreibung erlaubt CASE-Tools,
automatisch Adapter und Wrapper zu generieren, die die aus den Klassen abgeleiteten
Objekte miteinander verbinden. Eigenschaften (Properties) der Beans werden dabei
über spezielle Operationen konfiguriert. Beans kommunizieren über Ereignisse, die als
Event-Objekte gestaltet sind. Die Event Objekte werden von Event Sources erzeugt
und über einen Kommunikationsmechanismus an Event Listner weitergegeben. Die
Verknüpfung von Ereigniserzeugern (Event Source) und Ereignisempfängern (Event
Listner) erzeugt einen großen Anteil der Applikationslogik. Diese Verknüpfungen
können automatisch mit zugehörigen Adapterobjekten von CASE-Tools erzeugt werden. Allerdings läßt sich mit diesem Ansatz in der Regel nicht die volle Funktionalität
einer Applikation implementieren. An dieser Stelle müssen noch von Hand weitere
Objekte hinzugefügt werden.
Die Ähnlichkeit zu LONWORKS und LONMARK-Objekten ist auffallend. In beiden
Systemen wird die Applikationslogik durch die Verknüpfung von Funktionseinheiten
mit klar definierter Schnittstelle durch ein CASE-Tool generiert. Die Kommunikation
erfolgt über Ereignisse, die als Multicast von einem Sender an mehrere Empfänger
weitergeleitet werden.

3 Kopplung über Netzwerkvariablen

Für die Verknüpfung von LONWORKS-Netzen mit Java gibt es eine Reihe von verschiedenen Möglichkeiten. Die sicherlich einfachste Möglichkeit ist, einfache Java-
Applets zu schaffen, die direkt mit einer Netzwerkvariable im LONWORKS-Netz über
einen Interface-Knoten kommunizieren [3]. Auf dem Interface-Knoten ist ein Server
installiert, der ein proprietäres Protokoll auf TCP/IP-Basis implementiert. Die Kommunikation zwischen dem Interfaceknoten und dem Applet erfolgt über dieses Protokoll. Die Applets übernehmen durch ihre Verknüpfung mit nur einer oder wenigen
Netzwerkvariablen Funktionen der Visualisierung oder der Steuerung wie z.B. eine
Analogeingabe über einen virtuellen Drehknopf. Auf Grund der Sicherheitsmechanismen von Java muß das Applet entweder vom Interfaceknoten geladen worden sein,
oder lokal auf dem Rechner mit dem WWW-Browser installiert sein, um die TCP/IP-
Kommunikation zu ermöglichen. Da für die Installation auf dem Interfaceknoten ein

vollständiger WWWW-Server notwendig ist, wird im Normalfall das Applet lokal installiert. Sicherheitsmechanismen müssen vom Programmierer dieser Lösung selber implementiert werden.

Da die Struktur der Kopplung auf Netzwerkvariablenebene erfolgt, ist ein grundlegendes Verständnis über die semantische Bedeutung der einzelnen verbundenden Variablen notwendig. An dieser Stelle tritt das gleiche Problem auf, das bereits oben bei der Vernetzung einer dezentralen intelligenten Automatisierung beschrieben worden ist. Die Zahl der zu verknüpfenden Variablen, deren semantische Bedeutung unklar ist, führt zu zusätzlichen Aufwänden in der Entwicklung und Wartung der Anwendung.

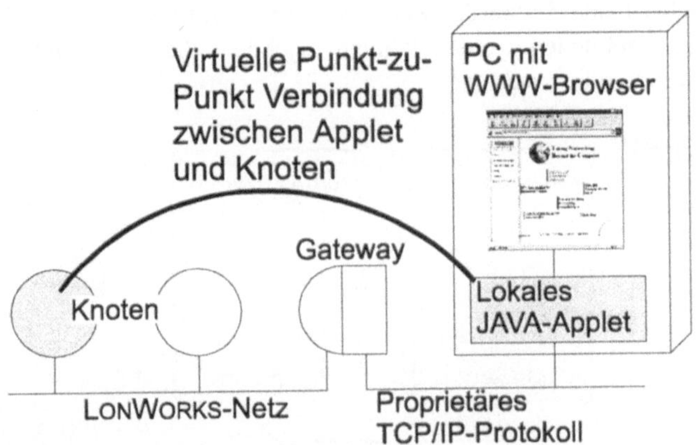

Fig. 3. Architektur einer einfachen Kopplung

4 Kopplung auf der Basis von Objekten

Um diesen Aufwand zu reduzieren, benötigt man ein konsistentes Objektmodell auf beiden Seiten der Schnittstelle. Die Konsistenz des Objektmodells zwischen der LON-WORKS-Ebene und der Java-Ebene erreicht man, wenn man auf LONWORKS-Ebene LONMARK-Objekte einsetzt, die im Netz verteilt operieren und auf Java-Ebene die Beans-Komponentenarchitektur benutzt.

Die RMI-Architektur basiert darauf, daß sich Nachrichtenempfänger bei den Sendern in einer Liste registrieren. An die Liste der Empfänger wird bei der Erzeugung einer Nachricht ein Nachrichtenobjekt verschickt. Wenn der Empfänger auf einem anderen Rechner installiert ist, wird die Nachricht über Stub-Objekte, welche die Stellvertreterfunktion für die Empfänger übernehmen und RMI-Server, die auf jedem Rechner installiert sind, weitergeleitet.

Das Interface zwischen beiden Netzen muß möglichst transparent sein. Die Unterschiede in den einzelnen Architekturen führen dazu, daß eine transparente Abbildung nur in eine Richtung mit vertretbarem Aufwand möglich ist. Um die Integration der Feldebene in übergeordnete Systeme zu unterstützen, werden hier LONMARK-Objekte auf JavaBeans abgebildet. Dabei stützt sich das Interface auf eine parametrierbare Host-Applikation, deren Schnittstelle zum LONWORKS-Netz und zur verteilten Java-Anwendung konfiguriert werden [4].

Für die Kommunikation bedeutet die gewählte Kopplung, daß die Netzwerkvariablen, die im Netz versandt werden, durch Ereignisobjekte modelliert werden. Durch die Verwendung einer Klassenbibliothek lassen sich mit relativ wenig Aufwand alle SNVT's (Standardnetzwerkvariablentypen) modellieren.

Darauf aufbauend läßt sich eine Klassenbibliothek schaffen, die sich äußerlich sehr LONMARK-Objekten ähnelt. Die Netzwerkvariablenschnittstellen werden durch Ereigniserzeuger und -empfänger modelliert, die Konfigurationsparameter durch Properties, die durch das CASE-Tool eingestellt werden.

Die Abbildung kapselt LONMARK-Objekte in JavaBeans. Dabei bildet das Bean einen Wrapper, der in direktem Kontakt mit dem zugehörigen LONMARK-Objekt steht. Ziel ist, daß möglichst beide Seiten des Interfaces durch graphische CASE-Tools an die jeweiligen Netze gebunden werden können, nachdem die in der Java-Applikation benötigten Daten definiert wurden.

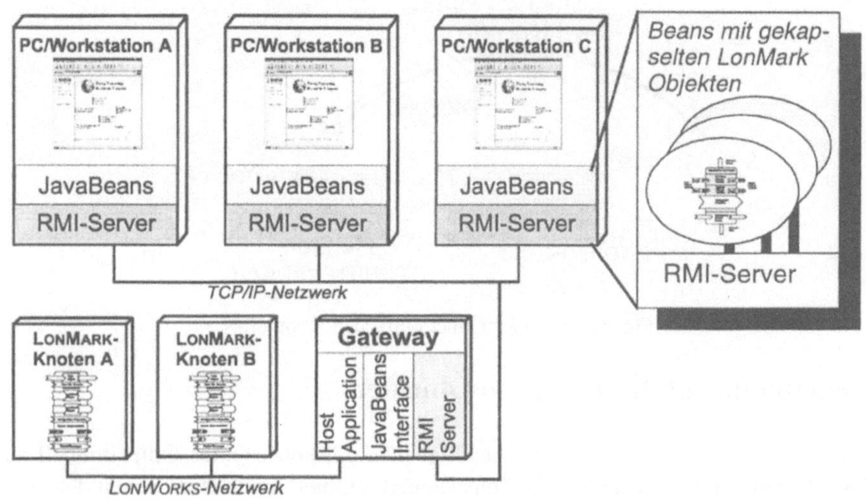

Fig. 4. Kommunikation mit RMI

5 Anwendungsbeispiel

Beispiel einer Kopplung zwischen einem LONWORKS-Netzwerk und Java-Applikationen ist die Integration der LONWORKS-Steuerung einer Förderanlage in eine komplexe Fertigungssteuerung.

Die Steuerung der Förderanlage ist dezentral aufgebaut und wurde bereits an anderer Stelle vorgestellt [2]. Die Förderanlage basiert auf schienengebundenen Fahrzeugen, die in geschlossenen Schleifen kreisen. Wichtig für die Integration ist, daß jeder Haltepunktation, jede Abzweigung und jeder Werkstückträger (Shuttle) mit einem Knoten versehen ist.

Die Förderanlage besitzt verschiedene Schnittstellen zu anderen Systemen im Fertigungsverbund:

- Die Transportaufträge werden von der Fertigungsplanung generiert und durch die Ablaufsteuerung an das Fördersystem übergeben.
- Bei der Ausführung der Aufträge müssen Fertigmeldungen und Störmeldungen zwischen Bearbeitungsstationen (Handhabungsgeräte) und dem Fördersystem in beiden Richtungen ausgetauscht werden. Durch die Anzahl und Ausführung der Stationen müssen unter Umständen eine ganze Reihe verschiedener Schnittstellentypen unterstützt werden.
- Vollzugsmeldungen der Transportaufträge werden an die Fertigungssteuerung übergeben, um weitere Bearbeitungsschritte auszulösen.

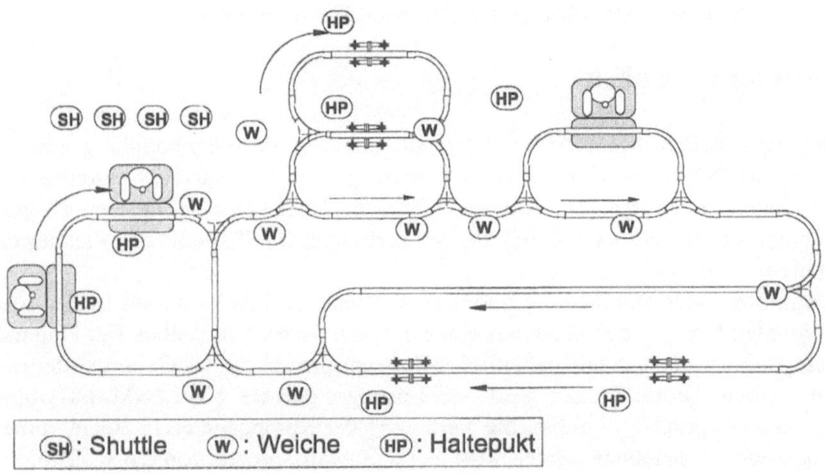

Fig. 5. Beispiellayout der Förderanlage mit Steuerungsknoten

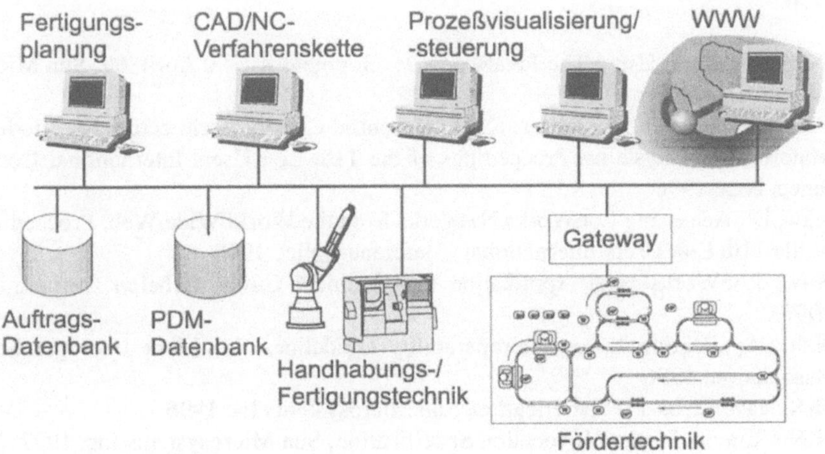

Fig. 6. Struktur des Gesamtsystems

Das überlagerte Fertigungsleitsystem ist auf einem Workstation/PC-Netzwerk implementiert. Um die verschiedenen Bestandteile der Fertigungssteuerung zu koppeln, wird eine Kommunikationsmiddleware verwendet. Werden die relevanten Teile der Anwendung in Java programmiert, bietet sich an, das RMI-Protokoll zu verwenden. Eine komplexe Fertigungssteuerung mit der ressourcenintensiven Fertigungsplanung ist mit heutigen Mitteln aus Performance-Gründen nicht komplett in Java realisierbar, jedoch bietet sich gerade im heterogenen Netz an, insbesondere die Applikationsteile mit Benutzungsoberflächen in Java zu programmieren. Dadurch läßt sich das Leitsystem praktisch von jedem verfügbaren Rechner weltweit bedienen. Die Verwendung von JavaBeans führt dabei zu einer deutlichen Verringerung des Entwicklungsaufwandes der Benutzungsschnittstellen und der internen Kommunikation.

6 Fazit und Ausblick

Mit der vorgestellten Methode wird der Aufwand der Prozeßankopplung erheblich reduziert. LONMARK-Knoten lassen sich ohne großen Aufwand in verteilte Java-Anwendungen integrieren. Der Anwender braucht dabei keine besonderen Programmierkenntnisse, sondern kann durch die ihm vertrauten CASE-Tools die Verknüpfung vornehmen.

Die Definition einer Schnittstelle und die nachfolgende Konfiguration beider Netze mit getrennten CASE-Tools kann nur einen Zwischenschritt darstellen. Für eine nahtlose Integration muß auch ein einheitlicher Entwurfsprozeß für das Gesamtsystem verwendet werden. Dieser Prozeß greift auf Lösungselemente wie LONMARK-Objekte und JavaBeans zurück und nimmt die Verteilung der Komponenten in einem vernetzten System vor. Unterstützt werden wird dieser Entwurfsprozeß von CASE-Tools, die eine automatisierte Verteilung von Objekten vornehmen und die benötigten Gateways ohne Eingriff des Benutzers erzeugen.

Literatur

1. De Soto, Alden: Using The JavaBeans Development Kit 1.0 April '97, Sun Microsystems Inc. 1997
2. Förste, S., Gehnen, G. Gerdes, K.-H.: Decentral Control Architecture For Modular Material Flow Systems, Proceedings of the 11th Lon Users International Conference, Nice 1996
3. Gaw, D.: Accessing LonWorks Networks from the World Wide Web, Proceedings of the 11th Lon Users International Conference, Nice 1996
4. N.N.: LonWorks Host Applikation Programmers Guide, Echelon Corporation, 1993
5. N.N.: Application Layer Interoperability Guidelines, LonMark Interoperability Association, 1996
6. N.N.: JavaBeans 1.0 Specification, Sun Microsystems Inc 1996
7. N.N.: Remote Method Invocation Specification, Sun Microsystems Inc. 1997

Einbettung von Feldbussen in das Internet-Management

T. Sauter, M. Knizak, M. Manninger

Institut für Computertechnik, Gußhausstr. 27-29, A-1040 Wien

Kurzfassung. Die Vernetzung in Bürokommunikation und Prozeßautomatisierung haben sich zwar unabhängig voneinander entwickelt, können aber in Teilbereichen günstig kombiniert werden. Die Anbindung von Feldbussen an klassische LANs und damit an das Internet ist beispielsweise in den Bereichen Fernüberwachung und -wartung interessant. Eine derartige Vernetzung bedarf aber eines globalen Managementkonzepts. Wir schlagen daher die Erweiterung des erprobten Internet-Standards SNMP auf Feldbusse vor und präsentieren eine Lösung für PROFIBUS.

Abstract. Although network concepts for process and office automation evolved independently, they can be combined with benefit. The primary advantage of such an interconnection is the possibility of remote access to fieldbus systems and the use of standard networking tools for this purpose. It is evident that this idea is feasible only with a unified and globally accepted management approach. Hence we propose to extend the proven Internet standard SNMP to fieldbus systems and present an implementation for PROFIBUS.

1 Motivation

Um die Vorteile von Feldbussen voll ausschöpfen zu können, ist es notwendig, sie nicht nur als Insellösungen anzusehen. Zum Protokollieren der Vorgänge auf einem Feldbus und zum Visualisieren genügt ein einfacher PC mit einer entsprechenden Feldbus-Einsteckkarte, den nötigen Treibern und einer Standardsoftware. Ein Nachteil ist dabei allerdings die Konzentration auf einen einzelnen Rechner. Sinnvoller wäre ein direkter Zugriff auf den Feldbus aus einem übergeordneten Netz, womit die Kopplung an klassische LANs und in weiterer Folge an das Internet auf der Hand liegen [1].

Die Vorteile einer solchen Anbindung sind einerseits eine räumlich stark erweiterte Zugriffsmöglichkeit auf Feldbusdaten, was zum Beispiel ein entscheidender Vorteil für eine zentrale Betriebsdatenerfassung ist. Wesentlicher ist aber, daß dafür eine vorhandene Infrastruktur genutzt werden kann - sowohl in Form von Hardware (der Übertragungswege) als auch Software (einer Fülle von Internet-Werkzeugen). Allerdings ist für die Vernetzung eine einheitliche Strategie nötig, und daher liegt es nahe, Methoden aus der Internet-Welt zu verwenden. Eine optimale Möglichkeit dazu bietet zum Beispiel das vom *Internet Architecture Board* standardisierte SNMP (Simple Network Management Protocol), das eine erprobte Struktur, die sich bei der Verwaltung von

Routern und Gateways bewährt hat, besitzt. Wenn ein Gerät über einen Zugang zum Internet (d. h. eine IP-Adresse) verfügt und SNMP unterstützt, sind die von ihm angebotenen Daten über SNMP weltweit abrufbar.

Nachfolgend werden zuerst die zum Verständnis wichtigen Grundlagen von SNMP und der ihm zugrundeliegenden Notation ASN.1 kurz zusammengefaßt, bevor ein Ansatz für die Kopplung zwischen LAN und Feldbus und eine Realisierung für PRO-FIBUS beschrieben werden.

2 Grundlagen

2.1 SNMP

Das *Simple Network Management Protocol* (SNMP) ist ein einfaches Protokoll zur Netzwerkadministrierung [2]. Dabei können sogenannte *Manager*-Stationen auf die ihnen zugeordneten *Agents* zugreifen. Die SNMP-Pakete zwischen Manager und Agent werden über das *User Datagram Protocol* (UDP) und das darunterliegende *Internet Protocol* (IP) übertragen [3].

SNMP umfaßt 5 verschiedene PDUs (Protocol Data Units): Mit *GetRequest* und *GetNextRequest* holt eine Management-Station einen oder mehrere Werte von einem Agent; *SetRequest* setzt einen Wert in einem Agent. Die restlichen PDUs, *GetResponse* und *Trap*, werden vom Agent zum Manager gesendet. *GetResponse* ist die Antwort auf eine *Get-*, *GetNext-* oder *Set*-Aktion des Managers. Die *Trap*-PDU gibt einem Agent die Möglichkeit, dem Manager Ereignisse auch ohne dessen Aufforderung anzuzeigen.

Über die bloße Kommunikationsverbindung hinaus stellt SNMP jedoch auch eine eindeutige Benennung der Datenobjekte zur Verfügung. Die dabei zugrundeliegende Datenstruktur ist die *Management Information Base* (MIB) [4]. Sie enthält in einer Baumstruktur alle über SNMP erreichbaren Objekte. Die Adressierung innerhalb der MIB erfolgt über sogenannte *Object Identifier* (OID). Der größte Teil der MIB ist weltweit einheitlich und daher fixiert, es gibt jedoch private Bereiche, in denen neue Strukturen angelegt werden können. In einem solchen Bereich können beispielsweise Feldbussysteme definiert werden.

2.2 ASN.1

Die Struktur der SNMP-Nachrichten ist in der Datenbeschreibungssprache *Abstract Syntax Notation 1* (ASN.1) definiert [5], wobei nur eine kleine Teilmenge. des Sprachumfanges verwendet wird. Für die Abbildung der abstrakten ASN.1-Konstrukte auf übertragbare Bitfolgen gibt es entsprechende Regeln (Basic Encoding Rules nach ISO 8825). Demnach ist ein ASN.1-Ausdruck in dreigeteilt: Das erste Feld gibt den Datentyp an, das mittlere Feld enthält die Länge des nachfolgenden Datenfeldes in Bytes, und zuletzt folgt der Wert selbst. Gruppen von solchen Ausdrücken können in *Sequence of*-Ausdrücken codiert werden, die selbst wieder dieser dreiteiligen Struktur folgen.

sequence of	Länge: 14	Daten

OID	Länge: 7	43.6.1.2.1.1.1
INTEGER	Länge: 1	16
NULL	Länge: 0	

INTEGER	0x02
STRING	0x04
NULL_TYP	0x05
OID	0x06
SEQUENCE	0x30
GETREQUEST	0xA0
GETNEXTREQUEST	0xA1
GETRESPONSE	0xA2
SETREQUEST	0xA3
TRAP	0xA4

Abb. 1. Beispiel für ASN.1 und die Typbezeichner

Abb. 1 zeigt als Beispiel eine ASN.1-codierte SNMP-Sequenz bestehend aus einer OID, dem Wert 16 und einem Nullwert. Außerdem sind die Codes für die einzelnen Datentypen und PDUs wiedergegeben, da sie das Verständnis der in Kapitel 5 behandelten Beispiele erleichtern.

3 Kopplung zwischen LAN und Feldbus

3.1 Gateway als Proxy Agent

Die SNMP-Anbindung von an sich nicht SNMP-fähigen Geräten erfolgt über sogenannte *Proxy Agents*, die auf verschiedene Art konzipiert werden können [1, 6]. Obwohl der Lösungsansatz mit Hilfe eines Gateways (Abb. 2) der aufwendigste ist, wurde er aus Gründen der Flexibilität gewählt. Der Anwender kann somit auf die Feldbus-Objekte über das einfache SNMP-Protokoll zugreifen, ohne den Feldbus selbst kennen zu müssen, da die Protokollumsetzung vollständig im Gateway erfolgt. Außerdem muß bei Änderungen, die den Feldbus betreffen, nur die Gateway-Applikation angepaßt werden.

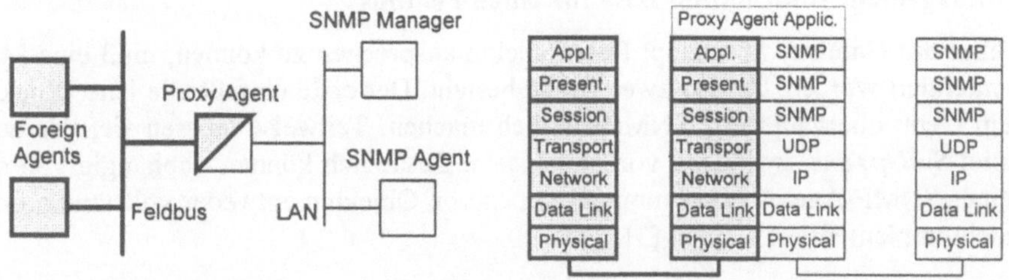

Abb. 2. Position eines Proxy Agents und Protokollstack eines Gateways

3.2 Aktualisierungsstrategie

Der Agent muß auf Anfragen des Managers im allgemeinen Daten vom Feldbus liefern. Dafür sind prinzipiell zwei Vorgangsweisen denkbar.

1. Bei der *Aktualisierung auf Anfrage* wird jede ankommende SNMP-Anfrage in Feldbusbefehle konvertiert und am Bus abgesetzt. Nach Eintreffen der Antwort wird diese an den Manager weitergeleitet. (Abb. 3a). Dabei muß allerdings sicher-

gestellt werden, daß keine Verwechslung zu knapp aufeinander folgender Anfragen möglich ist oder die Anzahl ausstehender Aufträge zu groß wird. Die Einrichtung von Befehls-Queues ist daher unumgänglich. Weiters muß auf eine Unterbrechung der Feldbusseite geeignet reagiert werden.

2. Die *zyklische Aktualisierung* schafft ein vereinfachtes Abbild der für SNMP relevanten Feldbusdaten im Gateway (Abb. 3b). Dies sorgt für eine schnellere Reaktionszeit auf SNMP-Anfragen, aber die Aktualität der Daten kann nicht mehr garantiert werden. Demzufolge müssen die Daten mit einem Zeitstempel versehen werden. Weiters ist zu bedenken, daß der Datenverkehr auf Feldbusseite durch ein zyklisches Polling stark erhöht wird.

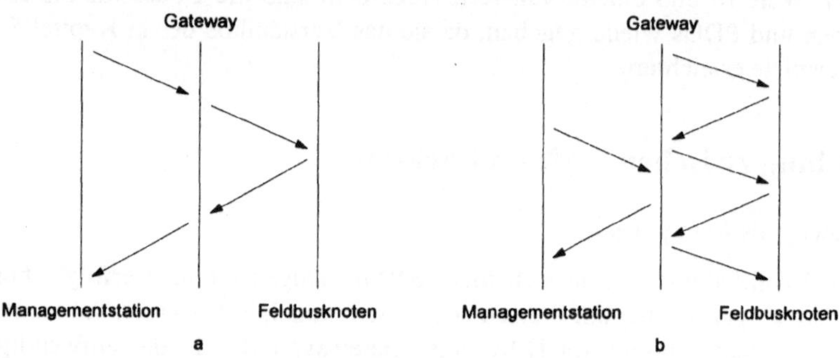

Abb. 3. Aktualisierungsstrategien für ein Gateway

Der Gewinn an Schnelligkeit und die genauere Information über den Zustand der Feldbusseite sprachen für eine Implementierung nach der zweiten Methode.

3.3 Management Information Base für einen Feldbus

Um über das Gateway überhaupt Datenobjekte ansprechen zu können, muß eine MIB implementiert werden, die aus zwei Teilen besteht. Der erste umfaßt alle jene Objekte, die ein Gerät überhaupt erst SNMP-tauglich machen. Teilweise müssen sie, wie zum Beispiel *SysUpTime,* zwingend vorhanden sein. Zusätzlich können - abhängig von der Art eines SNMP-Agents - bestimmte Gruppen von Objekten entweder vollständig oder gar nicht implementiert werden [3].

Zum Unterschied von den für SNMP notwendigen Objekten, die im globalen Bereich der MIB definiert sind und unverändert übernommen werden müssen, ist die feldbusspezifische Erweiterung in einem privaten Bereich der TU Wien angesiedelt. Bei der Definition dieser Feldbus-MIB wurde darauf geachtet, nur die wichtigsten Werte eines Knotens aufzunehmen, um den Umfang der MIB zu begrenzen. Außerdem wurde Wert auf eine leicht erweiterbare Struktur gelegt.

Wie Abb. 4 zeigt, gliedert sich die Feldbus-MIB in drei Unterbäume. Der erste, *fBTypeDescr,* definiert die Arten der im Feldbus vorhandenen Sensoren und Aktoren.

Über diese OIDs können keine konkreten Daten angesprochen werden, es handelt sich um reine Typdeklarationen.

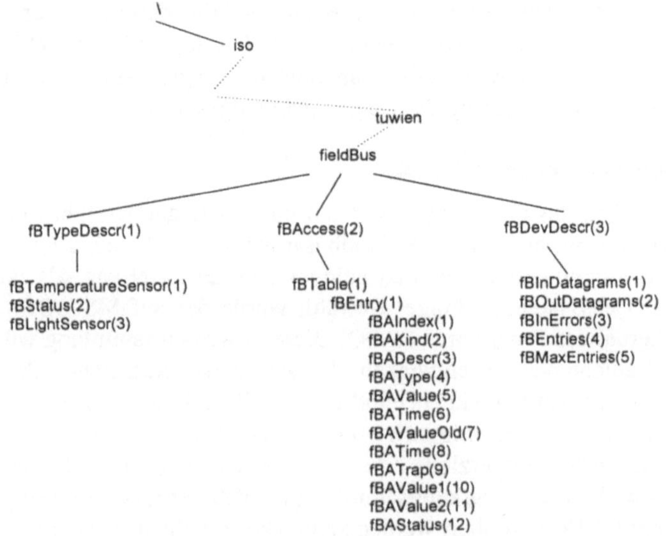

Abb. 4. Struktur der Feldbus-MIB

Der zweite Zweig, *fBAccess*, enthält die Daten der über SNMP ansprechbaren Feldbusknoten. Die Daten sind in einer Tabelle angeordnet, die leicht an die Zahl der vorhandenen Knoten angepaßt werden kann (was bei einer Baumstruktur nicht der Fall wäre). Das zweite Feld jedes Tabelleneintrags (*fBAKind*) gibt an, um welche Art von Gerät es sich handelt, und ist als Pointer auf den ersten Unterbaum zu verstehen. In der Tabelle werden auch die *Trap*-Bedingungen verwaltet. Der Zweig *fBDevDescr* schließlich stellt statistische Werte über das Gateway zur Verfügung und kann daher nur gelesen werden.

4 Ausführung eines PROFIBUS-Gateways

4.1 Protokollumsetzung

Neben der Definition der Feldbus-MIB ist die Protokollumsetzung der wesentlichste Punkt bei der Kopplung. Der Proxy Agent muß ja jedes SNMP-Kommando mit Hilfe eines oder mehrerer Feldbusbefehle durchführen. Beispielhaft für andere verbindungsorientierte Protokolle sei hier auf die Umsetzung von SNMP auf PROFIBUS eingegangen. PROFIBUS beinhaltet sowohl verpflichtende als auch optionale Dienste. Verpflichtend für den Kommunikationsaufbau sind: *Initiate*, *Reject* und *Abort*, sowie *Get OV*, *Identify* und *Status* [7]. Zusätzlich erscheint es zweckmäßig, auch die Basisdienste *Read* und *Write* in die Betrachtungen einzubeziehen. Die Dienste für das Kommunikationsmanagement laufen im Hintergrund und wirken sich nur auf das Zustandekommen der Kommunikationsbeziehung am Feldbus aus.

Von den angeführten Befehlen haben nur *Read* und *Write* ein direktes Äquivalent in SNMP (*Get* bzw. *GetNext* und *Set*, wobei die Antwort jeweils mit einem *Response*-Befehl geschickt wird). Alle anderen PROFIBUS-Befehle müssen, wenn man sie explizit über SNMP verwenden möchte, über eigens dafür definierte Variablen in der MIB angesprochen werden. Auf der anderen Seite hat der SNMP-Befehle *Trap* kein PROFIBUS-Äquivalent und wird deshalb im Gateway ausprogrammiert, was durch die gewählte zyklische Aktualisierung der Daten leicht möglich ist.

4.2 programmtechnische Realisierung

Die Implementierung des Gateways wurde auf einem PC durchgeführt, da für diesen sowohl Internetzugang als auch ein Feldbusanschluß verfügbar sind. Wegen der gleichzeitigen Kommunikation über den Feldbus und das LAN kam als Basis nur ein Multitasking-Betriebssystem in Frage. Gewählt wurde der auf MS-DOS aufsetzende Multitasking-Kernel des Programms KA9Q. Diese Programmsammlung wurde für den Gebrauch bei Radioamateuren entwickelt (KA9Q ist das Rufzeichen des Funkamateurs, auf den das System zurückgeht) und enthält alle gebräuchlichen TCP/IP-basierten Dienste wie *FTP*, *finger*, *ping* sowie einen UDP-Teil, auf den SNMP aufgesetzt werden kann. Für nicht kommerzielle Zwecke ist der Quellcode (in C++) frei erhältlich Ein weiterer Vorteil dieser Programmsammlung ist, daß der Quellcode so gestaltet ist, daß er auch unter UNIX compiliert werden kann. Dies erhöht auch die Einsatzmöglichkeiten dieses Systems.

Auf diese Umgebung wurde eine Datenstruktur aufgesetzt, die von allen Tasks verwendet werden kann und in der sich alle Daten befinden, die sowohl für die Internet- als auch für die Feldbusseite von Bedeutung sind.

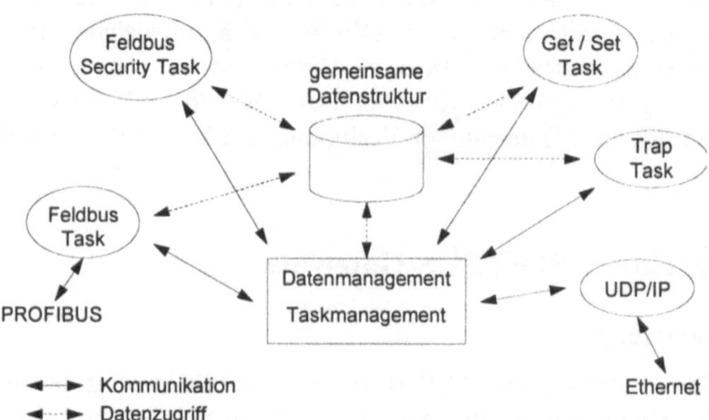

Abb. 5. Struktur der Gateway-Applikation

Folgende Tasks wurden definiert.

- Der Get/Set-Task hat die Aufgabe, hereinkommende Pakete auf ihre Richtigkeit zu untersuchen und die SNMP-PDUs *Get*, *GetNext* und *Set* an das für den jeweiligen

MIB-Unterbaum zuständige Unterprogramm weiterzuleiten, welches dann den Wert zurückliefert.

- Der Feldbus-Task aktualisiert die in der gemeinsamen Datenstruktur eingetragenen Werte zyklisch durch Befehle an den jeweiligen Feldbusknoten. Bei PROFIBUS-FMS werden eine Umsetzung in KBL-Nummer und OV-Nummer und anschließend der Datenaustausch mit dem PROFIBUS-Partner (siehe nächstes Kapitel) durchgeführt, bei PROFIBUS-DP und -PA ist nur eine Verschiebung der Daten innerhalb des Speichers notwendig, da die Feldbusdaten hier schon im Speicher vorhanden sind.

- Der Trap-Task durchsucht periodisch die gemeinsame Datenstrukur nach aktiven Trap-Bedingungen und löst gegebenfalls eine *Trap*-PDU an den Manager aus.

- Der Feldbus-Security-Task durchläuft ebenfalls periodisch die gemeinsame Datenstrukur, überprüft nach noch festzulegenden Algorithmen die Zuverlässigkeit der Feldbusdaten und setzt dementsprechend den Status der Feldbusknotens (FeldBusMIB.fBAccess.fBAStatus).

Das Konzept wurde so weit modularisiert, daß eine Portierung auf ein anderes Feldbussystem durch den Austausch des Feldbus-Tasks einfach möglich wird.

5 Der Weg einer Anfrage

Als Beispiel verfolgen wir eine SNMP-Anfrage nach einem bestimmten Wert eines PROFIBUS-Gerätes (in diesem Fall ein Zähler) auf ihrem Weg durch das Netz. Um den Wert des Zählers auszulesen, schickt man den *GetRequest*-Befehl mit dem passenden OID von derManagementstation zum Gateway. Dieses sucht in der gemeinsamen Datenstruktur nach dem gewünschten Wert und schickt ihn mit einem *Get-Response* zur Managementstation zurück. Für das zyklische Polling der PROFIBUS-Knoten, über das die Datenstruktur aktualisiert wird, werden im Gateway die Kommunikationsbeziehungsliste und das Objektverzeichnis benötigt. Abb. 6 zeigt ein typisches PROFIBUS-Antwort-Telegramm, Abb. 7 das zugehörige SNMP-Paket, mit dem dieser Wert vom Proxy zur Managementstation übertragen wird. Die Strukturen der unterlagerten Protokolle (IP, UDP und Ethernet-Protokoll) sind der Einfachheit halber weggelassen.

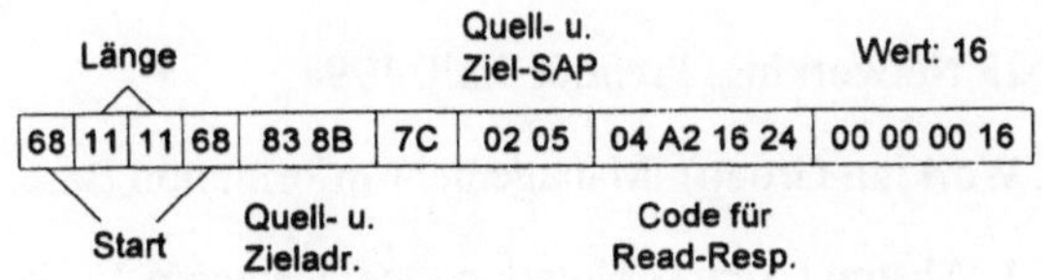

Abb. 6. typisches PROFIBUS-Telegramm

In den einzelnen Feldern des SNMP-Paketes erkennt man die dreiteilige ASN.1-Codierung. Die jeweiligen Nutzdaten sind grau unterlegt.

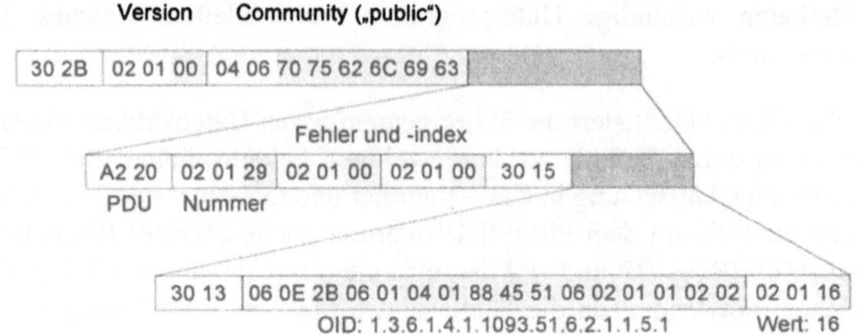

Abb. 7. typisches SNMP-Paket (*GetResponse*). Das *Community*-Feld ermöglicht eine einfache Gruppenbildung, im Feld *Nummer* steht ein fortlaufender Index.

6 Résumé

Wir konnten zeigen, daß das aus der Internet-Welt stammende Netzwerkmanagementprotokoll SNMP ein gut geeignetes Werkzeug zum Ansprechen von Feldbussen aus übergeordneten Netzwerken ist. Dazu muß im wesentlichen ein Gateway zwischen SNMP und dem Feldbus aufgesetzt werden. Wesentliche Punkte der Gatewayarchitektur sind die Wahl der Aktualisierungsstrategie, die Definition der MIB sowie die Protokollumsetzung, die für PROFIBUS ausgeführt wurde.

Natürlich ergeben sich durch die Anwendung dieser Möglichkeit zur Fernwartung von Feldbussen auch gewisse Risiken. Vor allem wenn das Gateway über das Internet öffentlich zugänglich ist, müssen geeignete Sicherheitsmaßnahmen den Zugriff unberechtigter Personen bzw. Geräte auf den Feldbus verhindern. Ein solcher Schutz muß auf Zertifikaten, also auf asymmetrischer Kryptographie basieren und kann rein softwaremäßig oder, um den gebotenen Schutz zu maximieren, unter Verwendung von Smart Cards realisiert werden.

Literatur

1. Knizak, M., Manninger, M., Sauter, Th.: Einbindung von Feldbussen in das Internet via SNMP. e&i 114, 258-262 (1997).

2. Miller, M. A.: Managing Networks with SNMP. New York: M&T Books, 1993.

3. Martin, J.: TCP/IP Networking. Prentice Hall, 1994.

4. IETF (Network Working Group): Management Information Base, RFC1156, 1990.

5. Gora, W.: ASN.1, Abstract Syntax Notation One. Datacom-Verlag, 1992.

6. Knizak, M.: Feldbus-Netzwerkmanagement mit SNMP, Diplomarbeit, TU Wien, 1997.

7. CENELEC: EN50170, 1996.

Das Producer/Consumer-Modell in Feldbussystemen der nächsten Generation

Viktor Schiffer

Fa. Rockwell Automation, Abt. EDC
Düsselberger Str. 15, D-42781 Haan

Abstract. Im folgenden wird das Producer/Consumer-Netzwerkmodell vorgestellt und mit herkömmlichen Systemen verglichen. Dabei werden die grundsätzlichen Neuerungen dieses Systems gezeigt. Die beim Producer/Consumer-Modell möglichen Datenauslösungsmechanismen ermöglichen eine schnelle Übermittlung von zeitkritischen und nicht zeitkritischen Daten an einen oder mehrere Busteilnehmer und erhöhen damit in erheblichem Maße die Leistungsfähigkeit eines Feldbussystems. Diese Leistungssteigerung wird mit Beispielen gezeigt und es wird ein Ausblick auf die nach diesem Prinzip entstandenen oder entstehenden Systeme gegeben.

Abstract. This paper presents the new producer/consumer network model and compares it with existing systems. All new features of this network model are being shown in detail. The new data production triggers found in the producer/consumer model allow fast transmission of data to one or multiple bus participants. This yields a dramatic increase of efficiency for such a field bus system. The increase in performance is shown in a couple of examples and a short look is taken at the fieldbus system already designed or being designed with this network model.

1. Einleitung

Im letzten Jahrzehnt sind die Anforderungen an Steuerungssysteme laufend gestiegen und damit auch an die Feldbusse, die solche Systeme verbinden. Der Benutzer verlangt verbesserte Diagnosemöglichkeiten über den Bus bei verringerten Ausfallzeiten und Installations- und Unterhaltungskosten. Gleichzeitig wird aber höhere Funktionalität, Zuverlässigkeit und Geschwindigkeit gefordert.
Die höhere Funktionalität hat mehr Verkehr und Daten auf dem Netzwerk als Folge. Heutige Netzwerke, die mit einer Quelle/Ziel-Adressierung arbeiten, können nicht gleichzeitig höhere Funktionalität und größere Datenmengen verkraften. Dadurch sind die Systemmöglichkeiten und Produktivitätsverbesserungen begrenzt. Diese erhöhten Anforderungen haben daher die Entwicklung eines neuen Netzwerksmodells forciert, bei dem höhere Funktionalität mit besserer Ausnutzung der Bandbreite auf dem Netzwerk kombiniert wird.

2. Beurteilungskriterien für Feldbusse

Leider konzentriert sich die Diskussion über Feldbusse oft nur auf Baudraten, Protokolleffizienz, und physikalische Eigenschaften (z.B welches Kabel wird benutzt). In Wirklichkeit ist die Situation jedoch weit komplexer. Wichtigere Kriterien bei der Auswahl eines Feldbusses sind jedoch Diagnosemöglichkeiten, Nachrichtenarten und Durchsatzgeschwindigkeit.

Den größten Einfluß auf diese Eigenschaften hat das verwendete Netzwerkmodell. Das in den letzten zwei Jahrzehnten verwendete Quelle/Ziel-Modell ist nicht mehr in der Lage, die heutigen Anforderungen an Feldbusse zu erfüllen. Das Producer/Consumer-Prinzip hingegen ist derzeit das einzige Modell, das auch noch zukünftigen Anforderungen gerecht wird. Das zeigt sich in folgenden Kriterien:

2.1 Diagnosemöglichkeiten

Feldbusse bieten leicht zu nutzende Möglichkeiten, Diagnoseinformationen aus den Geräten abzurufen. Gerätespezifische Informationen, wie zum Beispiel verringerte Funktionssicherheit einer Lichtschranke auf Grund einer verschmutzten Linse können während des Betriebs dem Steuerungssystem mitgeteilt werden. Der Feldbus kann diese Information dann einem Bediengerät mitteilen und damit das Maschinenpersonal auf dieses Problem hinweisen. Mittlerweile fordert man Gerätewartung, Auslesen von Fehlercodes etc. ohne Einfluß auf die Echtzeiteigenschaften des Systems.

2.2 Explizite Mitteilungen und E/A-Mitteilungen

Explizite Mitteilungen zur Gerätekonfiguration und -diagnose sind außerordentlich flexibel, dadurch, daß das Datenfeld Protokollinformationen und Anweisungen zu dem durchzuführenden Dienst enthält. Eine Anweisung könnte z.B. Timer-Werte in einer Steuerung einstellen oder einen Selbsttest veranlassen. Explizite Mitteilungen können zum Hinauf- und Herunterladen von Programmen verwendet werden, zur Gerätekonfiguration oder auch zur Datenaufzeichnung, Trendanalyse oder Diagnose. Der jeweilige Feldbusknoten muß derartige Nachrichten interpretieren, die angeforderten Aufgaben durchführen und die notwendigen Antworten absetzen. Diese Art Mitteilungen sind sehr variabel, sowohl in Länge als auch in Häufigkeit.

E/A-Mitteilungen dagegen sind von Natur aus implizit. Das Datenfeld enthält keinerlei Protokollinformationen, lediglich Echtzeit-Steuerungsdaten. Die Bedeutung dieser Daten wurde vordefiniert, so daß die Durchlaufzeit im Feldbusknoten minimiert werden kann. Als Beispiel einer E/A-Mitteilung sei hier eine Steuerung genannt, die Ausgangsdaten an einen E/A-Block sendet und der E/A-Block, der als Anwort seine Eingangsdaten sendet. Solche Mitteilungen sind kurz und häufig und verlangen kurze Verarbeitungszeiten.

In der Vergangenheit wurden für diese beiden Nachrichtentypen verschiedene Netzwerke eingesetzt, da ein Feldbus zur E/A-Steuerung nicht die Variabilität von expliziten Mitteilungen verträgt. Beispiele solcher Netzwerke sind DH+/RIO von Allen-Bradley oder Profibus FMS/DP. Heutzutage fordern die Benutzer beide Funktionen auf dem gleichen Bussystem. Dazu kommt, daß die hohe Funktionalität heutiger Geräte auch beide Nachrichtentypen im gleichen Netzwerk verlangt. Damit ist das herkömmliche Quelle/Ziel-Modell überfordert.

2.3 Datendurchsatz

Letztendlich legt der von der Applikation geforderte Datendurchsatz fest, welcher Netzwerktyp benötigt wird. Mit Datendurchsatz wird die Geschwindigkeit beschrieben, mit der die Eingangsdaten all den Knoten zur Verfügung gestellt werden, die sie benötigen, und die Ausgangsdaten (Entscheidungen) allen notwendigen Ausgangsgeräten zur Verfügung gestellt werden. Die beteiligten Knoten können Sensoren, Bediengeräte, Steuerungen, Datenaufzeichnungsgeräte, Alarmgeräte, Aktoren etc. Sein. Der Datendurchsatz wird durch Baudrate, Protokolleffizienz und insbesondere durch das Netzwerkmodell bestimmt.
Die <u>Baudrate</u> ist sozusagen „Brutto"-Geschwindigkeit. Leider wird sie am häufigsten zur Bestimmung der Leistungsfähigkeit herangezogen, dabei ist diese Angabe oft stark irreführend. Darüber hinaus ist diese Größe bei den heutigen neuen Netzwerken die unwichtigste der drei.
Die <u>Protokolleffizienz</u> - also Nutzdaten zu Gesamtdaten im Nachrichtenpaket - wird üblicherweise in Prozent ausgedrückt und ist ein Maß für den Protokoll-Overhead des Netzwerks. Die Protokolleffizienz ist zwar wichtig, aber bei weitem nicht so wichtig wie die Methode der Ablieferung und des Austauschs der Daten. Wenn zum Beispiel der Austausch von Informationen in zwei statt in einem Paket auf dem Bus stattfindet, wird eine um 25% höhere Protokolleffizienz des ersten Protokolls bedeutungslos.
<u>Netzwerkmodell.</u> Jeder Steuerungshersteller hat seine eigenen Vorzugsnetzwerke, egal, ob es sich dabei um Data Highway Plus, Remote I/O, Profibus FMS, Profibus DP, Interbus-S, ASI, Modbus Plus, GeniusLan oder Lonworks handelt. Alle diese Systeme haben die gleiche Eigenschaft gemeinsam: Sie gehorchen dem herkömmlichen Quelle/Ziel-Modell. Ein typisches Nachrichtenpaket wird hier gezeigt:

Quelle	**Ziel**	**Daten**	**CRC**

Abb. 1 Producer/Consumer-Netzwerkmodell

In Master/Slave-Anwendungen ist das „Quelle"-Feld üblicherweise nicht vorhanden, da der Master die einzige Nachrichtenquelle ist und da alle Antworten der Slaves für den Master bestimmt sind. Dieses Master/Slave-Pollingverfahren ist damit vom Grundsatz her ein Datenaustausch zwischen jeweils nur zwei Teilnehmern. Typischerweise werden hiermit nur Echtzeit-Steuerungsdaten ausgetauscht (E/A-Mitteilungen). Weitere Funktionen sind meist nicht vorhanden, um notwendige Wiederholbarkeit und Datendurchsatz zur Steuerung zu erreichen.
Beim Datenaustausch zwischen gleichartigen Teilnehmern (Peer to Peer) entwickelt man sich über Master/Slave hinaus und schafft dadurch bedeutend mehr Flexibilität. Dabei gehen die meisten Netzwerke zu expliziten Nachrichten über. Als Beispiele seien hier die Programmierung und Konfiguration sowie Bediengeräte auf PC-Basis genannt. Zusätzliche Geräte belasten den Bus im allgemeinen erheblich, da

üblicherweise jedes einzelne Gerät die gleichen Variablen aus den einzelnen Steuerungen ausliest. Somit liegen die gleichen Daten mehrmals hintereinander auf dem Bus. Erst dadurch wird sichergestellt, daß die gleichen Alarme, Trendmeldungen und graphischen Darstellungen an mehreren Bedienstellen zur Verfügung stehen. Damit nun nicht einzelne Geräte den Bus dauernd belegen, gehen die meisten Peer-to-Peer-Netzwerke zu einem Token-Passing-Prinzip über. Diese Algorithmen haben sich zwar als sehr „fair" den einzelnen Teilnehmern gegenüber erwiesen, es hat sich aber herausgestellt, daß die grundsätzliche Flexibilität, die dieses Verfahren attraktiv macht, auch gleichzeitig dazu führt, daß die Synchronisation von Steuerungen untereinander mit diesem Verfahren sehr problematisch ist. Die Antwortzeiten für eine vorgegebene Mitteilung variieren erheblich, abhängig davon, wie hoch die Buslast ist und wie weit „weg" man vom Token-Inhaber ist, wenn die Nachricht zur Übermittlung ansteht.

In vielen Fällen findet man einfache elektronische Bedien- und Anzeigegeräte auf E/A-Netzwerken, die einfache Schalte/Taster, Anzeigelampen und Zeigergeräte ersetzen. Hier führt nun das Hinzufügen von jedem einzelnen Gerät dazu, daß der Bus zusätzlich mit Daten belastet wird, die im allgemeinen gleich sind, aber an unterschiedliche Ziele gesendet werden. Dies verursacht zwar keine stark variierenden Buslasten, der erhöhte Busverkehr verlangsamt aber die Reaktionszeiten für alle Busteilnehmer, auch die der „echten" E/A-Geräte. Es sind jedoch nicht nur diese Geräte, die den Bus „verstopfen". Mittlerweile haben die E/A-Geräte immer mehr Intelligenz und die dadurch entstehenden zusätzlichen Diagnose- und Konfigurationsdaten können auch erhebliche Buslast verursachen.

Ob es sich nun um Master/Slave- oder Peer-to-Peer-Betrieb handelt, bei datenzielorientierten Netzwerken wird erhebliche Netzwerk-Bandbreite dadurch verschwendet, daß die gleichen Datensätze an verschiedene Knoten geschickt werden. Dazu kommt noch, daß die Synchronisation mehrerer Geräte, z.B. Übermittlung eines neuen Sollwerts an mehrere Frequenzumrichter, dadurch erschwert wird, daß die Daten bei jedem einzelnen Gerät zu einem anderen Zeitpunkt ankommen.

3. Vorteile des Producer/Consumer-Modells

Neue Netzwerke, die lediglich die Baudrate oder die Anzahle der Knoten heraufsetzen, um mit dem Problem der vergrößerten Datenmengen, der intelligenteren Geräte und der verbesserten Steuerungsmöglichkeiten fertig zu werden, verschieben nur das was sowieso unvermeidlich ist. Was wirklich nötig ist, ist ein völlig neues Netzwerkmodell, das auf die heutigen Probleme zugeschnitten ist.

Identifier	Daten	CRC

Abb. 2 Producer/Consumer-Netzwerkmodell

Dieses neue Modell ist das Producer/Consumer-Modell. In einem Producer/Consumer-Netzwerk, werden Nachrichten über ihren Inhalt identifiziert. Wenn ein Knoten diese

Daten benötigt, akzeptiert er diesen „Identifier" und „konsumiert" die Daten. Durch diese Methode ergeben sich einige neue Möglichkeiten.

3.1 Multicast

Da die Daten durch ihren Inhalt identifiziert werden, können mehrere Knoten zeitgleich die Daten des gleichen Produzenten konsumieren. Dadurch können Knoten präziser synchronisiert werden bei gleichzeitig effektiverer Ausnutzung der verfügbaren Bandbreite. Zusätzliche Bediengeräte können installiert werden, ohne den Busverkehr zu erhöhen, da sie ja die gleichen Nachrichten konsumieren. Weiterhin können die Knoten verschiedene Datensätze produzieren, die sich über unterschiedliche Identifier unterscheiden lassen.

In einem herkömmlichen Quelle/Ziel-Netzwerk ist ein Multicast vom Prinzip her nicht möglich, obwohl es Versuche dazu gegeben hat. Einige haben ein weiteres Feld für ein Gruppenziel aufgemacht und dann Knotennummern für diese Gruppenziele reserviert. Bei anderen kann ein Knoten mehr als eine Knotennummer haben. Im Grunde genommen handelt es sich aber nur um Flicklösungen. Seltsamerweise hat man dies meist bei Netzwerken angewandt, die auf expliziten Mitteilungen basieren, nicht bei E/A-Netzwerken, bei den dies am stärksten benötigt wird.

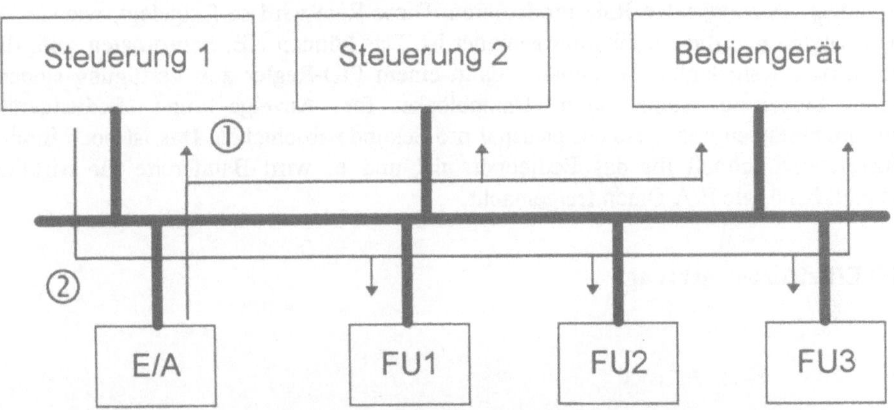

Mitteilung 1: Die Positionsangabe des E/A-Geräts geht per Multicast an beide Steuerungen und an das Bediengerät
Mitteilung 2: Die Geschwindigkeitsvorgabe geht als Multicast an alle drei Frequenzumrichter und an das Bediengerät

Abb. 3 Producer/Consumer-Multicast

3.2 Auslösemechanismen

Darüber hinaus erlaubt das Producer/Consumer-Modell zwei neue Auslösemechanismen für E/A-Mitteilungen zusätzlich zum traditionellen Polling. Das Polling hat seinen Ursprung im traditionellen Quelle/Ziel-Modell und ist vom Prinzip her eine bidirektionale Transaktion bestehend aus zwei Nachrichten (der Auslöser sendet Ausgangsdaten, der empfangende Knoten antwortet mit Eingangsdaten). Diese

Transaktionen werden in möglichst schneller Folge wiederholt, damit die Totzeit zwischen dem Auftreten einer Eingangsänderung und der Ablieferung dieser Information bei der Steuerung möglichst kurz wird. Die meisten Pollingzyklen werden so mit den gleichen alten Ein- und Ausgangsdaten ausgefüllt und verschwenden somit unnötig Übertragungsbandbreite.

Das Producer/Consumer-Modell bietet hier zwei neue Auslösemechanismen an, die effektiver sind und nur eine Nachricht in einer Richtung benötigen: ereignisgesteuert und zyklisch.

Die *ereignisgesteuerte Nachrichtenauslösung* führt dazu, daß ein Teilnehmer nur dann seine Daten sendet, wenn diese sich geändert haben; als Ereignis kann z.B. eine Datenänderung angesehen werden. Bei diesem Prinzip gibt es keine Verzögerung durch den Pollzyklus des Netzwerks. Die Daten werden gleichzeitig an alle Konsumenten verschickt, sobald eine Änderung vorliegt. Wenn keine Ereignisse vorliegen, läßt ein „Herzschlag" im Hintergrund die Konsumenten der Nachrichten ein einwandfreies Gerät ohne „Ereignisse" von einen gestörten Gerät unterscheiden. Diese Methode erlaubt eine drastische Reduktion des Netzwerkverkehrs und der einzelne Netzwerkknoten benötigt weniger Ressourcen, da er nicht immer wieder die gleichen alten Daten nach dem Pollingprinzip verarbeiten muß.

Die *zyklische Nachrichtenauslösung* ist für Knoten gedacht, die Daten in einer vom Benutzer festzulegenden Rate produzieren. Diese Rate wird so festgelegt, wie sie für den Knoten und die Applikation geeignet ist. Das können z.B. Sensordaten sein, die zu präzise festgelegten Zeitpunkten dann einem PID-Regler zur Verfügung stehen. Eine Steuerung kann dann Datenblöcke für Anzeige- und Bediengeräte zusammenfassen und diese ein paarmal pro Sekunde abschicken. Das ist noch immer ausreichend schnell für das Bedienpersonal und es wird Bandbreite für wirklich schnell benötigte E/A-Daten freigemacht.

3.3 Effizienzbetrachtung

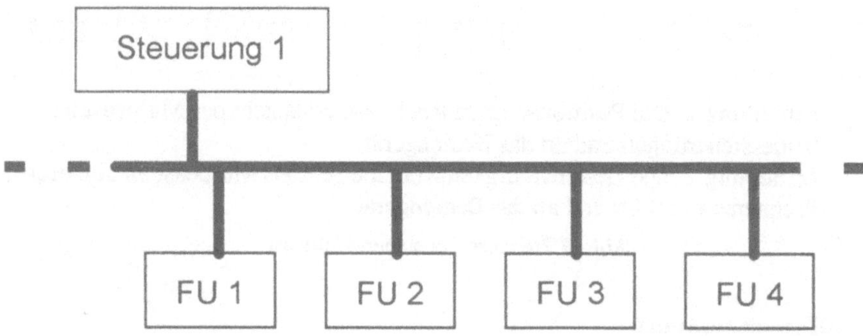

Abb. 4 Beispiel für die Effizienz des Netzwerks

Für das in Abb. 4 dargestellte Beispielsystem sollen drei Kommunikationsmethoden verglichen werden, die einmal pro Sekunde einen neuen Sollwert an die Frequenzumrichter übermitteln.

Polling (Quelle/Ziel-Modell, Scanzyklus 20 ms):
- 8 Nachrichten zur E/A-Datenübermittlung benötigt
- 200 Nachrichten pro Sekunde (50 Scans/Sekunde, 8 Nachrichten/Scan)
- max. 20 ms Netzwerkverzögerung

Zyklischer Datenaustausch (Producer/Consumer-Modell, Zyklus 20 ms)
- 1 Nachricht zur E/A-Datenübermittlung benötigt
- 50 Nachrichten pro Sekunde (50 Zyklen)
- max. 20 ms Netzwerkverzögerung

Ereignisgesteuerter Datenaustausch (Producer/Consumer-Modell, Herzschlag alle 200 ms)
- 1 Nachricht zur E/A-Datenübermittlung benötigt
- 5 Nachrichten pro Sekunde (4 mal Herzschlag und einmal „echte" Daten)
- wenige ms Netzwerkverzögerung

Sowohl Nachrichten vom „Peer-to-Peer"-Typ als auch Nachrichten zwischen Steuerungen und E/A-Geräten lassen sich in Producer/Consumer-Netzwerken mit zyklischer und ereignisgesteuerter Datenauslösung effizienter handhaben. Der Datenbedarf von Bediengeräten läßt sich dadurch mit nur minimaler Erhöhung der Netzwerkbelastung decken.
Auf der anderen Seite lassen sich gleichzeitig die flexiblen Anforderungen der Gerätekonfiguration und der Programmierung erfüllen. Typischerweise werden bestimmte Identifier für solchen Busverkehr reserviert und die einzelnen Knoten wissen, daß in diesen Nachrichten Informationen zu Ziel und Diensten eingeschlossen sind. Die Identifier werden so mit der Netzwerkszugriffsmethode kombiniert, daß die expliziten Nachrichten mit ihrem zugehörigen größeren Protokollanteil mit deutlich geringerer Priorität auf dem Netzwerk transportiert werden. Große Uploads oder Downloads, veränderte Konfigurationsparameter oder Diagnosemitteilungen werden somit in den Hintergrund geschoben und lassen sich dadurch zwischen höherprioren E/A-Nachrichten unterbringen

Als Folge davon braucht der Anwender nicht mehr zwei Netzwerke - eins für E/A-Betrieb und eins für explizite Nachrichten - durch seine Anlage zu ziehen. Auch die Gerätehersteller sind nicht mehr gezwungen, sowohl einen E/A-Anschluß als auch einen Konfigurations/Bedien/Diagnose-Anschluß in die Geräte einzubauen. Auf Grund dieser Vorteile ist es daher kein Wunder, daß die neuesten offenen Steuerungsnetzwerke - DeviceNet, ControlNet und Foundation Fieldbus (noch in Arbeit) - auf eben diesem Producer/Consumer-Modell basieren.

4. Beispiele von Netzwerken und ihren Modellen

Quelle/Ziel-Modell

Master/Slave	Peer to Peer
Data Highway +	Remote I/O
Profibus DP	Profibus FMS
Interbus-S	Modbus Plus
ASI	Lon Works

Abb. 5 Netzwerke mit Quelle/Ziel-Modell

Producer/Consumer-Modell

DeviceNet
ControlNet
Foundation Fieldbus

Abb. 6 Netzwerke mit Producer/Consumer-Modell

5. Wird das Quelle/Ziel-Modell verschwinden?

Das Quelle/Ziel-Modell ist praktisch ein „Erbstück" aus der Computer- und Datenverarbeitungs-Industrie. Trotz Ihrer Begrenzungen sind solche Systeme durchaus noch für eine Reihe von Anwendungen geeignet, die keine komplexe Koordination und gemeinsame Nutzung von Daten benötigen. Allerdings lassen sich nur mit der Flexibilität des Producer/Consumer-Modells die erweiterten Funktionalitäten der heutigen Anwendungen ausnutzen. Darüber hinaus ist diese Modell sehr gut für die weit intelligenteren Geräte von morgen geeignet. Heutzutage, wo der Anwender mehr Funktionalität bei geringerem Aufwand fordert, muß ganz einfach eine intelligentere Netzwerkstrategie her, und die benötigt ein intelligenteres Netzwerkmodell.

Literatur

1. DeviceNet Specifications, Rel. 2.0, Open DeviceNet Vendors Association, 1997

2. ControlNet Specifications, ControlNet International, 1997

3. http://www.odva.org

4. http://www.controlnet.org

Client-Server-Architekturkonzept mit heterogenem Kommunikationsinterface - Basis für das echtzeitfähige Betriebs- und Prozeßdatenmanagement in einem verteilten Prozeß

P. Löber / L. Clausnitzer / R. Fischer / P. Jarosz / S. König / M. Rössel

TU Bergakademie Freiberg, Institut für Automatisierungstechnik,
Lessingstraße 45, 09599 Freiberg / Sa.

Abstract. The client-server-concept as an essential part of a heterogeneous communication interface for the process-, operation- and quality-data management in distributed process is introduced. The same basic functionality concerns also, so called channel concept, for different fieldbus systems and ethernet. All parts of the client-server-concept are object-oriented developed. Real-time capabilities of the system are based on a multilevel synchronous and asynchronous event management.

1 Einleitung

Eine Abstraktion der echtzeitfähigen Betriebs- und Prozeßdatenerfassung (BDE/PDE) bzw. Maschinen- und Qualitätsdatenerfassung (MDE/QDE) in einem verteilten Prozeß reduziert sich bei Realisierung eines Client-Server-Konzeptes auf das in Fig. 1 angegebene Problem [1].

Im Mittelpunkt der Realisierung stehen die Client-Stationen, die z.B. diskless ausgeführt, die dezentrale Datenerfassung über ein oder mehrere Feldbusse (hier CAN-Bus) realisieren [2]. Die Erfassung beinhaltet sowohl die Maschinen- bzw. Prozeßdatenerfassung, die über Anschaltungen an die Meß- bzw. Stellgrößen des Prozesses der Teilanlage oder über speicherprogrammierbare Steuerungen (SPS) direkt erfolgt, als auch die Betriebsdatenerfassung, die per manueller Eingabe an der Clientstation zu erfolgen hat (z.B. per Prozeßtastatur, über Touch-Eingabe usw).

Die Erfassung der Qualitäts- und Produktionsdaten ist je nach Prozeß entweder nach erfolgter Inspektion manuell einzugeben oder erfolgt automatisch über direkt meßbare Größen, die im Prozeß online erfaßbar sind [3].

Nach der Erfassung und Vorverarbeitung der dezentral anfallenden Prozeß- bzw. Betriebsdaten erfolgt die Übertragung in die zentrale echtzeitfähige Datenbank des Servers.

Nachfolgend wird das informations- bzw. automatisierungstechnische Architekturkonzept des dafür notwendigen Programmsystems vorgestellt.

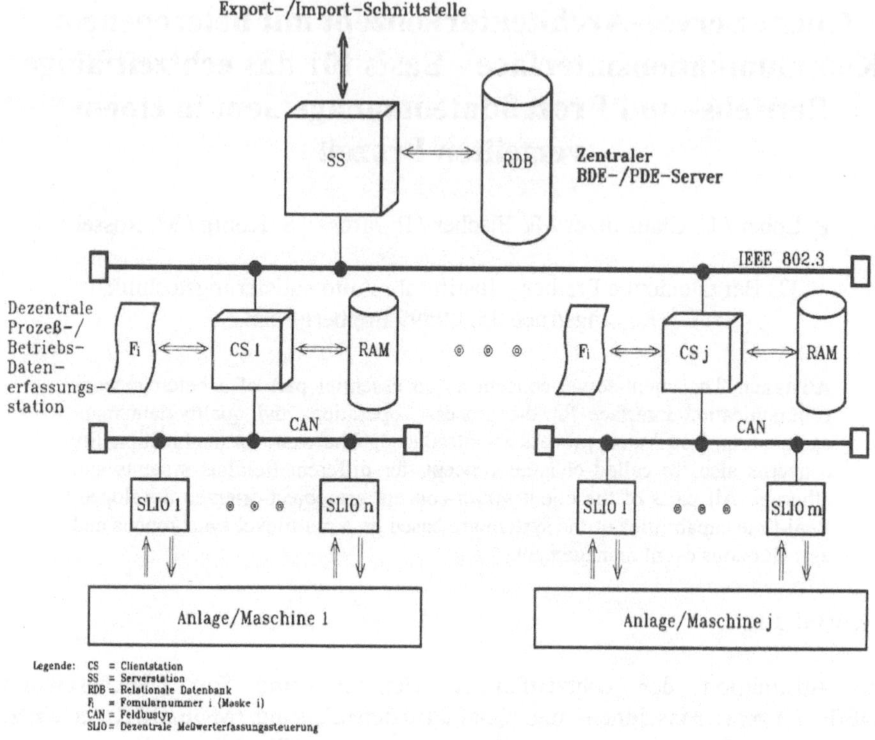

Fig. 1 : Client-Server-Basiskonzept für das dezentrale BDE-/PDE-Management

2 Hierarchiekonzept für die BDE/MDE-Erfassung in einem verteilten Prozeß

In Fig. 1 wird gezeigt, daß für die Realisierung des Problems das klassische hierarchische Ebenenmodell ebenfalls relevant ist. Insbesondere ist für die Realisierung derartiger automatisierungstechnischen Lösungen das Problem des echtzeitfähigen Kommunikationssystems bei Verwendung unterschiedlicher Feldbusse und den Ethernet-basierten Protokollen von besonderem Interesse. Das entwickelte Programmsystem nutzt z.Z. sowohl für die Clients als auch für die Server die Merkmale des Betriebssystems UNIX. Damit ist für die Clients prinzipiell der in Fig. 2 dargestellte Leistungsumfang (Applikationsumfang) notwendig und auch realisierbar.

Die für den Server notwendige Funktionalität unterscheidet sich von der des Clients, da hier das Kommunikationsinterface für die Datenbank vorhanden sein muß, dafür aber die direkte Kopplung zu den Feldbussen entfällt.

Im Punkt 3 wird gezeigt, daß das Client-Server-Konzept die prinzipiellen Voraussetzungen für die Kommunikation in einem verteilten System mit Mischkonfiguration besitzt.

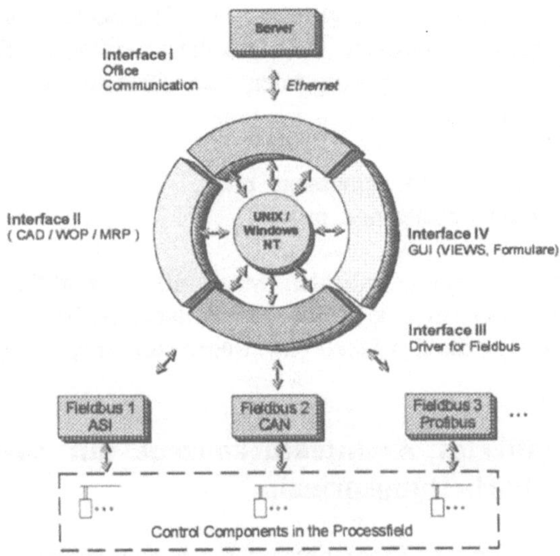

Fig. 2 : Kommunikationsschnittstellen eines Clients

3 Client-Server-Modell

Das Client-Server-Architekturkonzept ist eine klassische Kooperationsstruktur für die Erteilung und Abarbeitung von Aufträgen in einem verteilten Rechen- bzw. Steuersystem [5].

Gemäß Fig. 3 erfolgt in einem verteilten Prozeß die Auftragserteilung bei Anfrage durch die Clients. Diese Anfragen werden vom Server beantwortet (bedient).

Der wesentliche Vorteil einer Client-Server-Architektur liegt in der einfachen Arbeitsweise. Ein Client sendet eine Anfrage und der Server leistet einen Dienst, in dem er dem Fragenden eine positive, eine negative oder auch keine Antwort gibt.

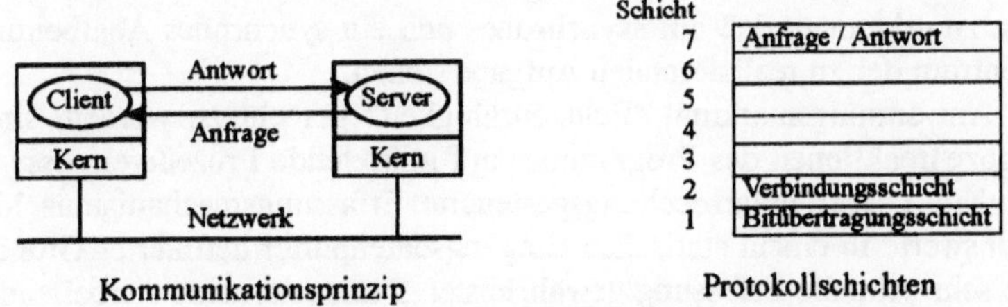

Fig. 3 : Client-Server-Modell

Client-Server-Konzepte benötigen für den Nachrichtenaustausch nicht den vergleichsweise hohen Aufwand der verbindungsorientierten Protokolle des OSI-Referenzmodelles. Von den dort vereinbarten 7 Schichten werden typischerweise nur die Bitübertragungsschicht (Layer 1), die Verbindungsschicht (Layer 2) und die

Anwenderschicht (Layer 7) verwendet. Die Protokollebenen 3-6 sind nicht vorhanden bzw. nicht ausgeprägt. Aufgrund dieser einfachen Struktur können die Kommunikationsdienste auf die folgenden zwei Dienste reduziert werden, die als Systemrufe ausgelegt sind :

- send   (destination, @message_ptr)
- receive   (addr, @message_ptr)

Es existieren viele Varianten der beiden Systemrufe. Sowohl für die Clients als auch für den/die Server werden eine Reihe von Parametern bzw. Konstanten in einer Headerdatei hinterlegt, die den Send- bzw. Receivedienst spezifizieren.

4 Objektorientiertes Architekturkonzept mit asynchronem und synchronem Abarbeitungsprinzip

Das Architekturprinzip des Programmsystems beruht auf einer konsequenten und durchgängigen objektorientierten Programmiermethodik, s.a. [6, 7 und 8].
Im Mittelpunkt der Anwendung steht die Deklaration einer globalen strategischen Basisklasse *K0Modul*. Der Präfix *K0* zeigt an, daß es sich um eine abstrakte Klasse handelt. Von dieser abstrakten Klasse werden Objekte abgeleitet, die als Module zur Anwendung zusammengefügt werden. Für die von der Basisklasse *K0Modul* abgeleiteten Module ist charakteristisch, daß sie alle Funktionen für die Programmablauforganisation, wie z.B. die Funktionen *idle, handle, act* (siehe auch Fig 4), für die Identifikation mittels der Funktion *get_name* und zur Anzeige des Zustand per *print_status*- Funktion von der Basisklasse ableiten bzw. erben. Die Verwaltung der Module erfolgt durch einen Modulhandler *KProgramm*.
Das Konzept einer konfliktfreien echtzeitfähigen Abarbeitungsweise eines Programmsystems für die Erfassung und Übertragung großer Prozeß- und Betriebsdatenmengen basiert auf einer zweistufigen Abarbeitungshierarchie, dessen prinzipielle Wirkungsweise in Fig. 4 dargestellt wird.
Daraus ist zu erkennen, daß ein asynchrones und ein synchrones Abarbeitungsprinzip im Zentrum der zu realisierenden Aufgabe stehen.
Um dem Anwendungsmerkmal "Echtzeitfähigkeit" gerecht zu werden, sind asynchrone Prozeßreaktionen des Programmes auf auftretende Prozeßereignisse (events) unverzichtbar. Dieser unterbrechungsgesteuerte Erfassungsmechanismus hinterlegt die Ereigniswerte in einem statischen Eingangsdatenpuffer definierter Größe. Damit wird eine sehr schnelle Erfassung gewährleistet. Dieses asynchron arbeitende Erfassungsprinzip nimmt keinen unmittelbaren Einfluß auf die übergeordneten komplexen Programmabläufe.
Die Weiterverarbeitung erfolgt dann in einem zweiten synchron arbeitenden Erfassungs- und Verarbeitungsprinzip, das von dem ersten Erfassungs- und Zwischenspeicherprinzip entkoppelt arbeitet.

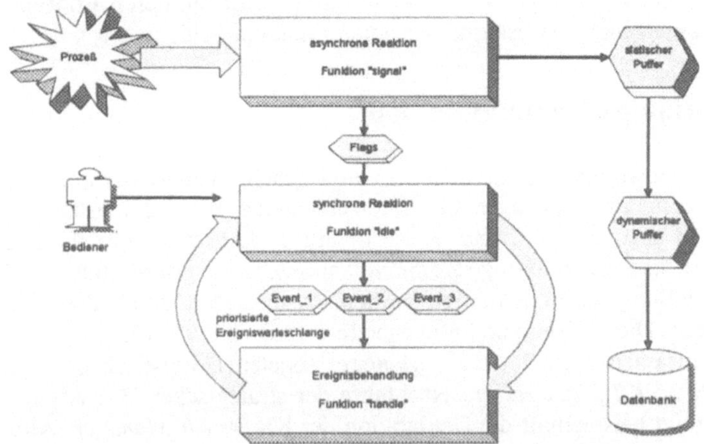

Fig. 4 : Objektorientiertes Architekturkonzept mit asynchronem und synchronem Abarbeitungsprinzip

Das Architekturkonzept arbeitet ereignisorientiert. Den Kern des synchron arbeitenden Programmteils bildet eine Ereignis-Schleife. Um die verschiedenen Systeme des Programms innerhalb einer Task quasiparallel arbeiten zu lassen, ermöglicht ein Scheduler, daß sich diese Ereignis-Schleife in mehreren Schichten schachteln kann. Innerhalb einer solchen synchronen Abarbeitungsschleife werden nacheinander folgende Funktionen zyklisch ausgeführt :

1. Ereignis holen
    Prioritätsorientierte Ereignisabholung aus der Ereignis-Warteschlange

2. *idle*-Funktion aller Programm-Module ausführen
    Sie übernimmt die Daten aus dem statischen Puffer , speichert sie dynamisch, generiert Ereignisse und setzt diese gemeinsam mit den Daten ab. D.h., die Ereignisse werden in die Ereignis-Warteschlange gemäß ihrer Priorität und zeitlichen Reihenfolge einsortiert.

3. Das aktuelle Ereignis wird, falls vorhanden, an alle Programmmodule weitergeleitet und dort ggf. mittels der (*handle*-Funktion) behandelt.

Im Programm muß sichergestellt werden, daß die Zykluszeit der Ereignisschleifen kleiner ist als die erforderliche Reaktionszeit. Bei länger dauernden Aktionen während der Bearbeitung eines Ereignisses (Laufzeit der entsprechenden Funktion ist zu testen) und beim Warten auf eingehende Daten, die innerhalb der *idle*-Funktion empfangen werden, muß der Scheduler Gelegenheit erhalten (Aufruf der *ps*-Funktion), die Abarbeitung unterbrechen und eine Ereignisschleife in einer höheren Schicht einschieben zu können. Der Scheduler sorgt dafür, daß in dieser höheren Schicht nur Ereignisse zur Ausführung kommen, deren Priorität höher ist als die des unterbrochenen Ereignisses auf der unteren Schicht.

5 Echtzeitfähige Kommunikation

Das Zusammenwirken von asynchronen und synchronen Abarbeitungsprinzip mit dem Kommunikationsmanager für unterschiedliche Informationsübertragungssysteme wird in abstrakter Art und Weise in Fig. 5 dargestellt. Für die Informationsübertragung sind echtzeitfähige Kommunikationsdienste erforderlich.
Die Bereitstellung dieser Kommunikationsdienste erfolgt durch den Kommunikationsmanager. Diese Kommunikationsmodule (z.B. KCAN_Manager) werden von einer gemeinsamen Basisklasse *K0_Manager* abgeleitet, wobei anzumerken ist, daß die Basisklasse *K0_Manager* ein Nachfahre der strategischen Basisklasse *K0Modul* ist. Fig. 6 zeigt beispielhaft die Deklaration der Klasse *K0_Manager*. Alle Kommunikationsmanager besitzen folgende Grundfunktionen :

1. Funktion zum Senden von Daten
2. Funktionen und Variablen zum Verwalten eines Sendepuffers (Füllen, Leeren)
3. Funktionen zur Empfangspufferverwaltung

In industrieller Erprobung befinden sich derzeitig folgende Kommunikationsmanager :

1. KCAN_Manager   für den CAN-Bus
2. KRS232_Manager   für die RS232-Schnittstelle
3. KTCP_Manager   für das TCP-Protokoll über IEEE802.3
4. KUDP_Manager   für das UDP-Protokoll über IEEE802.3

Der für den Datentransport notwendige Ablauf wird im wesentlichen über die Funktionen *idle* und *act* gesteuert . In der *idle*-Funktion prüft der Kommunikationsmanager seinen Empfangspuffer auf eingegangene Daten. Diese werden, wenn vorhanden, mittels einer spezifischen Aktion (*act*-Funktion) an alle Module gemeldet. Das Modul, für den diese empfangenen Daten bestimmt sind, läßt sich diese dann vom Kommunikationsmanager übergeben.
Müssen Daten gesendet werden, wird vom sendenden Modul die *send*-Funktion des Kommunikationsmanagers aufgerufen, der die Daten in seinen Sendepuffer schiebt. Der Sendepuffer wird dann in der *idle*-Funktion vom Kommunikationsmanager sukzessive geleert.

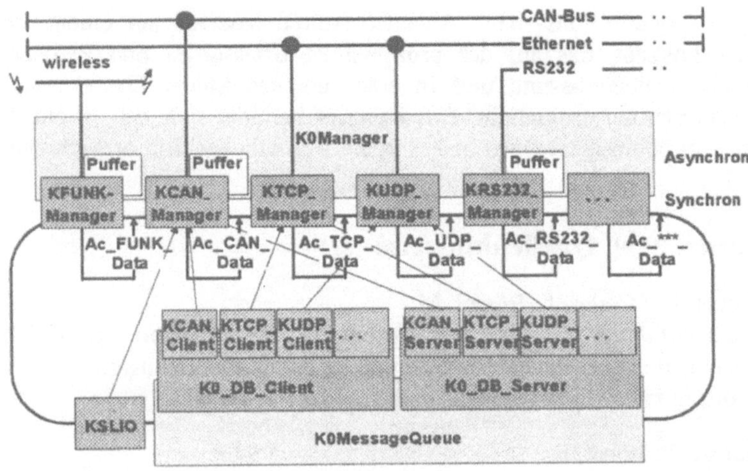

Fig. 5 :   Kommunikations- und Client-Server-Objekte

```
class K0_Manager : public K0Modul
{
protected :       int       sendIn ;
                  int       sendOut ;
                  int       sendSize ;
                  int       sendLen ;
                  void**    Sendepuffer ;


public :          K0_Manager(int=100) ;
                  ~K0_Manager() ;


protected :       BOOL      sendpush(void*) ;
                  void      sendpop() ;
                  void      *sendget() ;

                  virtual void sendflush(BOOL towait=FALSE)=0 ;
};
```

Fig. 6 :   Deklaration der Klasse **K0_Manager**

6 Zusammenfassung

Der Vortrag befaßt sich mit der Vorstellung eines Client-Server-basierten Architekturkonzeptes mit heterogenem Kommunikationsinterface. Das Konzept stellt die
Basis für den Entwurf eines echtzeitfähigen Betriebs- und Prozeßdaten-
erfassungssystems für verteilte Prozesse dar. Der programmtechnische Entwurf ist
konsequent und durchgängig objektorientiert ausgelegt. Die für die Umsetzung not-

wendigen Organisations- bzw. Ablaufstrukturen basieren auf einem zweistufigen Hierarchiekonzept, das auf der prozeßnahen Erfassungs- und Bedienebene eine asynchrone Datenerfassung und in einer übergeordneten Ebene eine synchrone Weiterverarbeitung ermöglicht. Das Konzept befindet sich z.Z. in einer großtechnischen Erprobungsphase, wo bisher keine wesentlichen Einsatzprobleme aufgetreten sind.

Literatur- bzw. Quellenhinweise

1. Löber, P. ; Geiler, I.; Rössel, M.
   Echtzeitfähiges Zellenrechnerkonzept für das Objektorientierte Prozeßdatenmanagement bei Beachtung heterogener Feldbusanbindungen.
   Vortrag zur EchtZeit '95 Karlsruhe, 22.Juni 1995

2. Jarosz, P; Löber, P.
   Microcontroller Based CAN-Bus-Network For Small And Medium Industrial Applications
   TU Bergakademie Freiberg, Institut für Automatisierungstechnik, Juni 1995

3. Fischer, R.; Rössel, M.
   Objektorientiertes Entwurfswerkzeug zur Erstellung von fensterorientierten Bedienoberflächen im Textmodus
   Dokumentation
   TU Bergakademie Freiberg, Institut für Automatisierungstechnik, Juli 1995

4. Jarosz, P.; Löber, P.
   The Link Module (Ver. 2.0) Hardware And Softwarespecification
   TU Bergakademie Freiberg, Institut für Automatisierungstechnik, Juni 1995
   CiA News 3/95

5. Tanenbaum, Andrew S.
   Moderne Betriebssysteme
   Carl Hanser Verlag München Wien 1994

6. Pirklbauer, Klaus
   Objektorientierte und konventionelle Automatisierungssoftware im Vergleich
   Institut für Informatik, Johannes Kepler Universität Linz, Juni 1994

7. Schöbel, Eckhard
   Objektorientierte Softwaretechnologie in den Komponenten des Steuerungssystems VACNT
   Tagungsband der 28. Jahrestagung des SAK vom 06.05. - 07.05.1997 in Dresden

8. WinCC
   Dokumentation A04 V3.0
   Siemens AG 7/97

Smart Bus Nodes in Feldbussystemen

Helmut Beikirch

Hochschule Wismar (FH)
Fachbereich Elektrotechnik und Informatik

Abstract. Demands on electronic components in the near of sensors grow by distributed automation processes. Bus node or bus participant is the name of such components. The components are working in field busses or sensor actuator busses. It's besides necessary to execute an open communication (ISO Layer 7) with the increasing implementation to signal conditioning functions in to the bus nodes. Such electronic parts must have an effective hardware and a high software performance. The demand of the bus node electronic are small hardware expense and very low power supply because of the distribution in the industrial process field. Electronic components with such demands and performances are called „Smart Bus Node" (SBN). Concepts and solution ways will be shown based on general principles and real applications.

1 Problematik, Aufgabe

Industrielle Netzwerke für den sensornahen Bereich, die allgemein als Feldbussysteme oder Sensor-/Aktorbussysteme bezeichnet werden, sind in ihrer Leistungsfähigkeit sehr unterschiedlich konzipiert. Mit der breiteren Verfügbarkeit hochintegrierter mikroelektronischer Bausteine kann der Leistungspegel dezentraler Netzwerke erheblich vergrößert werden. Die Verteilung „intelligenter" Funktionen auf dezentrale Busknoten in Feldbusnetzen eröffnet neue Dimensionen bei der Erfassung, Verarbeitung und Verteilung von Informationen.

Während sich standardisierte Feldbusnetzwerke wie beispielsweise Profibus, CAN (Controller Area Network), Interbus-S, LON (Local Operating Network) oder PNET durch organisierter Kommunikation entsprechend dem OSI-Schichtenmodell der ISO etablieren, wird der Umfang an Funktionalität in Sensornähe durch die Busknotenelektronik (Hard- und Software) bestimmt. Es besteht nun die Aufgabe, entsprechend einer umfangreichen Reihe von Anforderungen das Leistungsspektrum sensornaher Busknoten zu definieren. Bei der Formulierung von Anforderungen an die sensornahe Busknotenelektronik müssen künftige Entwicklungstrends unbedingt berücksichtigt werden. Zu diesen Anforderungen zählen folgende Schwerpunkte:

a) Bearbeitung eines Protokolls zur Kommunikation über Schicht 7 oder vergleichbare Schichten nach dem OSI-Modell $\Rightarrow$ Realisierung von Kommunikationsfähigkeit auf höherer Ebene;

b) „intelligente" sensornahe Funktionalität (beispielsweise Signalverarbeitung, Autokalibration, Eigenüberwachung, Diagnose, Prognose, wissensbasierte Entscheidungen);

c) Minimierung des Stromverbrauchs und des Hardwareaufwands; Zielstellung ist Realisierung von Single Chip Bus Nodes mit einem Leistungsbedarf im unteren mW-Bereich;

d) Hardwareflexibilität; mit der gleichen Hardware werden durch angepaßte Konfiguration bzw. On-chip-Umprogrammierung unterschiedliche Anwendungen realisiert; Nutzung analoger und digitaler PLDs bzw. FPGAs.

Diese sich teilweise widersprechenden Anforderungen (High Speed Signal Processing → Low Power Consumption) sind nur durch stufenweiser Realisierung umsetzbar. Umfangreiche Erfahrungen und langjährige Untersuchungen auf dem Gebiet des Entwurfs sensornaher feldbusvernetzter Komponenten führten zur Spezifikation von Busknoten, die für die Realisierung der genannten Anforderungen geeignet sind. Aufgrund der Leistungen und Funktionen werden solche Busknoten als *„Smart Bus Nodes" (SBN)* bezeichnet.

2 Smart Bus Nodes

Eine hohen Leistungsfähigkeit bei gleichzeitiger Effizienz an Hardwareaufwand und Stromverbrauch sind typische Kennzeichen für *Smart Technology*, nach dessen Regeln *Smart Bus Nodes* entworfen werden. Die geforderte hohe Leistungsfähigkeit kann nur durch einen geeigneten Mikrocontroller erreicht werden. Dieser Mikrocontroller darf aber bei der hohen Verabeitungsgeschwindigkeit nur einen sehr geringen Strom verbrauchen. In umfangreichen wissenschaftlichen Untersuchungen wurden geeignete Mikrocontrollerschaltungen analysiert und für die Eignung in *Smart Bus Nodes* bewertet. Die Kriterien spezifiziert durch Forderungen nach

- Verlustleistungsminimierung (wenige mW im aktiven Betrieb);

- Low Voltage-Technologie (3V oder geringer); steuerbare Taktteiler und Dual Clock-Systeme; über Software steuerbare Abschaltung prozessorinterner Systemkomponenten;

- kurze Zykluszeiten für die interne Befehlsbearbeitung (Bedingung für hohe Signalverarbeitungsgeschwindigkeit);

- ausreichender Programmspeicher (> 16kB ROM für Implementierung einer höheren Kommunikationsschicht und Anwendungen); ausreichender Arbeitsspeicher und nichtverlierbarer Parameterspeicher;

- implementierte I/O-Controller (Feldbuscontroller u.a.);

- Verfügbarkeit und Kosten für Bauelemente und Entwicklungswerkzeuge.

Mit diesen Schwerpunkten stehen hohe Ansprüche an die Gestaltung der Busknoten. Da besonders der Energieverbrauch der Teilnehmer in dezentralen Strukturen so gering wie möglich gehalten werden muß, können die dafür geeigneten Mikrocontroller nur unter bestimmten Restriktionen betrieben werden. Untersuchungen und Analysen der

Leistungsfähigkeit von Mikrocontrollern zeigten, daß nur wenige Typen in 16-Bit-Technik derartigen Ansprüchen gerecht werden [8]. Der Vergleich und die Auswahl geeigneter Mikrocontroller erfolgte auf der Basis typischer Parameter. Eine besondere Wichtung wurde dem Power Efficiency Factor (PEF - Verhältnis Verarbeitungsgeschwindigkeit zu der dabei aufgenommenen Leistung) gegeben, der für die in Frage kommenden Controller zwischen 130 und 560 MIPS/Ws lag [3] [8].

Bei künftigen Anwendungen spielen Netzübergänge zwischen Netzen des Automatisierungsbereichs und Netzwerken des Kommunikationsbereichs (z.B. Intranet, Internet) eine ebenso wichtige Rolle wie alternative und hybride Stromversorgungskonzepte (Lichtenergie, Bewegungsenergie, Wasser, Luft, biologische Energie) der angeschlossenen Netzwerkelektronik. Alternative Stromversorgungen sind effektiv erst möglich, wenn Busknoten über eine extreme Betreibsleistungsreduzierung verfügen.

2.1 Aufbau

Der systematische Entwurf von *Smart Bus Nodes* basiert hardwareseitig auf einer strukturierten Anordnung und Verknüpfung von Funktionsbausteinen. Fig. 1 zeigt die beiden Grundvarianten, einmal mit externer Stromversorgung und zum anderen mit der busintegrierten Stromversorgung. Es ist leicht zu erkennen, daß die Anzahl der Klemmen für die Kontaktierung der Interfaceschaltungen mit der busintegrierten Stromversorgung weiter minimiert wird. Besonders für Mikrosysteme und miniaturisierte Hardware ist dieser Aspekt bedeutend.

Kern der Struktur bildet ein leistungsfähiger Mikrocontroller mit einer ausreichenden Speicherbestückung. Das Prozeßinterface beinhaltet den Sensor-/Aktoranschluß mit Digitalisierung und Signalanpassung. Hier ist ein programmierbares Array (analog/ digital; softwaremäßig ladbar und umkonfigurierbar) zur flexiblen Kontaktierung unterschiedlicher Sensoren bzw. Aktoren vorgesehen. Das Businterface beinhaltet den Controller für das jeweilige Feldbussystem sowie zugehörige Leitungtreiber. Der Buscontroller wird vorzugsweise auch als mikrocontrollerinterne Interfaceschaltung bereitgestellt.

2.2 Funktionalität

Der Gesamtumfang an Funktionalität von *Smart Bus Nodes* wird von den Prozessorleistungen und letztendlich von den implementierten Anwendungen bestimmt. Grundsätzlich müssen für die Implementierung entsprechender Software Voraussetzungen geschaffen werden.

Neben herkömmlichen Lauf- und Überwachungsfunktionen, die Prozessor- und Speicherkontrolle gewährleisten, ist ein Management zu installieren, das die Komponenten Kommunikation, Processing/Control, Signal Acquisition/Signal Output und Recources Administration organisiert. Das Timing bzw. die interne Zeitverwaltung wird meist vom angeschlossenen Feldbus restriktiert oder zumindest mitbestimmt. Mit dem Feldbustiming muß die Sensor-/Aktordynamik synchronisiert werden. Gute Ergebnisse erreicht man mit Echtzeitkernen oder auch mit einfachen Dispatchern.

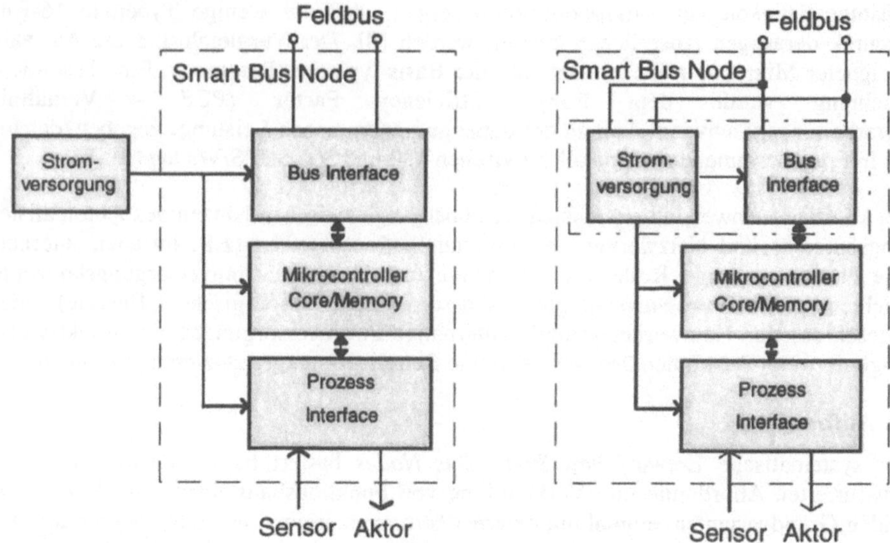

Fig. 1. Aufbau von Smart Bus Nodes (SBN) mit externer Stromversorgung und mit busintegrierter Stromversorgung (Prinzip)

Der Forderung nach extremer Betreibsleistungsreduzierung kann durch ein Spektrum von Maßnahmen nachgekommen werden. Durch die ereignisabhängige Reduzierung der internen Verarbeitunggeschwindigkeit kann neben der bekannten Absenkung in den Sleep-, Idle- oder Stop-Betrieb ein günstigeres internes „Switching" zu beachtenswerten Leistungsreduzierungen führen. Zeitverzögerungen für die Reaktivierung müssen dabei minimal gehalten werden. Das Spektrum der Maßnahmen umfaßt neben der anwendungsabhängigen Taktumschaltung (kontinuierliche Taktteiler oder Dual Clock-Betrieb $\Rightarrow$ z.B. fließende Taktumschaltung zwischen 10MHz und 38kHz) die Abschaltung interner nicht benötigter I/O-Komponenten (z.B. Timer, ADU, Counter). Ein gut organisierter Switch-Betrieb (Umschaltung zwischen Run und Sleep) kann zu Leistungsabsenkungen von bis zu 95% führen. Schnell rechnende Prozessoren haben für die gleiche Verarbeitungsfunktion längere Pausen als langsame Prozessoren. Das ist der Hauptgrund für die Auswahl von 16-Bit-Mikrocontrollern für diese Aufgabe. 8-Bit-Prozessoren sind zu langsam uns 32-Bit-Prozessoren verbrauchen derzeitig noch zuviel Energie.

3 Anwendung in einer fluidischen Anlage

Anhand eines Referenzobjekts wird der Entwurf eines *Smart Bus Nodes* auf der Basis eines Low Voltage-Konzepts (Mikrocontroller Toshiba TMP93CM41) für das CAN-Bussystem SDS (Smart Distributed System von Honeywell) durchgeführt. Fig. 2 zeigt die wesentlichen Baugruppen des Smart Bus Nodes im Blockschaltbild. Diese Implementierung, die von prinzipieller Bedeutung für weitere Entwurfsarbeiten ist, wurde für eine Anwendung in der Fluidik vorgesehen (gefördertes Forschungsprojekt). Fig. 3 zeigt die Anordnung der einzelnen Baugruppen zur Wegmessung und Steuerung eines Hydraulikzylinders in einer Anwendung der Fertigungstechnik. Hohe Anforderungen werden an den Busknoten durch eine große Sensordynamik und Präzision bei der Wegmessung gestellt.

Der Hydraulikzylinder ist nach einem auswählbaren Fahrprogramm zu steuern. Die Erfassung des Bewegungsablaufs muß mit einer hohen Auflösung erfolgen. Für die praktisch geforderte Kolbenpositioniergenauigkeit von 0,1mm...0,2mm und einer Kolbenlänge von 200mm (Wegänderung) ergibt sich eine notwendige Auflösung von 0,05%... 0,1%. Es muß also mit einer Auflösung der Digitalisierungsschaltung von mindestens 12 Bit gearbeitet werden.

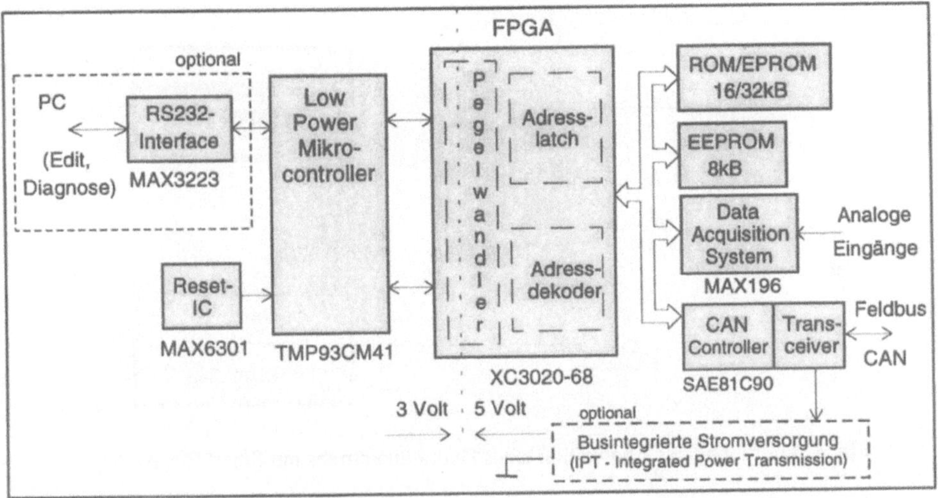

Fig. 2. Blockstruktur des SDS Busknoten

Der ausgewählte Prozessor TMP93CM41 kann mit 3V Betriebsspannung beschaltet werden. Im Entwurfsstadium wurden alle anderen Komponenten in 5V-Technik eingesetzt, da der CAN-Controller (Siemens SAB 81C90) nur mit 5V betrieben werden kann. Die Pegelanpassung zwischen den Komponenten sowie die Integration der notwendigen digitalen Zusatzschaltungen wurde durch ein FPGA (Field Programmable Gate Array) XC3020 vorgenommen. Diese FPGAs sind in CMOS-Technik ausgeführt und verbrauchen nur sehr geringen Ruhestrom. Mit PLDs oder CPLDs ist diese Schaltungstechnik nicht möglich (höhere Betriebsströme). Ebenfalls ist durch die einfache Tristate-Steuerung und die wahlweise Pegeleinstellung (TTL/CMOS) eine leichte Pegelanpassung zwischen 3V und 5V ohne großen Zusatzaufwand möglich. Für eine spätere Implementierung der Speicherfunktionen in den Controllerchip entfallen einige Baugruppen. Die Signalerfassung wurde bei diesem Entwurf mit einer separaten Signalerfassungbaugruppe (MAX196) realisiert. Sie weist einen sehr niedrigen Strombedarf auf und kann softwaremäßig gesteuert (Leistungsabsenkung) werden. Der Programmspeicherumfang von 16kB wird bei der Implementierung einer höheren Kommunikationsschicht sowie weniger Anwendungsprogramme schnell ausgeschöpft.

Der interne RAM von 2kB ist hat sich als ausreichend erwiesen. Das EEPROM arbeitet als Parameterspeicher und kann auch seriell ausgeführt sein.

Die Implementierung der SDS-Funktionen erfolgten über eine systematische Modellbildung [3]. Sie kann aufgrund des allgemeinen Ansatzes für nachfolgende Anwendungen erweitert bzw. leicht angepaßt werden.

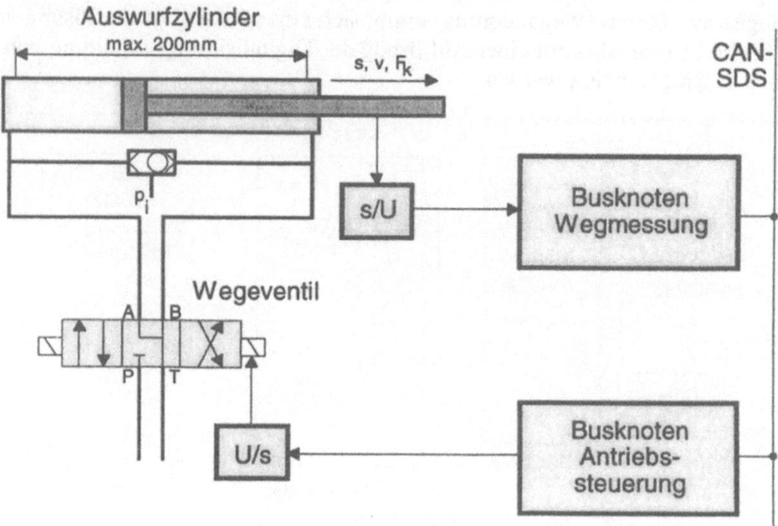

Fig. 3. Anordnung der Applikation eines Hydraulikantriebs mit Smart Bus Nodes

Bei praktischen Tests des Prototypen zeigte sich, daß bei Verarbeitungsleistungen (z.B. Gleitkommaarithmetik) große Prozessorzeitbelastungen auftraten. Die Leistungsbilanz wird durch diesen Faktor besonders beeinflußt. Hier ist durch bessere Software und optimierte Compiler mit den nächsten Prozessorgenerationen eine bessere Leistung zu erwarten (beispielsweise Mitsubishi M16C).

Fig. 4 zeigt die oszillographierten Bilder der Stromaufnahme des Busknotens unter verschiedenen Bedingungen im Switch-Betrieb. Für den 1. Fall bewegt sich das Eingangssignal nicht und es kommt zu einer sehr kurzen Verarbeitungszeit. Bei einer Prozeßzeitkonstante (max. zyklischer Verarbeitungsabstand) von 1s kommt es zu 203µs Verarbeitungszeit. Steigen die Berechnungsleistungen (Normierung, Skalierung, Triggerbewertung, ...), wird entsprechend dem 2. Fall (Fig. 4) das Verhältnis Verarbeitung zu Pause immer ungünstiger. Die Energieeinsparung bewegt sich von 90% im 1. Fall abwärts bis auf 0% bei ständiger Verarbeitung. Wird durch den ausbleibenden Switch-Betrieb keine Verlustleistung eingespart, so liegen die Meßwerte der max. Stromaufnahme beim Entwicklungsmuster bei 190mW. Die Benutzung des internen ROMs führt dann bei der Serienausführung zu einer weiteren Reduzierung auf max. 70mW.

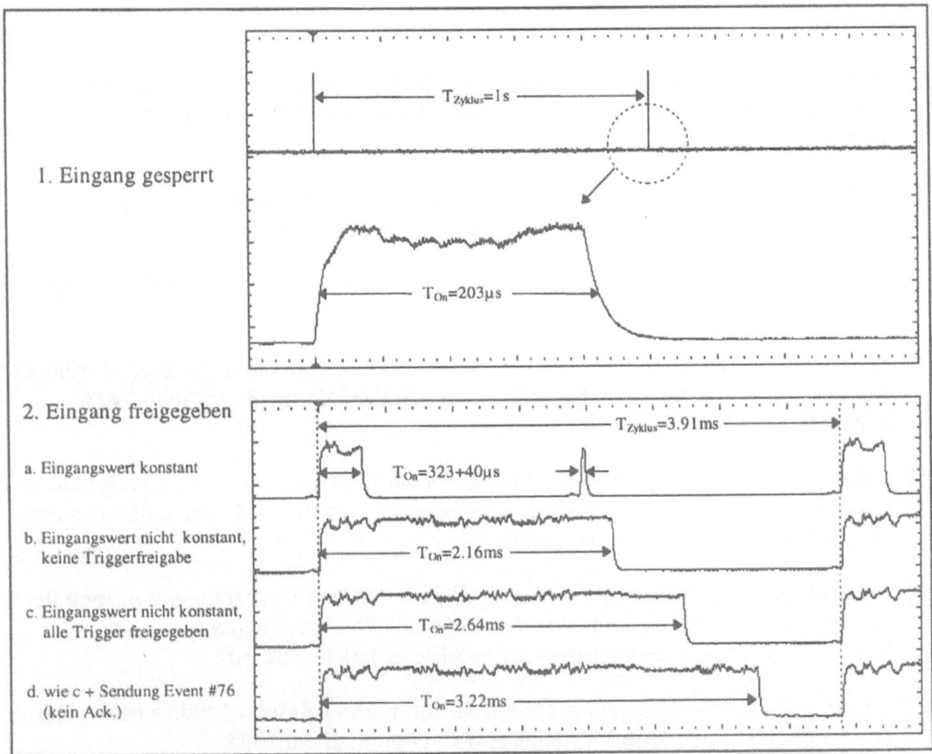

Fig. 4. Meßtechnisch ermittelte Stromaufnahme bei Switch-Betrieb [3]

4 Trend, Entwicklungsweg

Das vorgestellte Beispiel zur Realisierung von Smart Bus Nodes zeigt nur andeutungsweise die technologischen Möglichkeiten, die in den nächsten Jahren die Elektronikkomponenten in Feldbusnetzen beeinflussen werden. Hochintegrierte Schaltkreise mit Mikrocontrollerfunktionen werden für hohe Signalverarbeitungs- geschwindigkeiten bei niedrigen Betriebsleistungen (Betriebsspannung bis unter 1V) sorgen. Als Prozeßinterface werden flexible softwaremäßig konfigurierbare ASICs mit analogen und digitalen Schaltungen einsetzbar sein. Derartige Busknoten werden als Mikrosysteme in Sensorköpfe oder an Aktoren einfach installiert werden können. Über ein intelligentes Kommunikationsinterface lassen sich leicht Inbetriebnahme, Echtzeitbetrieb und Diagnose abwickeln. Industrienetzwerke werden über Gate Ways oder Bridges nahtlos in die Kommunikation der globalen Netze eingebunden werden können.

254

Literatur

[1] *Wehrmann, S.:* Untersuchung des Kommunikationsverhaltens in fluidischen
Anlagen. Diplomarbeit, Otto-von-Guericke-Universität Magdeburg, IPE,
Diplomarbeit, 1996

[2] *Beikirch, H.:* Sensornahe serielle Bustechnik. Wismarer Automatisierungs-
symposium, 17./18. Sept. 1996, Tagungsband, S. B3-1 bis 3-9

[3] *Voß, M.:* Smart-Busknoten in Feldbusnetzen. Diplomarbeit, Universität Rostock,
Fachbereich Elektrotechnik, 1997

[4] *Seifart, M.; Beikirch, H.; Rauchhaupt, L.:* Anforderungen und Leistungen serieller
Bussysteme im sensornahen Prozeßbereich. iNet'93, Kongreß-Vorträge 1993,
S. 59-66

[5] *Schultze, H.:* Schaltungskonzepte für Mikrocontrollerbaugruppen zum Anschluß an
bussysteme mit integrierter Hilfsenergieübertraung. Otto-von-Guericke-Universität
Magdeburg, IPE, Diplomarbeit, 1995

[6] *Beikirch, H.; Rauchhaupt, L.; Schultze, H.:* Low-Power Busknoten mit dezentraler
„Intelligenz". Kongreß „Embeddet Intelligenz '96", Sindelfingen,
14.-16. Febr. 96, Konferenzband Design&Elektronik S. 302-307

[7] *Lipsdorf, T.:* CAN-Applikationssystem für fluidische Antriebe. Studienarbeit, Otto-
von-Guericke-Universität Magdeburg, IPE, Diplomarbeit, 1995

[8] *Beikirch, H.:* Kommunikationsfähige Low-Voltage-Mikrocontrollerstruktur.
Kongreß „Embeddet Intelligenz '97", Sindelfingen, 19.-21. Febr. 1997,
Kongreßband S. 351-358

Codegenerator für Neuron-C

R. Pühringer, R. Schmit, M. Manninger

Institut für Computertechnik, Gußhausstr. 27-29, A-1040 Wien

Kurzfassung. Aufgrund der immer weiteren Verbreitung von LONWORKS-Anwendungen stellen wir eine Möglichkeit vor, verteilte Systeme, die auf dem Neuron-Chip basieren, mit einer SDL-Entwicklungsumgebung erstellen und warten zu können. SDL ist eine formale Beschreibungssprache, die für den Kommunikationsbereich entwickelt wurde und mittlerweile durch die ITU international genormt wurde. Mit dem von uns entwickelten Werkzeug werden LONWORKS-Anwendungen mit Hilfe eines SDL-Editors konzipiert, wobei sich auch die Netzwerkstruktur planen läßt. Die SDL-Beschreibung der Anwendung wird dann dem Codegenerator übergeben, der daraus Neuron-C-Programme für die einzelnen Knoten und eine Netzliste zum Binden des Netzwerks erstellt.

Abstract. As a result of the growing industrial acceptance of LONWORKS applications we present an SDL development kit for design and maintenance of distributed control networks based on the NEURON chip. SDL is an ITU recommendation of a formal description language that has originally developed for communication applications. With the use of our tool it is possible to develop LONWORKS-applications with a graphical SDL editor and to plan the structure of the network, too. The SDL description of the application is then passed to the code generator, which builds NEURON-C programs for each single node and a netlist for the binding of the network.

1 Einführung

Nicht zuletzt durch die breite Anwendung der Feldbussysteme im Bereich der Automatisierungstechnik sind die Vorteile dezentraler Systeme allgemein bekannt. Entlastung der übergeordneten Steuerungsebene, Fehlertoleranz und Flexibilität bedingen jedoch einen erhöhten Aufwand und eine neue Denkweise bei der Anwendungsentwicklung. Besonders kommt dieser Effekt beim Feldbussystem LONWORKS zum Tragen, da dieses total dezentral konzipiert wurde.

Der Entwurf, die Programmierung und die Wartung verteilter Systeme können nur mit einem entsprechenden Werkzeug gut erledigt werden. Diese Werkzeuge zwingen zu einer sauberen Spezifikation der Abläufe und bewirken dadurch die Vermeidung vieler Designfehler, die ohne ihre Anwendung fast zwangsläufig entstehen. Ein solches Werkzeug wurde auch am Institut für Computertechnik entwickelt. Codegeneratoren erzeugen aus den spezifizierten Abläufen automatisch Programmcode und vervollständigen somit das Design-Tool zu einem einfachen CASE-Tool.

Die hier vorgestellte Arbeit gibt dem Entwickler die Möglichkeit, mit den Methoden der formalen Beschreibungssprache SDL (siehe 2.1) Anwendungen für ein spezielles verteiltes System, das Feldbussystem LONWORKS, zu erstellen. Der Vorteil von SDL, eine Definition eines Systems nur aus der Beschreibung seines Verhaltens heraus zu erstellen, wird dabei auf die verteilte Knotenprozessorarchitektur von LONWORKS übertragen.

2 Basiskomponenten

Bevor die Architektur unseres Tools dargestellt wird, beschreiben wir nun kurz die beiden wesentlichen Gebiete, auf denen diese Arbeit beruht, nämlich SDL und LONWORKS.

2.1 SDL als Entwicklungsmethode

SDL (Specification and Description Language) wurde für Anwendungen im Telekommunikationsbereich entwickelt und besitzt mittlerweile einen mächtigen Sprachumfang, der von der ITU genormt [1] wurde. SDL ist eine formale Spezifikations- und Beschreibungssprache, mit der ein System nur durch die Beschreibung seines Verhaltens entworfen werden kann. Neben der textuellen Form SDL/PR (*Phrase Representation*), besitzt SDL als einzige formale Beschreibungssprache eine vollständige graphische Darstellungsform (SDL/GR). Dieser Umstand hat maßgeblichen Anteil an der großen Akzeptanz dieser Sprache.

SDL basiert auf dem Modell erweiterter endlicher Automaten, die miteinander asynchron über möglicherweise verzögernde Signalwege und unbegrenzte Puffer kommunizieren. In einem SDL-System gibt es einen oder mehrere Prozesse, deren Anzahl sich im Laufe der System-Lebenszeit verändern kann. Im allgemeinen kommuniziert nicht jeder Prozeß mit jedem anderen, sondern die Kommunikation ist durch die statische Struktur des SDL-Systems vorgegeben. Statische SDL-Diagramme zeigen also die einzelnen Prozesse und die zwischen ihnen verlaufenden Signalwege, wobei der Übersichtlichkeit halber mehrere Prozesse zu einem Block zusammengefaßt werden können.

Dynamische SDL-Diagramme sind Prozeßgraphen die das innere Verhalten eines Prozesses beschreiben. Dabei kann jeder Prozeß, ausgehend vom Zustand Start, mehrere Zustände einnehmen. Wesentlich ist nun, wodurch und wie ein Übergang von einem Zustand zu einem anderen geschehen kann. Während eines Zustandsüberganges, der immer durch eine Eingabe ausgelöst wird, können verschiedene Aktionen, wie etwa Zuweisungen und Ausgaben, durchgeführt werden.

Eine gute Einführung in SDL findet sich in [2]. Für eine weitere Vertiefung wird auf [3] und [1] verwiesen.

2.2 Das Feldbussystem LON

LON (Local Operating Network) ist ein Feldbussystem, das auf dem Neuron-Chip basiert. Je nach Anwendung enthält ein Knoten neben dem Neuron-Chip noch die Schnittstelle zur angeschlossenen Hardware und die Netzwerkanbindung. Die Programmierung der einzelnen Knoten erfolgt in Neuron-C, das ist ANSI-C, erweitert um die speziellen Befehle für den Neuron-Chip [4]. Durch das im Chip implementierte Netzwerkprotokoll LONTALK muß sich der Programmierer nicht um die Verteilung von Nachrichten an die restlichen Chips kümmern. Die Kommunikation geschieht über sogenannte Netzwerkvariablen, die im Quelltext definiert werden und deren Verbindung beim Netzwerkbinden am LONBUILDER festgelegt wird.

Auf eine weitere Beschreibung wird hier verzichtet, zur Information wird auf [5] verwiesen.

2.3 Aufbau der CASE-Umgebung

Als Arbeitsumgebung dient ein SDL-Tool (Abb. 1), das am Institut für Computertechnik entworfen wurde und bei der FeT'95 erstmals präsentiert wurde [6]. Dieses enthielt bisher neben den Grundkomponenten SDL/GR-Editor, SDL/PR-Import bzw. -Export und der Umsetzung in einen speziellen Zwischencode nur einen Codegenerator für C, wird aber nun mit einem weiteren Codegenerator für Neuron-C ausgestattet.

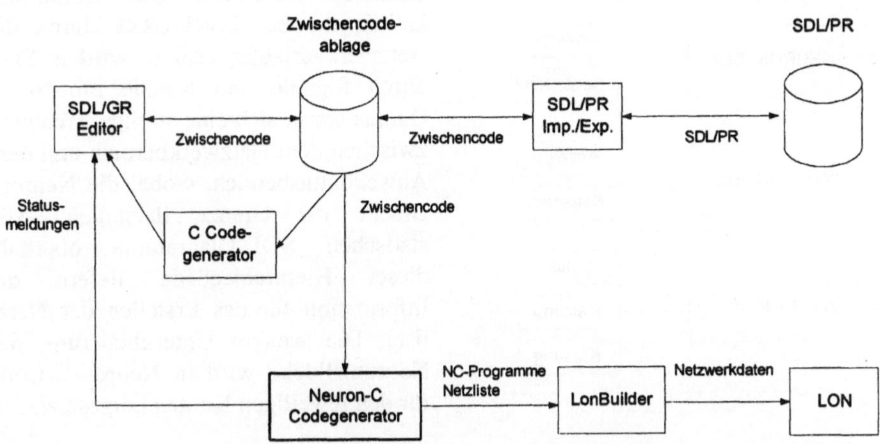

Abb. 1. Arbeitsumgebung des Codegenerators

Dieser Codegenerator muß alle SDL-Symbole, die zur vollständigen Beschreibung einer LONWORKS-Anwendung notwendig sind, in Neuron-C übersetzen. Weiters muß die Information über den Aufbau des Netzwerks dem LONBUILDER zur Verfügung gestellt werden, damit dieser das Netzwerk automatisch binden kann.

3 Spezifikation einer LONWORKS-Anwendung in SDL

Herkömmliche, kommerzielle Tools sind dafür ausgelegt, ein möglichst breites Spektrum von Zielsystemen und Anwendungen zu unterstützen. Der generierte Code ist meist zu allgemein gehalten und kann vielfach in zeitkritischen Anwendungen nicht eingesetzt werden. Unser Ziel ist es, einen möglichst schnellen und kurzen Code zu generieren, der speziell für den Neuron-Chip optimiert wird.

3.1 Spezifikation der Netzstruktur im SDL-Diagramm

Da einige SDL-Sprachelemente auf LONWORKS-Anwendungen nicht anwendbar sind, erfolgte eine Beschränkung auf eine Teilmenge, mit dem sich eine Anwendung sowohl vollständig und SDL-konform beschreiben läßt, als auch die Umsetzung in Neuron-C ermöglicht.

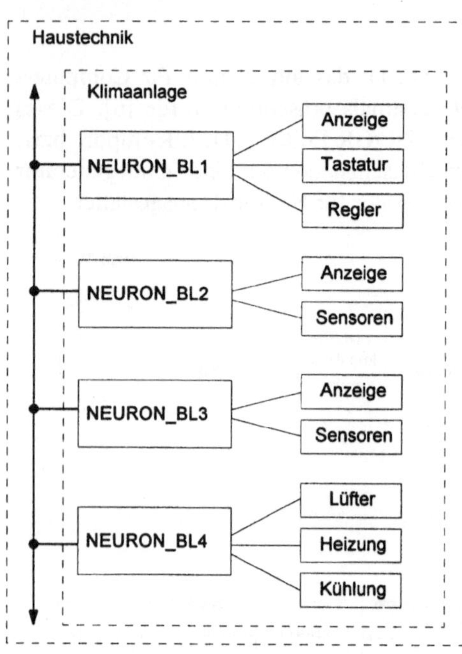

Abb. 2. Ausschnitt aus einer LONWORKS-Anwendung

Die Struktur des LONWORKS-Netzwerks spiegelt sich im SDL-Diagramm wider. D.h., die Hardwareknoten entsprechen Blöcken, die durch ein Schlüsselwort (*NEURON_*) gekennzeichnet sind, und die Kanäle repräsentieren das verbindende Netzwerk. Die Kommunikation, die bei LONWORKS durch die Netzwerkvariablen erfolgt, wird in SDL durch Signale und Kanäle modelliert. Daraus ergibt sich eine strenge Trennung zwischen dem Netzwerkbereich und dem Anwendungsbereich, wobei die Neuron-Blöcke die Grenze darstellen. Alle statischen SDL-Diagramme oberhalb dieser Hierarchieebene liefern die Information für das Erstellen der Netzliste. Die weitere Untergliederung der Neuron-Blöcke wird in Neuron-C-Code für den jeweiligen Knoten umgesetzt.

Als Beispiel siehe dazu eine LONWORKS-Anwendung im Bereich der Gebäudetechnik. Abbildung 2 zeigt als Ausschnitt die Klimasteuerung, wobei NEURON_BL1 dem Benutzerterminal, NEURON_BL2 und NEURON_BL3 den Meßknoten und NEURON_BL4 der Regelung entspricht. Die SDL-Darstellung dieses Sachverhalts sieht man in den Abbildungen 3 und 4. Auf der Systemebene befindet sich der übergeordnete Block *Klimaanlage*, der über *Kanal1* mit dem Rest des Systems kommuniziert und dessen Spezifikation Abb. 4 zeigt. Aus Platzgründen ist hier nur ein Neuron-Block vollständig beschrieben. Die Hardwareeinheiten aus Abb. 3 sind hier als Environment-Blöcke dargestellt.

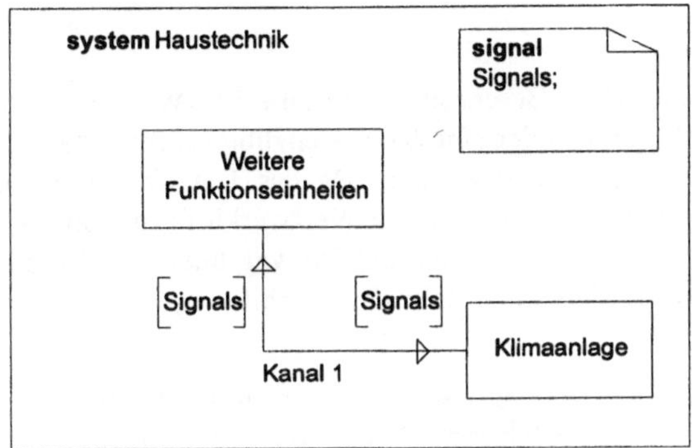

Abb. 3. SDL-Repräsentation von Abb. 2

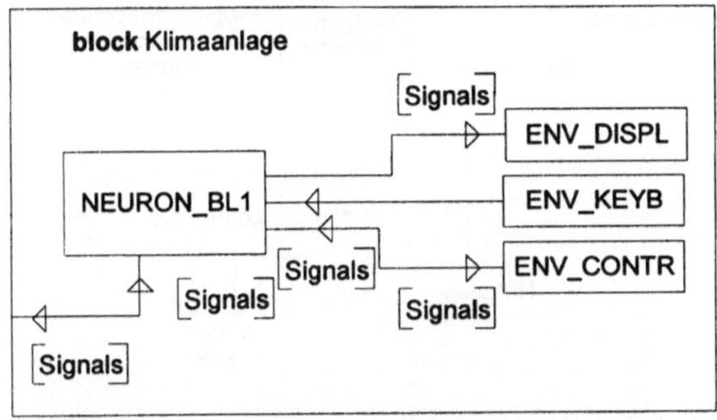

Abb. 4. Verfeinerung der Klimaanlage aus Abb. 3

3.2 Schnittstelle zur Umgebung

Das Environment stellt ein wesentliches Abstraktionselement von SDL dar. Man versteht darunter diejenigen Teile, die nicht im eigentlichen SDL-System spezifiziert werden müssen. Es genügt hier die Festlegung und die Zuordnung der Signale. In unserem Fall ist aber das Environment ein wesentlicher Bestandteil des zu realisierenden Gesamtsystems, und vor allem ist es sehr hardwarenahe. Es muß also die Möglichkeit geben, angeschlossene Hardwareeinheiten anzusprechen und mit Parametern zu versorgen. Da aber im SDL-Diagramm kein Sourcecode eingegeben werden soll bzw. kann, findet die Adressierung der angeschlossenen Hardware über Environment-Blöcke statt (Abb. 4). Ein solcher ist über einen Kanal direkt an einem Neuron-Block angeschlossen und trägt zur Identifikation durch den Codegenerator das Schlüsselwort *ENV_*. Aufgrund der besonderen Gegebenheiten einer LONWORKS-Anwendung kann davon ausgegangen werden, daß viele Knoten identisch aufgebaut sind, oder zumindest identische Hardwareeinheiten in Verwendung sind. Die zur Ansteuerung dieser Einheiten notwendigen Programmrümpfe werden in einer Funktionsbibliothek abgelegt und im SDL-Diagramm über den vorgegebenen Namen angesprochen, z. B. eine Anzeige durch *ENV_DISPL_xxx*, wobei xxx für den entsprechenden Bezeichner steht.

4 Realisierung

Die Übersetzung einer SDL-Beschreibung in eine LONWORKS-Anwendung geschieht in zwei Teilen. Zu Beginn findet eine Konsistenzüberprüfung statt. Danach erfolgt für jeden Neuron-Block die Codeerzeugung. Da das Dateiformat nicht offengelegt ist, besteht derzeit noch keine Möglichkeit, die Netzwerkinformation für den LONBUILDER automatisch zu erstellen. Es wird daher ein Umweg über eine Netzliste gegangen, die händisch eingegeben werden muß. Siehe dazu Abbildung 5.

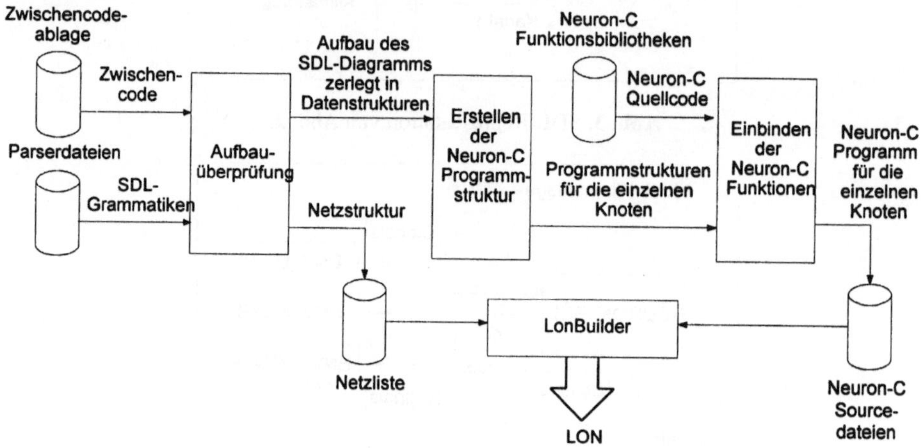

Abb. 5. Entstehungsprozeß einer LONWORKS-Anwendung

4.1 Aufbauüberprüfung und Erstellung der Netzliste

Wie bereits unter 2.1 erwähnt, spiegelt sich die Struktur des Netzwerks im SDL-Diagramm in der Anordnung der Kanäle in der Blockhierarchie oberhalb der Neuron-Blöcke wider. Vom Codegenerator wird diese Kanalstruktur nun in mehreren Schritten durchsucht, die kurz beschrieben werden sollen.

- Von der Systemebene bis zur Prozeßebene muß genau ein Neuron-Block vorkommen.
- Oberhalb jedes Prozesses muß ein Neuron- oder Environment-Block liegen.
- Ein Environment-Block ist über einen Kanal direkt mit einem Neuron verbunden. Sind mehrere gleiche Hardwareeinheiten an einem Neuron-Block angeschlossen, so müssen sich deren Namen unterscheiden.
- Signale, die hier den Netzwerkvariablen entsprechen, müssen einen Startblock und einen Endblock haben.

Um die Signalnamen und Variablendeklarationen aus der SDL-Spezifikation zu erhalten, werden die Textsymbole und Signallisten mit einem Parser durchsucht. Die generierte Netzliste enthält die Programmzuordnung zu den einzelnen Knoten und die Information, welche Netzwerkvariablen miteinander verbunden gehören. Daraus können später die Dateien für den LONBUILDER erstellt werden, falls das benötigte Dateiformat bekannt gegeben wird.

4.2 Realisierung des Environments

Die notwendigen Funktionen zur Ansteuerung externer Hardware werden in Funktionsbibliotheken zusammengefaßt und über einen vorgegebenen Namen angesprochen. Für eine einfache Anzeige sind dies folgende Funktionen.

- Init()   Einstellung der Anzeigenbetriebsart.
- Clear()   Anzeige löschen.
- Key(Param)   Den Buchstaben *Param* ausgeben; Überprüfung auf Zeilenüberlauf.
- Text(Param)   Den Text *Param* ausgeben; ebenfalls mit Zeilenüberlaufprüfung.

Die Adressierung der externen Hardware erfolgt über SDL-Task-Anweisungen. Dazu kann ein spezielles Feature des verwendeten Editors eingesetzt werden. Es ist somit möglich, mit jedem SDL-Symbol einen Kommentar zu übergeben, der vom Codegenerator direkt als C-Funktionsaufruf ausgewertet wird. In Abb. 6 ist dies im Task-Symbol durch ein vorangestelltes „C:" angedeutet.

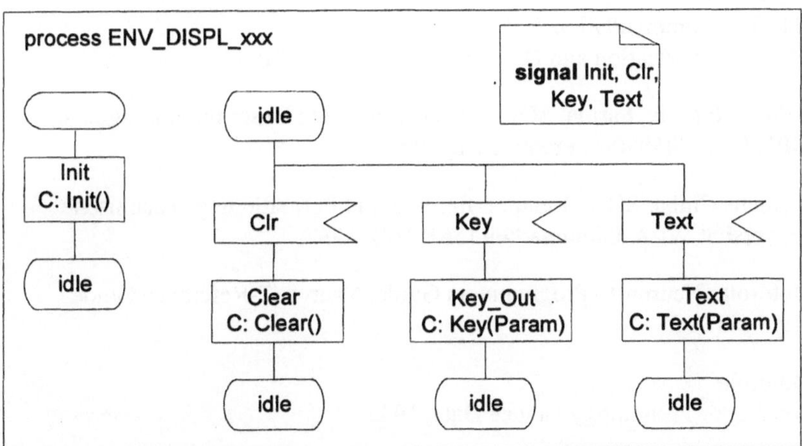

Abb. 6. Ein Environment-Block für eine Anzeige

4.3 Messagequeue und Scheduler

SDL basiert auf einem ereignisgesteuerten Ablaufsystem, wobei jeder Prozeß seine eigene lokale Messagequeue besitzt, in der die Ereignisse in der Reihenfolge ihres Auftretens eingetragen werden. Dieses Konzept galt es auf einem Neuron-Chip zu realisieren. Die wichtigste Forderung war die Gleichbehandlung von Signalen außerhalb der Neuron-Blöcke (den Netzwerkvariablen) und Signalen innerhalb der Neuron-Blöcke.

Realisiert wird dies durch eine globale Messagequeue. Jeder Neuron enthält eine Queue, in die sowohl die vom LONWORKS-Scheduler übernommenen Netzwerkvariablen als auch die internen SDL-Signale eingetragen werden. Jeder SDL-Prozeß wird als Neuron-C-Funktion realisiert, die in einer statischen Variablen den aktuellen Prozeßzustand hält. Die Queue übernimmt den Aufruf der einzelnen Funktionen und übergibt dabei das eingetroffene Signal als Parameter. Im Fall einer SDL-Output-

Anweisung wird das Signal in die Queue gestellt, die dann die weitere Verteilung (als Netzwerkvariable oder als internes Signal) übernimmt.

5 Resümee

Mit dem hier besprochenen Werkzeug ist es möglich LONWORKS-Applikationen in der Beschreibungssprache SDL zu entwerfen und automatisch einen entsprechenden Neuron-C-Code zu generieren. Die Netzliste zum Binden der Netzwerkvariablen wird ebenfalls automatisch erzeugt, wobei diese bis zur Bekanntgabe des vom LONBUILDER geforderten Dateiformats noch händisch übertragen werden muß. Das Projekt wird voraussichtlich im Herbst 97 abgeschlossen werden.

Literatur

1. ITU-T Recommendation Z.100
    CCITT Specification and Description Language, 3/1993

2. Belina, Hogrefe, Sarma: SDL - Specification and Description Language
    ISBN 0-13-785890-6, Prentice Hall 1991

3. Clemens Chiba: SDL : Benutzerführung und Formatierung, Technische
    Universität Wien, Diplomarbeit (561.982), 1997

4. Motorola: Neuron-C-Programmers Guide, Neuron-C-Reference Guide
    Motorola 1995

5. Motorola
    LonWorks Technology Device Data, 1995

6. R. Schmit: Systementwicklung mit SDL, FeT'95 Tagungsband
    ISBN 3-85133-004-8, ÖVE Schriftenreihe Nr. 9

Simulationsunterstützung für den Entwurf von Feldbussystemen

U. Donath[1], D. Hartenstein[2], K. Kabitzsch[2], P. Schwarz[1]

[1]Fraunhofer Institut für Integrierte Schaltungen, Außenstelle Dresden
Zeunerstraße 38, D-01069 Dresden
[2]Technische Universität Dresden, Institut für Informationssysteme
D-01062 Dresden

Abstract. Automatisierungssysteme der Feldebene zeichnen sich durch eine Verteilung der Automatisierungsfunktionen auf räumlich getrennte Geräte aus. Die Kommunikation der Subsysteme erfolgt zunehmend über Busse. Beim Entwurf derartiger Systeme ist die Funktionalität nachzuweisen und eine Leistungsbewertung vorzunehmen. Wir schlagen dafür einen gemeinsamen Modellierungs- und Simulationsansatz vor, der das Protokollverhalten der einzelnen Teilnehmer, ihre Interaktionen und die Prozeßdynamik einschließt. Als Basis für die Gesamtsystemsimulation wählen wir VHDL-Simulatoren, die vielfach beim Entwurf digitaler Systeme eingesetzt werden. Der Modellierungsansatz wird am Beispiel eines LON-basierten Systems erläutert. An einer Aufgabenstellung der Prozeßsteuerung werden die Möglichkeiten der Simulation zur Entwurfsunterstützung demonstriert.

Automation systems on the field level are characterized by a distribution of the control functions between separate devices. The communication of the subsystems is more and more performed by buses. During the design of such systems the funtionality and the performance have to be proved. For both we propose a common modelling and simulation approach which includes the protocol behaviour of the participants, their interactions, and the system dynamics. As the basis of the simulation of the whole system we choose VHDL simulators which are often used in design of digital systems. The modelling approach is detailed on a LON-based system. With a task of process control the features of a simulation supported design are demonstrated.

1 Einleitung

Wesentliche Schritte beim Entwurf von Feldbussystemen sind Spezifikation, Konfiguration, Test und Leistungsbewertung. Wichtige Fragestellungen [1, 2] dabei sind:
- Nachweis der Funktionalität
- Ermittlung der Antwort- und Reaktionszeiten
- Systemverhalten bei wechselndem Kommunikationsaufkommen
- Systemverhalten bei Parameteränderungen
- Empfindlichkeit gegenüber Störungen
- Aufdeckung der Fehlerursachen
- Auslastung des Übertragungsmediums.

Als Unterstützung bei der Bearbeitung dieser Fragestellungen schlagen wir eine vergleichweise genaue Modellierung und Simulation des Bussystems vor, die das Protokollverhalten der einzelnen Teilnehmer, ihre Interaktionen über den Bus und die Prozeßdynamik einschließt. Dadurch unterscheidet sich dieser Ansatz z.B. von bedienungstheoretischen Methoden oder der Verwendung von stochastischen Petri-Netzen. Diese Verfahren liefern Angaben über Mittelwerte oder Verteilungsfunktionen wichtiger Systemkennwerte, beabsichtigen aber keine detaillierteren Untersuchungen zu dynamischen Problemen, Fehlerursachen u.dgl. Außerdem ist auch bei diesen Methoden die Modellbildung recht kompliziert und aufwendig. Die simulationsgestützte Leistungsanalyse wird ausführlich in [3, 4] behandelt und in [5] auf verteilte, busgekoppelte Systeme angewendet.

Der von uns gewählte Ansatz geht von VHDL-Modellen aus, um weitverbreitete VHDL-Simulatoren nutzen zu können. Diese Simulatoren [6, 7] werden im Schaltkreis- und Systementwurf eingesetzt. VHDL gestattet eine kombinierte Beschreibung der Struktur und des Verhaltens diskreter Systeme. Das Abstraktionsniveau reicht von der Logikebene bis zur algorithmischen Ebene. Verhaltensmodelle in der Programmiersprache C können einbezogen werden. Weitere Gesichtspunkte für die Wahl von VHDL-Simulatoren sind die laufenden Arbeiten zur Erweiterung auf "Analog"-VHDL (VHDL-AMS) und bereits verfügbare Kopplungen zu Analogsimulatoren [8]. Damit wird auch die Behandlung hybrider (gemischt diskret-kontinuierlicher) Systeme einfach möglich. In der Framework-Einbindung der Simulatoren stehen zusätzliche ingenieurgerechte Beschreibungsmittel wie Blockschaltbilder, Flußdiagramme oder Automatengraphen zur Verfügung.

2 Modellierung

2.1 Modellansatz

Das Modell des Bussystems wird aus vorgefertigten Baublöcken zusammengesetzt. Jedes Automatisierungsgerät (Abb. 1) wird dabei als Knoten betrachtet, der eine Signalverarbeitungsleistung vollbringt. Da die Operationen bezüglich des Informationsaustauschs über den Bus einheitlich beschrieben werden können, sich jedoch bezüglich der Einflußnahme auf den Prozeß wesentlich unterscheiden, wird jeder Knoten so betrachtet, als ob er in einen Protokollprozessor (PP) und einen Applikationsprozessor (AP) geteilt sei [9]. Dem Protokollprozessor wird die Organisation der unterschiedlichen Übertragungsdienste einschließlich der Steuerung des Buszugriffs übertragen. Der Applikationsprozessor agiert mit dem Prozeß via Sensoren oder Aktoren und führt diskret-kontinuierliche Signalverarbeitungen aus.

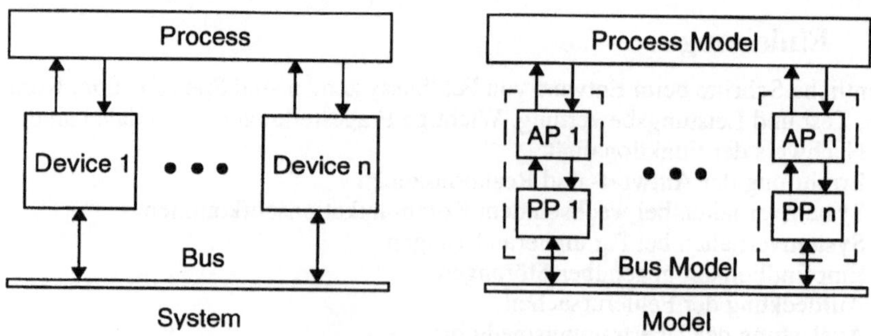

Abb. 1. Modellansatz

Die Modellbibliothek des VHDL-Simulators wird durch folgende Modelle erweitert:
- Bus
- Protokollprozessor

- ausgewählte Applikationsprozessoren
- Signalgeneratoren
- Monitore.

Der Bus ist ein bidirektionales Transferelement, das die Nachrichten im Telegrammformat überträgt. Der Protokollprozessor und diverse Applikationsprozessoren realisieren die o.g. Funktionen, wobei der Vorrat der Applikationsprozessoren vom Anwender durch neue VHDL-Modelle ergänzt werden kann. Die Signalgeneratoren ersetzen Subprozesse, die nicht vollständig durch kausale Relationen beschrieben werden können. Sie speisen Spurverfolgungsdaten oder Zufallsdaten ein. Sie werden genutzt, um beliebige Last- bzw. Verkehrssituationen zu erzeugen. Monitore zeichnen den Nachrichtenverkehr auf.

2.2 Detaillierung am Beispiel LON

Als Beispiel werden die Knoten eines Local Operating Network (LON) betrachtet. LON ist ein typisches Feldbussystem mit Multi-Master-Eigenschaft, das vorwiegend in der Gebäudeautomation, Industrieautomation und Heimvernetzung eingesetzt wird.

Der Protokollprozessor erfüllt Sende- und Empfangsfunktionen. Abbildung 2 zeigt die vom Modell erfaßten Knoteneigenschaften. Ausgangspunkt ist das OSI-Referenz-Modell. Ausgewählt wurden die "logischen" Merkmale eines Peer-to-Peer-Transfers innerhalb eines Simple-Channel-Netzwerks. Die Folge der Steueroperationen wird aus dem LonTalk-Protokoll [10] abgeleitet. Ihre Darstellung erfolgt zunächst als hierarchisches zeitbewertetes Zustandsdiagramm oder zeitbewertetes höheres Petri-Netz. Beides sind grafisch/textliche Formen, die manuell in VHDL- oder C-Verhaltensmodelle übersetzt werden. Die Zeitvorgaben resultieren aus den in der Produktbeschreibung [11] angegebenen Nominalwerten. Tools wie ExpressV-HDL oder SpeedCHART ermöglichen eine automatische VHDL-Code-Generierung aus Zustandsdiagrammen.

Protocol Processor

- explicit message handling
- request-reponse service
- acknowledged and unacknowleged unicast
- common odering and duplicate detection
- address recognition
- p-persistent CSMA
- optional priority
- optional collision detection

Abb. 2. Funktionen des Protokollprozessors

Abbildung 3 zeigt das Zustandsdiagramm des Protokollprozessors auf der obersten Hierarchieebene. Auf dieser Ebene werden alle Funktionen aufgeführt, die durch externe Ereignisse und interne Uhren ausgelöst werden. Die Funktionen korrespondieren mit gleichbenannten Zuständen, die durch Subautomaten untersetzt sind. Externe Ereignisse sind: Nachrichten vom Applikationsprozessor, Nachrichten vom Bus, Bus wird "idle" und Kollision von Nachrichten auf dem Bus. Interne Uhren steuern: Warten auf das Senden einer Nachricht, Sendezeit entsprechend der Nachrichtenlänge und der Übertragungsrate, Zeitpunkt der Wiederholung einer Sendung und Zeitpunkt des Löschens einer Nachricht aus dem Empfangspuffer. Zur Konstruktion der Automaten folgen wir den von Harel [12] definierten Regeln zur Bildung von Statecharts.

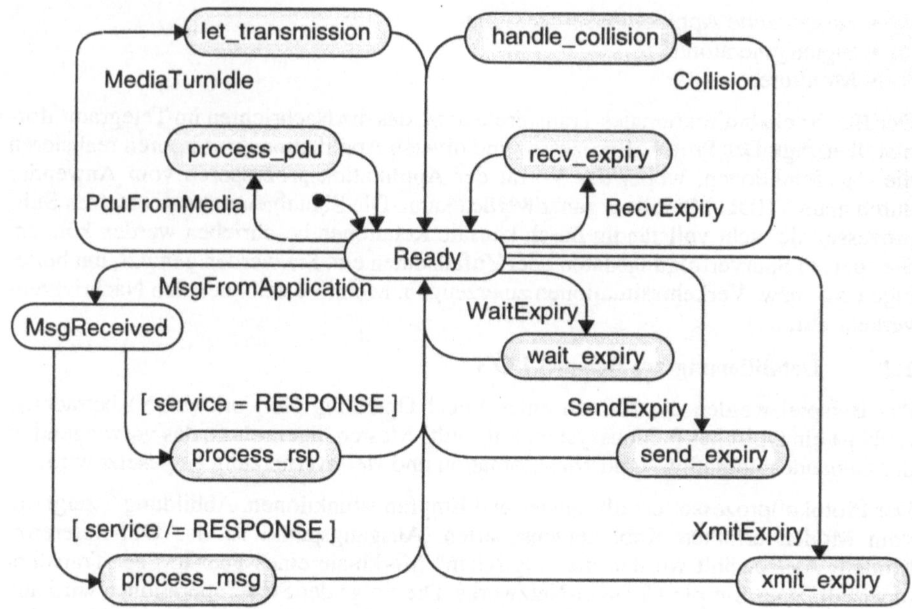

Abb. 3. Zustandsdiagramm des Protokollprozessors

Das Zeitverhalten des Bussystems wird im wesentlichen durch den Zeitbedarf der Signalverarbeitung, das p-persistente CSMA-Verfahren und den Zeitbedarf des Nachrichtentransfers bestimmt. Als Beispiel wird das Zeitdiagramm für den Acknowledged Service zwischen einem Knoten AP1 / PP1 und einem Knoten AP2 / PP2 betrachtet (Abb. 4). Die Signalverarbeitung durch die Applikationsprozessoren wird mit $\alpha 1$ und $\alpha 2$ Zeiteinheiten bilanziert. Zur Bestimmung des Sendezeitpunktes verwenden die Protokollprozessoren die Konstanten $\beta 1$ und $\beta 2$. Beide Konstanten werden aus Parametern der realen Prozessoren und des Übertragungsmediums bestimmt. $\beta 1$ ist identisch mit dem minimalen zeitlichen Abstand aufeinanderfolgender Nachrichten. $\beta 2$ ist die Dauer der Zeitscheiben innerhalb eines vorgegebenen Verzögerungsintervalls; n und m sind

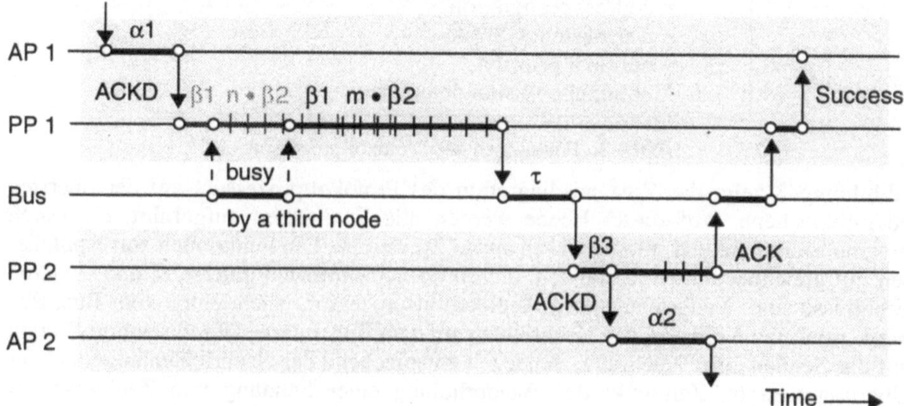

Abb. 4. Zeitdiagramm für den Acknowledged Service

die Nummern der zufällig ausgewählten Zeitscheiben. Zur Übertragung werden τ Zeiteinheiten benötigt, berechnet aus der Nachrichtenlänge und der Bitrate des Systems. Für den Empfänger werden für das Dekodieren der Nachrichten, Adreßerkennung usw. $\beta3$ Zeiteinheiten veranschlagt.

Die Modelldefinition der Applikationsprozessoren erfolgt in der Regel unmittelbar in VHDL. Die Signalverarbeitung wird durch Eingangssignaländerungen bzw. den Empfang von Nachrichten (WAIT ON signal, WAIT UNTIL condition) oder Abläufe interner Uhren (WAIT FOR time) ausgelöst. Typische Tasks der Applikationsprozessoren sind: Abfragen von Sensoren, Ereignismeldungen, Datenvorverarbeitungen und Stellgrößeneinstellungen. Komplexere Algorithmen wie z.B. PID-Regelalgorithmen werden als Differenzengleichungen implementiert. Die Applikationsprozessoren empfangen bzw. senden die Nachrichten als Record. Als Sender definieren sie u.a. Service, Adresse und Datum. Zur Messung der Antwortzeit wird ein Zeitstempel ergänzt.

Jeder Protokollprozessor kann die Übertragung einleiten, wenn er den Bus als "idle" erkennt. Ein Konflikt entsteht, wenn mehrere Prozessoren die Übertragung gleichzeitig beginnen. Vom Busmodell wird diese Situation durch das Setzen eines Zustandsflags den Sendern signalisiert, die daraufhin die Übertragung mit unterschiedlicher Verzögerung wiederholen. In der Terminologie von VHDL ist das Busmodell eine Resolution Function, die den jeweils aktiven Sender bestimmt und dessen Nachricht an alle angeschlossenen Prozessoren überträgt.

3 Simulation

3.1 LON-Applikation in der Prozeßsteuerung

Es soll geprüft werden, ob ein LON-Netz (Abb. 5) folgender Aufgabenstellung genügt: Teile werden in einem definierten zeitlichen Abstand einer Heizung zugeführt und entfernt. Zum Erzielen von bestimmten Materialeigenschaften wird die Erhitzung nach einer vorgegebenen Temperaturkurve gefordert. Nach dem Verlassen der Heizung gelangen die Teile in eine Kühlstrecke, für die vergleichbare Forderungen gelten.

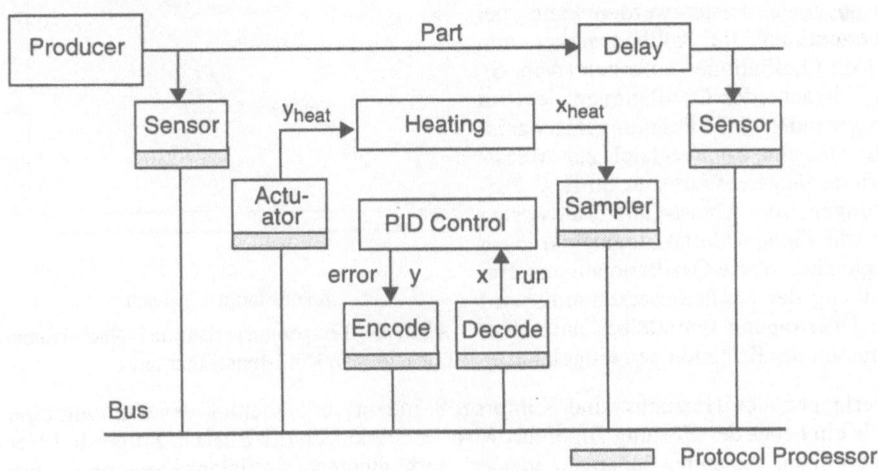

Abb. 5. Modell der Prozeßsteuerung

Das Eintreffen jedes Teils wird von einem *Sensor* registriert, der dies mit einem Signal run der *PID Control* meldet. Vom PID-Controller wird mit der Periode T die Regelgröße x erfaßt, die Führungsgröße w bestimmt und die Stellgröße y berechnet. Die Werte der Regelgröße x werden von einem *Sampler* geliefert, der den Temperaturverlauf x_{heat} der Heizung mit der Periode T abtastet und signifikante Änderungen dem PID-Controller meldet. Bis zum Empfang des neuen Signalwertes benutzt der PID-Controller den zuvor übertragenen Signalwert. Falls die Regeldifferenz eine vorgegebene Schranke überschreitet, wird vom PID-Controller ein Fehlersignal error ausgelöst. Die berechneten Werte der Stellgröße y werden einem *Actuator* zugeführt, der diese in das Stellsignal y_{heat} umformt. Dem PID-Controller sind zwei Glieder *Decode* und *Encode* vor- bzw. nachgeschaltet, die das Entpacken bzw. Packen der Telegramme und Signalkonvertierungen vornehmen. Die Regelstrecke *Heating* ist als Subsystem mit T1-I-Charakteristik modelliert. Die Glieder Sensor, Sampler, Actuator, Decode und Encode sind jeweils mit einem *Protocol Processor* gekoppelt, der den Nachrichtentransfer über den Bus organisiert.

3.2 Untersuchte Problemstellungen

Entscheidend für die Erfüllung der o.g. Funktionen sind folgende Parameter: Abtastperiode, Bitrate, Übertragungsdienst, Kollisionserkennung, Wiederholungen, Timer für die Wiederholung.

Die den Regelkreis betreffenden Signalübertragungen werden zunächst mit dem Acknowledged Service ausgeführt. Fehlermeldungen bezüglich der Einhaltung der Temperaturkurve benutzen den Unacknowledged Service. Abtastperiode, Bitrate und Timer wurden in Voruntersuchungen bemessen.

Heizung und Kühlung werden zunächst jeweils nur von einem Teil durchlaufen, so daß beide Prozesse abwechselnd ablaufen. Die Dynamik der Regelung zeigt, daß bei Nutzung der Kollisionserkennung die geforderte Temperaturkurve gewährleistet werden kann, bei Verzicht auf die Kollisionserkennung jedoch Oszillationen auftreten (Abb. 6). Die Ursache der Oszillationen liegt im einsetzenden Wiederholungsmechanismus, der eine im Vergleich zur Abtastperiode längere Wartezeit umfaßt. Pufferungen von Abtast- und Stellwerten sind die Folge. Übertragungsfehler lösen in gleicher Weise Oszillationen aus. Bei Nutzung der Kollisionserkennung wird die Übertragung unmittelbar nach dem Erfassen der Kollision neu eingeleitet.

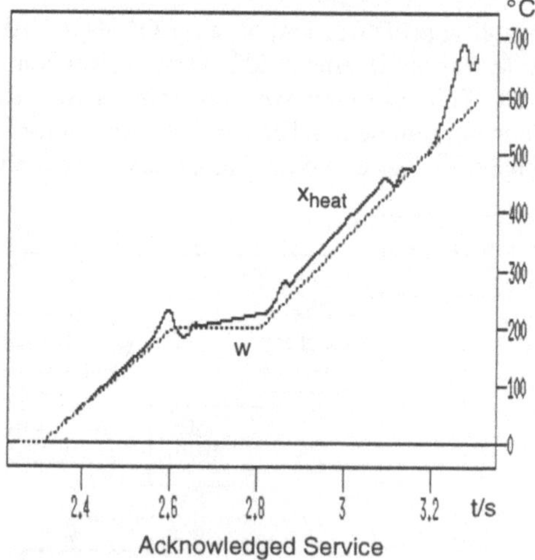

Abb. 6. Temperaturverlauf bei Nachrichtentransfer ohne Kollisionserkennung

Überlappen sich Heizungs- und Kühlprozeß, indem vor Abschluß der Kühlung eines Teils ein neues der Heizung zugeführt wird, verstärkt sich die Buslast. Folge sind Verzögerungen, die unter sonstigen idealen Bedingungen (Kollisionserkennung, keine Übertragungsfehler) zu wesentlich verstärkten Amplituden der Stellgrößen führen. Bei längeren Überlappungszeiträumen treten Instabilitäten auf.

Verwendet man an Stelle des Acknowledged Service den Unacknowledged Service ohne Wiederholungen, ist der Regelvorgang stabil. Die durch Kollisionen oder Übertragungsfehler verlorenen Werte der Regel- und der Stellgröße führen lediglich zu Abweichungen, die im Toleranzbereich liegen. Weitere Experimente zeigen, daß Heizung und Kühlung parallel für zwei Teile genutzt werden können (Abb. 7).

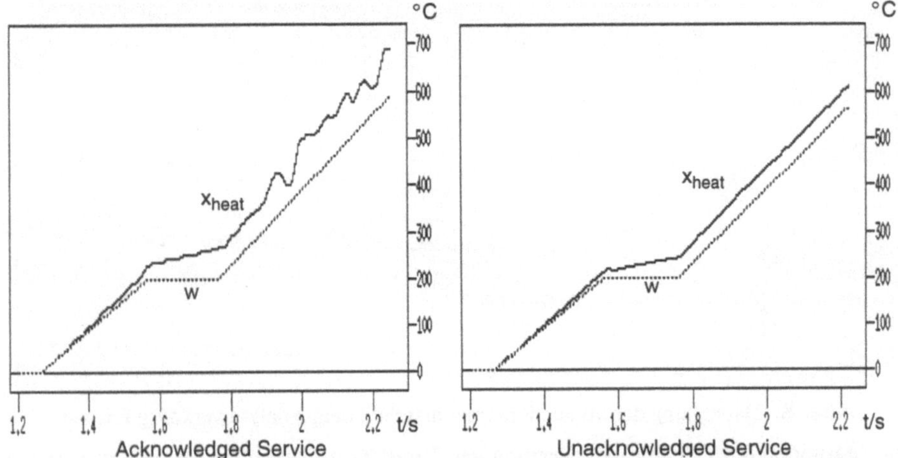

Abb. 7. Simulationsergebnisse bei Überlappung von Heizung und Kühlung

Das Experiment am Modell zeichnet sich durch ein erweitertes Monitoring aus. An mehreren Punkten wird der Nachrichtenverkehr von Monitor-Elementen protokolliert. Außerdem kann er vom Waveform-Tool des Simulators angezeigt werden. Ergänzungsmöglichkeiten sind On-line-Darstellungen der Busauslastung und von Reaktionszeiten. Desweiteren zeichnet jeder Protokollprozessor seine Aktivitäten auf. Dazu gehören der Übertragungsdienst, Zustandsfolgen, Füllstände der Puffer, Kollisionen und Wiederholungen.

4 Off-line-Auswertung

Zur Off-line-Auswertung werden die aufgezeichneten, chronologisch geordneten Simulationsergebnisse in eine Datenbank übertragen. Dabei wird jede dieser "Spuren" in eine Tabelle transformiert. Die Untersuchung der Spuren erfolgt mit dem MS-Excel-basierenden Analysewerkzeug Extrakt. Zu den Auswertungsmöglichkeiten gehören: 2D- oder 3D-Darstellungen zum Zeitverlauf, Erstellen von Statistiken und Filterungen. An die Datenbank können Anfragen nach bestimmten Prozeßabschnitten, deren Dauer und Häufigkeit gerichtet werden. Das Kommunikationsverhalten der Knoten kann u.a. durch die Sender-Empfänger-Beziehungen aufgezeigt werden.

Zur Auswertung der Simulationsspuren des Applikationsbeispiels sind zunächst die Ereignisse zu definieren, welche die Subprozesse Heizung und Kühlung begrenzen. Das sind die Ereignisse "Start der Heizung", "Ende der Heizung", "Start der Kühlung" und "Ende der Kühlung". Die Ereignisdefinition bezieht sich auf die Telegramme, die die entsprechenden run-Signale enthalten. Trägt das run-Signal den Wert "1" startet der Prozeß; mit "0" endet der Prozeß. Abbildung 8 zeigt in den beiden oberen Zeilen die Heizungs- und Kühlperioden. In den unteren Zeilen sind die korrespondierenden Start- und Ende-Ereignisse aufgezeichnet. Weiterhin werden die Ereignisse "Fehler Heizung" und "Fehler Kühlung" dargestellt. Sie werden den Telegrammen mit dem entsprechenden error-Signal entnommen. Die error-Signale kennzeichnen das Überschreiten der zulässigen Regeldifferenz. Die Ursachen sind oben aufgezeigt.

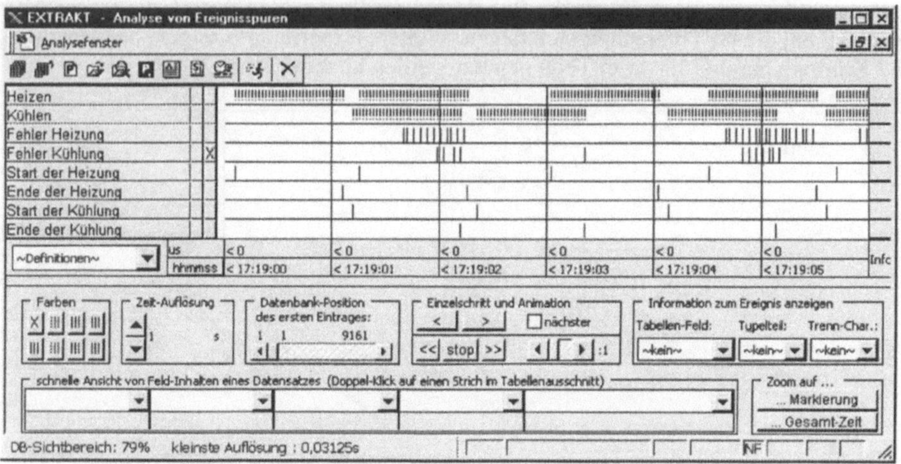

Abb. 8. Darstellung der Simulationsergebnisse mit dem Analysewerkzeug Extrakt

Die Methoden der Off-line-Auswertung der Simulationsergebnisse können in gleicher Weise zur Auswertung von realen Prozeßdaten benutzt werden.

Referenzen

1. Bonfig, K. W.: Feldbus-Systeme. expert-Verlag, Renningen-Malmsheim, 1995

2. Kabitzsch, K.; Hartenstein, D.: Fehlersuche in Automatisierungssystemen mit Feldbusvernetzung. e&i, 114.Jg. (1997), Heft 5, S. 274-279

3. Law, A.M.; Kelton, W.D.: Simulation Modeling and Analysis. McGraw-Hill, 1991

4. Schoen, J.M. (Ed): Performance and Fault Modeling with VHDL. Prentice-Hall, Englewood Cliffs, 1992

5. Schwarz, P.; Donath, U.: Simulation-based Performance Analysis of Distributed Systems. 5th Intern. Workshop on Parallel and Distributed Real-Time Systems (WPDRTS'97), Geneva, April 1997, will be printed by IEEE Computer Society

6. Navabi, Z.: VHDL Analysis and Modeling of Digital Systems. McGraw-Hill, 1993

7. Lehmann,G.; Wunder, B.; Selz, M.: Schaltungsdesign mit VHDL. Franzis-Verlag, Poing 1994

8. ANACAD Electrical Engineering Software GmbH: MixVHDL User's Manual. Revision 1.0. 1995

9. Altmann, S.; Donath, U.: Methodical Aspects of VHDL-based Field Bus Modelling with a Controller Area Network. SIG-VHDL Spring'96 Working Conference, Dresden, May 1996, pp 85-96

10. Echelon, Corp.: LonTalk Protocol Specification. Version 3.0. Palo Alto, 1994

11. Motorola, Inc.: LONWORKS Technology Device Data. 1995

12. Harel, D.: Statecharts: A Visual Formalism for Complex Systems. Science of Computer Programming 8 (1987), pp. 231-274

Softwaretest vernetzter Echtzeitsysteme

K. Kabitzsch[1], D. Hartenstein[1], U. Donath[2]

[1]Technische Universität Dresden, Fakultät Informatik D-01062 Dresden

[2]Fraunhofer Institut für Integrierte Schaltungen, Außenstelle Dresden, Zeunerstr. 38, D-01069 Dresden

Abstract. Some software problems like race conditions or deadlocks are caused by fieldbus errors. Rapid prototyping is a comprehensive method to find the root cause in a fast manner. This paper describes a hardware in the loop environment, a real time simulator and measurement techniques for distributed systems under test. The presented monitoring tools have filter and trigger capabilities. The event trace is stored in a data base and the user can analyze the data with pattern matching and statistical algorithms.

1 Fehler in komplexen Systemen

Viele Störungen und Fehler in vernetzten Echtzeitsystemen treten nur sporadisch auf. Tabelle 1 zeigt, auf welche Weise sich Störungen der Feldbusübertragung in den vom Anwender auf OSI-Schicht 7 aufgesetzten Automatisierungsalgorithmen als Fehler bemerkbar machen können. Übertragungsfehler entstehen z. B. nach Reflexionen durch falsch verlegte, fehlerhaft terminierte und zu lange Leitungssegmente. Die Fehlerbehandlung der Protokollmaschinen „behebt" solche Störungen durch Wiederholungsmechanismen, es entsteht aber zusätzlich eine oft beträchtliche Performance-Überlast sowie ein gelegentlicher Telegrammverlust.

Fügen dann im Empfänger stark kausale Algorithmen die als Ereignismeldungen übertragenen Wertinkremente zusammen, so führt der Ereignisverlust zu bleibenden Fehlern, die sich als Drift, Schlupf usw. bemerkbar machen. In schwach kausalen Algorithmen kann nach einem Ereignisverlust das Weiterschalten in den nächsten Automatenzustand unterbleiben und fatale Folgen auslösen. Enthält der Signalfluß stark kausaler Algorithmen Rückführungen, z. B. in geschlossenen Regelkreisen, so können übertragungsbedingte Totzeiten zu Instabilitäten führen. Für schwach kausale Algorithmen kann durch den Verlust einer Ereignismeldung auch ihre Lebendigkeit verlorengehen, d. h. das Weiterschalten in solchen Synchronisationskreisen ist auf Dauer unmöglich (Verklemmung). Entstehen Fehlfunktionen, weil das Rechnersystem die bei der Algorithmenkonstruktion von ihm erwarteten Rechtzeitigkeitsforderungen nicht erfüllen kann und seine Ergebnisse zu spät bereitstellt, so spricht man von Echtzeitfehlern. In der Praxis nimmt der Rechner dann z.B. einen kurzen Sensorimpuls nicht mehr wahr, schließt einen Robotergreifer zu spät, erzeugt Schlupf an einem Antrieb oder liest inkonsistente Prozeßabbilder von der Sensor-Peripherie ein. Die vom Feldbus fehlerhaft verursachten Zusatz-Totzeiten können grundsätzlich solche Echtzeitfehler hervorrufen.

Fehler der beschriebenen Art sind durch rechnergestützte Syntheseverfahren nicht völlig vermeidbar. Im folgenden werden Testwerkzeuge beschrieben, welche die Suche nach solchen Fehlern unterstützen.

Systemschicht	fehlerhaftes Verhalten und	seine Fortpflanzung in folgende Schichten
OSI-Schicht 1	Performanceüberlast	Reflexionen, Bitraten-Drift, EMV-Störungen
OSI-Schichten 2-6 (Telegramme)	Performanceüberlast	Fehlerbehandlung der Protokolle wandelt Fehler teilweise in Wiederholungen um
	Telegramm-Verzögerung	Telegramm-Verlust (selten -verdopplung, -verfälschung)
OSI-Schicht 7 (Events)	Event-Totzeit	durch Überschreiben im Empfangspuffer ⟶ Event-Verlust
ereignisorientierte Algorithmen: - stark kausal	bei Closed-Loop-Algorithmen: Instabilitäten	beim Addieren inkrementeller Größen: Drift-Fehler
- schwach kausal		kein Fortschalten zum nächsten Zustand bei Closed-Loop-Algorithmen: Lebendigkeits-Verlust (Verklemmung)
bedingungsorientierte Algorithmen: - stark kausal	Verlust von Gleichlauf (Jitter) Closed-Loop-Algorithmen: Instabilität	
- schwach kausal	Integritätsfehler durch Race Conditions	

Tab. 1: Störungen der Feldbusübertragung und ihre möglichen Auswirkungen auf Automatisierungsalgorithmen

2   Prüf- und Testmethoden im Ablauf des Rapid Prototyping

Beginnt die Diagnose bereits mit Prototypen, deren Realitätsnähe schrittweise gesteigert wird, so hat das viele Vorteile. Die Untersuchungen können beginnen, wenn die reale Anlage noch gar nicht existiert. Im Vergleich zur Messung am Originalsystem sind zeitgeraffte Untersuchungen möglich. Der Prototyp erlaubt teilweise Einblick in Funktionszusammenhänge, die später am realen Objekt gar nicht mehr zugänglich sind. Die Erprobung des Prototypen vollzieht sich ohne Produktionsausfall, Ausschußproduktion und anderweitigen Betriebsaufwand. An realen Anlagen dürfen viele Störungen und Gefahrenzustände gar nicht absichtlich herbeigeführt werden.   Am Prototypen sind solche Tests dagegen ohne Risiko möglich.

2.1   Formale Spezifikation und Verifikation

Hier geht man von einer formalen Beschreibung des Steuerungsprogramms sowie von formalen Annahmen über das Verhalten des Prozesses und seiner Umgebung (Bedienpersonal, Leitrechner, Umwelt usw.) aus. Bezüglich dieses Systems kann der Prüfer nun einzelne Thesen über konkrete Verhaltenseigenschaften formulieren. Ein Beweiskalkül weist dann entweder diese geforderten Eigenschaften nach oder es liefert ein Gegenbeispiel. *Vorteile:* Soweit alle formalen Annahmen das Gesamtsystem korrekt und vollständig beschreiben, sind die im einzelnen geforderten Eigenschaften exakt nachweisbar. *Nachteile:* Für komplexe Systeme lassen sich kaum hinreichend adäquate, formale Beschreibungen finden. Das Beweiskalkül beschränkt seine Aussagen auf solche Eigenschaften, nach denen es vom Benutzer explizit gefragt wurde. Es hängt also weiterhin von dessen Akribie ab, daß keine wichtige Frage „vergessen" wird.

2.2   Offline-Simulation

Hier implementiert man eine Nachbildung des Steuerungsprogramms, ein Modell des Prozesses sowie das Verhalten der Umgebung (Bediener, Leitrechner, Umwelt, Störgrößen) auf einem Rechner. Danach läßt man beide Modelle aufeinander einwirken, wobei die Zeitverläufe aller Modellgrößen aufgezeichnet werden, um sie später auszuwerten. *Vorteile:* Es ist eine fast beliebige Ähnlichkeit zum Verhalten des Originals erreichbar. So lassen sich auch die von kontinuierlichen Teilsystemen stammenden Zeitabhängigkeiten gut nachbilden. *Nachteile:* Das Modell benötigt meist lange Rechenzeiten. Da aus Zeitgründen niemals alle theoretisch möglichen Betriebszustände erprobt werden können, bleiben eventuell gerade die unbeachtet, die verborgene Fehlermechanismen auslösen könnten. Im Gegensatz zur formalen Verifikation läßt sich mit Simulationsmethoden zwar die Existenz von Fehlern zeigen, aber niemals die Abwesenheit bestimmter Fehler beweisen.

2.3   Online-Simulation

Hier besteht der Prototyp teilweise bereits aus originalen Baugruppen, während andere Systemteile noch durch Simulationsrechner nachgebildet sind, die mit den Originalkomponenten (hardware in the loop)   über definierte Schnittstellen zusammenwirken. Da an diesen Schnittstellen kein Unterschied zwischen Echtzeit und Modellzeit mehr zulässig ist, muß sich der Simulator an die Echtzeit der Originalkomponenten anpassen. Eine solche Simulationsunterstützung ermöglicht die Präzisierung korrekter Lastenhefte zwischen allen Beteiligten, umfangreiche Tests und sogar erste Softwareabnahmen zu einem frühen Zeitpunkt. Später werden im Prüffeld schrittweise Modellteile durch Originalaggregate substituiert. *Vorteile:* Die Steuerungshardware weist das originale Zeitverhalten auf, so daß kritische Effekte hier endgültig untersucht werden können. Für exakte Unterbrechungszeitpunkte von Betriebssystem-Interrupts, die Auflösung von Zugriffskonflikten vor gemeinsamen Ressourcen oder an CSMA-Medien, Race Conditions und Priority-Inversion-Effekte sowie ähnlich zeitabhängige Zusammenhänge bietet erst die Online-Simulation eine hinreichende Testumgebung. *Nachteile:* Für Simulatoren und Aufzeichnungstechnik ist ein großer Hardwareaufwand nötig.

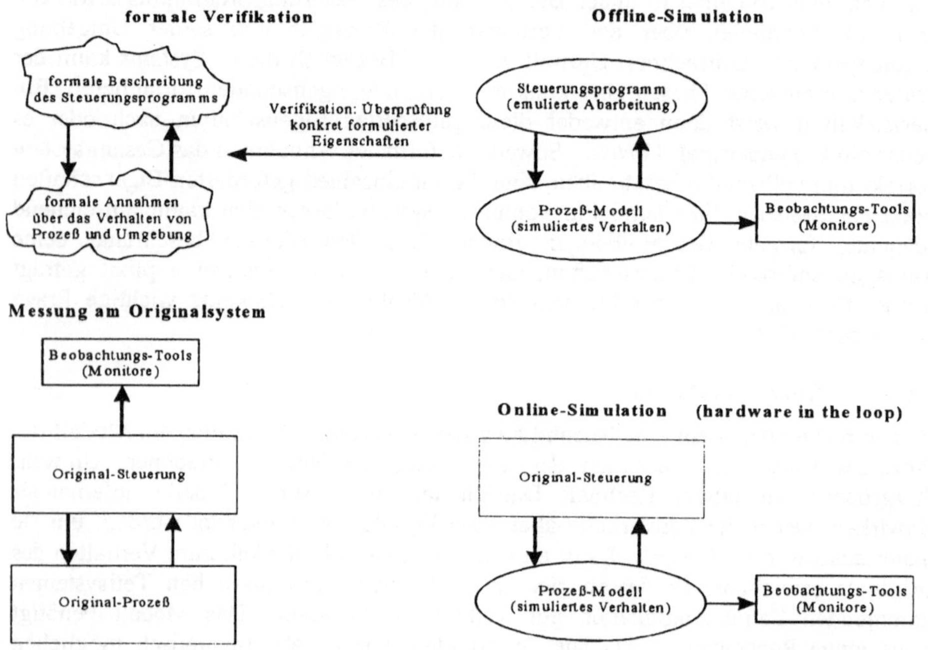

Bild 1: Methoden zur prototypischen Diagnose komplexer Automatisierungssysteme

2.4 Messung am Originalsystem

Die Online-Simulation geht im Prüffeld schrittweise in die Messung am Original über, nachdem die letzten Simulationskomponenten durch Originalbaugruppen substituiert wurden. Ähnliche Tests sind bei der Inbetriebnahme am Montageort, beim Einfahren sowie bei der späteren Wartung erforderlich. *Vorteile:* Montagefehler durch Abstimmungsmängel auf einer Baustelle lassen sich nur durch Messung vor Ort finden. Messungen sind vor allem zur Entdeckung von hardwarenahen Fehlermechanismen unverzichtbar (Haarrisse, Reflexionen, EMV). Erst in dieser

Testphase steht auch die Robustheit der Software gegenüber diesen Störeinflüssen auf dem Prüfstein. So muß sie auch mit unerwartetem Zeitverhalten einer Maschine fertig werden, das durch Fertigungs- und Montagetoleranzen entstanden ist. Bei der Wartung kommt man nur mit meßtechnischen Mitteln solchen Softwareproblemen auf die Spur, die z. B. von veränderten Impulszeiten an Sensoren herrühren, wie sie durch Alterung, Ermüdung, Verschleiß oder Korrosion entstehen. *Nachteile*: Werden Fehler erst in dieser Phase entdeckt, ist ihre Beseitigung besonders teuer.

3   Testbett zur Online-Simulation

3.1   Multivendoranlage

Zur Entwicklung und Erprobung von Testmethoden und -werkzeugen durch Messung am Originalsystem und Online-Simulation bauen 19 Firmen seit 1996 gemeinsam mit der Fakultät Informatik an der TU Dresden eine LonWorks-Multivendoranlage auf, die aus einer Stahlkonstruktion (B=600xH=295xT=60cm) mit vorgehängten Tafeln besteht. Die LonWorks-Vernetzung aller Tafeln erfolgt über einen abgedeckten Medienkanal, der ein schnelles Backbone (1,25 Mbit/s) enthält, von dem über rückseitige Router zahlreiche Medien abgehen. Insgesamt sind gegenwärtig Segmente mit folgenden Eigenschaften verfügbar:

1,25 Mbit/s Twisted Pair (TP/XF-1250)   78 Kbit/s Twisted Pair (TP/XF-78)
78 Kbit/s Free Topology (PT/FT-10)   4,8 Kbit/s Power Line (PL-21)

Entwurf, Installation, Konfiguration und Management der logischen Netzstrukturen erfolgen mit marktüblichen grafischen Werkzeugen.

Zum Bereich der Gebäudeautomation gehören die Funktionsblöcke Steuerung der Raumbeleuchtung und Jalousien, Sicherheitsbeleuchtung in Raumdecke und Tür, Personenzutritt (Automatik-Türmanagement, Chipkartensystem), Heizungsregelung, Lüftungssteuerung (Zuluft, Abluft, Lufttemperatur, CO-Sensor), Energiemanagement für Raumheizungen und Elektroverbraucher sowie Elektronische Geldbörse mit Chipkarten-Leser. Die Arbeitswand ist mit weiteren Komponenten im Laborraum verbunden, die Multivendoranlage wächst damit „in den Raum hinein". Mit den Spots der Arbeitswand und den dimmbaren Deckenleuchten lassen sich von der Tafel „Raumbeleuchtung" Szenenprogramme im Zusammenwirken mit anderen Raumkomponenten (z.B. elektrisch betriebene Jalousien bzw. Tür) gestalten. Die elektrisch angetriebene Zugangstür zum Laborraum führt in Verbindung mit der Tafel „Personenzutritt" das auf Chipkarten gestützte Zugangskontroll- und Türmanagement aus.

Zum Bereich der Industrieautomation gehören die Funktionsblöcke Erfassung und Anzeige von Störmeldungen sowie Identifikation von Personen und Gegenständen mit Transpondern. Auch diese Tafeln sind mit weiteren Laborkomponenten der Industrieautomation im Raum verknüpft. Ein 240cm x 140cm großes Modell einer automatischen Fabrik, das mit Fischertechnik aufgebaut ist und aus sechs

Bearbeitungsmaschinen, einem Roboter, verschiedenen Transportsystemen und einem Hochregallager besteht, wird von einem ABB-Fertigungsleitstand auf einer DEC-Alpha-Workstation gesteuert, der auch Auftragseingabe, Produktionsplanung und Visualisierung ausführt. Die Vewaltung der Arbeitspläne, Lagerbestände und Aufträge übernimmt eine Oracle-Datenbank unter UNIX. Sie wird vom Transpondersystem der Tafel „Identifikation" unterstützt, durch welches die in der Fabrik umlaufenden Werkstücke gekennzeichnet, Produktdaten mit den Werkstücken mitgeführt und so CIM-Konzepte im Fertigungsprozeß umgesetzt werden. Eine CNC-gesteuerte 3D-Fräsmaschine zur Fertigung von Teilen aus Automatenaluminium wird zur Zeit erprobt und soll mit ihren CAD-Systemen, den CAM-Komponenten zur Konturenherstellung sowie den Baugruppen zur Betriebsdatenerfassung künftig in die Multivendoranlage integriert werden.

Das modulare Konzept erlaubt eine ständige Modernisierung sowie einen zielgerichteten Ausbau der Anlage. Diese fest installierten, branchentypischen Applikationen sollen künftig auch als Testbett für Interoperabilitätstests nutzbar sein. Dazu werden neue Testlinge in die ihnen adäquate Umgebung eingebracht und auf korrektes Zusammenwirken mit den vorhandenen Komponenten untersucht. Parallel dazu sollen geeignete Werkzeuge entwickelt werden.

3.2 Echtzeitsimulator

Die Anlage enthält auch freie Montageflächen für Aufbau und Erprobung neuer Prototypen. Zur Nachbildung der zugehörigen Prozesse im Sinne der Online-Simulation dient ein leistungsfähiger Echtzeitrechner als Hardware-In-The-Loop-Simulator. Verbindet man ihn durch jeweils gespiegelte Ein-, Ausgangssignale mit den Automatisierungskomponenten der Arbeitswand, so lassen sich auf flexible Weise z.B. geschlossene Regelkreise aufbauen und erproben. Um die Vorteile herstellerunabhängiger, offener Hardware und Software nutzen zu können, wurde auf den Kauf eines Komplettgerätes (Tab. 2) verzichtet und ein System aus Standardkomponenten konfiguriert (VME-Bus, OS-9, MATLAB/SIMULINK).

Laufzeit-Plattform	Laufzeit-Betriebssystem	Entwurfs-plattform	Entwurfswerkzeuge	Produkt-beispiele
PC	Windows	PC	IEC-1131-Tool	WinMOD
VME-Bus (Multiprozessor)	OS-9, VxWorks	PC, Workstation	MATLAB/SIMULINK	CARTS MESA VERDE
Alpha-Chip-Cluster	Eigenentwicklung	PC	MATLAB/SIMULINK	dSPACE
Transputer-Cluster	Eigenentwicklung	PC	Eigenentwicklung	ASCET

Tab. 2: Charakteristische Typen marktüblicher Hardware-in-the-Loop-Simulatoren

3.3   Testwerkzeuge

Im Gegensatz zu üblichen Tools geben verteilte Debugger, wie sie z. B. für das LonWorks-Netzkonzept verfügbar sind, besondere Unterstützung beim Test verteilter Systeme. Die LonBuilder Developers Workbench umfaßt ein Multiknoten-Entwicklungssystem für 6 Emulatoren, wobei bis zu 4 LonBuilder kaskadiert und so gleichzeitig maximal 24 Knoten emuliert werden können. Ein solcher Debugger bietet beim Test verteilter Systeme gegenüber herkömmlichen Debuggern den Vorteil, daß sich Haltepunkte im gesamten System „verteilen" lassen. Nachteil der beschriebenen Haltepunkttests ist, daß das reproduzierbare Verhalten der Programme mit und ohne Haltepunkt nicht mehr vorausgesetzt werden kann, sobald mehrere parallele Tasks bzw. mehrere Knoten am Test beteiligt sind, jeder Haltepunkt das zeitliche Gefüge zwischen den Tasks und ihrer Umgebung erheblich verzerrt und die Reihenfolgen der Taskabarbeitung verändert.

Wesentlich vorteilhafter sind hier Testmethoden, die den freien Lauf des Testobjektes kaum behindern und deshalb mit dem Begriff der freien Aufzeichnung (Tracing) umschrieben werden. Die ablaufenden Programme, Netzprotokolle und Hardwareaktivitäten senden stattdessen über spezielle Diagnoseschnittstellen Ereignismeldungen (Events) an die Testumgebung. Diese werden dort in chronologischer Reihenfolge gesammelt und stehen für eine spätere Offline-Auswertung zur Verfügung (History-Datei auf Festplatte). So können beispielsweise Post-Mortem-Fehleranalysen nach Systemabstürzen und Verklemmungen vorgenommen werden. Damit wird das Entstehen von Fehlern kontinuierlich beobachtbar, denn die History enthält meist den gesamten Vorgang von der Fehlerursache bis zur Manifestation der Störung. Jeder einzelne Monitor muß dazu über folgende Komponenten verfügen:

- Zeitstempelung
- Temporäre Zwischenspeicherung im Ringpuffer
- Permanente Aufzeichnung auf der Festplatte
- Reduzierung der Datenmengen durch Online-Selektion mit Triggern und Filtern

Zur Fehlersuche in komplexen Systemen muß deren Verhalten durch mehrere Monitore auf möglichst verschiedenen Ebenen (Hardware, Betriebssystem, Anwendersoftware, Netztelegramme, physikalische Prozeßgrößen) beobachtet werden. Die Palette marktüblicher Monitore wie z. B. Protokollanalysatoren für Feldbusse, Transientenrecorder und Logikanalysatoren wurde deshalb durch Neuentwicklungen der TU Dresden ergänzt, welche erweiterte Möglichkeiten zur Triggerung, Filterung und Fokussierung der Aufzeichnungen auf fehlerverdächtige Zeitabschnitte bieten:

SPS-Analysator AutoSPy: Zur Beobachtung von Steuerungen der SIMATIC-Familie verbindet man ohne zusätzlichen Hardwareaufwand einen PC über die serielle Schnittstelle mit der SPS. Danach werden Variable ohne Unterbrechung des laufenden SPS-Anwendungsprogramms zyklusgenau gemessen, ihre Verläufe online im PC angezeigt und auf dessen Festplatte aufgezeichnet.

Feldmonitor HWDS /2/: Diese kompakten, programmierbaren Recorderbausteine werden während der Montage-, Inbetriebnahme- und Einfahrphase an die Rechner und Feldbusknoten angeschlossen und zeichnen interne Hard- und Softwareabläufe auf. Sie enthalten eine kleine, wechselbare Festplatte (PCMCIA). Ihr Ringpuffer kann alle 100 ns ein Datenwort aufnehmen. Als Filter und Trigger enthält jeder Diagnosebaustein einen ASIC. Erkennt dieser Event-Prozessor eines der dort abgelegten Muster online im gemessenen Datenstrom, so übernimmt er das entsprechende Stück der Aufzeichnung in den PCMCIA-Massenspeicher.

Integriertes Diagnosewerkzeug EXTRAKT: Bei komplizierten Fehlern reicht es oft jedoch nicht mehr aus, diese Geräte unabhängig voneinander einzusetzen und ihre Aufzeichnungsergebnisse getrennt auszuwerten. EXTRAKT ist ein an der TU Dresden entwickeltes Auswertewerkzeug, das die Resultate aller Geräte aufnimmt und den Vergleich verschiedener Signalklassen erlaubt. Dazu werden die aus verschiedenen Meßeinrichtungen kommenden Datenströme parallel nach chronologischer Ordnung in einer Datenbank zusammengeführt. EXTRAKT bietet für alle Signale Auswertungs- und Diagnosemethoden an:

- Abfragemechanismen (intelligente Suchstrategien, Filterung, Clusterung) nach Fehlerszenarien, Inspektion, Recherche
- Visualisierungsalgorithmen (Raum-Zeit-Diagramme, 3D- und Gantt-Diagramme)
- Berechnung von Lauf- und Totzeiten sowie anderer sekundärer Kenngrößen
- Statistische Auswertungen

Darauf setzen verschiedene Verfahren zum Aufspüren von Unregelmäßigkeiten durch Mustervergleich auf. Methoden zur Signalanalyse, statistischen und dynamischen Systemanalyse schließen aus der Beobachtung des gestörten Prozeßverlaufs auf gegenseitige Abhängigkeiten und decken so versteckte Funktionszusammenhänge im Inneren des Systems auf. EXTRAKT ist über OLE nach außen offen und besitzt einen Zugang zu anderen Datenbanksystemen über die ODBC-Schnittstelle.

Referenzen

[1] Kabitzsch, K. (Hrsg): Tagungsband "Automatisierungskonzepte mit verteilter Intelligenz". TU Dresden, Fakultät Informatik, Oktober 1995

[2] Kabitzsch, K.; Hartenstein, D.; Wurlitzer, T.: Integrierte Diagnose von Anlagen mit vernetzten Automatisierungssystemen. Vortrag, GMA-Kongreß'96, Baden-Baden, Sept. 1996, VDI-Berichte Nr. 1282, S. 517 - 526, VDI-Verlag Düsseldorf 1996

[3] Isermann, R.: Überwachung und Fehlerdiagnose - Moderne Methoden und ihre Anwendung bei technischen Systemen. VDI-Verlag Düsseldorf 1994

[4] Klar, R., u. a.: Messung und Modellierung paralleler und verteilter Rechensysteme. B. G. Teubner Stuttgart 1995

[5] N.N. LonWorks Technology Device Data. Firmenschrift DL 159/D, Motorola Inc. 1995

Verhalten der physikalischen Übertragungsschicht von Feldbussen bei elektromagnetischen Störungen

Jörg Neumann *), Harald Schumny *), Petr Kocourek **), Norbert Zisky *), Zdenek Štepka **), Jirí Novák **)

*) Physikalisch-Technische Bundesanstalt, Fürstenwalder Damm 388, D-12587, Berlin
**) Czech Tech. University, Dept. of Measurement, Technicka 2, CZ-16627, Prague 6

Zusammenfassung: Ein Kriterium bei der Auswahl von Feldbussen ist die Störfestigkeit der Datenübertragung. Die verschiedenen Feldbusse unterscheiden sich nicht nur in ihren Übertragungsprotokollen, sondern auch in den physikalischen Übertragungsschichten. Der Beitrag beschäftigt sich mit der Frage, ob die Störfestigkeit von der verwendeten physikalischen Übertragungsschicht abhängt oder im größeren Maße durch die Konstruktion der Feldbuskomponenten bestimmt wird. Die Untersuchungen wurden an einer Reihe von Schnittstellen durchgeführt. Die erhaltenen Ergebnisse werden am Beispiel der RS-485 und des M-Busses demonstriert. Anhand eines Modells werden mögliche Ursachen von Störungen der Datenübertragung erläutert.

Abstract. The immunity of data transmission is one criterion in the selection of field busses. The different fieldbusses differ not only in their protocols but also in the physical layers. The article deals with the question whether the immunity depends on the used physical layer or whether it is to a larger extend determined by construction of the fieldbus components. The investigations were carried out at different interfaces. The results are shown in examples of RS-485 and the M-Bus. Possible causes for disturbances of data transmission are explained with the help of a model.

1 Einführung

In einem gemeinsamen Forschungsprojekt der TU Prag und der Physikalisch-Technischen Bundesanstalt, gefördert durch die Volkswagen-Stiftung, wurde die EMV von Feldbussen untersucht [1]. Ein Schwerpunkt war der Einfluß der physikalischen Übertragungschicht (nach OSI-Referenzmodell DIN ISO 7498) auf die Störsicherheit der Datenübertragung. In diesem Beitrag wird die Störfestigkeit in Abhängigkeit von folgenden Faktoren betrachtet:

– Einfluß der elektrischen Eigenschaften der Übertragungsstrecke. Untersuchung von symmetrischen und unsymmetrischen Schnittstellen mit strom- oder spannungsmodulierten Signalen bei unterschiedlichen Signalpegeln.
– Einfluß der Schirmung der Übertragungsleitung.

– Einkopplungen von Störungen über die Signalleitungen in die Schnittstellenelektronik.

Schwerpunkt der Arbeiten war die Untersuchung der Störfestigkeit bei Einwirkungen von Transientenpaketen (Burst) auf die Übertragungsstrecke. Dadurch konnten sehr anschauliche, gut vergleichbare und reproduzierbare Ergebnisse gewonnen werden.

2 Versuchsaufbau

Für die Überprüfung der Störfestigkeit gegen Transientenpakete wurde ein Meßplatz entsprechend IEC 1000-4-4 aufgebaut (Abb. 1). Mit dem Burstgenerator wurden Impulse mit einer Amplitude bis zu 4 kV, einer Frequenz von 5 kHz bei einer Anstiegszeit von 5 ns erzeugt. Die Periode der Burst-Pakete betrug 300 ms. Die Einkopplung der transienten Störungen in die Signalleitung erfolgte mittels einer kapazitiven Koppelstrecke mit einer Länge von 1 m und einer Koppelkapazität von 50 pF..200 pF

Schnittstellenwandler für die verschiedenen Bussysteme bildeten die Datenübertragungseinrichtungen DÜE .

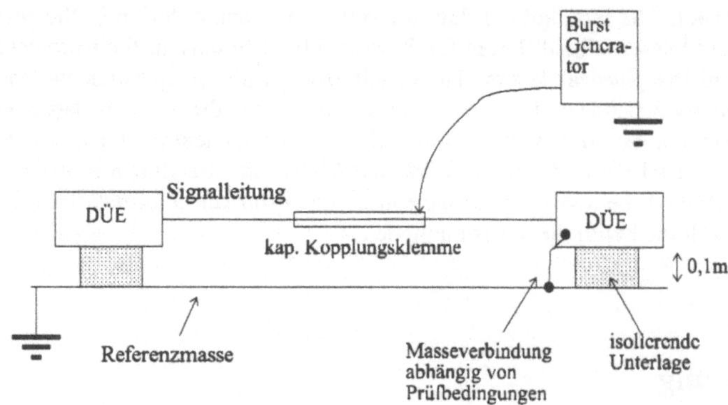

Abb. 1   Einkopplung in die Signalleitung mittels kapazitiver Kopplungsklemme

3 Untersuchung der Norm RS-485

Die Norm RS-485 [2] ist eine der am häufigsten verwendeten Schnittstellen bei Mehrpunktverbindungen. Feldbusse wie der Interbus-S, DIN-Meßbus, Profibus usw. verwenden in ihrer physikalischen Übertragungsschicht diesen Standard. Mit der RS-485 werden symmetrische Bussysteme aufgebaut, bei denen die Information in Form von Differenzspannungen übertragen wird.

Für die Versuche sind als DÜE handelsübliche PCs mit Schnittstellenwandlern eingesetzt worden. Somit konnte eine reale, in der Praxis häufig anzutreffende Übertragungstrecke nachgebildet werden. An dieser wurden in mehreren Versuchsreihen der Einfluß der Schirmung der Signaleitung auf die Störfestigkeit der Datenübertragung untersucht. Dazu wurden 50 kbyte große Datenblöcke mit einer Übertragungsgeschwindigkeit von 9600 bit/s übertragen. Der Schirm der Signalleitung wurde entweder gar nicht, auf die Referenzmasse, auf die Busmasse oder aber auf die Masse der DÜE aufgelegt.

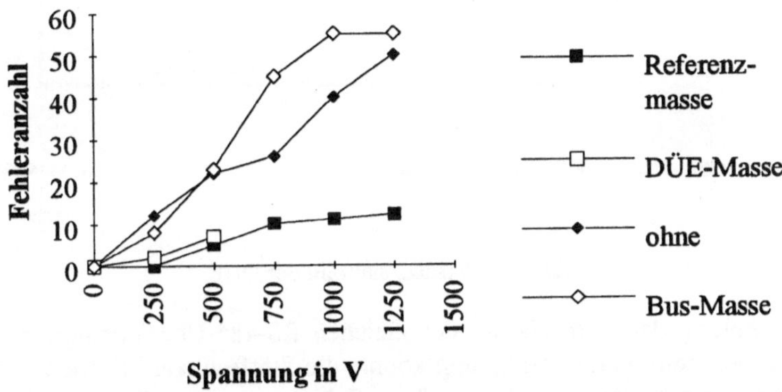

Abb. 2 Anzahl der Übertragungsfehler bei verschiedenen Schirmungsvarianten

Die erhaltenen Meßergebnisse (Abb. 2) zeigen, daß die Störfestigkeit stark von der Schirmung und deren Anschluß an die DÜE abhängt. Im ungünstigsten Fall wurde der Schirm direkt an die Masse der DÜE angeschlossen. Dabei kam es schon bei einer Burstamplitude von 500 V zu Fehlfunkionen der verwendeten PCs. Bei den anderen Schirmungsvarianten trat dieser Effekt erst bei höheren Spannungen auf. Um zu erklären, warum es zu Fehlfunktionen kam, wurde das in Abb. 3 dargestellte Modell für einen typischen Schnittstellenwandler im PC entwickelt.

Dieses Modell zeigt, daß Übertragungsfehler durch eine mangelnde Gleichspannungsfestigkeit der eingangsseitigen Differenzverstärker entstehen können. Oder es gelangen Störungen über die parasitären Kapazitäten des Optokopplers und des Gleichspannungsübertragers in die Elektronik des Schnittstellenwandlers und damit des PCs und verursachen somit nicht nur Übertragungsfehler, sondern führen auch zu Fehlfunktionen. Hauptursache für die schlechte Störfestigkeit sind nicht die Eigenschaften der Schnittstelle, sondern die ungenügende galvanische Entkopplung. Um diese Aussage zu bestätigen, wurden die Übertragungstrecke und die angeschlossene Elektronik getrennt. Dies erfolgte mit Hilfe von speziell konstruierten optischen Übertragungsstrecken, welche die Signalleitung und die DÜE vollständig galvanisch trennen.

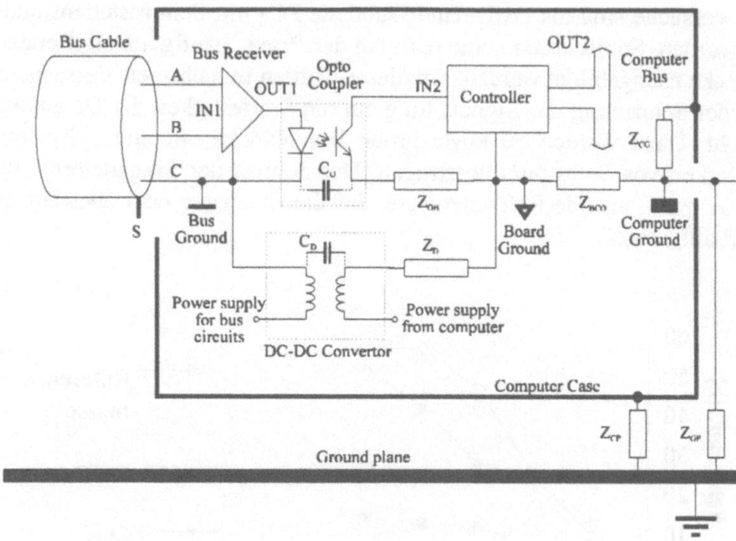

Abb. 3 Ersatzschaltbild der DÜE

Bei Wiederholung der Versuche an der gleichen RS-485-Übertragungsstrecke mit der angeführten optischen Entkopplung konnte die Störfestigkeit deutlich verbessert werden. Bei einem beidseitigen Anschluß der Schirmung und vollständigen galvanischen Trennung der Optokoppler von der Referenzmasse wurden bis zu einer Burstspannung von 4000 V keine Übertragungsfehler festgestellt. Sind die Optokoppler mit der Referenzmasse verbunden, so ist eine geringe Abhängigkeit der Fehleranzahl von der Art der Schirmung zu beobachten (Abb. 4).

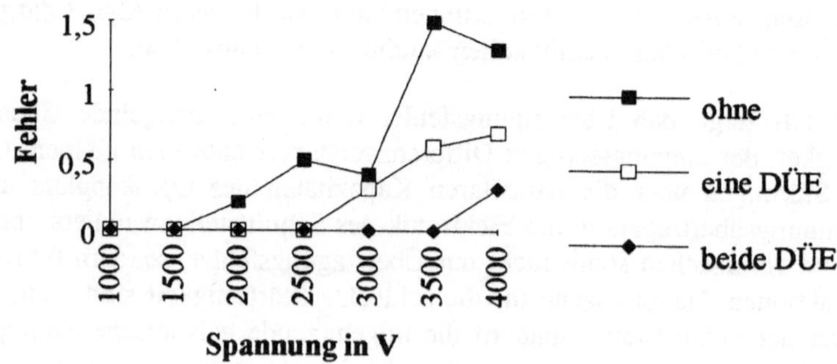

Abb. 4 Anzahl der Übertragungsfehler bei verschiedenen Schirmungsvarianten

Die Schirmung der Übertragungstrecke hatte nur noch einen sehr geringen Einfluß auf die Störfestigkeit gegenüber Burstimpulsen.

Es konnte gezeigt werden, daß die in die Übertragungsleitung eingekoppelten Störimpulse, unter Berücksichtigung üblicher Verfahren der Seriell-Parallelwandlung, zu keiner Störung der Datenübertragung führen. Somit konnten Störungen in der angeschlossenen Schnittstellenektronik als Hauptursache für Übertragungsfehler nachgewiesen werden.

4 Untersuchung des M-Bus

Der M-Bus [3] unterscheidet sich von anderen Feldbussen durch einige Besonderheiten der physikalischen Übertragungsschicht. Die Schnittstelle ist unsymmetrisch ausgelegt. Die Datenübertragung von einer Leitstation zu einem Teilnehmer erfolgt mittels der Modulation einer Spannung. In der entgegengesetzten Richtung erfolgt die Modulation eines Stroms. Somit unterscheidet sich die beim M-Bus verwendete physikalische Schicht grundsätzlich von der RS-485. Es wurden die gleichen Untersuchungen wie in Abschnitt 3 beschrieben auch für den M-Bus durchgeführt.

Messungen an einer M-Bus Übertragungsstrecke ohne zusätzliche optische Entkopplung ergaben, daß schon bei geringen Burstspannungen (500 V) eine maximale Fehlerrate von 5 % erreicht wurde (Abb. 5). Fehlfunktionen der DÜE wurden jedoch nicht beobachtet.

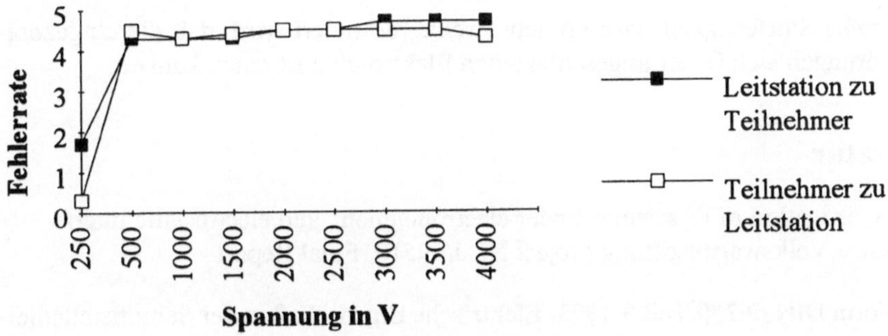

Abb. 5   Fehlerrate der M-Bus-Übertragungsstrecke

Eine einfache optische Entkopplung der Übertragungsstrecke wie in Abschnitt 3 war nicht möglich. Es wurde nur ein Teilnehmer mit entsprechender galvanischer Trennung aufgebaut. Messungen ergaben eine wesentliche Verbesserung der Störfestigkeit bis zu einer Burstspannung von 2000 V (Abb. 6). Die Ursachen für den folgenden starken Anstieg der Übertragungsfehler konnten noch nicht geklärt werden.

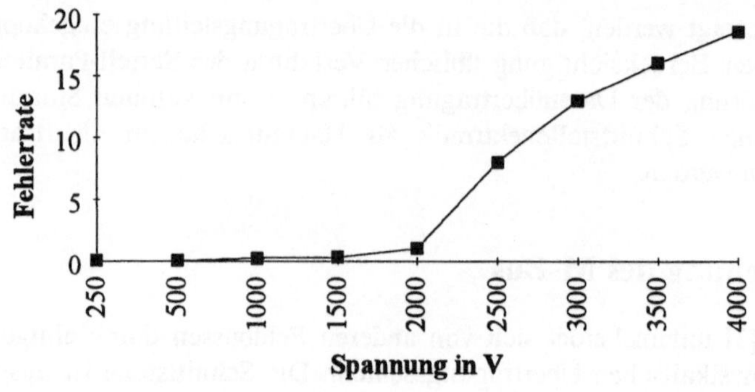

Abb. 6 Fehlerrate eines optisch entkoppelten Teilnehmers

5 Schlußfolgerungen

Bei der Betrachtung der Störfestigkeit der physikalischen Schicht von Feldbussen gegenüber Burstimpulsen kann festgestellt werden, daß der Typ der Schnittstellen einen geringen Einfluß auf die Störfestigkeit hat. Eine Schirmung der Übertragungsstrecke führt nur dazu, daß die Energie der eingekoppelten Störung verringert wird.

Die größte Störfestigkeit wird erreicht, wenn verhindert wird, daß die eingekoppelten Störungen sich in der angeschlossenen Elektronik ausbreiten können.

Literatur

1. Investigation of IT systems under electromagnetic and electrostatic interference, Volkswagenstiftung project Nr. I/69510, Final Report

2. Norm DIN 66259 Teil 3 1993. Elektrische Eigenschaften der Schnittstellenleitungen Doppelstrom, symmetrisch bis 10 Mbit/s

3. Norm prEN 1434-3 1994. Wärmezähler, Teil 3: Datenaustausch und Schnittstellen

Phasensynchronisation zyklisch getakteter Baugruppen an seriellen Bussystemen

M. Franke*, M. Hoffmann**

*TU Dresden, Elektrotechnisches Institut,
Lehrstuhl für Automatisierte Elektroantriebe
D-01062 Dresden, Germany
**SIEMENS AG, Automatisierungssysteme für
Werkzeugmaschinen, Roboter und Sondermaschinen
D-91050 Erlangen, Germany

Abstract. The state of the art to design automation systems is characterized by using fieldbus linked devices (e.g. sensors, drives, ...). Each single unit generates its own internal clock by means of quartz oscillators and timers. The relative phase shift of the sampling periods between different units is accidental and not constant, due to quartz oscillator tolerances and to an individual behavior after power-on. Given a possibility to read each actual phase position (timer value) repeatedly at common sampling points and to change the preset value, a selection of devices can be synchronized. The fieldbus system has to provide functions in order to sample data in different units at the same sample point, independent of its serial behaviour. This can be achieved by the Global Control Commands „Sync" / „Freeze" (PROFIBUS DP) or by similar broadcast mechanisms. The following publication shows an example for the synchronisation of digital drive units with practical measurements and it demonstrates the advantages in an application.

Feldbusgekoppelte digitale Baugruppen (z.B. Sensoren, Antriebe, ...) in verteilten Automatisierungssystemen kennzeichnen den Stand der Technik. Die einzelnen Baugruppen generieren ihren internen Takt mit Hilfe von programmierbaren Timern und Quarzoszillatoren. Aus dem Einschaltverhalten resultiert eine zufällige Phasenverschiebung der Abtastzyklen zwischen verschiedenen Baugruppen, die aufgrund der Quarztoleranzen variabel ist. Wenn es möglich ist, die aktuelle Phasenlage (Timeristwert) von verschiedenen Baugruppen zum selben Zeitpunkt zu erfassen, so können diese synchronisiert werden. Dazu muß das Feldbussystem Funktionen zur Verfügung stellen, die unabhängig von seinem seriellen Betriebsverhalten das zeitgleiche Abtasten von Daten in verschiedenen Baugruppen ermöglichen. Dazu können z.B. die Global Control Commands „Sync" / „Freeze" (PROFIBUS DP) oder ähnliche Broadcastmechanismen verwendet werden. Der Aufsatz stellt ein Beispiel für die Synchronisation von digitalen Antrieben und Messungen an einer realen Anlage vor und diskutiert anhand eines technologischen Beispiels die Vorteile der Synchronisation.

1 Einleitung

In räumlich ausgedehnten Automatisierungssystemen werden dezentrale Baugruppen, wie z.B. geregelte Antriebe oder Sensoren in zunehmendem Maße durch serielle Feldbusse untereinander bzw. mit einer zentralen Steuereinrichtung gekoppelt. Bedingt durch die serielle Arbeitsweise der Feldbussysteme können Informationen nicht zeitgleich zu den Teilnehmern übertragen bzw. in ihnen wirksam werden. Zusätzliche Dienste ermöglichen, daß zu den Teilnehmern zeitlich nacheinander übertragene Daten in ihnen zeitgleich übernommen bzw. von den Teilnehmern auszulesende Informationen in ihnen zeitgleich zwischengespeichert werden können. Ein Beispiel für solche Mechanismen sind die Dienste „Sync" bzw. „Freeze" beim PROFIBUS DP.

Sollen Daten zu definierten Zeitpunkten übertragen werden, so muß die Forderung nach deterministischem Verhalten erfüllt sein, was jedoch nur bei wenigen Feldbussystemen der Fall ist. Ist deterministisches Verhalten gegeben, so können verschiedene Busteilnehmer aufeinander synchronisiert werden.

Dezentrale Baugruppen, wie z.B. digital geregelte Antriebe arbeiten zyklisch nach einem intern generierten Takt im Bereich von wenigen Millisekunden. Aufgrund der sequentiellen Programmabarbeitung müssen Soll- bzw. Istwerte bis zu einem bestimmten Abschnitt des Abtastschritts im Antrieb vorhanden sein, um im aktuellen Abtastschritt berücksichtigt werden zu können. Bedingt durch Einschaltverhalten und Quarztoleranzen sind die Abtastzyklen selbst bei identischem Aufbau der Antriebe bezüglich ihres Startzeitpunktes nicht phasengleich und driften gegeneinander. Das führt dazu, daß selbst zeitgleich eintreffende Daten mit einer variablen Totzeit wirksam werden, die bis zu einem Abtastzyklus der Regelung betragen kann.

Im folgenden wird ein Verfahren zur Phasensynchronisation der Abtastzyklen in digital geregelten Antrieben vorgestellt, welches es ermöglicht, Baugruppen auch über nichtdeterministische Feldbussysteme zu synchronisieren.

Das Verfahren wurde mit Feldbus PROFIBUS DP experimentell verifiziert.

2 Beschreibung des Verfahrens

Die Abtastzyklen in digital geregelten Antrieben werden zumeist mit programmierbaren Timern, wie sie z.B. in Mikrokontrollern vorhanden sind, erzeugt. Die Zyklusdauer wird über entsprechende Timerregister vorgegeben. Nach erfolgter Initialisierung bestimmt der Timerstand die Phasenlage des Abtastzyklusses. Durch die Nutzung von sogenannten Global Control Commands (GC) wie z.B. „Sync" und „Freeze" ist es möglich, die Timerstände auf mehreren Antrieben zeitgleich zu erfassen und zu einer Auswerteeinrichtung zu übertragen [1]. Das Eintreffen eines GC muß von der Antriebssoftware mit hoher Priorität bearbeitet werden, um die Totzeit zwischen dem Eintreffen eines GC und dem Auslösen der entsprechenden Reaktionen im Antrieb (z.B. Timerstand abspeichern) gering zu halten. Eine Auswerteeinrichtung arbeitet einen Algorithmus ab, welcher die Timerstände miteinander vergleicht und

Korrekturwerte für die verschiedenen Antriebe ermittelt. Dabei ist es prinzipiell unerheblich, ob der Auswertealgorithmus in einer zentralen Systemsteuerung oder in einem der Teilnehmer abgearbeitet wird. Kriterien hierfür sind unter anderem die Möglichkeiten des Datenaustauschs zwischen Systemkomponenten über das Bussystem (z.B. Querverkehr, Multimasterbetrieb, ...) und Aspekte der Modularisierung von Funktionen. Im jeweiligen Antrieb bewirkt der Korrekturwert eine Modifikation der Zyklusdauer, was eine Verringerung des Phasenfehlers zur Folge hat. Mit diesem Mechanismus ist es möglich, die Phasenlage der Abtastzyklen digital geregelter Antriebe auf ein bestimmbares Toleranzband einzuregeln, so daß übertragene Sollwerte im selben Abtastschritt wirksam werden. Es muß sichergestellt werden, daß das Einlesen der Sollwerte erst außerhalb des Toleranzbandes erfolgen kann.

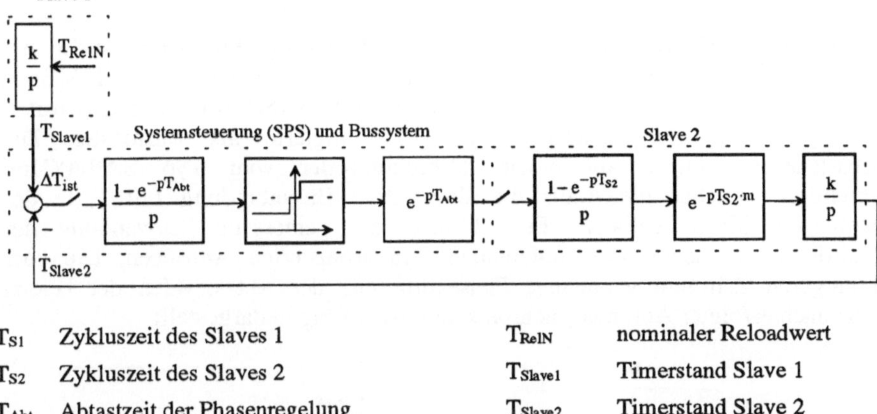

T_{S1}	Zykluszeit des Slaves 1	T_{ReIN}	nominaler Reloadwert
T_{S2}	Zykluszeit des Slaves 2	T_{Slave1}	Timerstand Slave 1
T_{Abt}	Abtastzeit der Phasenregelung	T_{Slave2}	Timerstand Slave 2
m	0...1	ΔT_{ist}	Phasendifferenz
k	Umrechnung Reloadwert $\rightarrow$ Periodendauer	T_{Rel}	Reloadwert (Stellgröße)

Fig. 1. Signalflußplan der Synchronisation zweier Antriebe

Die erreichbare Genauigkeit wird von der Auflösung der Timer in den Antrieben, von der Übertragungsgeschwindigkeit des Bussystems, der Drift der Quarze und der Verarbeitungsgeschwindigkeit der Auswerteeinrichtung bestimmt. Bussystem und Auswerteeinrichtung bestimmen die Abtastzeit der Synchronisation. Das Toleranzband ergibt sich folglich aus der Abtastzeit der Synchronisation und der Differenz der Abtastzeiten der aufeinander zu synchronisierenden Baugruppen. Die durch Modifikation der Abtastzeit einer Baugruppe hervorgerufene „künstliche" Drift muß größer als die zu kompensierende „natürliche" Drift der Quarze sein. Durch Softwaremodulation kann eine „künstliche" Drift erzeugt werden, die kleiner als die durch die minimale Timerauflösung vorgebbare ist.

Der Signalflußplan auf Fig. 1 beschreibt die Synchronisation von zwei Antrieben beispielhaft, wobei leicht zu erkennen ist, daß eine Synchronisation mehrerer Antriebe durch Mehrfachvorgabe des Sollwertes T_{Slave1} problemlos möglich ist. Praktisch durchlaufen die Timer einen endlichen Wertebereich von einer unteren zu einer oberen Grenze bzw. entgegengesetzt. Bei Erreichen einer Grenze erfolgt ein Rücksetzen auf den anderen Grenzwert (Reloadwert). Für die Systembeschreibung hingegen können die Timer als unendliche Integratoren eines der Timerauflösung entsprechenden konstanten Wertes über der Zeit modelliert werden. Der Regler kann sowohl linear, als auch nichtlinear entworfen und realisiert werden. Dem Grundprinzip nach handelt es sich bei dem Verfahren um eine Software-PLL, wobei ein Software-Dreipunktregler als Schleifenfilter verwendet wird [2].

Ausgehend von dem auf Fig. 1 dargestellten Signalflußplan wurde das Verhalten des Systems simuliert. Der Vergleich mit den im folgenden Abschnitt dargestellten Messungen zeigt eine gute qualitative und quantitative Übereinstimmung.

3 Experimentelle Ergebnisse

Das vorgestellte Verfahren wurde an einem Versuchsstand praktisch untersucht. Als Bussystem wurde der PROFIBUS DP mit einer Übertragungsrate von 12MBaud eingesetzt. Der Auswertealgorithmus wurde in einer SPS (Simatic S5) abgearbeitet, eine entsprechende Peripheriekarte (IM308C) fungierte als Busmaster. Die Antriebssteuerung wurde auf einem Mikrokontroller vom Typ SAB80C166 implementiert, welcher gleichfalls die slaveseitige Busanbindung mit Hilfe des Protokollasic's SPC3 realisiert. Fig. 2 zeigt die gemessene Schwankung der Phasendifferenz zwischen zwei aufeinander synchronisierten Antrieben. Die über einen längeren Zeitraum gemessene Phasendifferenz der Abtastzyklen der beiden Antriebe nach erfolgter Anfangssynchronisation ist auf Fig. 3 dargestellt.

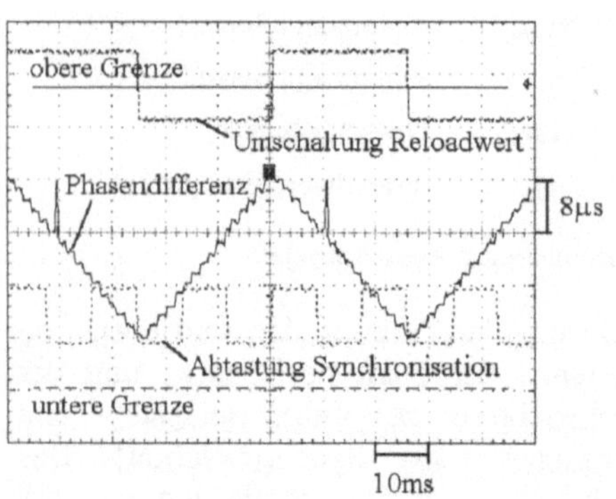

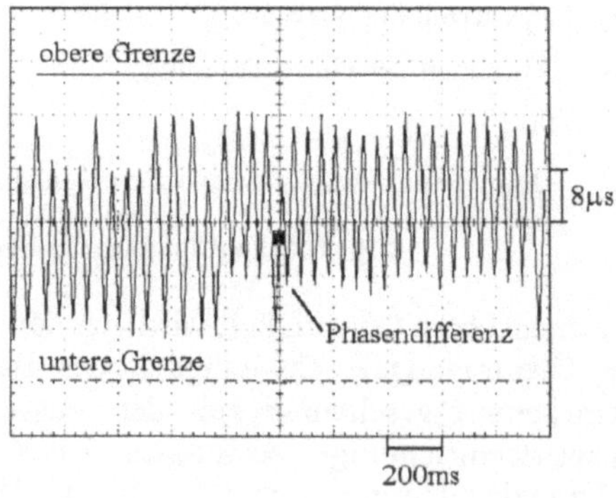

Fig. 2. Phasendifferenz der Abtastzyklen der Antriebe

Fig. 3. Phasendifferenz (längerer Zeitraum)

Die Schwankungen um den Nullpunkt resultieren aus der nicht äquidistanten Abtastung der Timerstände durch das Bussystem. In jedem zu synchronisierenden Teilnehmer muß im Kommunikationskanal ein Prozeßdatum für die Synchronisation zur Verfügung gestellt werden. Die ebenfalls abgebildeten Grenzen wurden aus der minimalen Timerauflösung (400ns) und der maximal auftretenden Abtastzeit der Synchronisation ermittelt. Die Meßergebnisse zeigen anschaulich, daß trotz einer bezüglich der Abtastfrequenz der Baugruppen relativ großen und nicht konstanten

Abtastzeit der PLL eine gute Synchronität der Baugruppen erreicht werden kann. Durch Erhöhung der Auflösung der Timer kann die Genauigkeit noch gesteigert werden.

4 Technologisches Beispiel

Die technologische Anordnung ist auf Fig. 4 dargestellt. Ein Werkstück (W) wird mit Hilfe einer Linearachse mit Absolutmeßsystem aus einer Bearbeitungsstation (B_Stat1) zu einer anderen (B_Stat2) transportiert. Während des Transportes soll ein für die weitere Bearbeitung erforderliches Teil eingelegt werden. Diese Aufgabe übernimmt ein Greifer auf einer Rundachse (R), der bei ca 180° die Teile aufnimmt und bei 0° paßgenau ablegt. Die Zuförderung der Teile zum Greifer wird in diesem Beispiel nicht betrachtet. Die Maschine ist dezentral aufgebaut. Alle Achsen sind untereinander und mit der übergeordneten Steuerung über den seriellen Feldbus PROFIBUS DP verbunden.

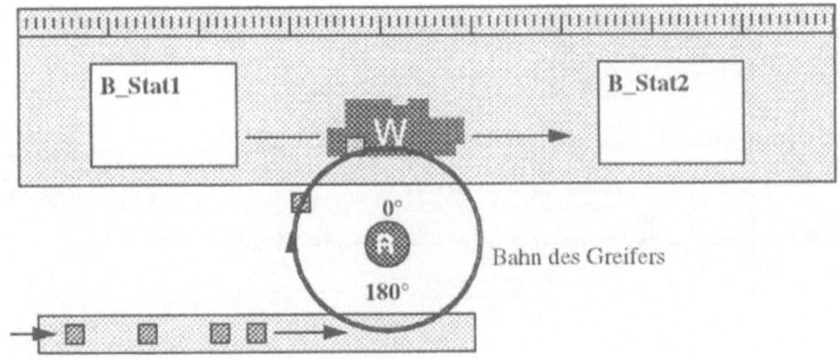

Fig. 4. Anordnung zum Fügen eines Teils in ein Werkstück

Nach den Meldungen der Bearbeitungsstation 1 „Werkstück abgesetzt" und des Greifers „Teil aufgenommen" müssen die Positionen an beiden Einzelachsen zur gleichen Zeit gemessen und daraus von der Steuerung die Geschwindigkeiten zum Erreichen des Rendezvous-Punktes berechnet werden. Dazu löst die Steuerung über das „Freeze" ein Meß-Kommando aus, das von allen Busteilnehmern gleichzeitig empfangen werden kann und sie veranlaßt, aktuelle Istwerte einzufrieren. Die Lageregler messen also in ihrem nächsten Takt die Position und geben den Wert zur Weiterverarbeitung (Berechnung neuer Fahrdaten) über den Bus an die koordinierende Einheit. Diese kann dann wieder neue Sollwerte an die Achse verteilen und mit dem „Sync"-Kommando auf allen betroffenen Busteilnehmern gleichzeitig, d.h. mit dem darauffolgenden Lageregelungstakt aktivieren.

Wenn die Regelungstakte der Einzelachsen untereinander nicht koordiniert sind, driften die Lageregler gegeneinander, sodaß die Synchronisationskommandos zwar zur gleichen Zeit empfangen, aber zu unterschiedlichen Zeiten ausgeführt werden. Das „Freeze" wird in der Achse$_1$ nach t$_{Freeze1}$ und in der Achse$_2$ nach t$_{Freeze2}$ wirksam.

290

Zwischen den Reglern besteht eine Differenz von t_d, die im worst-case unter der Annahme gleicher Abtastzeiten so lang ist wie der Lagereglertakt.

$$t_d \leq t_{LR1/2}$$

Dieser Zeitversatz begrenzt entweder die erzielbare Genauigkeit oder die mögliche Geschwindigkeit bei einem Rendezvous, denn durch die Unschärfe bestimmen beide Achsen ihre Positionen zu unterschiedlichen Zeitpunkten. Die Achse$_1$ liefert nach $t_{Freeze1}$ den Wert s_1 während die Achse$_2$ ihren Wert s_2 erst zum Zeitpunkt $t_{Freeze2}$ bestimmt, wenn Achse$_1$ um die Strecke s_d weitergefahren ist. Aus der Gleichung

$$s_d = v \cdot t_d \qquad \text{mit } v = \text{Geschwindigkeit}$$

lassen sich dann die Einschränkungen für Genauigkeit und / oder Geschwindigkeit berechnen.

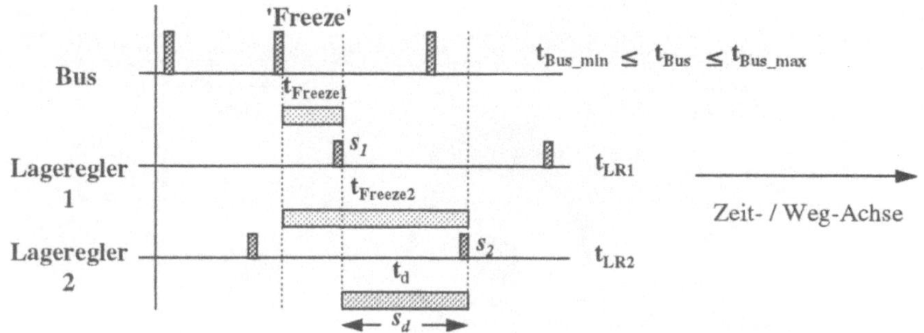

Fig. 5. Zeitdiagramm für die Istwerterfassung bei unsynchronisierten Achsen

Beispiel: $t_{LR1/2}$ =   1 ms

geforderte Genauigkeit   =   0,1 mm

Welche maximale Geschwindigkeit kann gefahren werden ?

$$v_{max\ unsync} = 0,1\ mm\ /\ 1\ ms = 0,1\ m/s$$

Synchronisiert man die Regler nach dem in Abschnitt 2 beschriebenen Verfahren, so liegt der Zeitfehler bei $t_{Jitter} = 50\ \mu s$. Daraus ergibt sich eine mögliche Geschwindigkeit von

$$v_{max\ sync} = 0,1\ mm\ /\ 0,05\ ms = 2\ m/s.$$

Durch die Synchronisation läßt sich die Genauigkeit der beiden Achsen um Größenordnungen verbessern.

5 Zusammenfassung

Im Beitrag wird ein Verfahren zur Synchronisation von zyklisch arbeitenden Baugruppen an seriellen nicht deterministischen Feldbussystemen vorgestellt. Das Systemverhalten wird modelliert und mit Meßergebnissen an einem Versuchsstand bei Verwendung des PROFIBUS DP verifiziert. Am Beispiel einer Werkstückübergabestation wird der Nutzen des Verfahrens durch eine überschlägige Rechnung nachgewiesen.

Das vorgestellte Verfahren wurde zum Patent angemeldet.

Literatur

1. Draft DIN E 19245 Part 3 Process Field Bus, Decentralized Peripherie (DP), PROFIBUS Nutzerorganisation, Karlruhe, 1994

2. Best, R.: Theorie und Anwendungen des Phase-locked Loops, 4. überarb. Aufl., AT-Verl., Stuttgart, 1987

Multi-Sensor-Controller mit CAN-Feldbus-Interface

S. Rietz, W.-J. Fischer

Fraunhofer-Gesellschaft, Institut für Mikroelektronische Schaltungen
und Systeme, Grenzstrasse 28, D-01109 Dresden

Abstrakt. Durch den Einsatz von Mikrosystemen wird eine Verteilung der
Intelligenz in Feldbussystemen der industriellen Automatisierung sowie der
Automobilbranche erreicht, die zur Freisetzung wertvoller Resourcen führt. Kern
solcher Mikrosysteme stellt ein am Fraunhofer-Institut für Mikroelektronische
Systeme und Schaltungen entwickelter ASIC dar, der die Anbindung von
Sensoren an einen CAN-Feldbus realisiert. Ein geeignetes Entwurfskonzept
garantiert eine flexible Konfiguration der on-Chip-Komponenten.

Abstract. A decentralisation of intelligence in field bus systems in the industrial
automation and the automotive branch can be reached by applying smart
microsystems. This saves precious resources. An ASIC developed by the
Fraunhofer-Institute of Microelectronic Circuits and Systems is the main part of
such microsystems. This ASIC links sensors to a CAN field bus. A flexible
configuration of the on-chip components could be reached by means of an
appropriate concept.

1 Einleitung

Eine wichtige Aufgabe in der heutigen Zeit stellt die Entwicklung von intelligenten
Mikrosystemen (Smart Microsystems) dar. In klassischen Feldbussystemen wird die
Datenverarbeitung von einem Prozeßrechner übernommen. Für komplexe Prozesse mit
einer Vielzahl von Sensor- und Aktordaten folgt daraus eine hohe Busbelastung sowie
ein hoher Rechen- und Speicheraufwand des Prozeßrechners. Durch den Einsatz von
Smart Microsystems kann sowohl der Feldbus als auch der Prozeßrechner entlastet
werden [1].

Am Fraunhofer-Institut für Mikroelektronische Schaltungen und Systeme Dresden
(IMS) wurde ein ASIC entwickelt, der das Kernstück eines intelligenten Mikrosystems
darstellt. Er ist in der Lage, analoge Sensorsignale aufzunehmen, in Digitalworte zu
wandeln, diese zu verarbeiten und zu speichern. Weiterhin besitzt der ASIC die
Fähigkeit, über einen Feldbus mit anderen Komponenten zu kommunizieren.

2 ASIC-Komponenten

Der ASIC-Entwurf fand unter Anwendung des „Baukasten-Prinzips" statt und stützte
sich auf Komponenten - sowohl HDL-Kerne (HDL - Hardware Description Language)
als auch Makrozellen - die am IMS entwickelt wurden. Mit dem „Baukasten-Prinzip"

ist es möglich, das System modular auf der Systemebene (oberste Ebene der Entwurfshierarchie) je nach Kundenwunsch zu konfigurieren. Das bedeutet einen schnellen, übersichtlichen, kostengünstigen, flexiblen und fehlerfreien Entwurf. Bild 1 zeigt die Struktur des entworfenen ASICs in der höchsten Ausbaustufe.

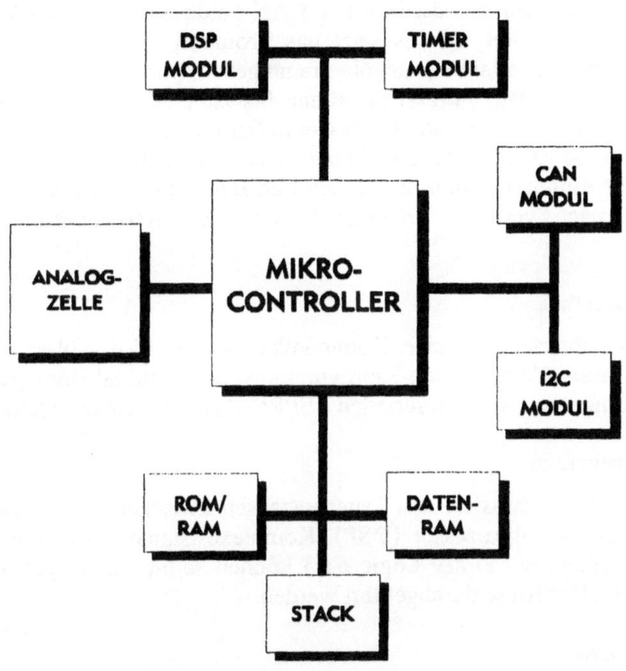

Bild 1: Systemstruktur des ASICs

2.1 Mikrocontroller

Ein 8-Bit-Mikrocontroller steuert den ASIC und ermöglicht eine einfache Signalverarbeitung wie beispielsweise eine Mittelwertbildung oder eine Bereichsüberwachung von Prozeßgrößen. Er weist eine RISC-Architektur auf und ist befehlskompatibel zu MICROCHIP's PIC16CXX. Das ermöglicht eine einfache und schnelle Entwicklung der Software unter Nutzung der Entwicklungstools von MICROCHIP. Der Mikrocontroller besitzt einen Watch-Dog-Timer sowie einen SLEEP-Mode, um die Stromaufnahme zu verringern.

2.2 Analogzelle

Zur Selektion von analogen Eingangssignalen wurde ein Analog-Multiplexer für bis zu 16 Kanäle implementiert. Eine Programmierung der Verstärkung erweist sich als sinnvoll für den Fall, daß unterschiedliche Sensoren mit verschiedenen Ausgangsspannungsbereichen an den Eingangsspannungsbereich des Analog-Digital-Umsetzers (ADU) angepaßt werden müssen. Beim konzipierten programmierbaren Verstärker lassen sich Verstärkungen von 1 bis 63 einstellen. Der A/D-Wandler dient der Transformation der analogen Sensorsignale in digitale Worte und arbeitet nach dem

Charge-Balancing-Prinzip [2,3]. Die Auflösung kann von 9 bis 15 Bit programmiert werden.

2.3 CAN-Feldbus-Interface

Zur Ankopplung an den standardisierten CAN-Feldbus wurde on-Chip ein CAN-Feldbus-Interface integriert. Dieses setzt das Protokoll des CAN-Busses nach dem Standard 2.0B um und erlaubt Datenübertragungen sowohl im Standard (11 Bit) als auch im Extended (29 Bit) Format bei Raten bis zu 1 Mbit pro Sekunde. Das Bus-Interface ist als Standalone-BASIC-CAN-Controller entworfen worden [5]. Das CAN-Bus-Protokoll wurde gewählt, da es nicht nur in der Automobilindustrie sondern auch in der industriellen Automatisierung zunehmend an Bedeutung gewinnt. Außerdem steht die Forderung nach verteilter Intelligenz, der das CAN-Bus-Protokoll sehr gut nachkommt.

2.4 I^2C-Interface

Mit Hilfe dieses Interface ist eine Kommunikation des ASICs über einen I^2C-Bus möglich. Programm-, CAN-Knoten-Konfigurations- sowie Initialisierungsdaten für den A/D-Wandler können in externen seriellen EEPROMs abgespeichert werden.

2.5 DSP-Interface

Die DSP-Schnittstelle realisiert den Datenaustausch zwischen dem ASIC und einem externen digitalen Signalprozessor (DSP). Komplexe Signalverarbeitungsalgorithmen (modellbasierte Diagnose, Fuzzy Logic o.ä.) können somit ausgelagert und auf einer leistungsfähigeren Hardware durchgeführt werden.

2.6 Timer Unit

Mit dieser Baugruppe werden die von den diversen ASIC-Komponenten benötigten Modultakte bereitgestellt (von 250 kHz bis 8 MHz). Außerdem können zur Anpassung an die jeweiligen Auflösungen die Zykluszeiten der Sensoren programmiert werden. Zusätzlich ist eine Einstellung per Software des Modultaktes für das CAN-Feldbus-Interface zur Generierung der nach [4] spezifizierten Bus-Frequenzen implementiert worden

Die einzelnen Komponenten wurden im Rahmen von Multi-Prjojekt-Chip-Fertigungen präpariert und gemessen. Das Bild 2 zeigt den Mikrocontroller-Kern und den CAN-Controller-Kern im Waferverband.

Für ein Industrieprojekt wurde eine abgerüstete Variante des Multi-Sensor-ASICs entworfen, wobei die DSP- und die I^2C-Schnittstelle nicht implementiert wurden. Der ASIC besitzt zwei Analog-Eingänge sowie zwei digitale I/O-Ports (5 und 8 Bit). Über einen Drei-Draht-Bus können Knotenkonfigurations- sowie Programmdaten des Anwenders aus externen seriellen EEPROMs eingelesen werden. Der Chip wird gegenwärtig in einer 1.0 mm-Technologie gefertigt. Meßergebnisse werden auf der Tagung vorgestellt.

Bild 2: Chip-Foto des Mikrocontroller-Kerns (links) und des CAN-Bus-Controller-Kerns (rechts)

3 Ausblick und Zusammenfassung

Marktprognosen [6,7] sagen für 8-Bit-Mikrocontroller mit integrierter CAN-Schnittstelle für das Jahr 2001 etwa eine Verdrei- bis Vervierfachung der CAN-Knotenanzahl gegenüber 1997 voraus. Die On-Chip-Integration eines im Fraunhofer-Institut für Mikroelektronische Schaltungen und Systeme entwickelten Signalprozessor-Kernes ermöglicht es, leistungsfähige Signalverarbeitungs-algorithmen bei verringertem Bauelementevolumen abzuarbeiten. Implementiert man anstelle des Analog-Digital-Wandlers einen Digital-Analog-Wandler, so wird es prinzipiell möglich sein, einen intelligenten Aktor-Knoten an ein Feldbus-System zu koppeln.

Mit der Entwicklung dieses ASICs wird ein weiterer Schritt in Richtung der Dezentralisierung der Intelligenz in Feldbussystemen getan. Sensordaten können bereits am Ort ihres Entstehens verarbeitet werden. Der entwickelte ASIC stellt eine integrationsfreundliche Komponente dar, weil durch den Entwurf auf der Systemebene bereits die Systemfunktion festgelegt wird und sich die Verifikation übersichtlich gestalten läßt. Durch die Anwendung des „Baukasten-Prinzips" (Modul-Bauweise) wird garantiert, daß kundenspezifische Wünsche erfüllt werden können. Hervorzuheben ist ebenfalls der Entwurf der digitalen Komponenten mit Hilfe von Hardware-Beschreibungssprachen, die eine wesentliche Zeit- und Kostenersparnis bedeuten. Die Integration von programmierbaren Baugruppen im System garantiert ein hohes Maß an Flexibilität für die Anwendungen. Spätere Änderungen der Anwender-Software können ohne Probleme vorgenommen werden. Durch die Vielzahl von Schnittstellen ist der ASIC offen für ein breites Spektrum von Anwendungen.

Literatur

1.   Rietz, S.: Entwicklung eines Multi-Sensor-ASICs mit CAN-Felbusinterface, Diplomarbeit, Technische Universität Dresden, 1996, S.1

2.   Hoeschele, David F.: Analog-to-digital and digital-to-analog conversion techniques, John Wiley & Sons, New York, 1994, S.35 f.

3.   Zander, H.: Datenwandler, Vogel, Würzburg, 1990, S.172 ff.

4.   Schneider, B.: Informationen zum CAN-Controller CAN_CTL, FhG-interne Publikation, Fraunhofer-Institut für Mikroelektronische Schaltungen und Systeme Dresden, 1996

5.   Köhler, C.: Entwurf eines synthesefähigen VERILOG-HDL-Modells für ein CAN-Bus-Interface

6.   Krappel, A. R.: Echtzeit-Vernetzung in Kraftfahrzeugen, Elektronik Heft10/1997, S. 85 ff.

7.   CAN Protocol Controller Chip Figures, CAN News Letter 2/1997

Smart-Card-Security für Feldbussysteme

M. Manninger*, B. Burgstaller**, H. Reiter**

* Institut für Computertechnik, Gußhausstr. 27-29, A-1040 Wien
** Institut für Automation, Treitlstr. 1, A-1040 Wien

Kurzfassung. Feldbussysteme benötigen besonders im Hinblick auf die immer wichtigere Anbindung an übergeordnete Netzwerke ein ausgereiftes Security-Konzept, und die Smart Card ist für diesen Zweck ein ausgezeichnetes Instrument. Wir besprechen die grundsätzlichen Überlegungen für ein umfassendes Konzept hinsichtlich nötiger Mechanismen, des Anwendungsbereichs des Schutzes und der Erfordernisse der Fehlertoleranz. Kernpunkte der Systemarchitektur sind die Verwaltung der Zertifikate und der Zugriffsrechte. Darüber hinaus werden die Erkenntnisse in das am Institut entwickelte Homenet Configuration Tool eingearbeitet.

Abstract. In the area of fieldbus systems, security is becoming more and more important, especially with respect to the growing number of systems connected to other networks like the Internet. Smart cards are an excellent instrument for this purpose, and we introduce a comprehensive concept including the necessary mechanisms, the area of application, and the requirements of fault tolerance. Crucial questions are furthermore the administration of the keys and the access rights. Finally, we apply our security concept to the Homenet Configuration Tool which has been developed at our institute.

1 Motivation

Feldbussysteme sind traditionellerweise offene Kommunikationssysteme, die nur über eingeschränkte Security-Mechanismen verfügen. Wichtige Anforderungen in der Automatisierung, wie etwa die Identifikation von Benutzern bzw. die Vergabe von Zugriffsrechten (z.B.: nicht jeder Arbeiter darf auf jeden Feldbusknoten zugreifen, der Schichtleiter verfügt über höhere Rechte, *Auditing* von Entscheidungen) können damit nicht abgedeckt werden. Noch wichtiger wird ein umfassendes Security-Konzept angesichts der derzeit entstehenden Ankopplungen von Feldbussystemen an das Internet.

Will man Feldbusdaten über das Internet übertragen (z.B. zur Fernsteuerung oder Visualisierung, siehe Abb. 1), so begibt man sich auf gefährliches Terrain. Das Internet bringt hier zusätzlich zu seinen hervorragenden Fähigkeiten der globalen Vernetzung ein nicht zu übersehendes Risikopotential mit sich. Es ist wichtig, erstens die Daten nicht unverschlüsselt zu verschicken, und zweitens sicherzustellen, daß empfangene Daten auch vom richtigen Sender stammen und nicht von etwaigen Angreifern verändert wurden.

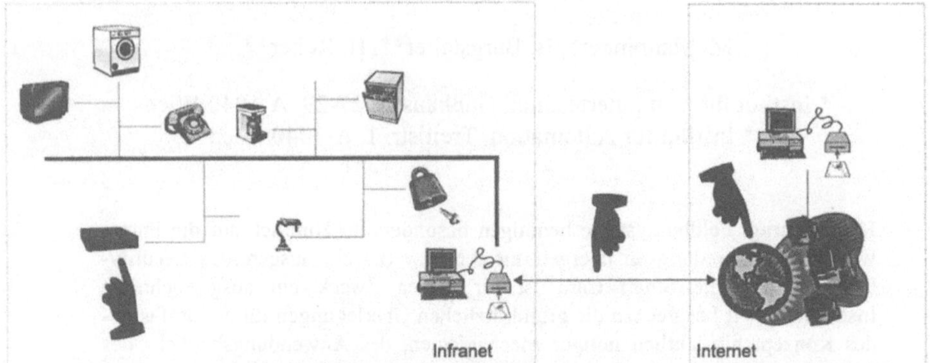

Abb. 1: Fernsteuerung eines Feldbussystems und Angriffe

2 Einsatz der Smart-Card-Technologie

Die Smart Card als sogenannte *Tamper-proof Hardware* ist eine anerkannt sichere Technologie, deren weitere Verbreitung wie im Fall des Internet rasant verläuft. Durch die geschlossene Struktur und die Ausführung aller sicherheitskritischen Operationen (Verschlüsselung, Authentisierung, Schlüsselmanagement) in der Karte selbst ist man gegen Angriffe bestmöglich geschützt. Systeme, die Sicherheit nur durch Software implementieren, haben demgegenüber einen entscheidenden Nachteil. Die Software, die die Sicherheit schaffen soll, ist dann selbst anfällig auf Veränderungen, und kriminelle Programme, die sich beispielsweise als Viren verbreiten können, führen diese Veränderungen bewußt her. Die Smart Card ist also das optimale Instrument, um die Datenübertragung über unsichere Netze (wie z. B. das Internet) sicher zu machen, und eignet sich somit auch für den Einsatz in Feldbussystemen.

Notwendige Voraussetzung dafür ist natürlich der Anschluß eines Chipkartenlesers an den beteiligten Netzknoten. Jede Übertragung von Daten kann von der Smart Card mit einer digitalen Signatur versehen werden, womit ihr Ursprung einwandfrei feststellbar ist. Eine Verschlüsselung zum Schutz vor dem Abhören ist ebenfalls möglich [5]. Insgesamt trägt die Smart-Card-Technologie zur Realisierung von Schutzmechanismen auf einem von konventionellen Lösungen nicht erreichbaren Sicherheitsniveau bei.

3 Security-Konzept

Zur Realisierung von Sicherheitsmechanismen in einem Feldbussystem ist es zunächst notwendig, mögliche Bedrohungen sowie die dazugehörigen Schutzmechanismen zu klassifizieren. Der Entwurf von auf die Möglichkeiten der Smart Card zugeschnittenen Algorithmen und Protokollen [2] steht dabei im Vordergrund, die Bedeutung von Fehlertoleranz im Hinblick auf Sicherheitsmechanismen wird ebenfalls untersucht.

3.1 Angriffe und Gegenmaßnahmen

Unbefugtes Abhören oder Verändern von Daten, Vortäuschen einer falschen Identität (*Masquerading*), Wiedereinspielung von Nachrichten (*Replay*), unautorisierter Zugriff auf Daten und Services sowie die Verleugnung der Teilnahme an einer Kommunikation (*Repudiation*) stellen Übergriffe auf das Kommunikationssystem dar, die es durch geeignete Maßnahmen zu verhindern gilt [6]. Schutzmechanismen gegen die genannten Übergriffe gliedern sich in drei Sparten:

Verschlüsselung: sowohl Ver- als auch Entschlüsselung in der Smart Card (Schutz vor Abhören und Verändern von Daten) unter Verwendung eines asymmetrischen Kryptosystems.

Authentisierung: Nachweis einer angegebenen Identität [1] einer Person oder eines Geräts durch die Smart Card (Schutz vor Masquerading). Nachdem der Benutzer seine Smart Card (z. B. mittels einer PIN) aktiviert hat, weist sie sich selbständig gegenüber dem System aus. Geräte haben eine Smart Card fix eingebaut.

Autorisierung: Vergabe von Rechten in bezug auf Personen, Dienste oder Geräte im System. Berechtigungen werden entweder direkt auf der Smart Card gespeichert oder nach Durchführung der Authentisierung von einer *Permission Authority* bzw. aus einem geräteinternen Speicher geholt.

3.2 Fehlertoleranz

Die implementierten Security-Mechanismen müssen darüber hinaus ein gewisses Maß an *Ausfallssicherheit* (Fehlertoleranz) aufweisen. Dies folgt aus der sog. *Fail-Secure-Property*: Im Fehlerfall kann das System Zutritt verwehren, den es erlauben hätte sollen. Es darf jedoch nicht passieren, daß Zutritt gewährt wird, der verboten werden hätte sollen. Für besonders kritische Anwendungen kann mitunter aber auch volle (*fail operational*) Fehlertoleranz gefordert werden. Zu diesem Zweck propagieren wir die Einteilung der Systemkomponenten in drei Klassen: *C1-Devices* sind Geräte, deren Sicherheitsmechanismen auch im allgemeinsten Fehlerfall (Ausfall des Kommunikationskanals oder Stromausfall) noch funktionieren (z.B. Türschloß mit Smart-Card-Leser und mechanischem Schlüssel). Die weniger mächtige Klasse der *C2-Devices* kann durch direkte Authentisierung am Gerät immerhin noch Ausfälle des Kommunikationskanals (nicht aber totale Stromausfälle) überbrücken. *C3-Devices* gehen im Fehlerfall lediglich in einen sicheren Abschaltzustand über.

3.3 Principals und die Vertrauensfrage

Konzeptionell zählen Feldbussysteme als Menge von mittels Kommunikationskanälen verbundenen Rechnerknoten zur Klasse der *verteilten* Systeme. Bedingt durch die Topologie eines solchen verteilten Systems ist es im Gegensatz zu *zentralisierten* Systemen unmöglich, alle Systemressourcen der Obhut eines einzigen zentralen Hosts zu unterstellen. Damit ist es nun aber auch nicht mehr möglich, sich einmalig „am System" anzumelden, um sodann gemäß den zugeteilten Rechten Zugriff auf alle Systemressourcen zu genießen.

Eine einfache, aber wichtige Frage betrifft daher den Anwendungsbereich des eingebrachten Schutzes. Wenn das Feldbussystem selbst in einer vertrauenswürdigen Umgebung steht und nur von vertrauenswürdigen Personen bedient werden kann, dann wird es genügen, die Verbindungen des Systems nach außen zu schützen. Dieser „Grenzschutz" entspricht dem Firewall-Konzept aus der Internet-Welt, wo die Firewall die Trennung zwischen „sicherem" und „unsicherem" Bereich darstellt und den Datenverkehr zwischen den beiden Bereichen einschränkt. In unserem Fall wäre dann typischerweise ein PC, der sowohl ein Feldbusknoten ist als auch am Internet hängt, diese Firewall. Nur dieser Knoten und der, der von außen auf den Bus zugreifen will, muß daher mit der Smart Card ausgestattet sein.

In einem dezentralen Feldbussystem ist es jedoch i.a. leicht möglich, daß Angreifer zusätzliche Knoten anschalten, oder daß sie einen bestehenden Knoten modifizieren. Befürchtet man dies, so kann der Bus nicht mehr als sicher betrachtet werden, und es muß ein vollständiger Schutz aller kritischen Knoten geschaffen werden. Dann muß jede zu sichernde Anwendung die Daten, die sie zu einem anderen Knoten schicken will, kryptographisch absichern. Dies bedingt aber in unserem Konzept auch den Einbau von Smart Cards in jedem der für sicherheitskritische Vorgänge benutzten Geräte.

Die Authentisierung in einem Feldbussystem kann daher jeweils nur für einen spezifischen Knoten erfolgen, was dann die Verwendung der auf diesem Knoten verfügbaren Ressourcen (Services) ermöglicht. Es ist dabei prinzipiell nicht notwendig, zwischen menschlichen Benutzern und Feldbusknoten als Clients anderer Knoten zu unterscheiden. Man spricht in diesem Zusammenhang von sog. *Principals* als Anbieter und Benutzer von Services im Feldbussystem.

Sicherheitsrelevante Überlegungen im Zusammenhang mit verteilten Systemen führen zwangsläufig auf die Frage, wem man im System vertraut bzw. mißtraut. Generelles Mißtrauen gegenüber allem und jedem im System impliziert zwar maximale Sicherheit, die Verwendbarkeit des Systems ist damit aber nicht mehr gegeben. Es ist daher notwendig, eine minimale, aber hochsichere Hard- und Softwarekonfiguration zu definieren, der alle Principals vertrauen, um auf dieser minimalen Vertrauensbasis (*Trusted Computing Base*, TCB [3]) die Implementierung der gesamten Sicherheitsmechanismen im System überhaupt erst zu ermöglichen.

4 Systemarchitektur

4.1 Schlüsselverwaltung

Die Aufnahme eines Principals in das System erfordert i.a. die Generierung eines Zertifikats, das fortan als Basis für die Authentisierung gegenüber anderen Principals im System dient. Dabei verschlüsselt eine sogenannte *Certification Authority* (CA) als Mitglied der TCB den Namen und den öffentlichen Schlüssel des neuen Principals mit ihrem privaten Schlüssel. Dieses Zertifikat übergibt der Principal bei der Authentisierung seinem jeweiligen Partner. Dieser entschlüsselt das Zertifikat mit dem öffentlichen Schlüssel der CA. Stimmt der im Zertifikat enthaltene Name nun mit dem Namen des Principals überein, so impliziert dies die ordnungsgemäße Anmeldung des Principals bei der CA. Die optionale Speicherung eines Zeitstempels im Zertifikat ermöglicht die temporäre Vergabe von Zertifikaten.

Zertifikate werden in der jeweiligen Smart Card des Principals gespeichert, sie befreien von der Notwendigkeit, für jedes Paar von Principals ein gesondertes Schlüsselpaar zu administrieren. Die CA ist in unserer Systemarchitektur aber kein Knoten, der die Zertifikate online vergibt. Da die Schlüsselverwaltung auf Smart Cards basiert, sind die Zertifikate bereits in der sogenannten *Personalisierungsphase* fix auf die Smart Card aufgebracht worden. Zu lösen ist nun noch die Frage der Zuordnung von Rechten zu diesen vorhandenen Zertifikaten.

4.2 Rechteverwaltung

Die Autorisierung von Principals erfolgt über die üblichen *Access Control Lists* (ACL). Eine solche Liste besteht aus Tripeln der Art (Principal$_X$, Service$_Y$, Principal$_Z$) mit der folgenden Semantik: Principal X darf Service Y auf Principal Z ausführen. Hält Principal Z seine ACL lokal, degeneriert natürlich das Tripel zu einem Tupel (Principal$_X$, Service$_Y$). Access Control Lists werden von einer *Permission Authority* (PA) generiert und signiert. Darüber hinaus wenden wir das verbreitete Konzept der Benutzergruppen an, welches eine einfachere Verwaltung von Rechten bringt. Anstatt Principal X kann also auch Group U in einer ACL vorkommen. Eine eigene *Group Member List* (GML) stellt dann die Verbindung zwischen den Principals und den Groups, denen sie angehören, her. Dabei kann jeder Principal mehreren Groups angehören.

Die Verwaltung der ACLs und der GMLs erfolgt nicht notwendigerweise in der TCB, da die Listeneinträge durch die Signatur mit dem privaten Schlüssel der PA gesichert sind. Dies ist schon deshalb notwendig, da Geräte der Klassen C1 und C2 die ACL (zumindest zusätzlich) direkt im Gerät halten müssen. Im Sinne einer erhöhten Sicherheit kann die PA somit auch offline gehalten werden.

5 Fallbeispiel aus der Gebäudeautomation

Im Rahmen des Forschungsprojekts HomeNet Configuration Tool[1] [4] wurde eine Java-basierte Software zur Fernsteuerung von Feldbussystemen im Bereich der Gebäude- und Heimautomation entwickelt. Neben einer komfortablen Benutzerschnittstelle galt es, eine offene Architektur für unterschiedliche Feldbustechnologien zu schaffen. Die Thematik der Security war dabei von entscheidender Bedeutung, denn eine Fernsteuerungssoftware ohne Authentisierung bzw. Autorisierung ist so gut wie wertlos.

---

[1] Dieses Projekt wird vom Fonds zur Förderung der wissenschaftlichen Forschung (FWF) finanziert (Projektnummer P-10699 ÖMA).

Abb. 2 zeigt, wie das HomeNet Configuration Tool (HCT) mit Hilfe von Lesegeräten für Smart Cards den Fernzugriff vor Mißbrauch schützen kann. Betrachten wir die Heizungssteuerung in einem Gebäude: Diese soll mittels der HCT-Software von jedem beliebigen Web-Browser aus manipuliert werden können. Die Sicherung gegenüber unautorisiertem Zugriff geschieht mit Hilfe eines Kartenlesers und einer Software, die die Kommunikation zwischen HCT-Software und Smart Card regelt. Diese Kommunikation (vom Web-Browser zum Smart Card Reader) ist auch durch jüngere Entwicklungen verschiedener Firmen[2] machbar geworden.

Ebenso wie auf der Client-Seite der Zugriff per Smart Card gesichert wird, muß man auch auf der Server-Seite (darunter versteht man jenen Rechner, der im Gebäude steht und das Java-Applet zum Browser schickt, siehe Abb. 2) die Autorisierung mittels Smart Card gewährleisten. Schließlich befindet sich auch in der Heizungssteuerung selbst eine Smart Card, die mit den anderen Karten im System den Kern der Trusted Computing Base bildet.[3] Die Heizungssteuerung hält übrigens als C2-Gerät ihre ACLs und die GMLs lokal.

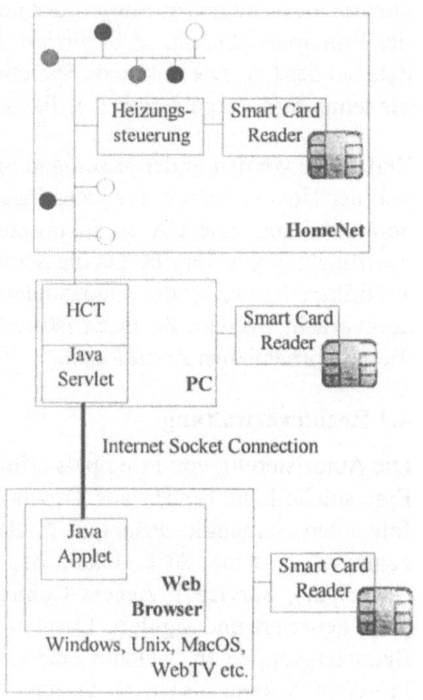

Abb. 2: Smart Card Security beim HomeNet Configuration Tool

Applet, Plug-In	Java Servlet	HCT	Feldbus-knoten
SSL	SSL	PC-Feldbus-Zugriff	
Internet Protocol	Internet Protocol		
Telefonleitung, Ethernet, ...		Feldbus	

Abb. 3: Protokollstack für sicheren Feldbuszugriff beim HomeNet Configuration Tool

Abb. 3 zeigt den beim HCT verwendeten Protokollstack. Dabei ist der Begriff *Plug-In* als Schnittstelle des Webbrowsers zum Kartenleser zu betrachten. Zur Verschlüsselung

---

[2] Auf der „1997 RSA Data Security Conference" wurde von den Firmen Consensus Development Corp., Gemplus, Hewlett-Packard, Litronic, Netscape Communications Corp. und VeriSign, Inc. eine Demonstration des „The Get Smartcard Demo" vorgeführt (siehe http://www.litronic.com/company/releases/rsa-rlsF.htm).

[3] Auch die PA als Erzeuger der signierten ACLs ist Bestandteil der TCB.

über Internet wird SSL verwendet, da die Verschlüsselung größerer Datenmengen in einer Smart Card lange Rechenzeiten verursacht. Die Smart Card erstellt die für diese Anwendung wesentlich kritischere Signatur des Zugriffskommandos. Der Vorgang des Fernzugriffs beginnt also mit einer gesicherten Kommunikation zwischen Client und HCT. Daraufhin besorgt HCT die Umsetzung des erhaltenen Kommandos auf den konkreten Feldbusbefehl, der wiederum in einer gesicherten Kommunikation zwischen HCT und der Heizung übermittelt wird. Da dieser Vorgang vom HCT stellvertretend für den Client ausgeht, ist damit i.a. eine Delegation der Rechte des Clients an das HCT verbunden.

6 Résumé

Security ist auch bei Feldbussystemen eine Eigenschaft mit wachsender Bedeutung, allgemein und speziell im Hinblick auf die Anbindung der Feldbusse an übergeordnete Netzwerke wie das Internet. Die Smart Card bietet ausgezeichnete Security-Mechanismen, mit denen die geforderte Verschlüsselung und vor allem die Authentisierung und die Autorisierung auf einem sehr hohen Sicherheitsniveau ausgeführt werden können. In der von uns vorgestellten Systemarchitektur kann die explizite Certification Authority entfallen, und die Permission Authority generiert ihre Access Control Lists offline. Verschiedene Geräteklassen bezüglich der geforderten Fehlertoleranz erfordern eine verteilte, weil teilweise nötige lokale Haltung der Access Control Lists. Diese Architektur kann auf das am Institut entstandene Homenet Configuration Tool, aber auch auf jeden anderen Feldbus mit intelligenten Knoten und optionaler Fernsteuerung durch Anbindung an ein übergeordnetes Netz angewendet werden. Bei Feldbussen mit unintelligenten Knoten ist nur die Grenze nach außen, also der Vorgang der Fernsteuerung absicherbar.

Literatur

1. Abadi et al: Authentication and Delegation with Smart-Cards, DEC Systems Research Center Report No. 67, 1992.

2. Burrows et al: A Logic of Authentication, Proceedings of the Royal Society of London A, Vol. 426, 1989, pp. 233-271.

3. Department of Defense: Trusted Computer System Evaluation Criteria, DOD 5200.28-STD, 1985.

4. Ochensthaler, Reiter, Redlein, Schildt: HomeNet Configuration Tool - ein Projektbericht, Proceedings of FACILITY '97, Wien, April 1997.

5. Rankl, Effing: Handbuch der Chipkarten, 2. Auflage, Carl Hanser Verlag, München 1996.

6. Schneier: Applied Cryptography, 2nd Ed., John Wiley & Sons, New York 1996.

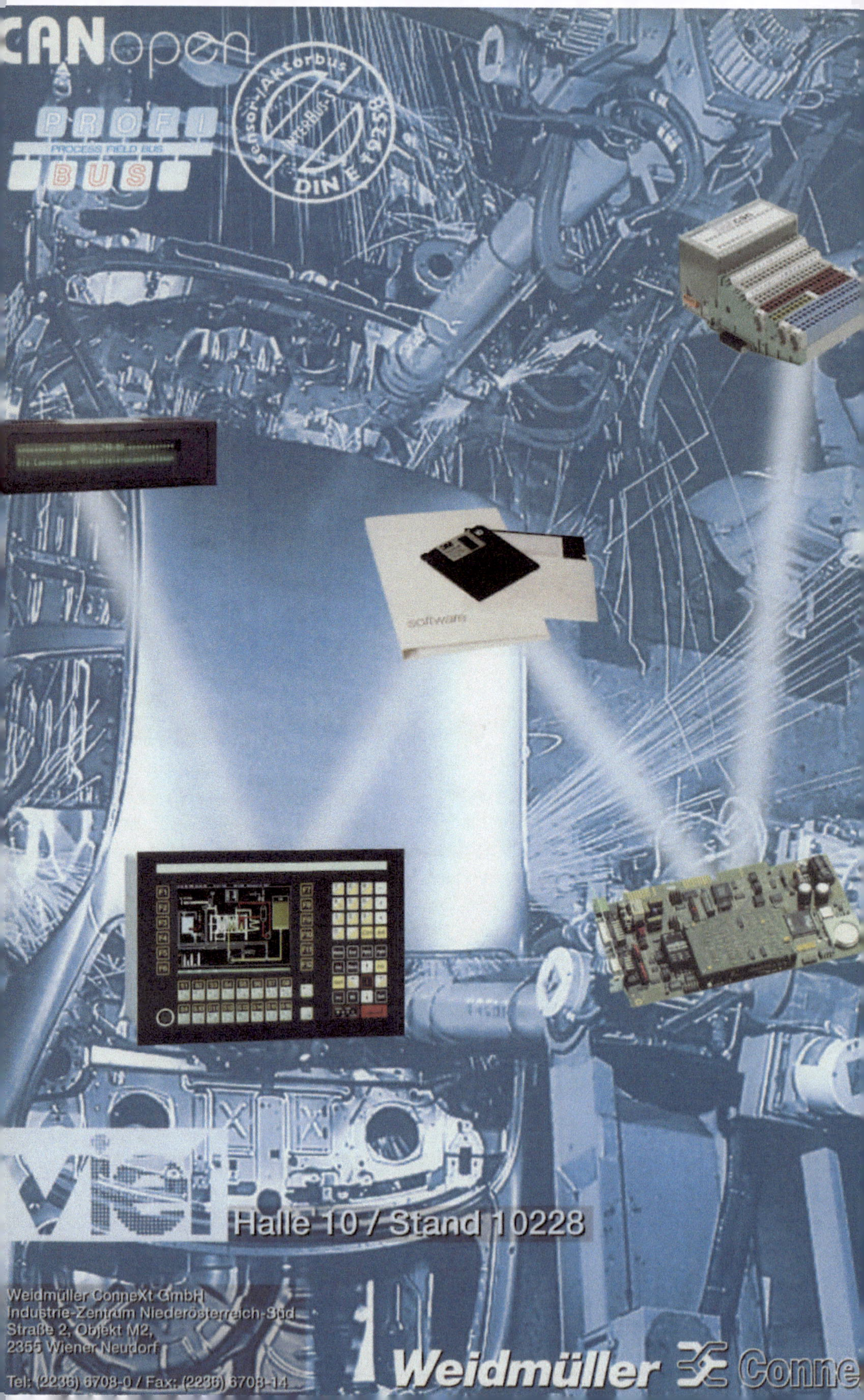
CANopen
PROFI
PROCESS FIELD BUS
BUS
Sensor-/Aktorbus
Interbus-S
DIN E 19258
Software
Halle 10 / Stand 10228
VIA
Weidmüller ConneXt GmbH
Industrie-Zentrum Niederösterreich-Süd
Straße 2, Objekt M2,
2355 Wiener Neudorf
Tel: (2236) 6708-0 / Fax: (2236) 6708-14
Weidmüller Conne

Teil 2 Produkt- und anwendungsorientierte Beiträge

Product- and application-oriented contributions

4 Industrieautomation

Industrial automation

Die Berichte zeigen deutlich, auf welch breiter Basis sich der Feldbus seinen Platz in der Industrie schon erobern konnte. Sie zeigen aber auch, daß erst durch die Feldbustechnologie eine durchgängige Automation ermöglicht wurde, die nun nicht mehr wie zur Zeit des Aufkommens von CIM ein Wunschdenken darstellt, sondern Realität geworden ist.

The reports demonstrate clearly the importance fieldbus systems were already able to gain in industry. They also show that fieldbus technology was the key to a comprehensive automation that is no longer wishful thinking like in the early days of CIM, but has become reality.

Projektierung, Inbetriebnahme und Diagnose von Feldbussystemen mit offenen PC-Plattformen

Jörg Böttcher

b-plus Meßsysteme GmbH, Pfleggasse 19, 94469 Deggendorf
Tel. ++49 991 340-856, Fax -858, E-mail: b-plus@t-online.de

Universität der Bundeswehr München, FB ET(FH) WE7, 85577 Neubiberg
Tel. ++49 89 6004-3993, Fax -3560, E-mail: joerg.boettcher@unibw-muenchen.de

Kurzfassung. Der vorliegende Beitrag beschäftigt sich mit Konzepten zur Projektierung, Inbetriebnahme und Diagnose von Feldbussystemen. Dazu werden zunächst die technologischen Grundlagen heute verfügbarer Konzepte erläutert. Wesentliches Element ist dabei die objektorientierte Softwareschnittstelle OLE 2, wie sie von Windows-basierten 32 Bit-Systemen zur Verfügung gestellt wird. Auf dieser bauen unter anderem auch zukünftig geplante Standards wie OPC bzw. aktuelle Projekte zur Vereinheitlichung des Feldbuszugriffs wie RACKS und NOAH auf. Schließlich wird anhand eines systemübergreifenden Tools gezeigt, wie bereits heute derartige Konzepte für die effiziente Installation von Feldbussystemen genutzt werden können.

Abstract. This lecture deals with concepts for designing, putting into operation and diagnosis of fieldbus systems. The technological fundamentals of concepts being available today are described. Among these are the object oriented OLE 2 interface which is supported by 32-bit Windows systems. Futural standards like OPC and actual projects for standardizing fieldbus access like RACKs and NOAH are working with OLE. Finally an overall tool is demonstrated which uses those concepts for the efficient installation of fieldbus systems.

1 Einführung

Bei der Arbeit mit Feldbus-basierten Systemen hängt der Aufwand an Zeit und Kosten für die Projektdurchführung im wesentlichen von der Verfügbarkeit einfach handhabbarer Werkzeuge für die Projektierung, Inbetriebnahme und Diagnose ab. Offene PC-Plattformen wie die unter modernen 32 Bit-Betriebssystemen (z.B. Windows 95/NT) vorhandenen unterstützen die Entwicklung leistungsfähiger Tools hierfür.

Wesentliche Motivationen für den Bedarf nach derartigen Systemen sind unter anderem folgende Sachverhalte:

- Der zeitliche Aufwand und die Systemkosten für die erfolgreiche Installation und

Inbetriebnahme eines Feldbussystems nehmen einen erheblichen Teil des Gesamt-aufwandes ein. Entsprechend einfach handhabbare Tools verringern diesen Auf-wand entscheidend.

- Durch die oftmals parallele Nutzung mehrerer Feldbussysteme innerhalb einer Anlage bzw. Firma entsteht der Wunsch nach einheitlicher Bedienung unabhängig vom im Einzelfall verwendeten System.

- Oftmals müssen Datenkopplungen zu bereits vorhandenen Standard-Software-Paketen (Datenbanken, Office-Anwendungen etc.) geschaffen werden, so daß ein-heitliche Software-Schnittstellen auch von der Feldbusanbindung unterstützt werden sollten.

2 Technologische Grundlagen

Moderne PC-basierte Tools für die Projektierung, Inbetriebnahme und Diagnose von Feldbussystemen ("Feldbus-Managementsysteme") arbeiten mit der in Bild 1 gezeich-neten Grundstruktur.

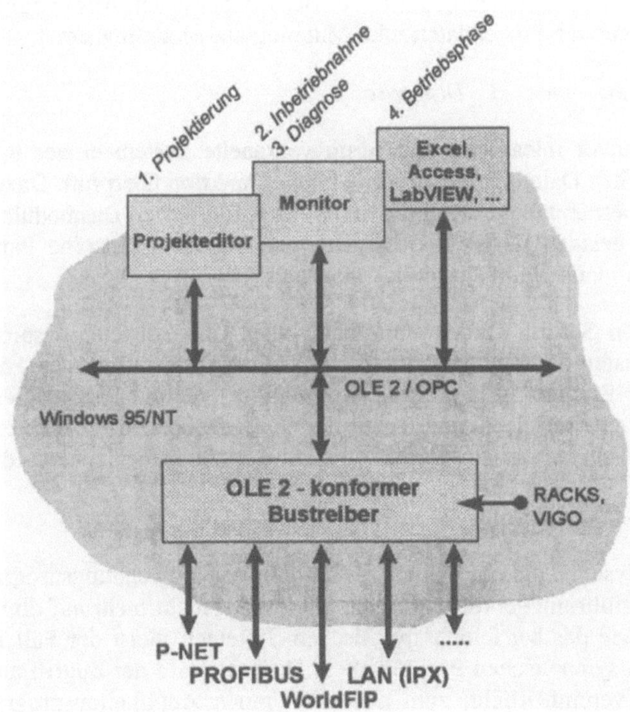

Bild 1. Grundstruktur moderner Feldbus-Managementsysteme

308

Während sämtliche Tools auf die standardisierte Windows-interne OLE 2 "Automatisierung"-Schnittstelle aufsetzen, stellen entsprechende Bustreiber einen dazu konformen Zugriff auf die Feldbusdaten (Parameter für die Buskommunikation, Prozeßdaten) zur Verfügung.

Im einzelnen sind folgende Teilaufgaben, denen meist einzelne Tools entsprechen, zu lösen:

1. Projektierung

Als günstig erweist sich die Verwendung sogenannter Projekteditoren, die den PC-unterstützten Entwurf von Feldbusinstallationen inklusive der dabei anfallenden Bus-modul-spezifischen Parametrierungen erlauben. In der Regel entsteht dabei folgendes Mengengerüst, das in einer Projektdatei abgelegt wird:

- Projektname
- verwendete Busmodule
- pro Busmodul Modulparameter wie Busadresse, Zugriffsrechte, prozeßrelevante Modulkonfigurationen (Grenzwerteinstellung, Parametrierung von Reglerparametern etc.)
- ggf. symbolische Bezeichnung der in der Betriebsphase von Anwendungsprogrammen benutzten Prozeßdaten inkl. Mapping auf Modulregister

2. Inbetriebnahme und   3. Diagnose

Hier wird das zuvor installierte und fertig verkabelte System eingeschaltet und in seinen wesentlichen Datenpfaden auf prinzipielle Funktion überprüft. Dazu werden in einem ersten Inbetriebnahmeschritt die unter 1. konfigurierten Busmodulparameter in die Busmodule geladen; je nach Bussystem müssen dabei einzelne Parameter wie Busadresse noch manuell am Busmodul eingestellt werden .

In einem zweiten Schritt werden sämtliche später vom Anwendungsprogramm benutzten Prozeßdaten auf Ihre grundsätzliche Übertragbarkeit getestet. Hierzu bieten sich Busmonitor-Programme an, die in systemnaher Art und Weise einen schnellen Aufruf unter Benutzung der unter 1. definierten symbolischen Prozeßbezeichnungen erlauben. Beide Schritte werden oftmals durch ein gemeinsames Tool erledigt.

4. Betriebsphase

Nach erfolgter Systemdiagnose können die endgültigen Anwendungsprogramme geladen und zur Ausführung gebracht werden. Sie greifen nicht mehr auf direkte Busmodulregister zu, wie das bei feldbusspezifischen Systemen meist der Fall ist, sondern arbeiten mit den symbolischen Prozeßnamen. Dadurch wird der Zugriff auf den Feldbus zum einen vereinheitlicht, zum anderen können Applikationsprogramme vollkommen unabhängig vom konkret verwendeten Feldbustyp erstellt werden.

Mit dem Trend, den Feldbuszugriff zu standardisieren, einher geht eine weitere Entwicklung: Die Verwendung applikationsspezifisch konfigurierter Standardprogramme für die Prozeßvisualisierung und -automation anstelle individuell programmierter Systeme. Die meisten der derzeit am Markt befindlichen Standardpakete, wie sie im Produktionsumfeld eingesetzt werden, unterstützen OLE 2 Automatisierung (z.B. LabVIEW, Excel, Access, Visual Basic etc.)

Folgende Kriterien werden dabei an alle Tools gestellt:

- intuitive Bedienung
- fehlerredundante Bedienung
- kein feldbusspezifisches Know-How erforderlich
- offene Softwareschnittstelle wie OLE 2 Automatisierung
- 32 Bit-fähig
- multiprotokollfähig
- Interoperabilität der Komponenten untereinander

3 OLE 2 und OPC - zukünftige Standards für den Feldbuszugriff

OLE 2 Automatisierung (Object Linking and Embedding) arbeitet objktorientiert nach dem Client-Server-Modell. Dies bedeutet, daß es einerseits Clients gibt, die bestimmte Funktionen vom OLE 2-kompatiblen Feldbustreiber anfordern. Clients sind sämtliche Tools, die von "oben" (vgl. Bild 1) auf den Feldbustreiber zugreifen. Der Feldbustreiber selbst ist Server; er stellt den Clients Objekte und darauf definierte Funktionen zur Verfügung. Dadurch wird eine weitgehende Entkopplung des anwendungsorientierten Buszugriffs (Prozeßdaten lesen und senden) von der feldbusspezifischen Protokollstruktur erreicht.

Die typischerweise benötigten OLE-spezifischen Methoden sind im wesentlichen:

- Feldbus-Objekt erzeugen (Create Object)
- Objekt auf Prozeßdatum zeigen lassen (Set Property)
- Prozeßwert einlesen (Get Property) bzw. Prozeßwert setzen (Set Property)
- Objekt verwerfen (Close Object)

Durch das Setzen eines "Zeigers" (zweiter Schritt oben) wird der Bezug zum symbolischen Prozeßwert kreiert. Dieser Bezug kann dynamisch verändert werden, so daß beispielsweise nacheinander auf alle relevanten Prozeßwerte zugeriffen werden kann.

Bild 2 zeigt die wohl derzeit wichtigsten Standardisierungsprojekte auf europäischer Ebene, welche die hier geschilderten Konzepte auf OLE 2-Basis heranziehen.

4 Anwendungsbeispiel

Das in Bild 3 abgebildete Tool VIGO stellt ein heute bereits verfügbares Werkzeug dar, welches auf OLE 2-Basis systemübergreifend die Projektierung, Inbetriebnahme und Diagnose von Bussystemen erlaubt.

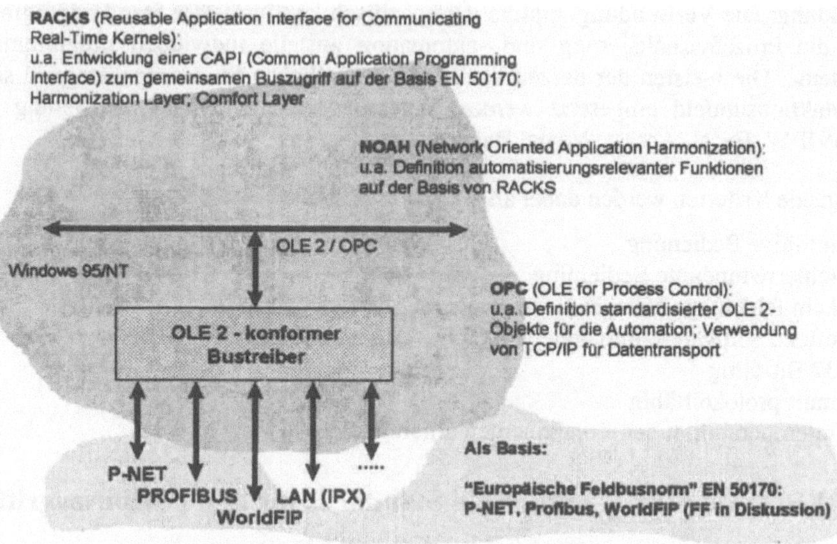

Bild 2. Standardisierungsprojekte mit OLE 2-Bezug

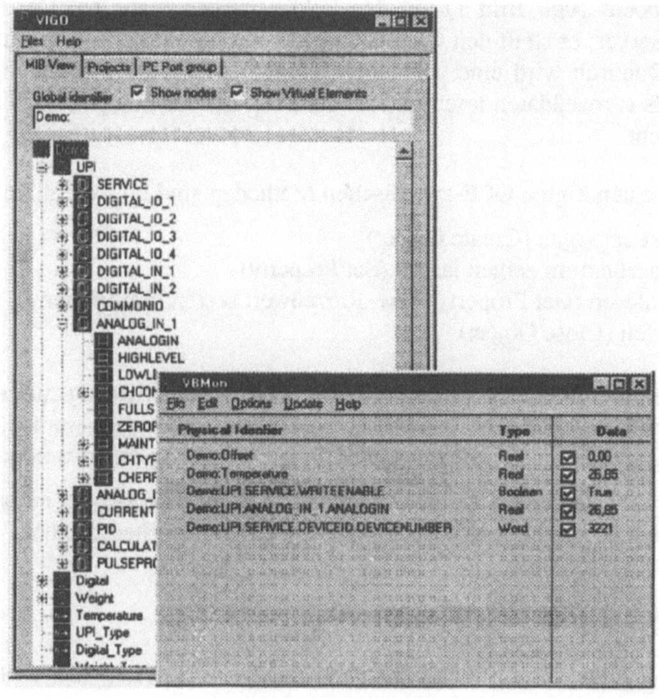

Bild 3. VIGO - Beispiel eines Tools auf OLE 2-Basis

VIGO integriert neben dem reinen Feldbuszugriff auch LAN-Zugriffe. Beispielsweise lassen sich über VIGO an einem PC, der selbst nicht direkt am Feldbus angeschlossen ist, sondern vielmehr über ein LAN eine Verbindung zu einem anderen PC mit lokalem Feldbus aufweist, sämtliche Feldbusdaten transparent bearbeiten ("Lokaler Fernzugriff").

Aufgrund der offenen Softwareumgebung OLE 2-basierter Systeme sind natürlich auch Datenaustausch und Interaktionen zwischen Tools mehrerer Hersteller möglich.

5 Von der Vor-Ort-Wartung zur Fernwartung

Offene PC-Plattformen erlauben ebenso einfach die Anbindung der lokalen Feldbussysteme an LANs und WANs. Insbesondere die TCP/IP-basierten Systeme können kompatibel an die OLE 2 - basierte Systemarchitektur, wie sie in diesem Beitrag erläutert wurde, adaptiert werden. Darauf aufbauend lassen sich Fernwartzugriffe über ISDN, Funktelefon oder Internet/Intranet bereits heute realisieren.

Folgende Systemerweiterungen zu dem bislang diskutierten OLE 2-Modell sind dazu sinnvoll und teilweise schon als Produkt verfügbar:

- OLE 2-kompatible Treiber für TCP/IP bzw. herkömmliche LAN-Techniken (IPX)
- OLE 2-kompatible Treiber für spezielle Übertragungsstrecken wie ISDN, GSM-Mobilfunknetz etc.
- Client-Server-Treiber für die OLE 2-transparente Verknüpfung zweier über TCP/IP bzw. Internet verbundener Rechner
- Host-Programme für den Fernzugriff inkl. Fernanwahl, Parameterferncheck, Fern-Debugging etc.
- Intranet-Applikationsprogramme für die Implementierung von Automatisierungs-Kreisstrukturen überhalb der Zellen-/Werksebene

Literatur

1. Schnell, G.: Bussysteme in der Automatisierungstechnik. Wiesbaden: Vieweg 1994.
2. Bretthauer, G. et. al.: GMA-Kongreß '96 Meß- und Automatisierungstechnik. Düsseldorf: VDI Verlag 1996.
3. Jamal, R.: VIP '96 Virtuelle Instrumente in der Praxis. München: National Instruments 1996.
4. Böttcher, J.: Einheitlicher Feldbuszugriff auf Basis europäischer Standards. IEE Nr. 3/1997, S. 73-74.
5. Böttcher, J.: EN 50170 - Die europäische Feldbusnorm (4 Teile). Elektronik Nr. 12/1996, S. 70-78, Nr. 14/1996, S. 63-67, Nr. 16/1996, S. 48-51, Nr. 17/1996, S. 54-57.
6. Böttcher, J.: Softwareanbindung von Feldbussen unter MS-Windows. e&i Nr. 5/1997, S. 241-244.

Methodischer Entwurf eingebetteter Software am Beispiel eines flexiblen Transportsystems

F. Kallmeyer, M. Miksic

Heinz Nixdorf Institut
Universität-GH Paderborn
Fürstenallee 11, 33102-D Paderborn

Abstract. Die Anforderungen an Automatisierungssysteme steigen. Gefordert werden Skalierbarkeit, geringe Reaktionszeiten sowie Unempfindlichkeit des Gesamtsystems gegenüber dem Ausfall von Subsystemen. Als Ansatz zur Beherrschung der resultierenden Komplexität bietet sich die dezentrale Bearbeitung der Steuerungsaufgaben an. Dabei ist Software zu entwickeln, die durch Parallelität und die Erfüllung von Echtzeitanforderungen gekennzeichnet ist. Um diese Herausforderungen zu bewältigen, ist ein methodischer, formaler Ansatz zur Entwicklung der Software notwendig, der einen eindeutigen Schwerpunkt in die frühen, implementierungsunabhängigen Phasen legt. Ein solcher Ansatz wird sowohl allgemein als auch anhand eines Beispiels ausführlich erläutert.

Abstract. Automation systems are increasingly facing demands concerning flexibility, short response times, and fault tolerance. Decentralized control systems are a promising approach to deal with those problems. On the other hand, such systems are characterized by a high degree of parallelism, which has to be handled during the development process. In order to successfully face this challenge, it is necessary to use a formal and methodical approach. Such an approach is characterized by focussing the early design stages, that are independent from implementation. A methodology that deals with the facts mentioned above will be described in general, and illustrated by an example.

1 Problematik

Die Automatisierungstechnik hat heutzutage stark gestiegene Erwartungen zu erfüllen. Typische Forderungen an die immer größer und komplexer werdenden Automatisierungssysteme sind z.B.: hohe Flexibilität bezüglich der Skalierbarkeit des Systems, geringe Reaktionszeiten und weitgehende Unempfindlichkeit des Gesamtsystems bezüglich des Fehlverhaltens einzelner Subsysteme.

Um diese Herausforderungen zu bewältigen, hat sich der Ansatz einer dezentralen Bearbeitung der resultierenden Steuerungsaufgaben als erfolgversprechend erwiesen. Dezentralisierung ermöglicht eine weitgehende Autonomie von Teilsystemen, erlaubt kurze Reaktionszeiten und stellt darüber hinaus einen Ansatz zur Beherrschung der Systemkomplexität dar. Es wird damit sowohl die Skalierbarkeit des Gesamtsystems als auch die Unempfindlichkeit gegenüber dem Ausfall von Teilsystemen unterstützt. Andererseits bedingt der dezentrale Ansatz einen erhöhten Aufwand an Kommunikation. Grundsätzliche Bedeutung kommt deshalb den Feldbustechnologien, wie beispielsweise den LON® (Local Operating Network) oder CAN® (Control Area Network) Feldbussystemen, als Kommunikationsmittel zu [1].

Der steigende Kommunikationsaufwand und die Notwendigkeit, in dezentralen, parallelen Abläufen zu „denken", stellen die Entwickler von Software für dezentrale Automatisierungssysteme in der Regel vor erhebliche Probleme. Begründet liegt dieser Sachverhalt in dem Umstand, daß die Auswirkungen von Entscheidungen im Entwicklungsprozeß aufgrund der vielfältigen Beziehungen zwischen weitgehend gleichberechtigten dezentralen Einheiten nur schwer zu überblicken sind. Diesen Schwierigkeiten bei der Softwareentwicklung ist nur mit einer geeigneten Software-Entwicklungsmethodik zu begegnen, die ein planmäßiges Vorgehen unter Berücksichtigung formaler Beschreibungsmittel zur Entwicklung von Software für Automatisierungssysteme unterstützt. Wesentlich ist dabei die Möglichkeit zur Modellierung von dezentralen Steuerungsabläufen auf einer abstrakten, d.h. implementierungsunabhängigen Ebene. Hierdurch rückt die Konzeption der Steuerungsaufgabe gegenüber der Implementierung in den Mittelpunkt der Betrachtung. In diesem Zusammenhang hat sich eine auf dem standardisierten Beschreibungsmittel SDL (Specification and Description Language) basierende Entwicklungsmethodik als geeignete Grundlage zur Softwareentwicklung für dezentrale Automatisierungssysteme herausgestellt.

2 SDL-basierte Entwicklungsmethodik für Echtzeitsysteme

Nach [2] besteht der Software-Entwicklungsprozeß im wesentlichen aus vier Entwicklungsphasen (Bild 1). Beim Durchlaufen jeder Phase werden phasenspezifische Tätigkeiten durchgeführt und bestimmte Aspekte mit Hilfe geeigneter Beschreibungsmittel modelliert, wodurch sog. *Aspektmodelle* entstehen. Durch die Beziehungen zwischen den Aspektmodellen sind die Abhängigkeiten zwischen den einzelnen Entwicklungsphasen gegeben.

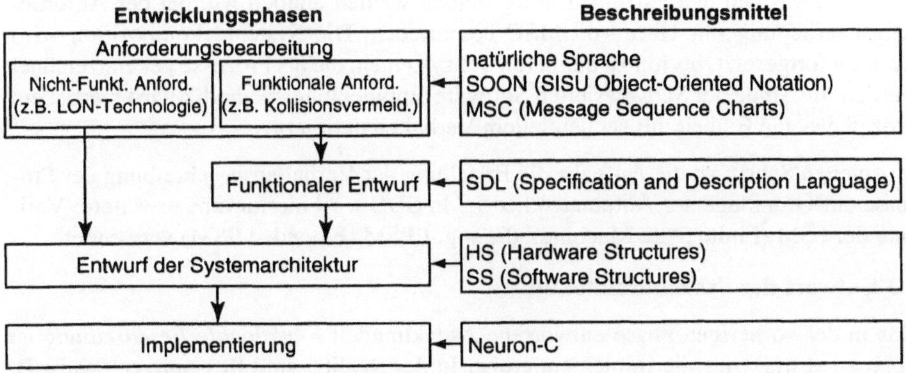

Bild 1. Entwicklungsphasen, Abhängigkeiten und verwendete Beschreibungsmittel der SDL-basierten Entwicklungsmethodik

2.1 Anforderungsbearbeitung

Das Hauptziel in dieser Phase besteht in einer möglichst genauen Beschreibung der Erwartungen, die das zu entwickelnde System erfüllen muß. Zu diesem Zweck wird als erstes Aspektmodell die *Aufgabenbeschreibung* erstellt. Als Beschreibungsmittel wird

314

hier die natürliche Sprache verwendet. Dabei wird empfohlen, sich auf das Wesentliche zu konzentrieren und die *Aufgabenbeschreibung* mit einfachen und verständlichen Aussagen präzise zu formulieren. Im nächsten Schritt wird ein *Begriffsverzeichnis* erarbeitet, in dem die wesentlichen in der Aufgabenbeschreibung verwendeten Begriffe definiert werden. Anschließend werden Systemgrenzen, eine grobe Systemarchitektur und sog. Rollen im Aspektmodell *Kontextbeschreibung* unter Verwendung sog. SOON Diagramme festgelegt. Die Rollen spezifizieren dabei die vom System erwarteten Funktionen. Abschließend werden in dieser Phase für jede Rolle Interaktionssequenzen mit der Umgebung im Aspektmodell *Protokollbeschreibung* beschrieben. Dabei werden sog. MSC (CCITT Z.100, 1993) als Beschreibungsmittel eingesetzt.

Alle bisher genannten Aspektmodelle dienen der Beschreibung Funktionaler Anforderungen. Nicht-Funktionale Anforderungenwerden werden hingegen als vorgegebene Entscheidungen in natürlicher Sprache dokumentiert.

2.2 Funktionaler Entwurf

Im Gegensatz zur Anforderungsbearbeitung - bei der in erster Linie das von außen sichtbare Systemverhalten modelliert wird - besteht beim Funktionalen Entwurf die Hauptaufgabe darin, das Systemverhalten von innen zu beschreiben. Dabei wird das Aspektmodell *Funktionale Beschreibung* mittels SDL entworfen. Der funktionale Entwurf erfolgt im wesentlichen in zwei Schritten.

Im ersten Schritt wird nach einem Top-Down Verfahren unter Berücksichtigung der Entwurfsregeln eine Systemdekomposition vorgenommen. Abstrakte Kommunikationskanäle werden zwischen den einzelnen Subsystemen festgelegt und Signale, die die beteiligten Kommunikationsteilnehmer austauschen, werden definiert. Interaktionssequenzen zwischen den einzelnen Subsystemen werden ähnlich wie bei der Anforderungsbearbeitung mit Hilfe von MSC beschrieben. Die Systemdekomposition wird rekursiv fortgesetzt, bis hin zu atomaren Subsystemen, die als Prozesse gekennzeichnet werden. Im Sinne der SDL-basierten Software-Entwicklungsmethodik wird hierbei ein Prozeß als eine Einheit mit sequentiellem Verhalten definiert.

Im zweiten Schritt ist die Aufgabe die Erstellung der Verhaltensbeschreibung der Prozesse auf Grundlage der Automatentheorie. In SDL wird hierfür eine erweiterte Variante der FSM (Finite State Machine), die sog. EFSM (Extended FSM) verwendet.

2.3 Entwurf der Systemarchitektur

Das in der vorherigen Phase entworfene Aspektmodell *Funktionale Beschreibung* ist noch zu abstrakt für die Implementierung. In der Realität sind Ressourcen - wie z.B. Speicher - sowie die Bearbeitungsgeschwindigkeit begrenzt. Das Ziel in dieser Entwicklungsphase besteht darin, unter Berücksichtigung der oben genannten Beschränkungen die Hardware- und Softwarearchitektur des realen Systems zu spezifizieren und dadurch die notwendigen Vorbereitungen für die Implementierung zu treffen. Als Resultat dieser Phase wird das Aspektmodell *Systemarchitektur* mittels HS (Hardware Structures) bzw. SS (Software Structures) Diagrammen entworfen.

2.4 Implementierung

In dieser Entwicklungsphase wird das Aspektmodell *Funktionale Beschreibung* unter Berücksichtigung des Aspektmodells *Systemarchitektur* auf der konkreten Hardware implementiert. Dabei entsteht das Aspektmodell *Code*. Die Implementierung erfolgt in der am besten geeigneten Programmierungssprache.

Die durchgängige Nutzung der aufgezeigten Methodik zur Entwicklung von Echtzeitsoftware wird im folgenden exemplarisch am Beispiel eines flexiblen Transportsystems beschrieben.

3 Anwendung der Methodik am Beispiel eines Transportsystems

Der vorgegebene mechanische und elektromechanische Aufbau des Transportsystems ist im Bild 2 dargestellt. Es handelt sich um ein System mit drei Stationen (Bestück - und Entnahmestationen) sowie mit elektrisch betriebenen Rollenförderer-, Kettenförderer und Umsetzer-Modulen. Der modulare Aufbau sichert dabei die für eine moderne Produktion notwendige Flexibilität der Anlage, da beliebige, von der dargestellten Konfiguration abweichende, Aufbauten möglich sind·[3].

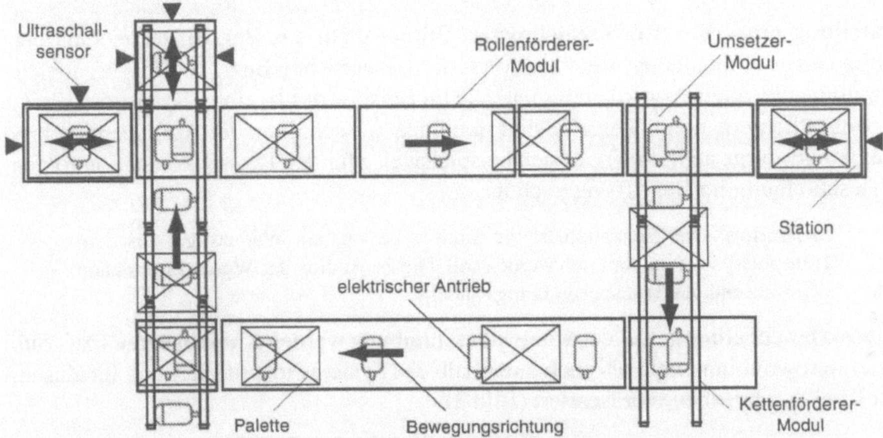

Bild 2. Mechanischer bzw. elektromechanischer Aufbau des Transportsytems

3.1 Anforderungsbearbeitung

Die durchzuführenden Tätigkeiten bei der Anforderungsbearbeitung sind - wie im Abschnitt 2 bereits angeführt - die Aufgabenbeschreibung, die Erstellung eines Begriffsverzeichnisses, die Kontextbeschreibung und die Protokollbeschreibung.

Aufgabenbeschreibung. Zu Beginn der Aufgabenbeschreibung soll in verständlicher Form die Gesamtfunktion des Systems geklärt werden. Für die Automatisierung des Transportsystems ergab sich vereinfacht folgende Beschreibung:

> „Es ist ein Softwaresystem zur Steuerung des Transportsytems nach Bild 2 zu entwickeln. Die Gesamtfunktion des Systems ist der Transport von *Warenträ-*

gern zwischen Stationen. Die Warenträger werden über *Stationen* in das System aufgenommen bzw. aus dem System entnommen. Damit erfüllt jede Station sowohl eine Bestückungs- als auch eine Entnahmefunktion. Eine *leere Station* kann zu einem beliebigen Zeitpunkt bestückt werden. Nach Abschluß des Bestückungsvorgangs erfolgt die Festlegung der *Zielstation* für den jeweiligen Warenträger. Anschließend startet der *Transportvorgang*, der mit der Ankunft des Warenträgers bei der Zielstation beendet wird. Von einer *bestückten Station* kann der Warenträger zu jedem beliebigen Zeitpunkt entnommen werden. Die Bestückungs-, Transport- und Entnahmevorgänge sowie die Festlegung der Zielstation sollen dabei jeweils individuell und unabhängig voneinander für jeden einzelnen Warenträger und jede einzelne Station im System erfolgen ..."

Hierbei können in weiteren Schritten einzelne Aussagen weiter konkretisiert werden, z.B. gilt für den Bestückungsvorgang:

„Der Bestückungsvorgang beginnt, indem eine leere Station von einem *Bediener* mit einem Warenträger bestückt wird. Befindet sich anschließend für mindestens fünf Sekunden kein Objekt (Bediener oder *Gegenstand*) im *Gefahrenbereich* der Station, so wird der Bestückungsvorgang als beendet betrachtet, und die Bestimmung der Zielstation kann vorgenommen werden. Andernfalls ist ein akustisches Warnsignal zu geben und die Räumung des Gefahrenbereichs abzuwarten ..."

Erstellung eines Begriffsverzeichnisses. Primär dient das Begriffsverzeichnis dem Aufbau eines einheitlichen Verständnisses der wesentlichen Begriffe der Aufgabenbeschreibung bei allen Projektbeteiligten. Die im Beispiel des flexiblen Transportsystems als wesentlich erkannten Begriffe sind in obigen Ausschnitten der Aufgabenbeschreibung kursiv dargestellt. So ergab sich beispielsweise für den Begriff Zielstation folgende Beschreibung im Begriffsverzeichnis:

Zielstation: Die Zielstation ist die Station, bei der ein Warenträger aus dem Transportsystem entnommen werden soll. Die Zielstation des Warenträgers kann sich während des Transportvorgangs ändern.

Kontextbeschreibung. Bei der Kontextbeschreibung werden Komponenten bzw. Subsysteme sowohl innerhalb als auch außerhalb des Systems identifiziert und ihr Zusammenspiel durch Rollen repräsentiert (Bild 3).

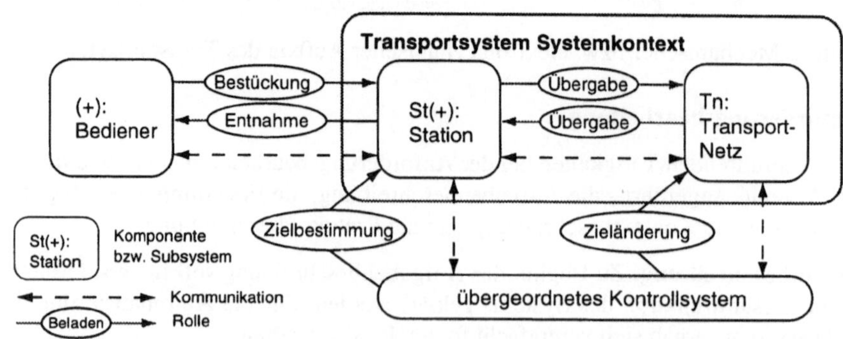

Bild 3. Vereinfachte Darstellung des Aspektmodells *Kontextbeschreibung*

Die beiden wesentlichen Einheiten des Transportsystems sind durch die Komponenten Bediener bzw. übergeordnetes System gegeben, wobei nur ein übergeordnetes System, aber mehrere Bediener gleichzeitig mit dem Transportsystem interagieren können (verdeutlicht durch das Symbol (+) bei der Komponentenbezeichnung, s. Bild 3). Da in dieser Phase noch keine detaillierte Systemdekomposition vorgenommen wird, sind alle Fördermodule außerhalb der Stationen durch das Subsystem Transportnetz abstrahiert worden. Die in der Aufgabenbeschreibung spezifizierten Vorgänge - wie z.B. der Transport- oder Entnahmevorgang - werden auf die entsprechenden Rollen abgebildet. Der zwischen den einzelnen Komponenten notwendige Informationsaustausch wird mittels Kommunikationsbeziehungen modelliert.

Protokollbeschreibung. Die abschließende Aufgabe der Anforderungsbearbeitung besteht in der Modellierung der Ablaufsequenzen der in der Kontextbeschreibung definierten Rollen mittels MSC. Bild 4 zeigt beispielhaft die Rolle *Bestückung* mit Hilfe zweier Ablaufsequenzen. Die gerichteten waagerechten Linien kennzeichnen Signale, die einen Informationsaustausch zwischen den beteiligten Komponenten darstellen. Die senkrechten Säulen repräsentieren die Zeitachse.

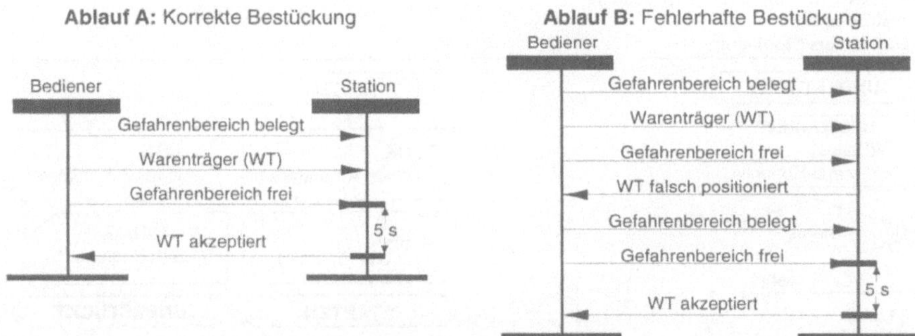

Bild 4. Protokollbeschreibung mittels MSC am Beispiel der Bestückung einer Station

3.2 Funktionaler Entwurf

Wie bereits in Kapitel 2 beschrieben, erfolgt der Funktionale Entwurf in den zwei Schritten Systemdekomposition und Verhaltensbeschreibung. Das hierbei entstehende Aspektmodell *Funktionale Beschreibung* wird mittels SDL beschrieben.

Systemdekomposition. Bei dieser Tätigkeit werden die im Aspektmodell Kontextbeschreibung identifizierten Komponenten und Subsysteme in einem Top-Down-Vorgehen bis hin zu elementaren Prozessen verfeinert. Beispielsweise wurde beim flexiblen Transportsystems das Subsystem Transportnetz in die Komponenten Umsetzer, Rollenförderer und Kettenförderer unterteilt, was auf einsichtige Weise aus dem mechanischen Aufbau der Anlage resultiert. Zwischen benachbarten Modulen wurden anschließend die entsprechenden Rollen mit den zugehörigen Interaktionssequenzen zur Übergabe der Warenträger spezifiziert. Als Beispiel für die Festlegung elementarer Prozesse kann die Komponente Station dienen. Entsprechend ihrer Aufgaben wurde sie in die drei Prozesse Bediener-, Ausrichtungs- und Stations-Kontrolle unterteilt. Die Bediener-Kontrolle dient dabei zur Überwachung des Gefahrenbereichs, die Ausrichtungskontrolle zur Überprüfung der korrekten Positionierung der Warenträger und die Stations-Kontrolle zur Koordination und Steuerung der Abläufe.

Verhaltensbeschreibung. Unter Verwendung von SDL wird in diesem Schritt das ereignisabhängige Verhalten der einzelnen Prozesse modelliert. Im Bild 5 ist beispielhaft ein Ausschnitt aus der Verhaltensbeschreibung des Prozesses Stations-Kontrolle wiedergegeben: Der Prozeß wartet im Zustand *UNBESTÜCKT* auf das Signal *BESTÜCKEN*, das vom Prozeß Bediener-Kontrolle gesendet wird. Anschließend wird der Prozeß Ausrichtungs-Kontrolle gestartet. Verläuft die Ausrichtungskontrolle erfolgreich, so erfolgt das Setzen eines Timers, um das spezifizierte Zeitintervall von fünf Sekunden abzuwarten. Anderenfalls wird der Bediener durch ein Warnsignal zur ordnungsgemäßen Bestückung der Station aufgefordert.

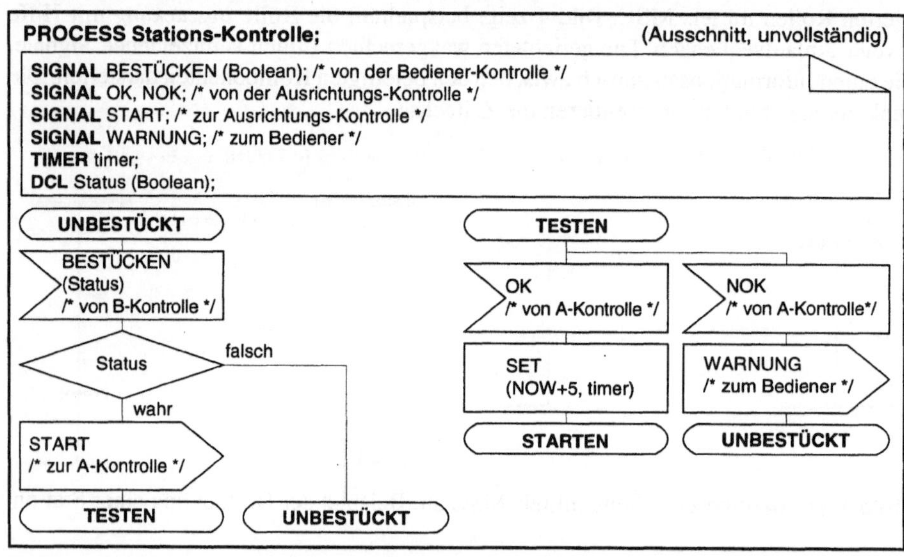

Bild 5. Ausschnitt aus der Verhaltensbeschreibung des Prozesses *Stations-Kontrolle*

3.3 Entwurf der Systemarchitektur

Beim Entwurf der Systemarchitektur (Abbildung von Prozessen auf Hardwarekomponenten) sind als wesentliche Sachverhalte die Bearbeitungsgeschwindigkeit sowie die vorhandenen Ressourcen zu berücksichtigen. Im Beispiel des flexiblen Transportsystems bezog sich der erste Aspekt vor allem auf die Reaktionszeit. So bestand die Nicht-Funktionale Anforderung als Kommunikationsprotokoll LONTalk® zu verwenden. Da dieses Protokoll ein indeterministisches Verhalten aufweist, mußte der Zuordnung der Prozesse auf die Hardware-Einheiten besondere Aufmerksamkeit geschenkt werden. Aus diesem Grund sollten Aktionen, die eine kurze Reaktionszeit erfordern, die Kommunikation über den Feldbus vermeiden. Als Beispiel für eine zeitkritische Aktion kann das unmittelbare, zuverlässige Ausschalten des Antriebs einer Station bei der Ankunft bzw. Detektion eines Warenträgers dienen. So waren zur Vermeidung der Kommunikation der Prozesse über den Feldbus die Prozesse Gefahrenbereichs-Überwachung und Stationskontrolle auf einer gemeinsamen Hardware zu implementieren. Als knappe Ressourcen haben sich die E/A-Anschlüsse der Steuerungshardware erwiesen, da ihre Anzahl herstellerseits fest vorgegeben war. So mußte beispielsweise das Subsystem Umsetzer aufgrund der großen Anzahl benötigter Aktoren und Sensoren auf zwei Hardware-Komponenten abgebildet werden.

3.4 Implementierung

Für die Implementierung wurde aufgrund Nicht-Funktionaler Anforderungen die LON® Technologie gewählt. Die Kodierung erfolgte in der Programmiersprache Neuron-C®, die als Standardwerkzeug zur Programmierung der LON®-basierten Mikrokontroller angeboten wird [4]. Diese Programmiersprache stellt aufgrund des ereignisorientierten Ansatzes eine geeignete Basis für die Implementierung der mittels SDL spezifizierten Verhaltensbeschreibung der Prozesse dar. Die Kommunikation zwischen den auf verschiedenen Hardwarekomponenten implementierten Prozessen wird dabei mittels sog. Netzwerkvariablen auf Basis eines für die LON®-Technologie spezifischen Kommunikationsverfahrens realisiert.

4 Ergebnisse und Ausblick

Es wurde an einem praktischen Beispiel die Stärke eines methodischen Software-Entwicklungsansatzes gezeigt, der sich besonders durch eine Verkürzung der Entwicklungszeit sowie eine geringere Fehleranfälligkeit der entwickelten Softwaremodule auszeichnet. Durch den Einsatz der dezentralen Feldbustechnologie konnte eine bessere Skalierbarkeit und räumliche Integration der mechanischen und elektronischen Komponenten erreicht werden. Validierungsmethoden zur Analyse des Systems in allen Entwicklungsphasen, eine stärkere Integration mit anderen Ingenieurbereichen (Regelungstechnik, Hardware) sowie die Einbettung in eine integrierte Entwicklungsmethodik für mechatronische Systeme sind Gegenstände unserer weiteren Forschung [5].

Literatur

1. Bonfig, K. W.: Feldbus-Systeme. Expert Verlag, Deutschland, 1992

2. Bræk, R.: Engineering Real Time Systems: An object-oriented methodology using SDL. Prentice-Hall, USA, 1993

3. Tüchelmann, Y.: Automatisierter Material- und Informationsfluß als CIM-Komponente. VDI-Bericht 580, VDI-Verlag, Düsseldorf, 1987

4. N.N.: Neuron-C Programmers Guide. Echelon Corporation, USA, 1993

5. Gausemeier, J., Brexel, D., Flath, M., Kallmeyer, F., Miksic, M.: Integrated Product Modeling - A contribution to Computer Aided Product Development. Tagungsband IMACS Symposium on Mathematical Modelling, Wien, 1997

Introducing an automation system
using SQL database software
and LON components

Eberhard Knechtel

structur Gesellschaft für Datenverarbeitung mbH
Ostfildern/Germany

Abstract. The paper describes the implementation of an automation system for a production line consisting transport elements, operator consoles, process I/O and testing systems using a distributed SQL database containing all production and quality data and LON components extending and integrating the given production environment. This was the first experience using LON components, so the implementation isn't an example for an ideal system, but shows how to manage the process of integrating LON elements and to substitute existing components.

1 Introduction

The product, for which the production line had been set up, is a high integrated part of a vehicle and sold to many different automotive companies with different technical specifications. So, the product has to be delivered in many different versions of hardware, software and adaption parts like cables and sockets. The production parameters (e.g. production steps, the transport sequences, electrical and geometric data, download software) and the testing methods should be hold in an existing SQL database of another production line, that has been installed two years ago using a PC based network.

The SQL database is maintained and searched by different users over the intranet of the company, e.g. the production coordinator knows the production parameters and supervises the whole production process, the quality engineer installs the testing rules and reads the results and the worker reads the given informations and sets error messages. The database environment consists of a central database unix server equiped with a dual processor and a redundant disk array and two local database unix servers with less capacities forming a distibuted database. The testing procedures ran on some unix workstations acting as database clients. The principle was to have the database as near as possible to the process to keep the data actual during all production steps.

2 Describing the existing environment

A few years ago, the first production line consisted of a PC network controlling the workstations for manual and automatic production steps and a transport system

between them. Each PC (Standard PC, VGA or LCD display) was equipped with a process interface card (digital I/O with opto inputs and relais outputs), controlling the workstation, and an ethernet card, connecting the PC to the local database host. Each PC started by a network boot mechanism and was connected to the transportation system via a serial interface. The number of installed PCs was more than 30.
The testing machine additionally had a second direct connection to a unix workstation for controlling the test equipment and doing the automated test procedure.

There are different reasons for changing this concept together with our customer: the new different production environment, the experiences with the stability and costs of a PC network and the suitability of the new automation system for prototyping a different configuration.

3 Why using LON for networking ?

For economic reasons first, a LON based operator console, a LON based serial interface and a LON based digital I/O module substitute each workstation controller PC. The price reduction was about 30%. In fact, the price reduction was higher, because the application handling with the LON nodes was much easier than planned.

Second, the LON based network using free topology transceivers and standard cable and connectors was much more robust and flexible and substitutes the 10Base-T cables and hubs. The smaller performance of the LON network was no problem, because the data rates are only about 10 kBit per second. The cable supports the components with network access and 24 volt power supply, no additional cabling was required and no active network controller (e.g. hub) need to be installed.

Third, there was no topologic difference between the PC network and the LON based network, so the database structures fit and need no redesign.

Fourth, the easy way to extent the functionality, e.g. by simply adding and connecting LON displays to visualize information at the right place, without changing software.

4 Implementing an interface between LON and SQL database

At start of the project there was no standard interface definition for accessing a remote SQL database from a LON based node. There were two ways to implement this functionality:

1. using a Windows NT PC with a LON interface (DDE or LON NSS based) running the SQL application and acting as the gateway, or,

2. connecting the LON through a LON ethernet gateway module or a LON serial module direct to the database host running the SQL application.

First, it was not possible to start with a NSS based software, because there were no

322

experiences and not enough budget for this software product. Over and above that, the NSS was just introduced by Echelon and we didn't manage to work with brand new software products. Second, the now available LAN gateway wasn't introduced so only the DDE server and the direct connected serial module could be tested.

4.1 Connecting through Windows NT running the LON DDE server

Echelon offers a DDE server [1] and Windows NT driver software [2], which should integrated into a PC application (see Fig.1).
The designed application uses "transaction definitions" (TD), which are triggered from LON nodes via network messages and network variables, the results are sent back to the LON. The TD consists of a unique name, a list of input/output parameters, a trigger definition and a database operation defined by using SQL. The list of input/output parameters are in fact network variables, and the trigger definition consists of a message with node information and the trigger name.

Because the PC working as a gateway is a very critical part of the system, we choosed the second version.

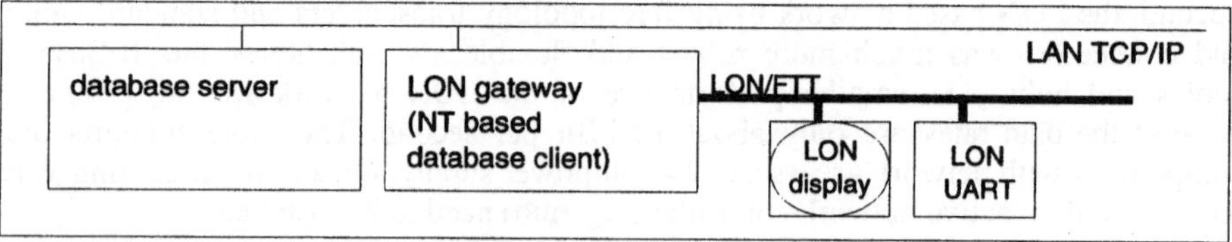

Fig. 1 connecting LON and database server via NT gateway

4.2 Connecting through a serial module directly to the unix system

The second approach implemented the application as a Neuron-C server program [3] for a LON based serial node and a client unix program. The application parts which handle the SQL database are located on the unix system. The LON nodes are connected to the serial gateway module and receive and transmit data by network variables and messages.

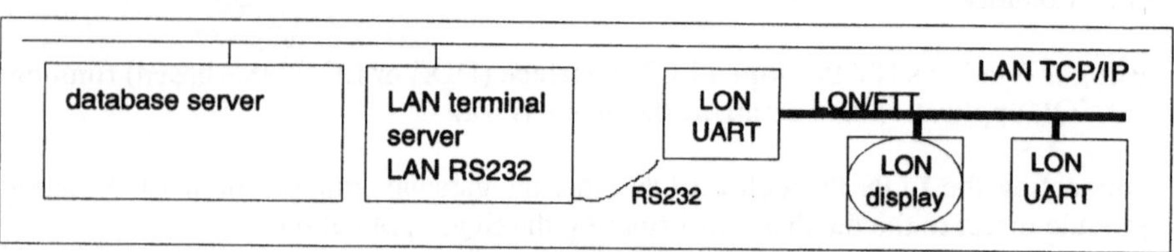

Fig. 2 connecting LON and database server via serial module

The server application frequently triggers the client application and transmits and receives data. Unlike the standard serial program, the application never transmits pure network traffic: The data consists of trigger events and results in an application near manner. But the configuration is done with standard tools located on a PC. The advantage of this strategy is the separation of managing a network and using the network as a process frontend.

Only the data of interest is transmitted at once and stored in the SQL database. All temporary data is stored in the LON nodes, so the transaction rates are low.

5 Summary

The described procedure shows a pragmatic way to integrate LON modules in an existing production environment by substituting some elements, which needed to be changed or by adding new functionality. The approach was to make a system, which conforms as much as possible to garantee interoperability. The given project frame forced this pragmatic way, but the given budget and time schedule was usual.

6 Final implementation

Figure 3 shows the final implementation using a serial controller acting as LON gateway. The offline configuration of the network is installed on a PC running Windows 95 using Echelon's products "Lonmaker" and a LON based PC card.

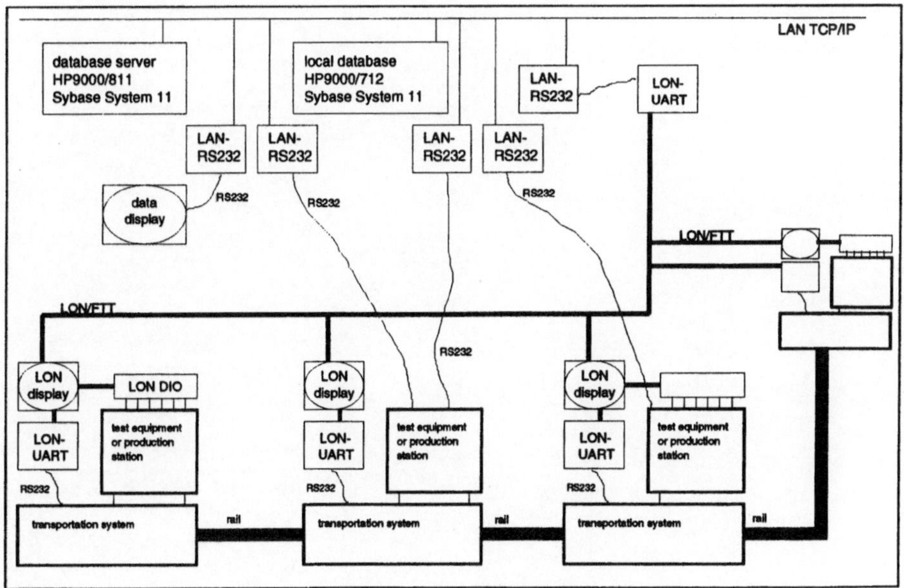

Fig. 3 final configuration

The final implementation consists of:
1. Echelon LON management tool "Lonmaker" for controlling and installing the network,
2. serial controller from SYSMIK (Dresden/Germany) with special application developed with Echelon's "Nodebuilder" acting as a gateway,
3. same serial controllers with another special application controlling the transport system of the production line installed at each workstation,
4. display modules from Beckmann & Egle (Kernen/Germany) with special application handling the user interface for the workstation operator and triggering the transactions as decribed above,
5. digital I/O modules with 8 relais output and 8 opto input lines from UNITRO (Backnang/Germany) controlling some automation systems, and
6. patch cables, sockets and T-connectors from VOGEL (Remchingen/Germany).

References

1. Echelon Corporation: Node Builder DDE Server User's Guide (1993-1995)
2. Echelon Corporation: LONWORKS PC LonTalk Adapter (PCLTA) Driver for Microsoft Windows NT (1995-1996)
3. Echelon Corporation: Node Builder User's Guide (1995)

Dezentrale Automatisierung
mit IEC 1131 und CANopen

Armin Walter[1], Harald Hellmann[2]

1): Weidmüller Interface GmbH, Ohmstraße 9, D-32758 Detmold
2): Weidmüller ConneXt GmbH, Paderborner Str. 170, D-32760 Detmold

Abstract: Der Trend zu dezentralen Automatisierungssystemen in der Fertigungsindustrie ist unverkennbar. Um dem Anwender die Vorteile dieser Systeme nutzbar zu machen, müssen geeignete Hardwarekomponenten, ein geeignetes Netzwerk und Softwarehilfsmittel für Projektierung, Programmierung und Inbetriebnahme zur Verfügung stehen. In diesem Beitrag wird ein solches dezentrales Steuerungssystem vorgestellt. Basis sind nach IEC 1131.3 programmierbare Module, die untereinander oder mit nicht programmierbaren Modulen über das Netzwerk CANopen kommunizieren. Die entwickelten o.g. Softwarewerkzeuge sowie der entwickelte Ansatz zur Integration von Netzwerktechnik und Steuerungstechnik werden erläutert.

In production engineering a trend to decentralized automation systems can be observed. These systems offer a lot of advantages to the user, if suited hardware components, a communication network and software tools for network configuration, control programming and debugging are available. Such a control system is described in this paper. The control system is based on IEC 1131.3 programmable modules which can communicate mutually or with remote-I/O-modules via the CAN based protocol CANopen. The above-mentioned software tools which were developped as well as the choosen approach for the integration of network communication and control technique are explained.

1 Einleitung

Die Entwicklung im Bereich der Produktionsanlagen verlagert sich in Richtung flexibler und modularer Komponenten und Systeme. Voraussetzung dafür ist die Verfügbarkeit von Automatisierungslösungen, die diese Modularität auf die Steuerungsebene abbilden können und die sich durch

- dezentrale, vernetzte Einheiten (Hardwarekomponenten),
- eine leichte Programmierbarkeit der Steuerungsfunktionen und
- eine weitreichende Bedienerunterstützung (Netzwerkprojektierung, Unterstützung bei Test und Inbetriebnahme der Anlage)

auszeichnen. Neben der hohen Flexibilität wird immer eine effiziente und kostengünstige Lösung auf Basis bestehender und verbreiteter Standards gefordert. Bei der Realisierung dezentraler Automatisierungssysteme sind die Kommunikationstechnik in der Feldebene und die Steuerungstechnik miteinander zu verbinden, d.h. die Komponenten beider Techniken müssen die Schnittstellen in die jeweils andere Welt zur Verfügung stellen. Dieser Beitrag beleuchtet die genannten Aspekte, insbesondere die erforderlichen Systemeigenschaften, und beschreibt das realisierte dezentrale Steuerungssystem.

2 Dezentrale Automatisierung

Die herkömmliche Steuerungsstruktur bestand i.a. aus einem zentralen Rechnersystem mit zentralen Ein-/Ausgabebaugruppen, an die die zur Prozeßsteuerung erforderlichen Sensoren- und Aktoren direkt angeschlossen sind, Abb. 1. Im nächsten Schritt wurden vornehmlich zur Verringerung des Verkabelungsaufwands Feldbussysteme eingesetzt, die auf einen schnellen Datenaustausch zwischen der zentralen Steuerung und den dezentralen Ein-/Ausgabemodulen hin optimiert sind. Bei dieser „Remote-I/O"-Struktur ist die Auswahl eines geeigneten Feldbussystems ein zentrales Problem. Flexible Kommunikationsstrukturen sind bei dieser Lösung nicht von Bedeutung.

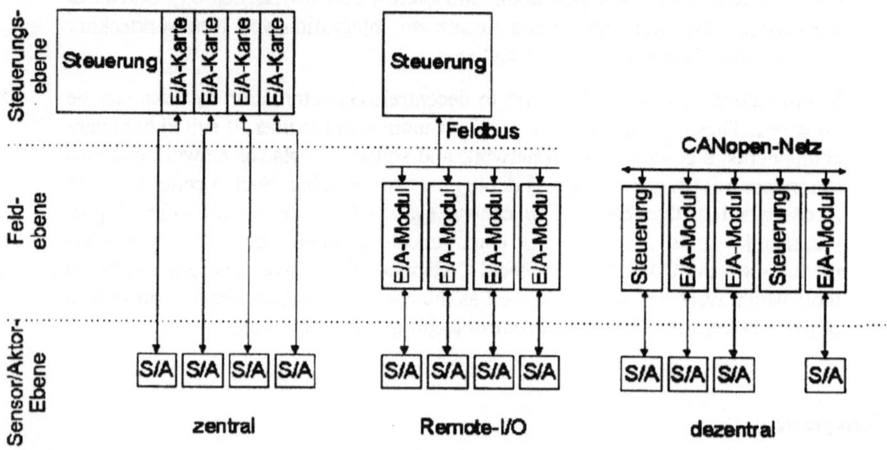

Abb. 1: Steuerungsarchitekturen

Bei der dezentralen Automation müssen diese Strukturen wesentlich erweitert werden. Kommunikationsbezogen sind folgende Eigenschaften zu erfüllen: *Vernetzung* der dezentralen Module, Konfigurierung der Netzwerkknoten, Projektierung des Netzwerks. Darüber hinaus müssen für die Steuerungsknoten mit/ohne E/A-Funktionalität Anwendungsprogramme erstellt werden, die auf Remote-I/O-Module zugreifen und mit anderen programmierbaren Modulen kommunizieren können. Die kommunizierte Datenmenge ist hierbei geringer, die Qualität des Informationsaustausches steigt aber.

Bei einer konsequenten Umsetzung dieser Idee kann die Steuerungsstruktur für eine Maschine oder Anlage komplett auf deren mechanische modulare Struktur abgebildet werden. Für den Anwender bedeutet dies, daß die Hardware bedarfsgerecht zusammengestellt werden kann. Bezüglich der Steuerungsprogramme ergeben sich ebenfalls kleinere Module, die problemlos wiederverwendet werden können. Diese Flexibilität und Skalierbarkeit des Steuerungssystems führt letztlich zu kostengünstigen und transparenten Lösungen von der Hardware bis hin zur Wartung einer laufenden Maschine oder Anlage.

3 Erforderliche Systemeigenschaften

Bei der Realisierung dezentraler Automatisierungssysteme nach Abb. 1 müssen die Steuerungstechnik und die Netzwerktechnik zusammengeführt und aufeinander abgestimmt werden. Es ist darauf zu achten, daß die gewählten Komponenten dieser Techniken jeweils für die Aufgabenstellung geeignet sind und auf möglichst breiter Basis stehen, d.h. standardisiert sind. Diese Aspekte sollen zunächst beleuchtet werden, bevor die Realisierung beschrieben wird, für die auch Ergänzungen dieser Standards notwendig waren.

3.1 Norm IEC 1131

Aufgrund verschiedener Randbedingungen ist die Steuerungslandschaft heute sehr heterogen [2]. Diesem Problem wurde durch die Einführung der Norm IEC 1131 begegnet [4, 7]. Insbesondere durch die Normung von textuellen und grafischen Programmiersprachen (Teil 3 der Norm) können Engineeringkosten gesenkt werden [3]. Diese Sprachen erlauben die Programmierung unterschiedlicher Steuerungen in der gleichen Syntax und Semantik und haben sich als wichtiger Standard in der industriellen Steuerungstechnik etabliert. Endgültige Zielsetzung ist die Portierung von Anwendungsprogrammen auf unterschiedliche Steuerungen [2].

3.2 Protokoll CANopen

Das Protokoll CANopen basiert auf dem Feldbus CAN, der aufgrund seiner Eigenschaften (Multimasterfähigkeit, kurze Reaktionszeiten, hohe Übertragungssicherheit) den Gedanken der dezentralen Automatisierung weitgehend unterstützt. CANopen ist ein europäischer Standard, der innerhalb der internationalen Anwender- und Herstellervereinigung CAN in Automation (CiA) entstanden ist. Das Kommunikationsprofil beschreibt die grundsätzlichen Mechanismen des Datenaustausches und umfaßt das Netzwerkmanagement, den Boot-up-Prozeß, die Kommunikationsobjekte für Service- und Prozeßdaten, einen Zeitstempel sowie weitere optionale Dienste. Es beschreibt ferner die Struktur des Objektverzeichnisses, das eine standardisierte Schnittstelle für den Zugriff auf Service- und Prozeßdaten darstellt. Applikationsspezifische Funktionen bestimmter Gerätetypen (z.B. E/A-Module, Antriebe, Regler) werden in Geräteprofilen festgelegt, die aber immer auf dem Kommunikationsprofil basieren müssen. Eine Übersicht über diese Thematik findet sich in [1]. In CANopen-Profilen ist immer ein bestimmter Bereich für herstellerspezifische Mechanismen vorgesehen.

3.3 Integration von Steuerungstechnik und Netzwerktechnik

Die Integration beider Techniken wurde erst in letzter Zeit in Angriff genommen, die Standardisierungsbemühungen waren zuvor auf das jeweilige Gebiet beschränkt. Die

IEC 1131 beinhaltet den Teil 5 „Kommunikation", der noch in Arbeit ist. Es gibt dazu verschiedene Arbeiten innerhalb und außerhalb der Normung, die vorrangig auf Systeme für die Remote-I/O-Architektur abzielen und nicht feldbusneutral sind [2].

Die Programmiersprachen nach IEC 1131 stellen dem Programmierer zwei Sprachkonstrukte für die Kommunikation zur Verfügung: Kommunikation über Funktionsbausteine und Datenzugriff über Zugriffspfade. Die erste Variante ist vorrangig für die Kommunikation zwischen Steuerungen geeignet, die zweite auch für die Kommunikation mit einem Feldgerät. Die Nachteile der zweiten Variante, insbesondere bei ereignisgesteuertem Datentransfer wie bei CANopen, sind in [8] dargestellt. Insbesondere wird auf die Abhängigkeit vom jeweiligen Zielsystem und von der herstellerspezifischen Realisierung verwiesen.

Für die Beschreibung und den Betrieb von „intelligenten" Geräten in einem Netzwerk sind über die grundlegenden hinausgehende Mechanismen erforderlich. Solche intelligente Geräte wie Steuerungen, Mensch-Maschine-Schnittstellen oder Softwarewerkzeuge (zur Programmierung, Visualisierung oder Netzwerkprojektierung) verlangen verschiedenen Dienste, von denen die wichtigsten sind:

 - dynamischer Aufbau von Servicedatenverbindungen zwischen einzelnen Geräten
 - dynamische Allokierung von Objektverzeichniseinträgen (z.B. für Prozeßabbild)
 - Mechanismus zum Herunter- und Hochladen von Daten (z.B. für Programme)
 - gleichzeitiges Beschreiben von Objektverzeichniseinträgen mehrerer Geräte.

Neben diesen Erweiterungen des CANopen-Kommunikationsprofils ist ein Geräteprofil erforderlich, in der IEC 1131-spezifische Funktionen beschrieben sind. Insbesondere müssen die Hierarchieobjekte „Konfiguration", Ressource" und „Programm" bzw. „Task" sowie Befehle an die Steuerung in das CANopen-Protokoll eingebracht werden, wobei die unterschiedlichen Datenstrukturen der einzelnen IEC 1131.3-Programmiersysteme zu berücksichtigen sind.

Letztlich ergibt sich aus der Integration beider Techniken ein praktisches Problem. Die Startbedingungen der Applikation auf der dezentralen Steuerung sind in Abhängigkeit des aktuellen Netzwerkzustands zu definieren.

4 Realisierung

Die Ausführungen im vorigen Abschnitt haben deutlich gemacht, daß die bestehenden Standards nicht für eine Realisierung von dezentralen Steuerungssystemen ausreichen. Es wurden wesentliche Ergänzungen des CANopen-Standards erarbeitet, was durch dessen offener Struktur begünstigt wird; diese wurden in die Gremien des CiA eingebracht [5, 6].

4.1 Entwicklungswerkzeuge

Die erforderlichen Funktionen der Entwicklungswerkzeuge wurden genannt. Sie sind in einer Software-Toolbox zusammengefaßt, die aus insgesamt 4 Werkzeugen besteht:

„DIAnet" für die Netzwerkprojektierung, „DIApro" für die Steuerungsprogrammierung, „DIAvis" für die Prozeßvisualisierung und „DIAmon" für die Busüberwachung. Der Schwerpunkt dieser Betrachtung liegt auf den beiden ersten Werkzeugen.

Mit „DIAnet" erfolgt die Netzwerkkonfiguration textuell oder grafisch, Abb. 2. Die einzelnen Netzwerkknoten werden parametriert und die Kommunikationsbeziehungen zwischen allen Knoten hergestellt. Fremdprodukte können eingebunden werden. Weiterhin können Netzvariable mit symbolischen Namen deklariert werden, die netzwerkweit zur Verfügung stehen und in einem Steuerungsprogramm verarbeitet werden können. Mit „DIApro" können in den IEC 1131.3 konformen Sprachen AWL, FBS, KOP und ST [7] Steuerungsprogramme erstellt und dokumentiert sowie für das Zielsystem übersetzt werden. Darüber hinaus beinhaltet „DIApro" umfangreiche Funktionen für die Test- und Inbetriebnahme: Starten und Stoppen von Programmen, lokale und Netzvariable lesen und setzen, Prozeßperipherie forcen, Programmstatus lesen.

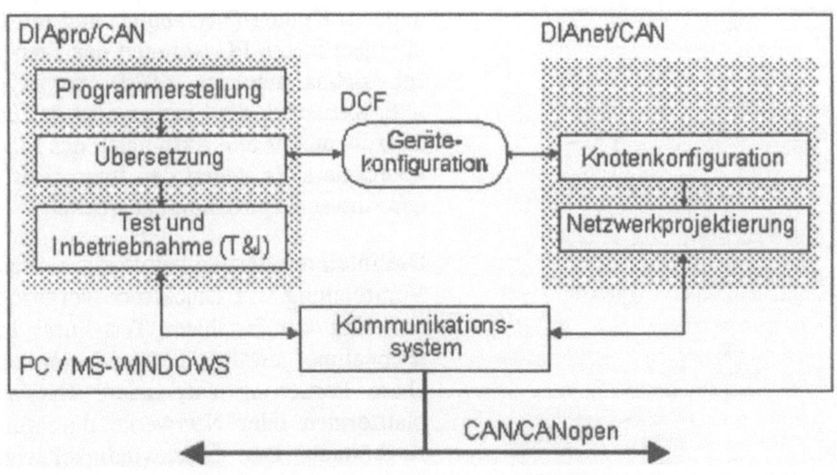

Abb. 2: Struktur der Entwicklungswerkzeuge

Die Kommunikation zwischen beiden Werkzeugen erfolgt über das in CANopen spezifizierte „Device Configuration File" (DCF), das von „DIAnet" erzeugt und von „DIApro" zur Übersetzung des Programms eingelesen wird. Es enthält unter anderem die für „DIApro" erforderlichen Informationen über die Zielhardware und eine Liste mit den symbolischen Namen der Netzvariablen, die in dem Steuerungsprogramm dieses DIAcan-Moduls verwendet werden sollen. Innerhalb des Programms werden die gleichen Namen im Variablenblock „VAR_EXTERNAL" deklariert.

4.2 Hardwarekomponenten

Die Hardwaremodule für die Signalverarbeitung und die Prozeßanbindung des Steuerungssystems basieren auf der modularen Gerätefamilie „WINbloc". Mit den ca. 20 verschiedenen digitalen und analogen Ein- und Ausgabemodulen kann eine flexible

Prozeßankopplung erfolgen. Diese Standardmodule arbeiten nach dem Remote-Prinzip. Aus dieser Palette wurde ein digitales Kombimodul (24 Ein-, 8 Ausgänge) zu einer dezentralen Steuerung mit Netzwerkfunktionalität erweitert („DIAcan"-Modul). Dieses Modul ist mit einer CPU vom Typ 80C320 (Taktfrequenz 24 MHz) sowie 32 KByte RAM, 64 KByte EPROM und 64 KByte Flash-EPROM ausgestattet. Die Funktionalität des Moduls wird über die entwickelte Firmware bereitgestellt.

4.3 Firmware des DIAcan-Moduls

Abb. 3 zeigt die Firmwarestruktur des DIAcan-Moduls. Das Modul muß alle Funktionen der Entwicklungswerkzeuge, insbesondere von DIApro unterstützen. Die Funktionalität umfaßt hinsichtlich der IEC 1131-konformen Datentypen und Operatoren mehr als den PLCopen Base Level II.

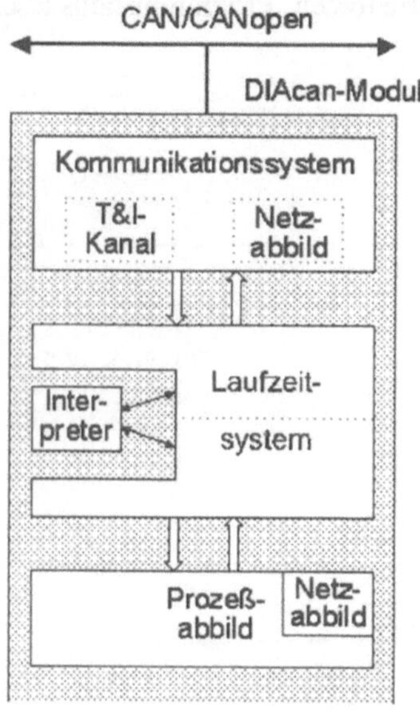

Das CANopen-konforme Kommunikationssystem ist für den Datenaustausch mit anderen Knoten (Netzabbild) und mit dem übergeordneten PC während der Test- und Inbetriebnahmephase (T&I) verantwortlich. Kernstück der Firmware ist das Laufzeitsystem, das alle Aktivitäten des Moduls koordiniert. Es steuert den Interpreter, der das Anwenderprogramm abarbeitet.

Die Interpreterlösung wurde einer direkten Verarbeitung von Objektcode vorgezogen, um eine komfortablere Test- und Inbetriebnahme gewährleisten und eine einfachere Portierung auf andere Hardwareplattformen oder Netzwerke durchführen zu können. Der Geschwindigkeitsverlust ist aufgrund der Parallelverarbeitung in der dezentralen Struktur tragbar.

Abb. 3: Firmwarestruktur des DIAcan-Moduls

Das Laufzeitsystem ist in eine obere und eine untere Schicht aufgeteilt. Während die obere Schicht mehr administrative Aufgaben wie Verwalten von Programmen oder Steuerung des Interpreters übernimmt, ist die untere Schicht für den Zugriff auf den Speicher und die Peripherie verantwortlich. Es sind asynchrone und zyklische Aktivitäten erforderlich: Fehlermeldungen und Steuersignale müssen asynchron verarbeitet werden, während zyklische Aktivitäten wie Steuerung des Interpreters und Auffrischen des Prozeßabbilds im sogenannten Systemkontrollpunkt durchgeführt werden. Das Netzabbild wird über andere (auch Remote-I/O-) Module zur Verfügung gestellt.

Es kann festgelegt werden, ob die Applikation nur bei bereitem Netzwerk, nur bei nicht bereitem Netzwerk oder immer startet.

4.4 Leistungsdaten

Wesentliche Leistungsdaten des DIAcan-Moduls sind die Zykluszeiten und die netzwerkbezogenen Einflüsse darauf. Die Zykluszeit liegt bei Verwendung von lokalen Operanden zwischen 30 ms (bei nur Sprungoperationen) und 55 ms (bei nur arithmetischen Operationen) für jeweils 1000 AWL-Zeilen.

Die Verwendung von Netzvariablen hat keinen Einfluß auf die Zykluszeit, da das Netzabbild wie das Prozeßabbild gehandhabt wird. Die Zeit für die ereignisgesteuerte Auffrischung des Netzabbilds wird jedoch von der Buslast bestimmt, die wiederum von einer Reihe von Faktoren abhängt. Als Beispiel sei genannt, daß ca. 1 ms vergeht, bis eine Ausgangsvariable eines DIAcan-Moduls am Ausgang eines Remote-Ausgabemoduls wirksam wird. Die umfangreichen Test- und Inbetriebnahmefunktionen verbrauchen einen Anteil der Prozessorleistung des DIAcan-Moduls. Dieser Anteil vergrößert die Zykluszeit um bis zu 7 %. Die zu übertragenden Daten werden als Servicedatenobjekte versendet und haben eine niedrige Priorität im Netzwerk.

Bei der Wertung aller Angaben ist zu beachten, daß die Komponenten eines dezentrales Steuerungssystems aufgrund der parallel arbeitenden Struktur nicht die Leistungsfähigkeit einer zentralen Steuerung besitzen müssen. Im Vergleich zu Remote-I/O-Systemen ist weniger Datentransfer erforderlich.

5 Zusammenfassung und Ausblick

Das beschriebene Steuerungssystem beinhaltet alle Komponenten, die für eine dezentrale Automatisierungslösung erforderlich sind. Aufbauend auf der modularen Produktfamilie WINbloc/CANopen wurde das DIAcan-Modul entwickelt, daß es nach IEC 1131.3 programmierbare Funktionen abarbeiten und in einem CAN/CANopen-Netzwerk variabel kommunizieren kann. Geeignete Softwarewerkzeuge für die Netzwerkprojektierung und Steuerungsprogrammierung sowie für Test- und Inbetriebnahmefunktionen liegen vor. Die dafür erforderlichen CANopen-Erweiterungen wurden erarbeitet und in die CANopen-Gremien eingebracht.

Zur Zeit wird dieser Ansatz auf die leistungsfähigere Hardwareplattform PC104 portiert, wobei diese Module zwei CAN/CANopen-Schittstellen besitzen und somit eine Gatewayfunktion zwischen einem übergeordneten und einem unterlagerten Netz wahrnehmen. Abb. 4 zeigt einen möglichen Aufbau.

Der jetzige Stand der CANopen-Standardisierung beinhaltet noch keine Mechanismen für die Behandlung hierarchischer Netzwerke. Dementsprechend muß jedes Teilnetz einzeln projektiert und programmiert werden. Zukünftige Arbeiten konzentrieren sich daher auf eine entsprechende Erweiterung des CANopen-Kommunikationsprofils.

Eine Portierung des Ansatzes auf ähnliche Module mit LON-Anschaltung wird zur Zeit bearbeitet. Dies wird durch die universelle und offene Struktur des Laufzeitsystems, des Interpreters und der IEC1131.3 begünstigt.

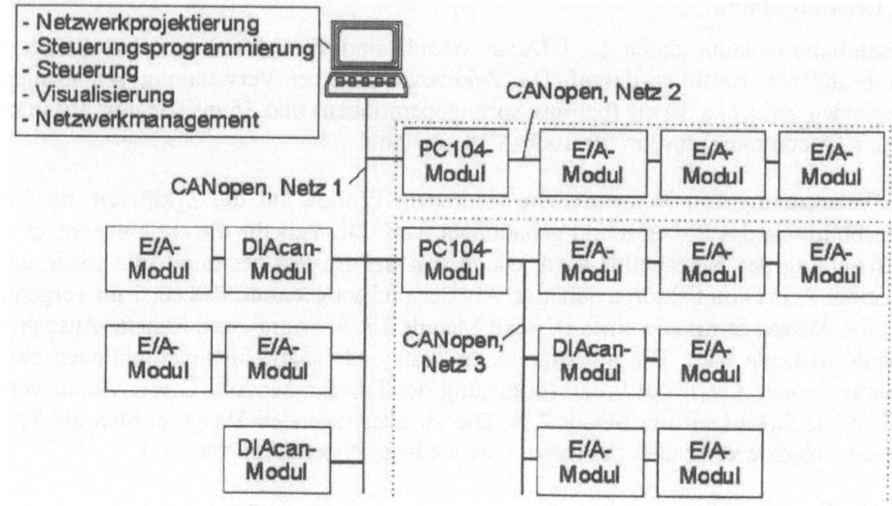

Abb. 4: Möglicher Aufbau eines Gesamtsystems

Literatur

1. Förster, J.; Zeltwanger, H.: Das offene CAN-Protokoll - Hinweise zur Implementierung von CANopen. Elektronik (1997) 10, S. 96-104

2. Fussel, B.: SPS-Programmierung und Offene Kommunikation auf der Basis IEC1131. atp 39 (1997) 6, S. 34-45

3. John, K.-H.; Tiegelkamp, M.: SPS-Programmierung mit IEC 1131-3. Berlin, Heidelberg, New York: Springer-Verlag, 1995

4. Neumann, P. et. al.: SPS-Standard: IEC 1131. München, Wien: R. Oldenbourg Verlag, 1995

5. N.N.: CANopen device profile for IEC1131 devices. CiA (Hrsg.): CiA Work Draft Standard Proposal 405, nicht veröffentlicht

6. N.N.: Framework for programmable CANopen devices. CiA (Hrsg.): CiA Work Draft Standard Proposal 302, Stand Juli 1997, Erlangen: CiA-Selbstverlag, 1997

7. N.N: Speicherprogrammierbare Steuerungen. DIN EN 61131, Berlin: Beuth-Verlag, 1995

8. Sperber, M.: Distributed IEC1131-programming and CANopen. In: CiA (Hrsg.): Proceedings 3rd International CAN Conference, Paris 1996. Erlangen: CiA-Selbstverlag, S. 87-90

Eigensichere Feldgeräte für PROFIBUS-PA und Foundation™ Fieldbus

Hans Endl

Softing GmbH
Richard-Reitzner-Allee 6
D-85540 Haar

Abstract. Die Feldbusstandards PROFIBUS-PA und Foundation™ Fieldbus nehmen für sich in Anspruch, speziell auf die Belange der Verfahrenstechnik zugeschnitten zu sein. Beiden gemeinsam ist, daß sie auf einer Busphysik gemäß IEC 1158-2 basieren, die für den Einsatz im eigensicheren Bereich geeignet ist. Der limitierte Energieverbrauch eigensicherer Geräte steht jedoch im Widerspruch zu den Leistungsanforderungen moderner Feldbusprotokolle, welche häufig die der eigentlichen Gerätefunktion um ein Vielfaches übersteigen. Zunächst werden die funktionalen Anforderungen, welche PROFIBUS-PA und Foundation™ Fieldbus an Feldgeräte stellen, benannt und ihre Auswirkungen auf die Gerätearchitektur aufgezeigt. Anschließend wird eine für beide Feldbusse geeignete Hardware-Architektur beschrieben, welche für den Einsatz in eigensicheren Feldgeräten geeignet ist.

Abstract. PROFIBUS-PA and Foundation™ Fieldbus claim to fulfill the specific requirements on fieldbusses in process control applications. Both fieldbus standards are based on a physical layer according to IEC 1158-2 which allows for intrinsic safety. The limited power consumption of intrinsically safe instruments is in contradiction to the performance requirements of state-of-the-art fieldbus protocols. The functional requirements on instruments to be used in PROFIBUS-PA or Foundation™ Fieldbus networks will be listed and the overall architecture of fieldbus devices will be shown. Finally a hardware design suitable for both fieldbusses will be described, which also allows for intrinsically safe applications.

1 Einführung

Viele der in den letzten Jahren entstandenen Feldbusse lassen aufgrund der Eigenschaften ihrer Busphysik den Bau eigensicherer Feldgeräte nicht zu. Dies mag einer der Gründe sein, warum Feldbusse bisher hauptsächlich in der Fertigungsautomatisierung, kaum jedoch in der Prozeßautomatisierung eingesetzt werden.

Beim Interoperable Systems Project (ISP) wurde erstmals der Physical Layer nach IEC 1158-2 verwendet. Nach dem Zusammenschluß von ISP und WorldFIP zur Fieldbus Foundation wurde der IEC Physical Layer sowohl vom Foundation Fieldbus als auch von PROFIBUS-PA, dem PROFIBUS-Profil für die Verfahrenstechnik, übernommen.

Für diese beiden Feldbusse erscheinen zur Zeit die ersten Feldgeräte mit IEC-1158-Anschluß auf dem Markt.

2 Busphysik nach IEC 1158-2 als Basis für eigensichere Feldbusse

2.1 IEC 1158-2

Die physikalische Schicht nach IEC 1158-2 kennt verschiedene Ausprägungen, welche alle die Fremdspeisung der angeschlossenen Geräte über den Feldbus zulassen. Für die Realisierung eigensicherer Bussegmente ist die Variante mit einer Übertragungsgeschwindigkeit von 31,25 kBit/s im „Voltage Mode" am besten geeignet. Dabei wird die vom Busspeisegerät zur Versorgung der Geräte bereitgestellte Gleichspannung (9V bis 32V) von dem jeweils sendenden Gerät durch ein Informationssignal (0,75V bis 1V peak-to-peak) überlagert. Busgespeiste Geräte können diese Spannungsmodulation nur über eine Veränderung ihres Strombezugs bewirken, was wiederum zu bestimmten Anforderungen an die Eingangsimpedanz und die maximale Flankensteilheit im Stromverbrauch dieser Geräte führt.

Der Aufwand für die Realisierung eines Busanschlusses nach IEC 1158-2 ist deshalb höher als beispielsweise für normale PROFIBUS-Anschaltungen auf Basis von RS-485 mit NRZ-Kodierung. Man benötigt eine analoge „Medium Attachment Unit" (MAU) zur sende- und empfangsseitigen Aufbereitung des Signals und zur Einkopplung der Speisespannung. Diese MAU kann diskret aufgebaut werden, es sind aber auch bereits von mehreren Herstellern integrierte Lösungen verfügbar.

Die Signalübertragung erfolgt über eine gleichstromfreie Manchester-Biphase-Kodierung, die eine Rückgewinnung des Taktsignals ermöglicht. Da gängige Microcontroller in der Regel keine Manchester-Encoder/Decoder mit FCS-Prüfung nach IEC 1158-2 integriert haben, ist auch hierfür eine separate integrierte Schaltung notwendig.

2.2 Eigensicherheit

Die Zündschutzart „Eigensicherheit" beruht darauf, daß die Energie innerhalb der im explosionsgefährdeten Bereich betriebenen Stromkreise so begrenzt wird, daß kein zündfähiger Funke entstehen kann. Dazu muß nicht nur die maximale Stromaufnahme, sondern auch die in Kondensatoren und Induktivitäten gespeicherte Energie beschränkt werden. Diese Restriktionen lassen sich aber nur sehr schwer mit den Anforderungen vereinbaren, welche komplexe Feldbusprotokolle an die Rechenleistung und damit an die Hardware-Ressourcen stellen.

Für die Versorgung busgespeister Feldgeräte in einem eigensicheren Bussegment stehen typischerweise etwa 110 mA bei einer Spannung von 12 Volt zur Verfügung. Dies limitiert die Zahl der in einem Segment betreibbaren Geräte. Sollen bis zu zehn Geräte angeschlossen werden, so bleiben pro Gerät nur etwa 10 mA, die für die eigentliche Gerätefunktion und den Feldbusanschluß aufgeteilt werden müssen. Es gibt nur wenige Low-Power-Versionen gängiger Microcontroller mit derart niedrigem Stromverbrauch.

3 Funktionale Anforderungen

3.1 Modell

Feldbusprotokolle basieren in der Regel auf einem reduzierten OSI-Schichtenmodell mit drei Schichten: Physical Layer, Data Link Layer und Application Layer. Auch bei kompakten Implementierungen spiegeln sich diese Schichten in der Systemarchitektur wider, wobei neben dem Physical Layer häufig auch Teile des Data Link Layers in Hardware realisiert sind.

Bild 1 zeigt links in allgemeiner Form die Hard- und Softwareschichten eines Feldgeräts mit Busanschluß. In der Mitte und rechts sind die jeweiligen Ausprägungen für PROFIBUS-PA und Foundation Fieldbus dargestellt. Während beide Protokolle auf demselben Physical Layer basieren und damit ähnliche Voraussetzungen an die Hardware stellen, unterscheiden sich die darüberliegenden Protokollschichten ganz erheblich (zumal das für PROFIBUS spezifizierte und auch beim Foundation Fieldbus verwendete FMS-Protokoll bei PROFIBUS-PA jetzt durch das einfachere DP-Protokoll ersetzt wurde).

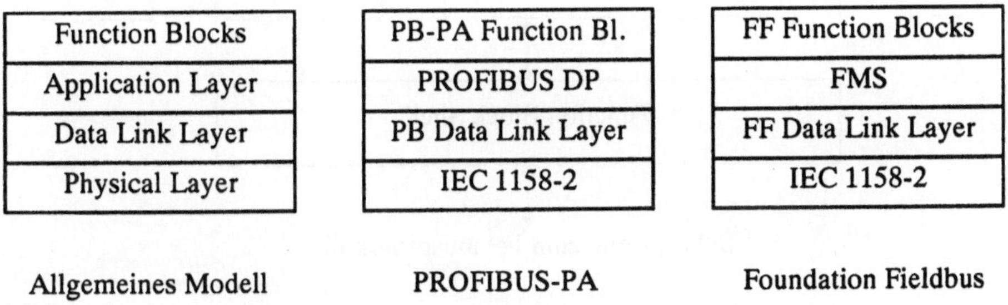

Function Blocks	PB-PA Function Bl.	FF Function Blocks
Application Layer	PROFIBUS DP	FMS
Data Link Layer	PB Data Link Layer	FF Data Link Layer
Physical Layer	IEC 1158-2	IEC 1158-2

Allgemeines Modell	PROFIBUS-PA	Foundation Fieldbus

Fig. 1. Schichtenmodell eines busfähigen Feldgeräts

Um über die Kommunikationsfunktion hinaus auch eine Standardisierung der (verteilten) Applikationfunktion zu erreichen, wurden sowohl für den Foundation Fieldbus als auch für PROFIBUS-PA standardisierte Funktionsblöcke definiert. Da beide hier betrachteten Bussysteme dabei auf die im ISP-Projekt geleisteten Vorarbeiten aufsetzen, sind ihre Funktionsblockdefinitionen sehr ähnlich.

Einfache Funktionsblöcke sind zum Beispiel *Analog Input* oder *Analog Output*, während als komplexerer Funktionsblock ein PID-Regler genannt werden könnte. Um diese Funktionsblöcke geräteunabhängig zu halten, erfolgt die eigentliche Abbildung der Gerätefunktion auf den Funktionsblock über sogenannte *Transducer Blocks*, welche ebenfalls standardisiert sind. So besitzen beispielsweise ein Temperaturtransmitter und ein Coriolis-Massendurchflußmesser zwar den gleichen Funktionsblock für Analog Input, aber zwei unterschiedliche Transducer Blocks, welche alle die Parameter enthalten, die für die jeweilige Gerätefunktion und das Meßprinzip spezifisch sind.

Darüberhinaus ist für jedes Gerät noch ein *Physical Block* (PROFIBUS-PA) oder *Resource Block* (Foundation Fieldbus) erforderlich, in dem Parameter zur Beschreibung des physikalischen Geräts abgelegt sind, wie zum Beispiel Herstellerbezeichnung, Modellbezeichnung, Seriennummer usw. Die Ankopplung der Funktionsblöcke an die Kommunikation und den (möglicherweise systemweit synchronisierten) Start der Funktionsblockalgorithmen erledigt eine *Function Block Shell*, die als eine Art Betriebssystem für Funktionsblöcke angesehen werden kann.

Bild 2 zeigt die Struktur der Geräteapplikation und deren Anbindung an die Kommunikationsschichten des Feldbusprotokolls am Beispiel eines einfachen Temperaturtransmitters.

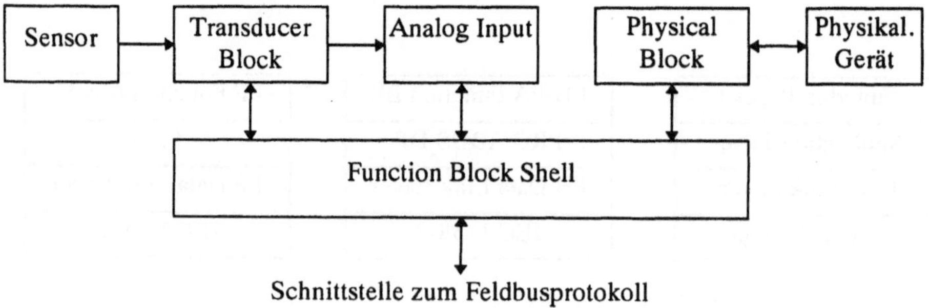

Fig. 2. Anbindung der Gerätefunktion an die Kommunikation

Die Anzahl der Parameter aller Blöcke kann selbst bei einem einfachen Sensor erheblich sein und stellt entsprechende Anforderungen an das Mengengerüst der Kommunikationsobjekte. Für einen Temperaturtransmitter mit je einem Function Block und Transducer Block ergeben sich minimal bei PROFIBUS-PA etwa 80, beim Foundation Fieldbus sogar über 100 Parameter. Die als statisch deklarierten Parameter müssen nichtflüchtig gespeichert werden, so daß ein EEPROM entsprechender Kapazität vorzusehen ist.

Die durch das Kommunikationsprotokoll bedingten Anforderungen unterscheiden sich für PROFIBUS-PA und Foundation Fieldbus, da beide Bussysteme unterschiedliche Protokolle im Data Link Layer und auch im Application Layer nutzen.

3.2 PROFIBUS-PA

Durch die im Herbst 1996 von der PROFIBUS Nutzerorganisation (PNO) beschlossene Abbildung des PROFIBUS-PA-Profils auf die Kommunikationsfunktionen des PROFIBUS-DP-Protokolls hat sich der Realisierungsaufwand für die Protokollsoftware eines PROFIBUS-PA-Feldgeräts etwas vereinfacht. Der zyklische Datenaustausch sowie die Alarmmeldung erfolgen über den zyklischen DP-Kanal, während der Zugriff auf die zahlreichen Parameter über azyklische Verbindungen erfolgt.

3.3 Foundation Fieldbus

Die Realisierung eines Feldgeräts für den Foundation Fieldbus gestaltet sich schwieriger. Hauptgrund ist das im Vergleich zu PROFIBUS-PA wesentlich komplexere Protokoll, welches mehrere Kommunikationsmodelle enthält: Neben dem für die Abwicklung des zyklischen Datenverkehrs vorgesehenen Publisher-Subscriber-Modell ist für die Parametrierung des Geräts ein Client-Server-Modell und für die Alarmabwicklung ein Report-Distribution-Modell zu implementieren. Darüberhinaus bietet der Foundation Fieldbus umfangreiche System-Management- und Network-Management-Funktionen, mit denen sich zum Beispiel die Ausführungszeitpunkte von Funktionsblöcken und die Übertragungszeitpunkte der zyklischen Daten systemweit vorgeben lassen, was für eine präzise und synchrone Abarbeitung verteilter Applikationen unerläßlich ist.

Die Profilklasse „Intelligent I/O-Device", die einem klassischen Feldgerät entspricht, erfordert bereits die Unterstützung eines wesentlichen Teils des möglichen Funktionsumfangs. Wenn ein Feldgerät zusätzlich in der Lage sein soll, bei Ausfall des zentralen Leitsystems im eigenen Bussegment die Rolle des *Link Active Schedulers* (LAS), also die zeitgesteuerte Koordination der Kommunikationsaufrufe zu übernehmen, dann muß der gesamte Funktionsumfang des Foundation Fieldbus implementiert werden.

In jedem Fall ist für den Protokollstack mit einem Codeumfang zwischen 64 und 128 KByte zu rechnen. Dazu kommen noch die Funktionsblöcke und die Anbindung an die Gerätefunktion, so daß eine Programmspeichergröße von 128 KByte oder 256 KByte vorzusehen ist.

4 Realisierungsbeispiel

Im folgenden wird am Beispiel einer Busanschaltung für eigensichere Feldgeräte aufgezeigt, welche Technologien heute für diesen Einsatzfall zur Verfügung stehen. Neben den im letzten Kapitel genannten grundlegenden funktionalen Anforderungen wurde die Realisierung durch folgende spezielle Anforderungen bestimmt:

- Die mechanischen Abmessungen durften höchstens 56 x 40 mm (22,4 cm^2) betragen.

- Der Gesamtstromverbrauch von Feldbusanschaltung und Feldgerät durfte 10 mA nicht überschreiten.

- Die Hardware sollte sowohl für PROFIBUS-PA als auch für Foundation Fieldbus geeignet sein.

Bild 3 zeigt die realisierte Lösung im Blockschaltbild. Der Feldbus wird über eine Medium Attachment Unit (MAU) angekoppelt, welche dem Bus auch die Energie für die gesamte Schaltung und den Sensor/Aktuator entnimmt. Das noch serielle aber bereits digitalisierte Signal wird dann über einen speziellen seriellen Schnittstellen-baustein (SIO) an den Microcontroller übergeben. Der Sensor/Aktuator (oder ein komplettes Feldgerät mit seriellem Interface) wird über eine der Schnittstellen des Microcontrollers angebunden.

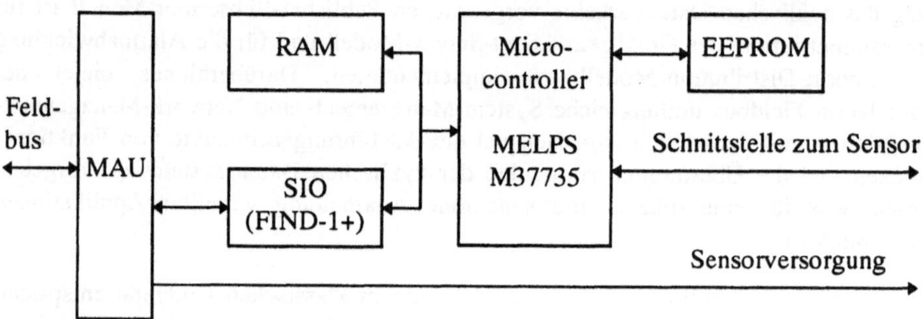

Fig. 3. Blockschaltbild der Feldbusanschaltung

Eine diskret aufgebaute Medium Attachment Unit schied aufgrund der Platzverhältnisse aus, stattdessen wurde der Baustein μSAA22Q von Yokogawa eingesetzt. Als Alternative ist hier noch der Baustein SIM-1 von Siemens zu nennen.

Einziger PROFIBUS-ASIC mit intergriertem Data Link Layer und synchronem Inter-face für IEC-MAUs ist der SPC4 von Siemens. Da die Busanschaltung jedoch auch für den Foundation Fieldbus geeignet sein sollte, konnte stattdessen nur ein synchroner Schnittstellenbaustein verwendet werden, und der Data Link Layer mußte für beide Protokolle vollständig in Software realisiert werden. Die Wahl fiel auf den FIND-1+ (YTZ440-F) von Yamaha, der neben der Manchester-Kodierung und -Dekodierung noch die Generierung und Überprüfung des CRC-Polynoms gemäß IEC1158-2 durchführt und einige nützliche Protokoll-Timer enthält.

Die Suche nach einem Microcontroller, der bei geringem Stromverbrauch eine akzeptable Rechenleistung anbietet und dabei noch einen Adreßraum größer als 64K erlaubt, gestaltete sich schwierig. Gegenüber dem in einem früheren Projekt eingesetzten Motorola MC68L11K1 erwies sich hier der Mitsubishi MELPS M37735 als gute Wahl, da er bei gleicher Taktrate weniger Strom aufnimmt und aufgrund seiner Architektur Adreßräume größer 64 KByte besser unterstützt.

Da es den Microcontroller M37735 maskenprogrammiert oder in einer OTP-Version gibt, ist kein externes FLASH-Memory notwendig. Die ROM-Größe von 128 KByte reicht für die Protokollsoftware und kleinere Applikationen aus. Als Datenspeicher wurde ein SRAM der Größe 128 KByte gewählt. Zum nichtflüchtigen Speichern der Kommunikations- und Funktionsblockparameter wurde ein EEPROM mit 32 KBit vorgesehen.

Für die Ankopplung der Sensor- oder Aktuatorfunktionen stehen im M37735 Ports, ein A/D-Wandler und serielle Schnittstellen zur Verfügung. Letztere bieten sich als Interface zu existierenden Geräten mit vorhandener serieller HART-Schnittstelle an, um diese HART-Geräte auf PROFIBUS-PA und Foundation Fieldbus umzurüsten, ohne das Gerät komplett neu entwickeln zu müssen. Die hier beschriebene Hardware wird als Gateway zwischen HART und PROFIBUS-PA bzw. Foundation Fieldbus eingesetzt. Bei einer Taktrate von 4 MHz beträgt die Stromaufnahme der gesamten Schaltung weniger als 5 mA, so daß einschließlich der Versorgung des angeschlossenen HART-Geräts mit 4 mA die Forderung nach einem Gesamtstromverbrauch von höchstens 10 mA eingehalten werden konnte.

5 Zusammenfassung

Sowohl PROFIBUS-PA als auch der Foundation Fieldbus erlauben die Realisierung eigensicherer Feldgeräte auf einer Busphysik nach IEC 1158-2. Die Beschränkung des Stromverbrauchs auf 10 mA für das gesamte Feldgerät mit Busanschluß schränkt die Auswahl an verwendbaren Komponenten stark ein. Trotzdem konnten für beide Feldbusprotokolle portierbare Software-Implementierungen realisiert werden, die auf der vorgestellten Hardware-Plattform lauffähig sind. Es zeigt sich, daß die für den Einsatz von Feldbussen in der Verfahrenstechnik notwendigen Technologien verfügbar sind.

Literatur

1. CENELEC: General Purpose Field Communication Systems - CENELEC Standard EN 50170. 1996

2. PROFIBUS Nutzerorganisation e.V.: PROFIBUS-PA, Profile for Process Control Devices. March 1997

3. Fieldbus Foundation: Foundation™ Specification, Rev. 1.1. February 1997

4. HART® Communication Foundation: HART® -SMART Communications Protocol, Rev 7.1. January 1997

Integration von PROFIBUS-PA in Systemen und Netzwerken

Pierre Kobes

Siemens AG, AUT 93 FBK, Karlsruhe

Abstract. PROFIBUS is very largely established in the European market mainly in the Manufacturing Automation. PROFIBUS offers an efficient fieldbus system with the variations FMS and DP as well in the cell level as in the field area. The extension PROFIBUS-PA opens the usage of PROFIBUS for Process Automation and provides an integrated fieldbus solution for all relevant applications.

Abstract. Ausgehend von der Fertigungsautomatisierung hat sich PROFIBUS als leistungsfähiges Feldbussystem im europäischen Markt durchgesetzt. Mit den Ausprägungen FMS und DP hat PROFIBUS bisher schon eine effiziente Lösung für die Kommunikation auf Zellen- und Feldebene ermöglicht. Die Erweiterung bei PROFIBUS-PA erlaubt den Einsatz in der Prozeßautomatisierung. PROFIBUS bietet damit eine universelle und durchgängige Feldbuslösung für alle marktrelevanten Anwendungen an.

1 Einleitung

PROFIBUS-PA ergänzt die bisherigen Protokolle PROFIBUS-FMS (Fieldbus Message Specification) und PROFIBUS-DP (Decentralised Periphery). Die FMS-Variante ist für anspruchsvolle Kommunikationsaufgaben im Zellbereich ausgelegt wie z.B. die Verbindung von Automatisierungssystemen. Die Variante DP dagegen ist für die Feldebene zugeschnitten und ist für schnelle und zeitkritische Steuerungsvorgänge optimiert. PROFIBUS-DP hat in der Fertigungsautomatisierung einen breiten Einsatz gefunden. Die Übertragungsgeschwindigkeit kann bei DP bis zu 12 Mbit/s betragen.

PROFIBUS-PA ist speziell für Anwendungen in der Prozeßautomatisierung ausgelegt. Die Norm EN 50170 Vol. 2 wurde an zwei Stellen erweitert. Die erste Erweiterung betrifft die Übertragungstechnik. In PROFIBUS-PA wird neben der Standard-Übertragungstechnik RS 485 auch die IEC-genormte Übertragungsschicht IEC 1158-2 übernommen. Die IEC-Übertragungstechnik ist für den Einsatz im explosionsgefährdeten Bereich zugelassen. Ähnlich zu der 4-20 mA Zweileitertechnik erfolgen hierbei Energiespeisung und Datenübertragung mit nur zwei Adern. Die RS

485- und die IEC-Übertragungstechniken lassen sich über DP/PA-Koppler oder DP/PA-Link verbinden [1, 3]. Damit ist eine nahtlose Erweiterung eines PROFIBUS-Netzes für Anwendungen in explosionsgefährdeten Bereiche gegeben.

PROFIBUS-PA beschreibt Standardfunktionalitäten in der Anwendung auf Basis von Function Blocks. Standardprofile legen das Verhalten der Feldgeräte am Feldbus fest und sind damit die Basis für die Austauschbarkeit von Geräten verschiedener Hersteller. Die Geräteprofile setzen auf dem Protokoll PROFIBUS-DP auf und verwenden die DP-Erweiterungen, die zusätzlich zu dem zyklischen Aktualisieren der Prozeßgrößen einen azyklischen Zugriff auf Parametrierdaten erlauben. Profile für Druck-, Temperatur-, Durchfluß- und Füllstandmeßumformer sowie Stellgeräte und binäre Ein/Ausgänge wurden von der PROFIBUS Nutzerorganisation (PNO) definiert. Weiterhin sind die Profile Wartengeräte, Einzelregler und Analysengeräte in Arbeit. Mit PROFIBUS-PA wird die Palette der PROFIBUS-Produkte um Komponenten für die Prozeßautomatisierung erweitert, so daß PROFIBUS den gesamten Bereich der Industrieautomatisierung abdeckt.

2 Anschluß von Feldgeräten über DP/PA-Koppler und DP/PA-Link

Für den Übergang der Übertragungstechnik von PROFIBUS-DP (RS 485) auf PROFIBUS-PA (IEC 1158-2) stehen die beiden Netzkomponenten DP/PA-Koppler und DP/PA-Link zur Verfügung (Abb. 1.). Ihr Einsatz richtet sich nach den Anforderungen an die Automatisierungstechnik.

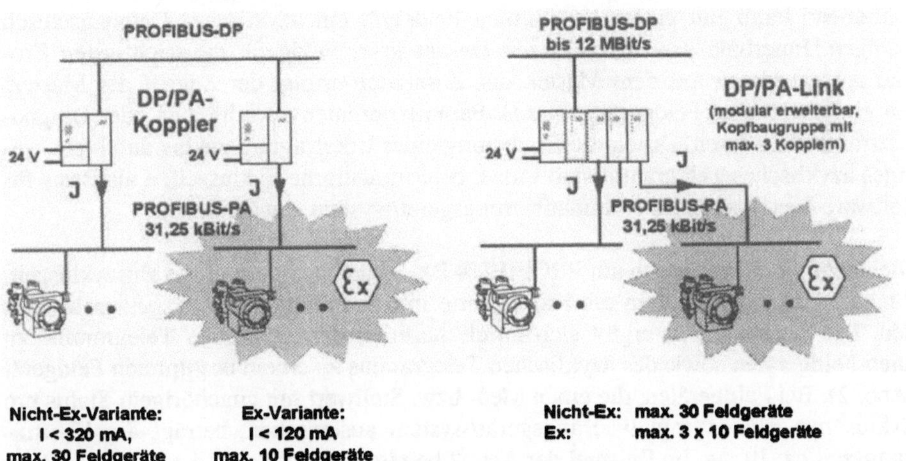

Abb. 1. Anschluß von Feldgeräten über DP/PA-Link DP/PA-Koppler oder DP/PA-Link

Der DP/PA-Koppler hat folgende Aufgaben:
• Umsetzen des Datenformats von asynchron (11 Bit/Zeichen) auf synchron (8 Bit/Zeichen) und damit verbunden eine Umsetzung der Übertragungsgeschwin-

digkeit auf 31,25 kBit/s. Der Koppler wirkt „wie ein Draht", wird nicht projektiert und ist aus Sicht der Busteilnehmer nicht zu erkennen.

- Speisung der Feldgeräte
- Begrenzung des Speisestroms durch Barrieren (für Ex-Anwendungen)

Es werden 2 Varianten des DP/PA-Kopplers angeboten: Eine Nicht-Ex-Variante mit Speisung für max. 30 Feldgeräte und eine von der Physikalisch Technische Bundesanstalt (PTB) zertifizierte Ex-Variante mit Speisung für max. 10 Feldgeräte für den Einsatz in Zone 1 und 2.

Das DP/PA-Link besteht aus max. 5 DP/PA-Kopplern (Ex- oder Nicht-Ex-Variante), die über eine Kopfbaugruppe als ein Teilnehmer an PROFIBUS-DP angeschlossen werden. Die Kopfbaugruppe ist ein Slave am übergeordneten PROFIBUS-DP (max. 12 MBit/s) und ein Master an den unterlagerten PA-Segmenten. Diese PA-Segmente bilden zusammen einen logischen Bus. Die Summe aller Feldgeräte an einem DP/PA-Link ist auf max. 30 begrenzt.

Das DP/PA-Link wird bei hohen Anforderungen an die Zykluszeit und hohen Mengengerüsten eingesetzt.

3 Das DP/PA-Link bietet Vorteile bei der Zykluszeit

Der Datenaustausch bei PROFIBUS-DP verwendet einen zyklischen Datenverkehr. In einem Zyklus werden alle Ausgangswerte (Stellbefehle) in die Feldgeräte geschrieben und alle Eingangswerte (Meßwerte) aus den Feldgeräten gelesen. Anschließend kann mit einem bestimmten Feldgerät ein azyklischer Datenaustausch erfolgen. Innerhalb der Buszykluszeit tauscht jedes Feldgerät die wichtigsten Ein- und Ausgabedaten mit dem Master aus. Zusätzlich erfolgt der Zugriff des Masters auf ein bestimmtes Feldgerät, um z.B. Parametrierdaten zu schreiben oder Diagnoseparameter zu lesen. Durch die Ergänzung eines Übertragungszyklus durch ein einziges azyklisches Telegramm sind kurze, deterministische Zykluszeiten als Basis für Software-Regelungen im Automatisierungsgerät/-system gewährleistet.

Die Anzahl der Feldgeräte am PROFIBUS-PA-Segment bestimmt die Buszykluszeit, d.h. das Zeitraster, in dem die Prozeßwerte mit den Feldgeräten ausgetauscht werden. Die Buszykluszeit ergibt sich durch Addition der zyklischen Telegramme zu allen Feldgeräten sowie des azyklischen Telegramms zu einem bestimmten Feldgerät (Abb. 2). Bei Feldgeräten, die einen Meß- bzw. Stellwert mit zugehörigem Status pro Zyklus mit dem Automatisierungsgerät/-system austauschen, beträgt die Übertragungszeit ca. 10 ms. Im Beispiel der Abb. 2 beträgt die Zykluszeit 4 x 10 ms + 10 ms = 50 ms.

Bei Einsatz des DP/PA-Link werden die zyklischen und azyklischen Daten in je einem Telegramm über den PROFIBUS-DP an das Automatisierungsgerät/-system weitergeleitet. Aufgrund der hohen Datenübertragungsrate von bis zu 12 MBit/s treten dabei nur unwesentliche Verzögerungen in der Datenübertragung auf (selbst bei 30 Feldgeräten pro DP/PA-Link nur max. 1 ms).

Das DP/PA-Link hat bis zu der max. anschließbaren Anzahl von Feldgeräten (30 Feldgeräte pro DP/PA-Link) das gleiche Zeitverhalten wie der DP/PA-Koppler. Entscheidende Vorteile ergeben sich bei Strukturen, bei denen die Feldgeräte auf mehrere DP/PA-Links aufgeteilt sind. Bei einer Übertragungsgeschwindigkeit von 12 MBit/s am übergeordneten PROFIBUS-DP treten lediglich Verzögerungen im Bereich von max. 1 ms auf, so daß die Zykluszeit von der Anzahl der Feldgeräte nahezu unabhängig bleibt. Bei 10 Feldgeräten pro DP/PA-Link beträgt die Zykluszeit ca. 100 ms, bei 30 Feldgeräten pro DP/PA-Link ca. 300 ms.

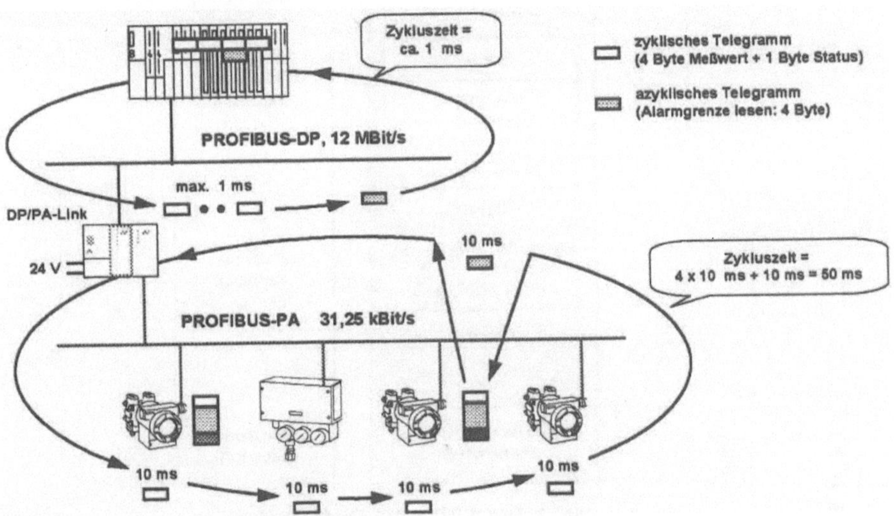

Abb. 2. Buszyklus mit DP/PA-Link

4 Mit den PA-Profilen ist die Interoperabilität gegeben

Die Parameter in einem Feldgerät können in 3 Gruppen eingeteilt werden (Abb. 3):
1. Prozeßparameter: Meßwert bzw. Stellwert und zugehöriger Status
2. Betriebsparameter: Meßbereich, Filterzeit, Alarmparameter (Meldung, Alarm- und Warngrenzen), Standardparameter (Meßstellenkennzeichen, TAG)
3. Herstellerspezifische Parameter: z.B. spezielle Diagnoseinformationen

Die Parameter der 1. Gruppe werden zyklisch oder azyklisch vom Automatisierungsgerät/-system gelesen bzw. geschrieben. Die Parameter Meßwert und Status sind in allen messenden Feldgeräten, die Parameter Stellwert und Status in allen stellenden Feldgeräten vorhanden und einheitlich kodiert (z.B. Meß-/Stellwert in 4 Byte IEEE-Format).

Die Parameter der 2. Gruppe können bei Bedarf vom Automatisierungsgerät/-system azyklisch gelesen bzw. geschrieben werden. Ein Teil dieser Parameter wird über die Funktionsbausteine im Automatisierungsgerät/-system mit den Feldgeräten ausgetauscht, um den Zugriff der Bedien- und Beobachtungssysteme zu ermöglichen (z.B. Visualisierung einer Alarmüberschreitung).

Die Parameter, d.h. die damit verbundenen Feldgeräte-Funktionen der 1. und 2. Gruppe sind im „PROFIBUS-PA Profile for Process Control Devices" der PNO festgelegt. Diese Feldgeräte-Funktionen sind teilweise Pflicht, teilweise wahlfrei. Sind optionale Funktionen im Feldgerät realisiert, müssen diese der Beschreibung nach dem PA-Profil entsprechen.

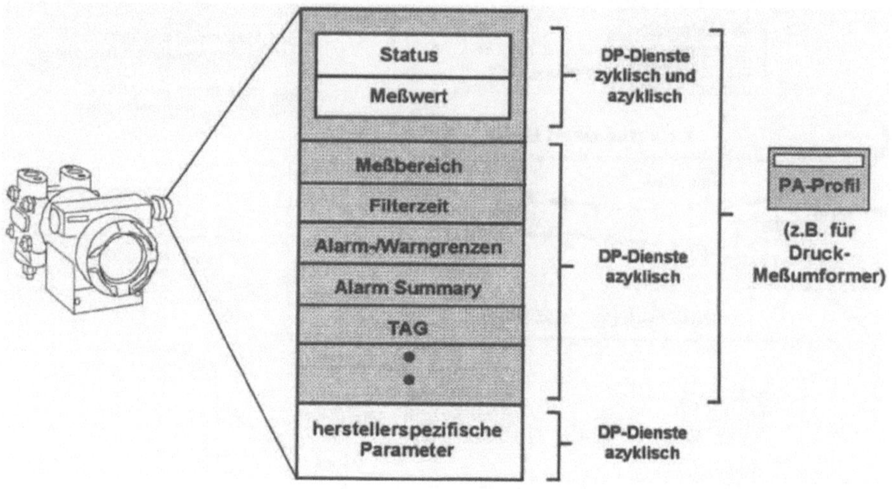

Abb. 3. PROFIBUS-PA-Profile

Die Parameter der 3. Gruppe sind herstellerspezifisch. Der azyklische Zugriff erfolgt in der Regel mit einem Personal Computer zu Diagnose- und Wartungszwecken. In Ausnahmefällen werden bestimmte Parameter aus dieser Gruppe auch vom Automatisierungsgerät/-system gelesen bzw. geschrieben.

Unter dem Begriff Interoperabilität versteht man das Zusammenspiel von Komponenten (hier: Leitsysteme und Feldgeräten) verschiedener Hersteller an einem offenen Bussystem auf Basis einer herstellerübergreifenden Festlegung der Geräte- und Kommunikationsfunktionen [2, 4]. Automatisierungsgeräte/-systeme und PC's unterschiedlicher Hersteller können über PROFIBUS-PA die im PA-Profil festgelegten Parameter aus allen Feldgeräten lesen/schreiben und damit die im PA-Profil definierten Feldgeräte-Funktionen beeinflussen. Die Austauschbarkeit der Feldgeräte unter Beibehaltung der Funktionalität („Interchangeability") ist gegeben, wenn die Funktionen nach dem PA-Profil genutzt werden.

5 PROFIBUS-PA ist integraler Bestandteil des Leitsystems

PROFIBUS-PA ist integraler Bestandteil des Prozeßleitsystems (Abb. 4). Die volle Integration beinhaltet u.a. die Parametrierung des DP/PA-Link durch die prozeßnahe Komponente (PNK), Projektierung der Stellvertreter-Bausteine sowie Feldgeräte-Projektierung durch das zentrale Engineeringsystem ES. Das Leitsystems bietet insbesondere die optimale Kommunikation über alle Leitsystemhierarchien und das zentrale Engineering für alle Systemkomponenten.

Die Kommunikation zwischen PNK und den Feldgeräten am PROFIBUS-PA erfolgt über sog. Stellvertreter-Bausteine in der CPU. Die Ein- und Ausgänge der Stellvertreter-Bausteine stehen „stellvertretend" für die Feldgerätedaten in der CPU, um den Zugriff des Bedien- und Beobachtungssystems OS über den Anlagenbus zu ermöglichen. Die Projektierung der Stellvertreter-Bausteine erfolgt über das zentrale Engineeringsystem ES. Dort werden auch die Konfigurationsdaten für das DP/PA-Link und die Feldgeräte erzeugt, die im Anlauf zuerst in das DP/PA-Link und anschließend in die Feldgeräte geladen werden.

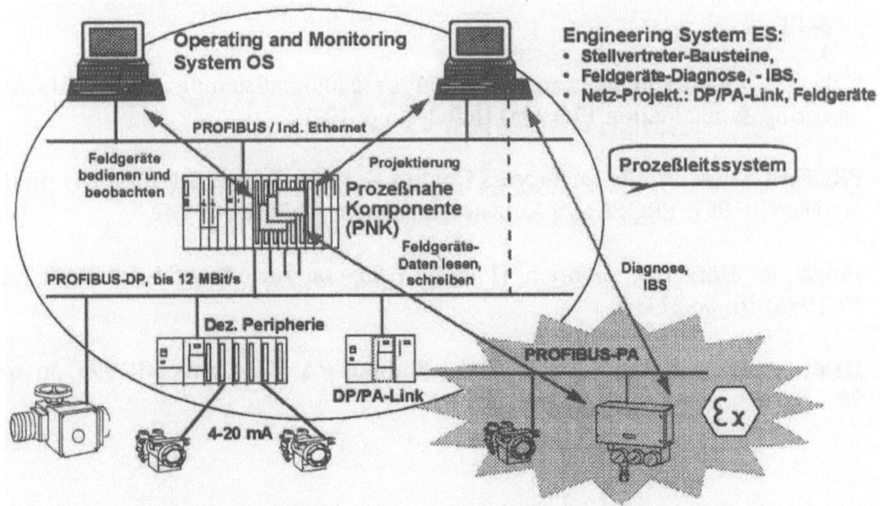

Abb. 4. PROFIBUS-PA als integraler Bestandteil des Leitsystems

Die Inbetriebsetzung sowie Diagnose der Feldgeräte mit Hilfe der Parametriersoftware kann bei Bedarf über das zentrale ES oder ein PC/PG am PROFIBUS-DP erfolgen (gestrichelte Linie zwischen ES und PROFIBUS-DP). Beim zentralen Feldgeräte-Engineering werden die Telegramme zwischen ES und Feldgeräten über einen transparenten Kanal in der CPU der PNK (am Anwenderprogramm vorbei) ausgetauscht.

Pro CPU können mehrere DP-Stränge für den Zugang zur Feldebene genutzt werden. Da alle Feldkomponenten, wie z.B. Antriebe, dezentrale Peripherie oder

DP/PA-Link gemeinsam am PROFIBUS-DP angeschlossen werden können, ist eine Anpassung der Feldinstrumentierung an die örtlichen Gegebenheiten der Anlage leicht möglich, z.B. ein DP-Strang pro Teilanlage.

6 Ausblick

Mit PROFIBUS-PA erfolgt die Integration der Feldgerätetechnik in die Prozeßleitsysteme. Dabei werden insbesondere Eigensicherheit (Zündschutzart Eexi), Feldgerätespeisung über die Datenleitung, Fernparametrierung sowie vorbeugende Wartung und Diagnose berücksichtigt. Bei der Entwicklung von neuen Prozeßleitsystemen werden die Anforderungen
- zentrales Engineering (Parametrierung und Diagnose)
- Verlagern der Automatisierungsfunktionen (Intelligenz) in die Feldgeräte
- Einbindung in vorhandene Feldbuslösungen (z.B. PROFIBUS-DP)
von vornherein berücksichtigt.

Literatur

1. Kobes, P.: Offener Feldbus auch für die Prozeßautomatisierung: PROFIBUS. engineering & automation 17(1995) Heft 3-4, pp. 10-12

2. PROFIBUS-PA. Profile for Process Control Devices. Second Draft Profile. Edition: March 1997. PROFIBUS Nutzerorganisation, Order No. 3.042

3. Heidel, R., Hörger, J., Döbrich, U.: Vom Feldgerät zur offenen Feldtechnik. atp 37(1995) 10, pp.21-33

4. Diedrich, C., Preuss, A.: Das Profil des Profibus-PA. Elektronik 20/1995, pp. 68-78

Das CANopen Antriebsprofil

G. Beckmann, S. Wagner, J.-U. Varchmin

Technische Universität Braunschweig
Institut für Elektrische Meßtechnik und Grundlagen der Elektrotechnik (emg)
Hans-Sommer-Str. 66, 38106 Braunschweig
Tel. +49 (0)531 / 391-3859, Fax +49 (0)531 / 391 - 5768

Abstract: Parallel zu laufenden Standardisierungsbemühungen im Bereich der Programmierumgebungen für Industriesteuerungen (IEC 1131) erstellen Arbeitsgruppen innerhalb der Anwendervereinigung CAN in Automation (CiA) u.a. auf der Basis des CANopen Kommunikationsprofils verschiedene Geräteprofile für CAN-vernetzte Systeme. Dadurch stehen dem Anwender standardisierte Funktionen für bestimmte Geräteklassen zur Verfügung. Dieser Beitrag gibt einen kurzen Überblick über die Aufgaben und die Einordnung eines Profiles innerhalb des Kommunikationssystems und stellt anschließend die Funktionen und Betriebsarten des Geräteprofils für drehzahlveränderliche Antriebe vor.

Abstract: Beside the efforts to standardize the programming environment for industrial controlsystems (IEC1131), some work groups inside the international users and manufactures group CAN in Automation (CiA) define different device profiles for CAN-connected systems, based on the communication profile CANopen. These profiles offer standardized functions and services for certain classes of devices. This presentation gives a general idea of the jobs and the functions of a device profile inside the communication system and introduce afterwards the functionality and modes of operation of the profile for drives and motion control.

1 Einleitung

Die weite Verbreitung des CAN-Busses, sowohl in der industriellen Automatisierungstechnik und der Gebäude-Installationstechnik, als auch im Kraftfahrzeug, ruft immer wieder die Diskussion um eine geeignete Protokollschicht auf. Wichtig ist ein einheitlicher Zugang zu den Funktionalitäten eines Gerätes oder einer Gerätefamilie durch die Dienste und Parameter der Anwendungsschicht.

Innerhalb der Automatisierungstechnik stellt das Kommunikationsprofil CANopen, das von der Anwendervereinigung CiA zur Verfügung gestellt wird, dem Anwender Funktionen zum Aufbau eines CAN-vernetzten Systems zur Verfügung. Um auch gerätespezifische Funktionen einheitlich nutzen zu können, werden in Arbeitskreisen unterschiedliche Geräteprofile erarbeitet, die sich auf das Kommunkationsprofil stützen.

1.1 Aufgaben eines Geräteprofils

Hauptaufgabe eines Geräteprofils ist es, die kommunkationsrelevanten Funktionen und Parameter einer Gerätegruppe zu beschreiben und sie einem Kommunikationssystem in standardisierter Form zur Verfügung zu stellen. Dabei wird sich das Geräteprofil der Dienste eines unterlagerten Kommunikationsprofils bedienen, wie Fig. 1 zeigt.

Ein Geräteprofil soll die Eigenschaften von Geräten verschiedener Hersteller für ein begrenztes Anwendungsgebiet umfassen. Da jeder Hersteller seine Geräte mit spezifischen Eigenschaften ausstattet, muß es eine Schnittmenge der vorhandenen Vielfalt bilden. Durch ein definiertes Kommunikationsverhalten bei Fehlern oder z.B. Parameterabfragen entsteht eine einheitliche Schnittstelle zur Umgebung eines Gerätes. Die Menge der Parameter, die ein bestimmtes Gerät aus dem Vorrat aller vorhandenen Parameter unterstützt, muß einem Netzwerk und den angeschlossenen Teilnehmern mitgeteilt werden können oder abfragbar sein.

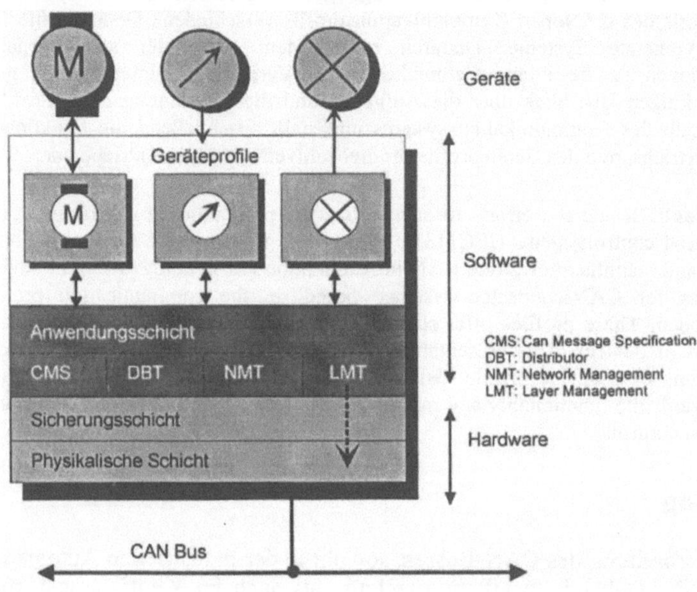

Fig. 1 Einordnung eines Geräteprofils

Die Realisierung von Gerätefunktionen geschieht über Zustandsmaschinen, die z.B. Betriebsarten oder -zuständen entsprechen können. Durch Kontroll- und Statusworte wird der aktuelle Zustand umgeschaltet oder abgefragt. Im wesentlichen ergibt sich so eine umfassende Parameterliste, welche die Eigenschaften eines Gerätes beschreibt. Die Einheitlichkeit der Schnittstelle wird über einen Standard-Gerätetyp mit Standard-Parametern erzielt, der von allen Herstellern unterstützt werden muß; herstellerspezifische Besonderheiten eines Gerätes werden durch optionale Parameter erfaßt, die aber kommunikationstechnisch trotzdem in definierter Art und Weise behandelt werden. So bleiben technische Leistungsmerkmale prinzipiell erhalten, und für Standardaufgaben entsteht eine gewisse Austauschbarkeit von Geräten unterschiedlicher Hersteller. Um zu optimalen Lösungen für spezifische Aufgabenstellungen zu kommen, wird es Kombinationen aus Geräten und zugeordneten Steuerungen bestimmter Hersteller geben, die auch die besonderen Geräteeigenschaften und -parameter ausnutzen.

Die Eingliederung in das CANopen Protokoll geschieht über Einträge in das Objektverzeichnis, an den dafür vorgesehenen Stellen:

Tabelle 1 Das CANopen Objektverzeichnis

Index		Objekt
0000		nicht belegt
0001	- 001F	statische Daten-Typen
0020	- 003F	komplexe Daten-Typen
0040	- 005F	herstellerspezifische Daten-Typen
0060	- 007F	statische geräteprofilspez. Daten-Typen
0080	- 009F	komplexe geräteprofilspez. Daten-Typen
00A0	- 00FF	reserviert
1000	- 1FFF	Kommunikationsprofil
2000	- 5FFF	herstellerspezifische Profile
6000	- 9FFF	standardisierte Geräteprofile
A000	- FFFF	reserviert

Für Funktionen, die weder vom Kommunikationsprofil noch vom Geräteprofil abgedeckt werden, stehen im Bereich von 2000h - 5FFFh herstellerspezifische Objekte zur Verfügung.

1.2 Aufteilung der Geräte in verschiedene Klassen

In einem Automatisierungssystem kommen verschiedene Geräte mit völlig unterschiedlichen Funktionalitäten zur Anwendung. Ein digitales E/A-Modul benötigt beispielsweise keine Dienste zur Übergabe eines neuen Sollwertes, wie er aber von einem Antrieb verlangt wird. Deshalb stehen Geräteklassen zur Verfügung, in die ein Gerät eingeordnet werden kann und innerhalb derer ein Geräteprofil mit spezifischen Funktionen und Betriebsarten definiert ist. Die Tabelle 2 zeigt bereits vorhandene Klassen und den Stand der Geräteprofile.

Tabelle 2 Geräteprofile in Bearbeitung oder bereits veröffentlicht

Geräteklasse	aktueller Stand		Bezeichnung
E/A-Module	DSP	Version 1.4	CiA DSP-401
Antriebe	DSP	Version 1.0	CiA DSP-402
Encoder	DSP	Version 1.0	CiA DSP-406
Regler und Meßtechnik	WD	nicht veröffentlicht	CiA WD-404
IEC 1131	WD	nicht veröffentlicht	CiA WD-405
Nahverkehr	WD	nicht veröffentlicht	CiA WD-407

DSP: Draft Standard Proposal, veröffentlichter Spezifikationsentwurf.
WD: Work Draft, CiA-internes Arbeitspapier.

2 Das Geräteprofil für drehzahlveränderliche Antriebe

Das hier beschriebene Geräteprofil behandelt Schrittmotoren, Frequenzumrichter und Drehstromservoantriebe.

Von der Profibus Nutzerorganisation (PNO) wird an der Realisierung eines Geräteprofils gearbeitet, das ebenfalls auf dem Mechanismus des Objektverzeichnisses
beruht. Eine Reihe von Herstellern aus dem Bereich der Antriebstechnik hat bereits auf
Basis des Interbus S das Geräteprofil der DRIVECOM realisiert. Aus dieser Situation
heraus wurde entschieden, das Geräteprofil für CAN an die DRIVECOM-Spezifikati-
onen anzulehnen. Dadurch ergibt sich für den Anwender der Vorteil eines einheitlichen
Zugangs zu den Funktionalitäten eines Gerätes unabhängig vom unterlagerten
Kommunikationssystem.

2.1 Die Zustandsmaschine

Das Verhalten und der momentane Status einer Antriebseinheit werden durch eine
Zustandsmaschine bestimmt, deren Übergänge durch das *<controlword>* oder
geräteinterne Ereignisse ausgelöst werden.

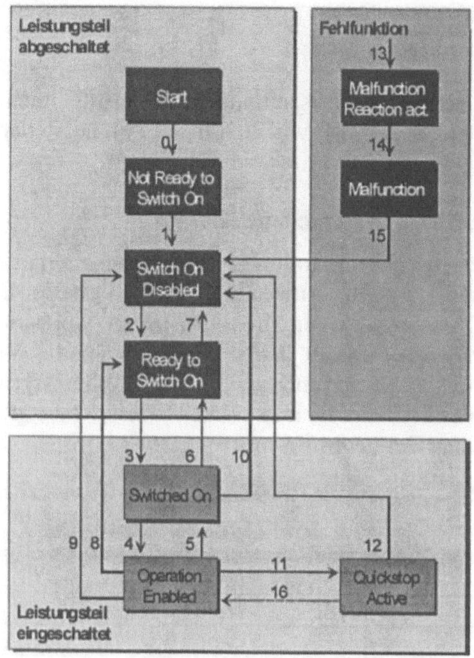

Fig. 2   Zustandsmaschine

Unterschieden werden drei Hauptzustände:

* die Initialisierung

Nach der Initialisierung und Anmeldung eines Gerätes in einem CANopen-Netzwerk
gemäß dem Kommunikationsprofil, müssen die Antriebsfunktionen zunächst
parametriert werden.

- der Betriebszustand

Der Übergang von *<Ready to Switch On>* zu *<Switched On>* ist gleichzusetzen mit dem Einschalten des Leistungsteils. Das erfordert allein aufgrund der damit verbundenen Gefährdungspotentiale ein Kommando von einem angeschlossenen Hostcontroller. Jetzt können Antriebsparameter verändert werden; diese werden allerdings erst im Zustand *<Operation Enabled>* an den Motor weitergegeben und führen damit zu Achsbewegungen. Der Zustand *<Quick Stop Active>* kann optional als stabiler Zustand eingerichtet werden. Dadurch besteht die Möglichkeit, ohne Abschalten des Leistungsteils und erneute Initialisierung des Gerätes einen Schnellhalt auszuführen und anschließend direkt in den normalen Betriebszustand zurückzukehren.

- Fehlerzustände

Ein Fehlerzustand kann zum einen aus jedem anderen Zustand heraus erreicht werden und kennzeichnet zum anderen eindeutig eine Fehlersituation. Das Verlassen des Funktionsblockes zur Fehlerbehandlung ist außerdem immer mit einer erneuten Initialisierung des Gerätes über die Zustände *<Switch On Disabled>* und *<Ready to Switch On>* verbunden.

2.2 Die Betriebsarten

Entsprechend der Gerätekategorie "drehzahlveränderliche Antriebe" ergeben sich die zu unterstützenden Betriebsarten und Geräteparameter. Die große Zahl relevanter Parameter verbietet an dieser Stelle eine Aufzählung, die im Geräteprofil nachgelesen werden kann. Wichtiger sind die vorgesehenen Betriebsarten und generellen Mechanismen des Geräteprofils auf die jetzt näher eingegangen wird.

Tabelle 3 Betriebsarten des Geräteprofils für drehzahlveränderliche Antriebe

Betriebsart	Funktion
Homing Mode	Referenzfahrt
Profile Position Mode	Positionierung der Achsen
Interpolated Mode	Interpolationsfunktionen
Profile Velocity Mode	Geschwindigkeitsprofile
Velocity Mode	Geschwindigkeitsregelung mit Frequenzumrichtern
Profile Torque Mode	Momentenregelung

Zum Auffinden einer Referenzposition bei Antriebssystemen ohne absolute Lagegeber existiert die Betriebsart *<Homing Mode>*. Es werden verschiedene Methoden vom Profil unterstützt, um den existierenden unterschiedlichen konstruktiven Lösungen dieses Problems in Form von Refernzfahrtschaltern, Endschaltern sowie Indexpulsen zu entsprechen. Neben der Auswahl der Referenzfahrtmethode ist die Einstellung einer entsprechenden Geschwindigkeit sowie zulässiger Beschleunigungen während der Referenzfahrt möglich.

Der *<Profile Position Mode>* behandelt die Positionierung von Achsen. Neben einem Fahrprofil zum Erreichen einer vorgegebenen Lage können die dazu gehörenden Geschwindigkeiten und Beschleunigungen gewählt werden. Die Übertragung von Sollwerten wird in einem 'handshake'-Verfahren zwischen einem Hostcontroller und dem Antriebscontroller ermöglicht. Es sind zwei Varianten vorgesehen:

- die Einzelwertvorgabe: Nachdem ein Sollwert erreicht ist, signalisiert die Antriebseinheit diesen Zustand über ein Bit im Statuswort und erhält anschließend einen neuen Sollwert vom Hostcontroller, was durch Setzen eines Bits im Kontrollwort angezeigt wird.

- die Satzvorgabe: Nachdem ein Sollwert erreicht wurde, steuert die Antriebseinheit unmittelbar den nächsten an.

Der *<Interpolated Position Mode>* beschreibt sowohl die zeitliche Interpolation von Sollwerten einer Achse als auch die räumliche Interpolation mehrerer Achsen. Für Interpolationsdaten sind angepaßte Datenpuffer einzurichten und zu verwalten, wofür geeignete Kommunikationsobjekte zur Verfügung stehen. Es wird fast immer ein Interpolationsdatenpuffer benötigt, der als Ring oder als FIFO angelegt sein könnte. Die Interpolationsalgorithmen sind herstellerspezifisch und daher nicht im Geräteprofil enthalten. Für die Kommunikationsschicht eines Antriebssystems ist es unerheblich, welche Form der Interpolation implementiert oder benutzt wird. Neben den beiden genannten Arten sind prinzipiell weitere Formen der verkoppelten Sollwertbearbeitung möglich. Für jede Form der Interpolation sind aber einige Grundelemente erforderlich, die im Profil definiert wurden.

Für die Interaktion mehrerer Achsen ist die Synchronisation der beteiligten Netzwerkkomponenten von großer Bedeutung. Hier werden die vom Kommunikationsprofil vorgesehenen Methoden angewendet (Sync-Object) oder ein Prozeßdatenobjekt, daß als Broadcast-Telegramm angelegt ist und über eine im Telegramm enthaltene Gruppennummer applikationsspezifische Gruppierungen von Antriebseinheiten gestattet.

Zusammen stellen die drei bisher beschriebenen Modi einen lageorientierten Betrieb einer Antriebseinheit dar. Daher sieht das Profil funktional einen gemeinsamen Lageregler für diese drei Betriebsarten vor. Die genaue Struktur der implementierten Regler ist wiederum sehr herstellerspezifisch und wird durch das Geräteprofil nicht weiter behandelt.

Neben der Lageregelung einer Achse ist ein Verfahren mit Geschwindigkeitsprofilen ein häufiger Anwendungsfall weshalb zwei Betriebsarten *<Profile Velocity Mode>* und *<Velocity Mode>* unterstützt werden. Der *<Velocity Mode>* ist verstärkt auf die Anwendung in einfachen und kostengünstigen Frequenzumrichtern zugeschnitten. Die Mehrzahl der Parameter ist auf 16 Bit begrenzt und gestattet die Nutzung einfacher Prozessoren in den Antriebscontrollern. Der *<Profile Velocity Mode>* ist stärker mit den anderen möglichen Betriebsarten verflochten und wirkt gegebenenfalls als zusätzliche Störgröße auf die Reglerstrukturen der anderen Betriebsarten (z.B. *<Profile Torque Mode>*) ein.

Zur Erzeugung von Profilformen aus vorgegebenen Sollwerten gibt es einen Profilgenerator (*<Trajectory Generator>*), der für alle möglichen Betriebsarten identisch sein kann. Der Algorithmus zur Berechnung von Stützstellen gemäß einer vorgewählten Kurvenform (linear, $\sin^2$, Splines) ist prinzipiell unabhängig davon, ob es sich um Stützstellen für die Lage- oder für die Geschwindigkeitsregelung handelt.

2.3 Zustandskontrolle und -überwachung

Die Zustandsübergänge der Fig. 2 werden überwiegend durch Veränderung des *<controlword>* initiiert. Das *<controlword>* nach Fig. 3 ist standardmäßig im ersten Datenwort in den vordefinierten Empfangskommunikationsobjekten (aus Sicht der Antriebe) für schnellen zyklischen Datenaustausch vorhanden (Process Data Object Receive PDOR). Umgekehrt wird das *<statusword>* zusammen mit der aktuell eingestellten Betriebsart *<modes_of_operation_display>* in das erste Datenwort der Default-Sendeobjekte gemappt (PDO Transmit PDOT). Bei Geräten, die ein remapping von PDO`s unterstützen, kann dies selbstverständlich mit SDO-Zugriffen (Service Data Object) geändert werden.

controlword (Index 6040h)

Bit	
0	Switch On
1	Disable Voltage
2	Quick Stop
3	Enable Operation
4	Operation Mode Specific
5	Operation Mode Specific
6	Operation Mode Specific
7	Reset Fault
8	Halt
9	Reserved
10	Reserved
11	Manufacturer Specific
12	Manufacturer Specific
13	Manufacturer Specific
14	Manufacturer Specific
15	Manufacturer Specific

statusword (Index 6041)

Bit	
0	Ready to Switch On
1	Switched On
2	Operation enabled
3	Fault
4	Voltage Disabled
5	Quick Stop
6	Switch On Disabled
7	Warning
8	Manufacturer Specific
9	Remote
10	Setpoint Reached
11	Limit Value
12	Operation Mode Specific
13	Operation Mode Specific
14	Manufacturer Specfic
15	Manufacturer Specfic

Velocity Mode	Profile Position Mode	Profile Velocity Mode	Profile Torque Mode	Homing Mode	Interpolated Position Mode
RFG Disable	New Setpoint	reserved	reserved	Start homing	Enable IP
RFG Stop	change_set_i	reserved	reserved	reserved	reserved
RFG zero	abs_rel	reserved	reserved	reserved	reserved

controlword

Velocity Mode	Profile Position Mode	Profile Velocity Mode	Profile Torque Mode	Homing Mode	Interpolated Position Mode
reserved	setpoint_ack.	Speed=0	reserved	Homing att.	IP-Mode act
reserved	following_err	reserved	reserved	Homing error	reserved

statusword

Fig. 3 Das Kontroll- und das Statuswort

Die Bit 4 bis 6 des Steuerwortes und die Bits 12 und 13 des Statuswortes haben betriebsartenabhängige Bedeutungen. Die anderen Bit sind zwingend zu implementieren, um das Geräteprofil zu erfüllen. Für herstellerspezifische Anpassungen sind im *<controlword>* fünf und im *<statusword>* 3 Bit vorgesehen. Die Inhalte des *<statusword>* spiegeln den aktuellen Zustand der Antriebseinheit zum Zeitpunkt der Telegrammgenerierung wieder und sind nicht gepuffert.

2.4 Multi-Achsmodule

Um Geräteherstellern die Integration von mehrachsigen Antriebssystemen zu erleichtern, wurde der Indexbereich des Geräteprofils im Objektverzeichnis in einen Basisbereich für Einachsmodule und in bis zu sieben Offsetbereiche mit einem Abstand von 800h eingeteilt. Damit sind die Kommunikationsobjekte der einzelnen Achsen lediglich durch einen Offset unterschieden. Mit Hilfe der höherwertigen Bit des Index erfolgt die Auswahl der anzusprechenden Achse.

3 Ausblick

Geräteprofile bieten Geräteherstellern den Vorteil, die von ihnen am Markt angebotene Komponente im Zusammenspiel mit verschiedenen Steuerungen und mit weiteren Peripheriebaugruppen einfacher vertreiben zu können. Steuerungsherstellern fällt es leichter, einheitliche und durchgängige Bedien- und Programmieroberflächen für die Geräte unterschiedlicher Hersteller zu erstellen, wenn Grundfunktionen der Geräte identisch sind. Außerdem muß die Prozeßschnittstelle nicht für jedes Gerät individuell angepaßt werden. Anwender erhalten die Möglichkeit, unterschiedliche Komponenten mehrerer Hersteller zu funktional optimierten Gesamtsystemen zusammenzustellen.

Das darf aber nicht mit dem idealisierten Ziel verwechselt werden, echte Austauschbarkeit einer Komponente des Herstellers "A" gegen die von Hersteller "B" zu erreichen. Für Basiseigenschaften der untersten Kategorie läßt sich so etwas erzielen; Systeme, die eine anspruchsvolle Automatisierungsaufgabe lösen müssen, benötigen immer einige spezielle Produkteigenschaften, die sicherlich nicht in allen vergleichbaren Geräten einer Gruppe verfügbar sein werden. So wird es zukünftig Kombinationen am Markt befindlicher Geräte und darauf abgestimmter Steuerungen bzw. umgekehrt geben, die auch ihre Spezialeigenschaften gegenseitig ausnutzen.

Mit Hilfe des Kommunikationsprofiles CANopen war es möglich auf der HMI '97 eine Multi-Vendor-Anlage aufzustellen, die Komponenten verschiedener Hersteller zu einem System vereint und über den CAN-Bus miteinander vernetzt. Die Basis für eine Interoperabilität ist damit hergestellt.

Literatur

1.   CAN Application Layer Spezifikation (CiA DS-201 bis DS-207), Version 1.1, Erlangen, 02/96

2.   CANopen Communcation Profile (CiA DS-301), Version 3.0, Erlangen, 10/96

3.   *Zeltwanger, H., Förster, J.*, Das offene CAN-Protokoll, Elektronik 10/1997

4.   CANopen Device Profile for I/O Modules (Cia DSP-401), Version 1.4, Erlangen, 12/96

5.   CANopen Device Profile for Drives and Motion Control (CiA DSP-402), Version 1.0, Erlangen, 05/97

6.   CANopen Device Profile for Encoder (CiA DSP-406), Version 1.0, Erlangen, 05/97

Österreichs modernste Walzwerkstechnologie setzt auf Profibus

Siegmund Scherney

Siemens AG Österreich, Automatisierungstechnik
Wolfgang-Pauli-Straße 2, 4020 Linz

Abstract. Im Jahre 1996 modernisierte die Firma VOEST ALPINE Stahl LINZ GmbH ihr Kaltwalzwerk 2 von Grund auf. Neben umfangreichen mechanischen Änderungen wurde die gesamte Automatisierungs- ebene erneuert. Als Feldbussystem wurde dafür PROFIBUS und PROFIBUS-DP ausgewählt. Vor allem durch die Durchgängigkeit und die hohe Flexibilität, sowie die optimale Einbindung in die verschiedenen Hardwarekonzepte konnten die Anforderungen dieser komplexen Aufgabenstellung bestmöglich realisiert werden.

Abstract. In 1996 Voest Alpine Stahl Linz GmbH has entirely modernized its cold rolling-mill KWW2. Besides extensive mechanical changes the complete automation level has been modernized. PROFIBUS-DP was chosen as a fieldbus system to manage the communication tasks. Most crucial for going with PROFIBUS was its high transparency and flexibility as well as the ability of an optimum integration of different hardware concepts. These features complied ideally with the tough requirements for Voests complex tasks.

1 Einleitung

In den vergangenen Jahren haben sich die Qualitätsanforderungen an die Endprodukte von Walzwerken ständig erhöht. Dementsprechend sind die Anforderungen an die Walztechnik gestiegen. Zielsetzung der Modernisierung des Kaltwalzwerkes 2 der VOEST ALPINE Stahl LINZ GmbH war die Abbildung der in den beiden bestehenden Kaltwalzwerken installierten Kapazität in *einem* Werk - dem modernisierten KWW2. Dies auf Basis des Einsatzes modernster Technologie. Realisiert wurde eine sogenannte Beize- Tandem- Kopplung, was bedeutet, daß das Walzgut kontinuierlich durch die Beizanlage und unmittelbar daran durch die Tandemstraße läuft. Der Umbau wurde in der Rekordzeit von 43 Tagen im Spätsommer 1996 durchgeführt und umfaßte folgende Arbeiten:
- Abbruch der Beizanlage
- Demontage der mechanischen Einrichtungen am Ein- und Auslauf der Tandemstraße
- Errichten einer neuen Flachtankturbulenz Beizanlage
- Einbau der neuen Mechanik für die Kopplung zur Tandemstraße
- Neuinstallation aller Versorgungssysteme wie Hydraulik etc.

- Erneuerung der gesamten Elektrik
wie Energieversorgung, Motoren, Antriebsspeisungen usw.
- Kompletter Tausch der Basisautomation samt Prozessrechner für die gesamte
gekoppelte Anlage
Die beeindruckende Dimension des Gesamtprojektes wird durch nachstehende technische Daten verdeutlicht.

Beize

Bandbreite	600 - 1640 mm
Banddicke	1,5 bis 6 mm
Einlaufgeschwindigkeit	max. 600 mm/min
Einfädelgeschwindigkeit	60 m/min
Bundgewicht	max. 30 to

Tandem

Einlaufgeschwindigkeit vor Gerüst 1	380 m/min
Haspelgeschwindigkeit	1200 m/min
Gesamtanschlußleistung	ca. 60 MVA
Leistung Hauptantriebe	49 MVA
Verbaute Fläche der Gesamtanlage:	mehr als 37.000 m2

Bild 1. Gesamtansicht

2 Automatisierungskonzept

2.1 Basisautomatisierung

Das reibungslose Zusammenspiel aller Steuer- und Regelsysteme ist natürlich Grundvoraussetzung für eine erfolgreiche Automatisierung des Gesamtprozesses und deshalb das ausgewählte Feldbussystem von grundlegender Bedeutung. Moderne Automatisierungskonzepte strukturieren den Gesamtprozess unter Beachtung der technologischen Anlagengegebenheiten. Dabei kommt es zur Funktionsaufteilung in verschiedene Automatisierungsebenen wie Feldebene, Gruppenleitebene und Prozessleitebene. Das wesentlichste Merkmal der einzelnen Ebenen ist heutzutage der in sich geschlossene Funktionsumfang. Das bedeutet, daß nach Installation jeder einzelnen Ebene eine voll funktionsfähige Anlage zur Verfügung steht. In solch komplexen Systemen wie einem Walzwerk wird als Mindestausbaugrad die Feld- und Gruppenleitebene angesehen, da mit der Feldebene allein nur Tippbetrieb möglich ist. Deshalb muß das Feldbussystem die Anforderungen der Feld- und der Gruppenleitebene abdecken.

2.2 Feldbus heißt mehr als dezentrale I/O

In der Feldebene werden alle Stellgliedregelungen und Steuerungen der einzelnen Funktionseinheiten zusammenfaßt.

Funktionseinheiten sind z.B.
- Einlaufteil, Auslaufteil, Haspeln
- Bandschweißmaschine
- Bandscheren, Besäumschere
- Antriebseinheiten
- Hilfsaggregate

Stellgliedregelungen sind z.B.
- Regelung für die Hydraulikaggregate
- Regelung des Beizprozesses
- Druck-Momenten-Zugregelung

Steuerungen sind z.B.
- mechanische Einzelbewegungen
- Ablaufautomatiken im Ein- und Auslaufteil der Straße

Jede Funktionseinheit der Feldebene wird als autarkes Gerät betrachtet, das zunächst als einzelnes mit einem zugeordneten Bediengerät oder Operator Panel betriebsbereit gemacht werden kann. Diese betriebsbereiten Funktionseinheiten melden sich bei der Gruppenleitebene an, die dann koordiniert Sollwerte und Steuerbefehle vorgibt. Ebenso werden alle übergeordneten Regelungen und Steuerungen sowie die Visualisierung in der Gruppenleitebene zusammengefaßt, z.B.

- Dickenregelung
- Leitsollwertvorgaben
- Bandverfolgung
- Betriebsartensteuerung
- Sollwertsteuerung

Mit den Funktionen der Feld- und Gruppenleitebene ist bereits ein Automatikbetrieb möglich, der sich, für ein Band betrachtet, vom Prozessrechnerbetrieb nur dadurch unterscheidet, daß die erforderlichen Sollwerte vom Bedienungsmann per Hand vorgegeben werden.

Diesem hierarchisch aufgebauten Konzept der Feld-, Gruppen- und Prozessebene stehen auf der Steuerungs- und Leittechnik dementsprechende Hard- und Softwaremodule gegenüber. Um eine eindeutige und klare Strukturierung der Gesamtanlage zu erreichen, wurde konsequent eine dezentrale Automatisierungslösung verwirklicht.

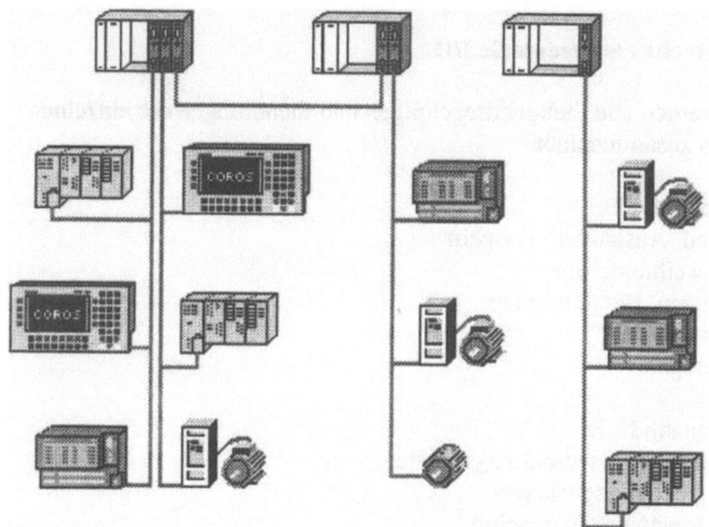

Bild 2. Dezentrale Automatisierungskomponenten KWW2

Als Steuerungssysteme sind Automatisierungsgeräte SIMATIC S5-155U mit jeweils einer oder mehreren Zentralbaugruppen CPU 948 eingesetzt,
zur Regelung hochdynamische Regelsysteme SIMADYN D.

Insgesamt sind an 30 PROFIBUS DP Masteranschaltungen
mehr als 650 Peripheriestationen angeschlossen, und zwar ca.

115   Kompakte Peripheriestationen ET 200B in Schaltschränken und Verteilern
300   Vor-Ort Kästen IP65 mit eingebauter Stromversorgung, Optischem Link
       Modul OLM, und Modularer Peripherie ET200U.

50   Bedienstationen , Anzeigen in Pulten
10   HMI, (Human Man Interface , COROS)
80   AC und DC Drives
80   Encoder

Verbunden sind alle Feldbusteilnehmer mit ungefähr 12 km Lichtwellenleiterkabel und 8 km elektrischer Profibus Leitung.

2.3 Maßgeschneiderte Vor-Ort Stationen

Neben kompakten Peripheriemodulen sind einzelne dezentrale Peripheriestationen als modulares System ausgeführt, wobei einerseits sehr einfach optimierte Standardkonfigurationen, und andrerseits aus einem Baugruppenspektrum von mehreren hundert Peripheriebaugruppen ausgewählt, 100%ig maßgeschneiderte Stationen realisiert wurden.

Durch die Festlegung auf einige wenige Standardperipheriestationen und damit verbunden die Erstellung von darauf abgestimmten Softwaremodulen konnte bei der notwendigen Flexibilität eine wirtschaftlich optimierte Lösung erreicht werden. In den dezentralen Peripheriestationen werden ausnahmslos alle Signale von und zur Peripherie, wie z. B. digitale Signale von Sensoren und Endschaltern, usw., analoge Signale wie Temperaturen, Meßwerte etc. erfaßt, und über Profibus-DP an die zugehörige Steuerung weitergeleitet.

Ein Vorteil von Profibus DP ist dabei die „Shared Input" Funktion, was bedeutet, daß diese Eingangsinformationen mehreren Mastern an einem Busstrang zur Verfügung gestellt werden können, und über die Möglichkeit eines gleichzeitigen Master/Slave Betriebs eine homogene Kommunikation zwischen den Mastern, und damit den SPS untereinander erreicht wird.

Damit wird auch die max. mögliche Telegrammlänge von 244 Byte zu den Stationen, bzw. zwischen den Steuerungen untereinander optimal ausgenützt und sichert größtmögliche Effizienz in der Kommunikation.

2.4 Höchste Anlagenverfügbarkeit

Neben der Effizienz ist auch die Sicherheit und Zuverlässigkeit der Datenübertragung ein wesentliches Kriterium. Hier vertraut man voll auf Profibus-DP. So sind alle Funktionen der gesamten Handebene, wie Tippbetrieb oder Einrichtbetrieb ebenfalls über das FBS realisiert, d.h. es wird nicht mehr wie bei früheren Konzepten üblich hardwaremäßig eine eigene Handebene aufgebaut.

In diesen Hand-Betriebsarten werden über busfähige Grafik Displays alle Schalthandlungen direkt Vor-Ort vorgenommen, wie z.B. Band Start-Stop, etc.

Dieses Konzept verlangt natürlich eine entsprechende Geschwindigkeit der Datenübertragung am Bus und die höchste Verfügbarkeit des Feldbussystems. Um größtmögliche Unempfindlichkeit gegenüber EMV Störungen zu erreichen wurden einzelne Bussegmente in Lichtwellenleitertechnik ausgeführt. Damit sind auch keine Einschränkungen in den Baudraten bei den notwendigen Segmentlängen gegeben. Damit und mit der gesicherten Übertragung der Einzeltelegramme wird die notwendige 100%ige Verfügbarkeit auf alle Fälle gewährleistet. Optional könnte man die dabei eingesetzten Optical Link Module zu einem redundanten optischen Ring verbinden, wodurch trotz eiern eventuellen mechanischen Unterbrechung des Lichtwellenleiters an einer beliebigen Stelle alle Teilnehmer erreichbar bleiben.

2.5 Parametrierung und Programmierung machen Profibus DP zum Systembus

Eine Schlüsselkomponente der gesamten Automatisierung ist die Antriebstechnik, ist sie doch ganz wesentlich für die Qualität des Endproduktes verantwortlich. Die hochdynamischen Regelungsvorgänge beim Walzen werden in eigenen speziellen Regelungssystemen strukturiert, die Kommunikation zur Peripherie erfolgt ebenfalls über Profibus-DP. Auch für die Kommunikation der Steuerungssysteme mit den zugeordneten Antrieben, z. B. im Bereich der Beize wird Profibus verwendet, wobei hier neben der zyklischen Kommunikation der Soll- und Istwerte im Betrieb ganze Parametersätze ausgetauscht werden können.

Ein sehr großer Vorteil des ausgewählten Bussystemes zeigte sich bei der Inbetriebnahme. Erstens ist es bei dermaßen ausgedehnten und umfangreichen Anlagen durchaus der Regelfall, daß verschiedene Komponenten zeitweilig nicht verfügbar sind, oder, schlimmstenfalls, überhaupt noch nicht geliefert. Das würde während der Inbetriebnahme ein permanentes Anpassen der Kommunikationsstruktur des Busses an die tatsächlichen Gegebenheiten bedeuten. Bei Profibus erlaubt das Zu-und Abstecken von Teilnehmern während des Betriebes, bei gleichzeitiger freier Zuordnung der Adresse zum Einbauplatz am Bus eine einfache und vor allem nur einmalige Busparametrierung. Dadurch werden bei der Inbetriebnahme lästige und aufwendige Doppeleingaben bei. notwendigen Änderungen vermieden. Wenn man bedenkt, daß in der Endphase mehr als 30 Inbetriebsetzungsingenieure auf dieser Anlage waren, kann man ermessen wie vorteilhaft eine eindeutige und immer gleichbleibende Adressierung ist.

3 Zusammenfassung

Bedingt durch die Komplexität der Anlage, damit verbunden dem umfangreichen Engineeringaufwand und letztendlich kürzesten Lieferterminen, vor allem aber einer extrem kurzen Inbetriebnahmephase war eine ganz klare Strukturierung und Dezentralisierung von vornherein Grundlage aller weiteren Planungsarbeiten. Profibus und Profibus DP erfüllen bestmöglich die Anforderungen, welche bei solchen komplexen Anlagen an die Kommunikation gestellt werden, nämlich
- Durchgängigkeit in der Feld- und Gruppenleitebene

- höchste Zuverlässigkeit in ausgedehnten Anlagen unter schwierigen
  Umweltbedingungen
- äußerste Flexibilität während der Inbetriebnahme und Hochlaufphase
- bestmögliche Einbindung in die Software- und Hardwarekonfigurationswerkzeuge
  der Steuerungshersteller
- Durchgängigkeit in der Projektierung und Dokumentation

Mit Profibus wurden Kosten gespart
durch   - drastische Reduzierung der Verkabelungskosten
          - Reduktion des Engineeringaufwandes zur Schnittstellenklärung
          - kürzere Engineering- und Lieferzeiten
          - verkürzte Inbetriebnahmezeiten
          - Fernprogrammierung
Profibus sichert höchste Anlagenverfügbarkeit
durch   - leistungsfähige Kommunikation können klare und übersichtliche
            Funktionseinheiten gebildet werden
          - die Sicherheit der Datenübertragung sowie möglichen redundanten
            Busstrukturen
          - einfachste Erweiterbarkeit , Zu- und Abstecken der Teilnehmer während
            des Betriebes
Profibus sichert einmal getätigte Investitionen und ist zukunftssicher
durch Standardisierung in der europäischen Feldbusnorm EN 50170 bei gleichzeitiger
Herstellerunabhängigkeit

Literatur

1. Profibus Nutzerorganisation e.V. (Hrsg.).: PROFIBUS
   Technische Kurzbeschreibung. April 1997

2. Popp Manfred.: Profibus DP Schnelleinstieg. PNO e.V: 1996

3. Literatur

1. Profibus Nutzerorganisation e.V. (Hrsg.): PROFIBUS. Technische Kurzbeschreibung, April 1997
2. Popp, Manfred: Profibus-DP Schnelleinstieg, PNO e.V. 1998

5 Gebäudeautomation

Building automation

Vergleicht man die Technik, die in den folgenden Berichten dargestellt wird, mit der konventionellen Technik der Gebäudeautomation, erkennt man, welch enormer Quantensprung in den letzten Jahren stattfand und welche Entwicklung in den nächsten Jahren noch erfolgen wird. Auch hier, wie in Teil 1, liegt wiederum das Hauptaugenmerk auf den Aspekten Engineering und Integration.

Comparing the techniques presented in the following articles with those of conventional building automation, one recognizes the tremendous quantum leap that occurred in the past few years as well as the evolutional potential for the future. Like in part 1, the focus is on the aspects of engineering and integration.

Der Markt für Feldbussysteme in der Gebäudeautomation

Winfried Brandt

VDMA, Frankfurt/Main

Abstract. Im Bereich der Gebäudeautomation hinkte der Einsatz und die Standardisierung von Kommunikationssystemen immer hinter den industriellen Bereichen hinterher. Dies zeigt sich auch in dem noch nicht weitverbreiteten Wissen über den Einsatz von Kommunikationstechnik in Anwendungen der Gebäudeautomation. Es gibt aber auch schon erste Reaktionen auf die vielfältigen neuen Möglichkeiten und Märkte, die sich durch Kommunikationsstandards ergeben. Als aktuelles Beispiel werden die Einsatzmöglichkeiten für standardisierte Feldbussysteme im Bereich Hausautomation und Hausmanagement vorgestellt.

Abstract. The use and the standardisation of communication systems for building automation and control systems is a bit behind the industrial one. Therefore the knowledge about using the communication technologies in applications of building automation and control systems is not yet wide spread. On the other hand there are some reactions to the many new possibilities and markets which result from the communication standards. As an example the new possibilities of using the standardised fieldbus systems for home automation and home management will be presented.

1 Von der ZLT-G zur GA

Zu Beginn der 70er Jahre begann die Entwicklung weg von der "Zentralen Leittechnik für Gebäude (ZLT-G)" mit analoger Regelungstechnik hin zu digitalen Leitsystemen mit Unterzentralen und schnellen bitseriellen Kommunikationssystemen. Gegen Ende der 70er Jahre gab es die ersten volldigitalen Regelungssysteme, die sogenannten DDC-Systeme. Durch den Einsatz von DDC-Systemen für die Meß-, Steuer- und Regeltechnik (MSR) wurden die bisher getrennten und nur über Klemmleisten verbundenen Systeme ZLT-G und MSR zusammengeführt, wobei einfache proprietäre Kommunikationssysteme eingesetzt wurden. Man erkannte sehr schnell, daß über die nun vorhandenen Schnittstellen verschiedene Geräte miteinander verbunden werden und untereinander ihre Informationen austauschen konnten. 1990 wurde in der VDI Richtlinie 3814, Blatt 1, der Begriff ZLT-G durch den Begriff "Gebäudeleittechnik (GLT)" ersetzt.

Wenn heute von Gebäudeautomation (GA) gesprochen wird, so sind dies verteilte Systeme mit verschiedenen Bedien- und Beobachtungseinrichtungen, die über die gesamte Liegenschaft verteilt sein können, vielen Automationsstationen und zum Teil kommunikationsfähigen Feldgeräten. Außerdem findet man sowohl Software-, als auch Kommunikationsschnittstellen, die die Kopplung zu Automationssystemen andere Gewerke außerhalb des HLK-Bereiches (Heizung, Lüftung, Klimatisierung) ermöglichen, wie z. B. Beleuchtungstechnik und Sicherheitstechnik.

2 Einsatz von Kommunikationsstandards

Während für den Bereich der industriellen Prozeßleittechnik Kommunikationsstandards schon länger vorhanden sind, wie z. B. MAP, begann die Standardisierung für den Bereich der Gebäudeautomation erst in der zweiten Hälfte der 80er Jahre, wobei in Deutschland mit den Standards FND (DIN V 32 735) und PROFIBUS (DIN 19 245) die ersten Normen für diesen Bereich spezifiziert wurden. Obwohl viele Hersteller beide Standards in ihren Systemen unterstützen, ist die Akzeptanz am Markt doch sehr gering. Sowohl für den FND, der für die Managementebene spezifiziert worden ist und der 1997 europäische Vornorm wurde (ENV 1805-2), als auch für den PROFIBUS FMS, der mittlerweile für den Feldbusbereich als europäischer Standard veröffentlicht ist (EN 50170), gibt es jeweils nur eine geringe Anzahl von Projekten, in denen mindestens zwei Systeme unterschiedlicher Hersteller miteinander über den jeweiligen Standard kommunizieren.

Die Gründe dafür sind vielfältig. Bei der Kommunikationstechnik handelt es sich um ein spezielles, anspruchsvolles und neues Wissensgebiet für alle am Projekt Beteiligten. Dies gilt sowohl für die Hersteller, als auch für die Planer und Betreiber. Bei der Realisierung kommunikationstechnischer Projekte ist ein interdisziplinäres Zusammenwirken dieser Gruppierung erforderlich; hinzu kommt die Umsetzung der jeweiligen funktionalen Anforderungen aus den unterschiedlichen Gewerken. Dieses in der Vergangenheit ungewohnte Zusammenspiel und das noch nicht ausreichend in der Breite vorhandene Know-how bei der Umsetzung von offenen Kommunikationsstandards sind derzeit wohl noch Gründe, die die offene Kommunikation nicht voll zur Wirkung kommen lassen.

3 Einsatz von Feldbussystemen in Gebäuden

Betrachtet man den Feldbusbereich in der Gebäudeautomation, gab es schon lange zwei "quasi-offene" Standards, nämlich die 4-20 mA-Schnittstelle und die 0-10 V-Schnittstelle. Derzeit stehen in der Gebäudeautomation als europäische Standards

- Batibus,

- EIB,

- EHS und

- LonTalk Protocol

zur Verfügung, die eine Kommunikation in der Feldebene zwischen anwendungsspezifischen Controllern und der Sensorik/Aktorik bieten.

Der klassische Anwendungsbereich für Feldbussysteme in der Gebäude- und, neuerdings auch, in der Hausautomation ist die Einzelraumregelung, die hauptsächlich bei HLK-Systemen, Fan-coil- und Induktionssystemen und Elektroheizungen eingesetzt werden.. Hier besteht zum einen die Anforderung, die Gegebenheiten des Raumes den individuellen Wünschen des oder der Benutzer anzupassen, wie z. B. Raumtemperatur, Raumfeuchte und Beleuchtung. Zum anderen hat man aber auch die Möglichkeit, alle Räume mit bestimmten Informationen zu versorgen oder Informationen aus ihnen abzurufen; als Beispiele seien hier die Nachtabsenkung der Raumtemperatur und die Abfrage der Fensterkontakte nach Ende der normalen Nutzungszeit der Räume. Ein solcher Raum ist in dem folgenden Bild (Fig. 1.) schematisch dargestellt. Der eingezeichnete Fensterkontakt ist in erster Linie vorhanden, um zu verhindern, daß die Klimatisierung des Raumes bei geöffnetem Fenster eingeschaltet ist. Weil der Kontakt aber als Sensor an einen Feldbus angeschlossen ist, steht diese Information nun auch anderen Gewerken zur Verfügung, wie z. B. einem Einbruchmeldesystem. Dieses erhält dieselbe Information wie das HLK-System, benutzt sie aber für andere Zwecke. Anwendungen die auf diese Weise in die Einzelraumregelung eingebunden werden können, sind z. B.

- Brandmeldetechnik,

- Zutrittskontrolle,

- Beleuchtungstechnik,

- Elektroverteilung,

- Sanitär- und Wassertechnik,

- Beschattungstechnik,

- Fördertechnik sowie

- Außenanlagen.

Wie die einzelnen Komponenten eines Feldbussystems genutzt werden, hängt also einzig und allein von der Anwendung ab. "Klassische" Anwendungen sind, wie oben bereits erwähnt, z. B. das Klimatisieren des Raumes nach den Sollwertvorgaben und das Einschalten des Lichtes. In modernen Gebäudeautomationssystemen werden die vielen vorhandenen Informationen genutzt, um den Raum zu "managen". Die Beleuchtung wird nicht nur ein- und ausgeschaltet, sondern es wird auch festgestellt, ob überhaupt jemand im Raum ist, ob die Helligkeit durch die Fenster ausreichend ist und wieviele Stunden die Leuchtmittel insgesamt schon eingeschaltet waren. Diese Raummanagement-Informationen spielen eine immer größer werdende Rolle in dem Kontext "Facility Management", in dem die kompletten Einrichtungen eines Gebäudes umfaßt und die gesamten betrieblichen und personenbezogenen Abläufe organisiert, überwacht, koordiniert und optimiert werden (siehe [1], [2]).

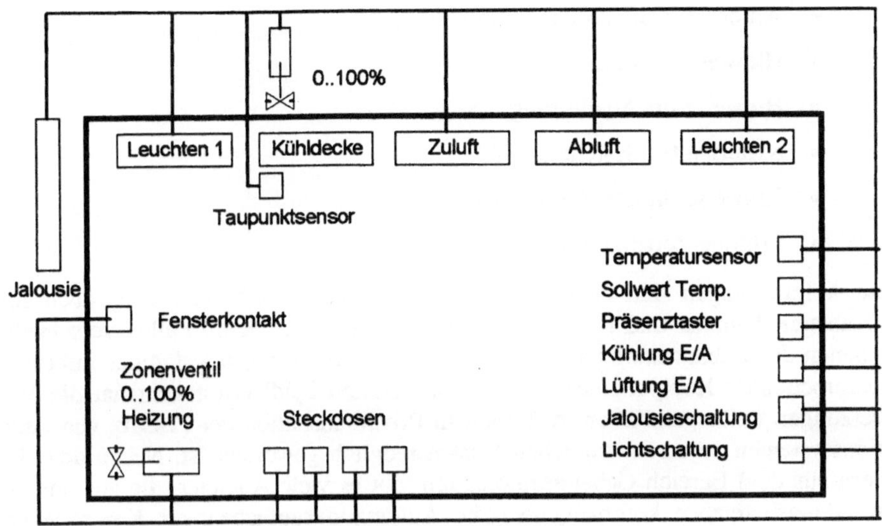

Fig. 1. Beispiel für den Einsatz eines Feldbusses in einem Raum

4 Neue Märkte durch die Hausautomation

Während in der Gebäudeautomation Feldbussysteme bereits häufig eingesetzt werden, ist dies in der Hausautomation noch nicht der Fall, die ihren Schwerpunkt im Wohn- und wohnähnlichen Bereich bis zu kleineren, gemischt genutzten Gewerbeobjekten hat. Produkte und Systeme der Hausautomation sind am Markt verfügbar, bzw. befinden sich vor der unmittelbaren Markteinführung. Die Funktionalität von Hausautomationssystemen erstreckt sich von HLK-Anwendungen über Einbruchschutz bis hin zur Einbindung von Hausgeräten sowie der Aufschaltung von Life-Safety-Funktionen und der Nutzung von Multimedia-Diensten als mögliche Bediener- und Benutzeroberfläche.

Die technischen Grundlagen für Hausautomationssysteme werden derzeit beim VDMA erarbeitet. Sie umfassen Festlegungen über

- Anwendungsbereiche,
- Begriffe,
- Systemstruktur,
- Funktionen,
- Intelligente Verknüpfungen,
- Informationselemente,
- Anforderungen (Geräte, Netze, Sicherheit),

- Anforderungen an Kommunikation,
- Hinweise für Planung,
- Hinweise für Ausführung,
- Kriterien für Systemauswahl,
- Hinweise für den Service und
- Prüfung/Zertifizierung.

Diese Anforderungen sind denen der Gebäudeautomation sehr ähnlich, jedoch sind die Anbieter, Kunden und Kosten der Hausautomation in vielen Fällen verschieden von denen der Gebäudeautomation. Erst die Entwicklung leistungsfähiger informationsverarbeitender Komponenten von standardisierten Feldbussystemen hat die Voraussetzungen dafür geschaffen, daß auch in Privathaushalten der Einsatz von Automationssystemen zu erschwinglichen Kosten möglich geworden ist. Neben den Herstellern aus dem Bereich Gebäudeautomation gibt es viele Anbieter, die aus anderen Anwendungsbereichen kommen, aber die Automationsaufgaben zur Kontrolle von energie- und umweltrelevanten Prozessen ebenfalls beherrschen.

Hierbei treten auch Synergieeffekte auf, wenn sich z. B. ein Betrieb aus dem Sanitär-, Heizung-, Klimabereich mit einem Betrieb des elektrotechnischen Handwerks zusammenschließt, um gemeinsam komplette Hausautomationssysteme anbieten, planen, installieren und warten zu können. Durch die Tätigkeiten Angebot, Planung, Installation und Wartung ergeben sich somit für die beiden Betriebe auch neue Betätigungsfelder und neue Märkte. Auch Hersteller von Rolladensteuerungen, Audio-/Videogeräten, Haushaltsgeräten, Sicherheitssystemen, Telekommunikationsanlagen, Personalcomputern usw. bieten bereits Hausautomationssysteme an oder zumindest die Anbindung an solche Systeme (siehe Fig. 2.).

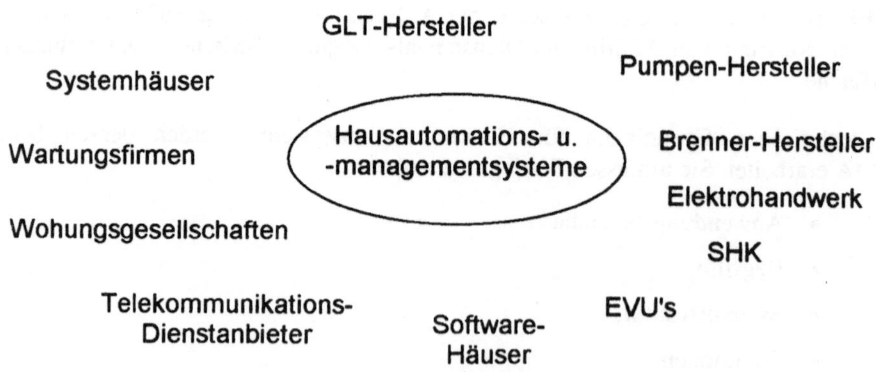

Fig. 2. Anbieter auf dem Markt für Hausautomations- und -managementsysteme

Parallel zur Durchsetzung der Hausautomation am Markt entwickelt sich ein weiterer neuer Markt - und zwar die Nachfrage nach Hausmanagement-Dienstleistungen.

Hiermit ist die Bewirtschaftung von Wohn- und wohnähnlichen Gebäuden bzw. kleineren Zweckbauten gemeint mit dem Ziel, die Wirtschaftlichkeit, die Sicherheit und den Komfort zu steigern [3]. Als Beispiel sei hier die Erfassung von Verbräuchen in großen Wohnanlagen mit vielen Wohneinheiten erwähnt. Der Verbrauch von Strom, Wasser, Gas, Heizleistung u.a.m. wird heute noch getrennt von verschiedenen Firmen erfaßt und den Nutzern in Rechnung gestellt. Eine Überwachung und Wartung der entsprechenden Anlagen findet dabei meistens nicht statt. Mit Hilfe eines Hausmanagementsystems kann dies alles aus einer Hand geschehen, mit kürzeren Erfassungszeiträumen (bisher häufig jährlich!), genaueren Verbrauchsmessungen und damit besseren Abrechnungen. Hausmanagement umfaßt also die Gesamtheit der technischen, infrastrukturellen und kaufmännischen Leistungen, die allesamt keine Technik-, sondern Dienstleistungen sind.

Ein neuer Dienstleistungsmarkt entsteht. Auch in diesem Fall können die Anbieter aus sehr unterschiedlichen Bereichen kommen, wie z. B. Wartungsfirmen, die die Heizungsanlagen warten und die verbrauchten Wärmemengen in den Wohnungen erfassen, und Versorgungsunternehmen, wie Gas-, Wasser-, Stadtwerke und Energieversorgunger. Auch Telekommunikations-Diensteanbieter können sich in Zukunft an diesem Markt beteiligen, wobei diese bereits über einen kommunikationstechnischen Zugang zum Haus verfügen.

Durch die Zunahme des Einsatzes von Feldbussystemen in Wohnbauten, werden die Grundlagen für neue Anwendungen und Dienstleistungen geschaffen. Neue Märkte entstehen mit neuen Szenarien bezüglich Anbieter und Nutzer. Durch das kommunikative Zusammenwirken an ein und demselben Feldbussystem treten die verschiedenen Gewerke nicht mehr einzeln in Erscheinung, sondern dem Nutzer eines Hausautomations- oder Hausmanagementsystems stehen alle Informationen und Dienste über eine Bedieneinheit zur Verfügung. So gesehen wird deutlich, welche positiven Auswirkungen die Standardisierung der Kommunikation in der Feldebene der Gebäudeautomation in Bezug auf neue Anwendungen (Hausautomation) und Dienstleistungen (Hausmanagement) nach sich zieht.

Literatur

1. Bügel, Ulrich: Gebäudeautomation auf dem Weg zum ganzheitlichen "Facility Management". atp - Automatisierungstechnische Praxis 39 (1997) 3, S. 9.

2. Kranz, Hans: Gebäudeautomation im Wandel. atp - Automatisierungstechnische Praxis 39 (1997) 3, S. 10-15.

3. VDMA-Einheitsblatt (Entwurf): VDMA 24 197. Hausmanagement - Begriffe und Leistungen. Juli 1997

Component-based Project Engineering

M. Goossens

The EIB Association (EIBA),
Tinklaan 5, B-1160 Brussels, Belgium

Abstract. The EIB Home and Building Electronic System offers a vendor-independent tool platform for design and commissioning of projects. The user is always presented with customisable, high-level components featuring a unified look. The development engineer constructs these components to fit the need of the application. The objective is to make project design accessible to people without any electronics or programming skills (such as the electrical contractor).
The EIB Tool Software suites for PC's share the common ETE framework, which provides both the tools and the underlying mechanisms to create, customise and distribute such Project Engineering Components. They are entirely data driven: the solution provider simply distributes a 'component catalogue' in electronic form to all users.

1. Preface

None of those involved in the configuration, commissioning and maintenance of building automation installations, underestimates the engineering effort required to design a project. Indeed, *project engineering* often has to be carried out by highly skilled system engineers, and in many cases is also a major cost factor. To meet this challenge, the EIB system [1] has pioneered the concept of Component-based Project Engineering.

The idea is to provide a Repository (or "product database") which contains re-usable, customisable software building blocks, called Components. The structure of each component should be rich enough, so that it can be handled off-line as a virtual device. The binding to actual hardware takes place "at installation time" through an active linking and downloading procedure (see also section 4.1). On the other hand, a high-level, graphical configuration interface has to enable the qualified user with limited engineering skills to design and commission installations.

2. Project Engineering for EIB

Precisely what an EIB Component achieves becomes clear only when looking at the 'global' dynamics of the EIB Component-based Installation Design paradigm as sketched in fig. 1. Components shift the engineering effort away from installation design and maintenance, towards product development. The Component hides the complexity of the underlying solution from the user. Importantly, this abstraction and presentation "layer", once created, is available for unlimited re-use.

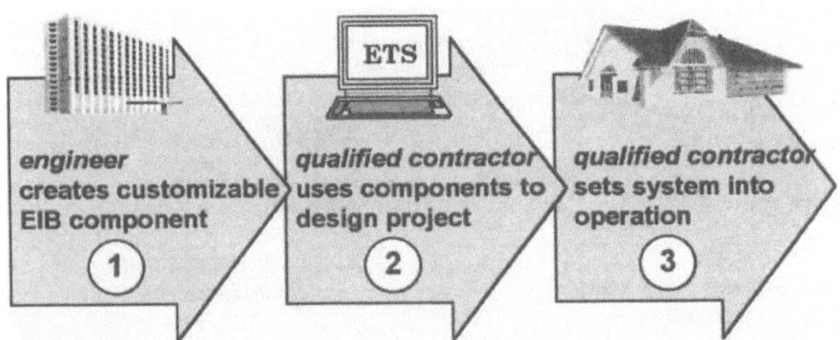

Fig. 1. Components shift the engineering effort to the development phase.

2.1 The Development Phase

Not bound to any particular processor, EIB does not prescribe any particular compiler or applet development environment. All commercially available tools, assemblers, compilers, emulators and debuggers can be put to use, though the EIB Integrated Development Environment offers a framework to facilitate this step.
Standardized file interfaces allow the resulting binary "applet" data to be imported into the ETE Component Repository. The component designer now uses the ETS Component Manager to create the component wrapper around the "raw" applet. The Component wrapper makes the EIB applet re-usable and customisable, with properties which can be tuned for each specific instance in every project. Section 4 analyses the resulting data structures.

2.2 Registration and Distribution

The development phase is rounded off by submitting the Component to the registration body in the form of (an extract of) the Component Repository. This step is important to ensure the integrity of the Component concept in a multi-vendor environment.
Fig. 2 shows the (public part of the) global model of distributing the components to the ETS users. The exchange of component data is handled via defined import/export file interfaces. This process is supported by dedicated tools and utilities for all 'agents' in the model.

2.3 Designing Installations

Finally, the ETS user normally receives Component Repository updates from various vendors in the form of *product database* diskettes which can be imported. In ETS, they appear as entries in an on-line catalogue. The user now selects, combines and customises the components per drag-and-drop.

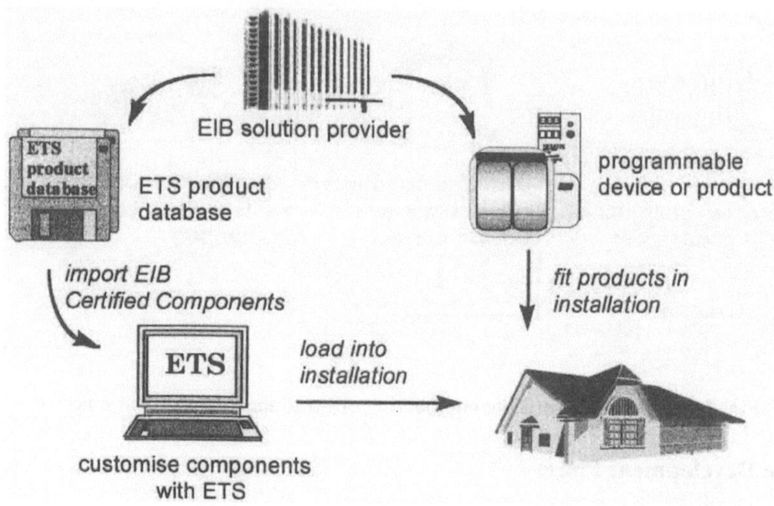

Fig. 2. EIB Components - from concept to installation.

3. The EIB Tool Environment Framework

Critical in an open, multi-vendor environment is the availability of standard tools, both for creating components and for project engineering. For EIB, this is realised with the EIB Tool Software (ETS) suites for developers and qualified contractors respectively. Both tool-suites use the EIB Tool Environment (ETE) Framework, which contains the vendor-independent Component Repository. This Repository plays an important role, because the EIB Project Engineering Components are entirely data-driven. Currently, it is implemented in the shape of relational database. Finally, the Repository also provides special infrastructure to deal with multiple language versions.

The ETS Developer's Edition comes with all necessary utilities and editors to create and handle components. The ETS Project Edition provides the user with browsers to navigate the repository, as well as various browsers and editors for project engineering, with the components as drag-and-drop building blocks. So, together with the ETE Framework, these tools constitute the universe where EIB Project Engineering Components live.

4. Project Engineering Components

4.1 Data Model

Now, let's look a bit more closely at the components themselves, and discover how they are built up and achieve their purpose. Fig. 2 shows how the EIB Component essentially equips the (invariant part of the) binary image (or Applet Code) which has to be downloaded (say into EEPROM), with two wrappers:

- The *configuration* information tells the tool about the capabilities of the application in terms of parametrisation, addressing and supported data formats.
- The *presentation* information allows the tool to generate the appropriate user interface which gives access to the configuration. It also embeds the "solution" into a navigation structure; with various browsing and search mechanisms, it can be retrieved quickly.

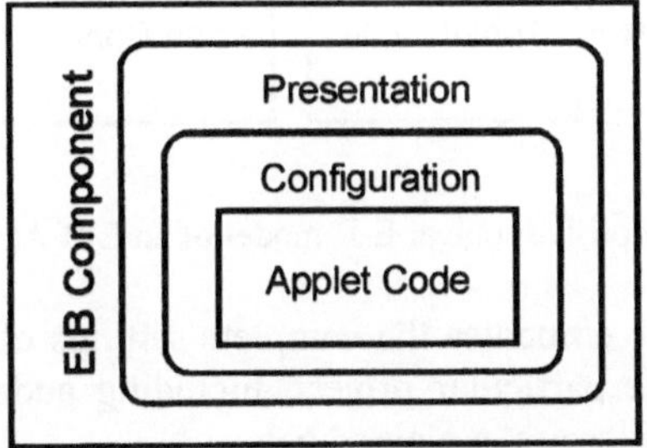

Fig. 3. Overall composition of an EIB Component

Drilling down further, we leave out the hardware, navigation and commercial information, concentrating instead on the Application entity which is at the heart of the component. Again, the following analysis illustrates that the concept of an Application which can be customised and downloaded into a device from a PC is central to EIB.

Remember that an EIB Project Engineering Component actually consists of data. The highly simplified Entity-Relationship (E-R) model of fig. 3 shows what this looks like in terms of the data model. Double arrows indicate a "to many" relationship; not all multiplicities or transitivity relationships are shown. Only the underscored entities are visible to the user: they contain attributes which allow the developer to control the appearance, and add texts, default values, admissible value ranges etc., i.e. the *presentation* wrapper. The remaining entities stay hidden.

The Application entity provides the common header for the binary Applet Code, around which the wrappers are constructed. To begin with, we distinguish:

- Communication Object: the EIB shared variables, via which different devices can be logically linked together using group addresses [2].
- Distributed Object: a structured, logical grouping of Properties of the application, which may be accessed via the EIB network [3].

- Parameter: a visible property which the user can set to tune or customise the Application's behaviour.

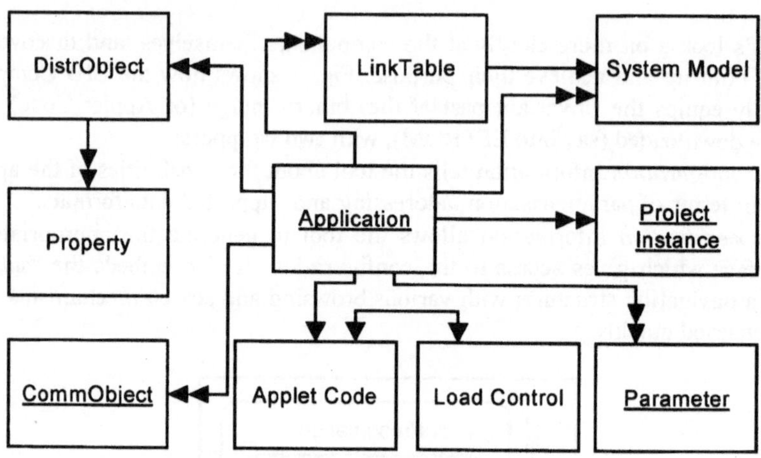

Fig. 4. Simplified logical E-R model of an EIB Application.

Next, the Project Instance embodies the complete settings of one particular instance of the Application in one particular project, including addressing, parameters and system flags. Further entities are explained below.

4.2 Customisation and System Version Independence

This section explains how the model also goes a long way towards uncoupling the application from the precise version and implementation details of the device's EIB system firmware. It should be stressed that these features, together with the Component Model as a whole and the software framework on which it resides, form an intrinsic part of the EIB system.

Active Linking. Basically, the tool can use the entries from the Link Table to link the Application's binary Applet Code for the particular System Model (or even the actual physical medium) encountered in the installation 'at download time'.

Dynamic Load Control. This entity provides the developer with direct, application specific, low-level control over the management of the device on download. A dedicated command syntax covers all standardised EIB Management Operations. Such a Load Control sequence can be appended to the Applet Code, so that the tool's default behaviour can be adjusted or overridden.

The *Product Snap-in* Call-back Interface. A call-back interface allows ETS default behaviour to be replaced by custom functions, wherever relevant. This functionality must be provided in the shape of Windows Dynamic Link Libraries (DLL).

5. A Universal Push-button Component

As an example of a component, the figures below show the parameter structure (German version) of a universal 4-fold push-button. At project design time, each button can dynamically be assigned a particular function, such as switching, dimming control, shutter control etc. The detailed behaviour for each function chosen may be controlled via a set of dedicated parameters.

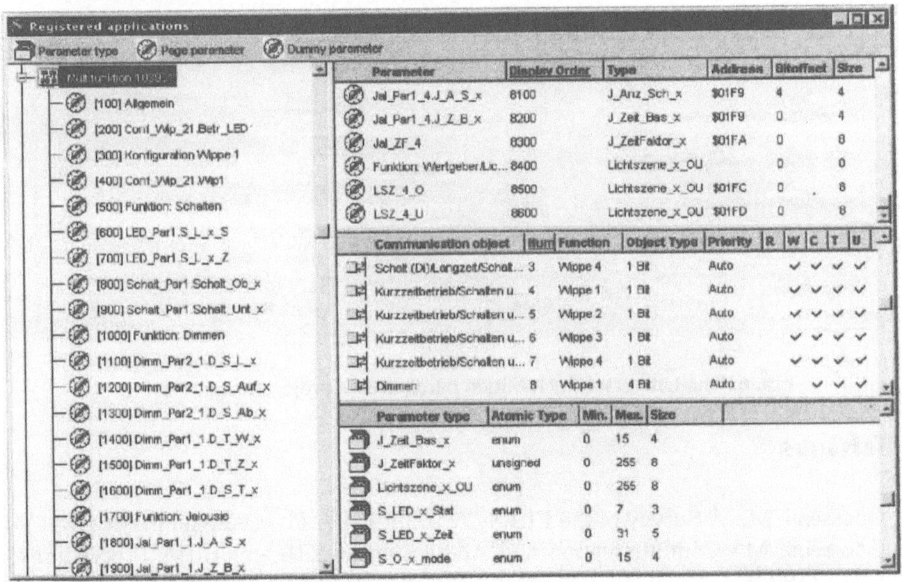

Fig. 5. Developer's view of parameter structure for a Universal Push-button application.

The editor which a developer will use to design this part of the component is depicted in fig. 5. Certain parameter attributes (such as visibility) may be made dependent on the value chosen for a parent parameter. The resulting user's view is shown in fig. 6.

Acknowledgements

The author wishes to thank S. Lüling (INSTA GmbH) for kindly providing the universal push-button example.

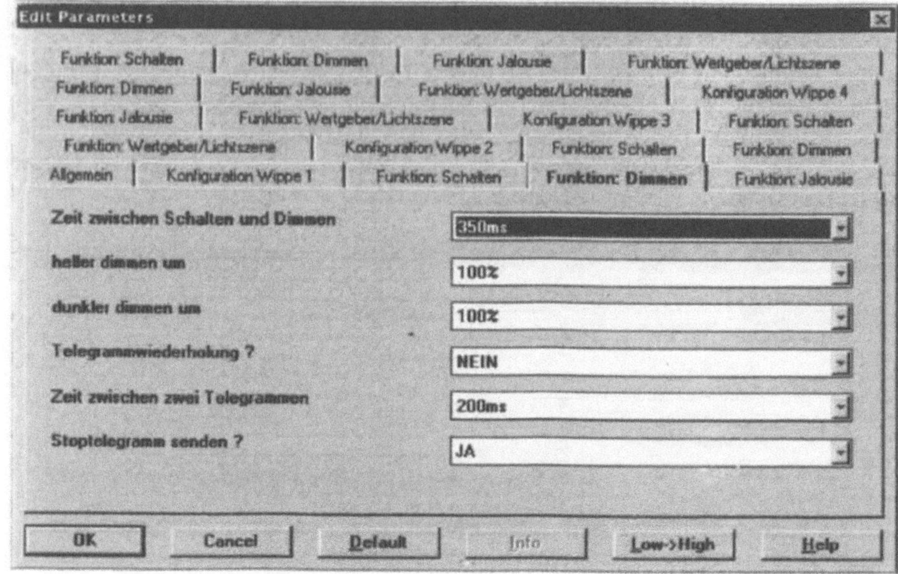

Fig. 6. The Universal Push-button parameters - project design view.

References

1. Goossens, M.: A Survey of the EIB System. In: EIBA Proceedings, EIBA (1997).
2. Goossens, M.: Communication and Addressing on EIB. In: EIBA Proceedings, EIBA (1997).
3. Goossens, M.: Object-based Distributed Application Design. In: FeT Proceedings 1997, Springer 1997.

Systemintegration in der Gebäudeautomation mit grafisch orientierter Software für LonWorks

Markus Gehnen

AEG Lichttechnik GmbH
Rathenaustr. 2-6
D-31832 Springe

Abstract. Die breiten Einsatzmöglichkeiten von LonWorks bergen die Gefahr der Unübersichtlichkeit für den einzelnen Anwendungsfall. Die Auswahl von Geräten und ihr sinnvoller Einsatz erfordert in der Regel eine Vertrautheit mit technischen Details, die man bei vielen potentiellen Anwendern nicht voraussetzen kann. Als Hilfestellung kommt daher PC-Software zum Einsatz, mit der sich der Nutzer auf die Erstellung der gewünschten Funktionalitäten konzentrieren kann, ohne sich mit Lon-spezifischen Dingen belasten zu müssen. Das strikt grundrißorientierte Softwarekonzept schafft Nähe zu vertrauten Planungsmethoden. Der notwendige Schulungsaufwand bleibt bei diesem Ansatz gering, wie Praxiserfahrungen bestätigen.

The wide possibilities of LonWorks may confuse the user. For an efficient application of Lon devices you need a deeper understanding of the technology than most potential users have. The solution is a graphical PC-Software. It helps concentrating on the desired functionalities while it protects from Lon-specific details. The floor plan oriented approach is similar to conventional planning methods and minimizes the necessary training effort.

1 Aufgabenstellung

LonWorks (Lon = Local Operating Network) ist eine offene Technologie, die auf preiswerten busfähigen Microcontrollern, den *Neurons*, basiert. Das zugehörige Kommunikationsprotokoll LonTalk besteht aus weitgehend standardisierten Informationseinheiten, den Netzwerkvariablen. LonWorks eignet sich für den Einsatz in zahlreichen Gewerken der Gebäudetechnik mit der Möglichkeit zur gewerkübergreifenden Kommunikation.

Diese breite Anwendbarkeit kann aber zugleich den Nutzer vor Schwierigkeiten stellen: Die mittlerweile sehr vielfältigen Kombinationsmöglichkeiten zwischen den Geräten verschiedener Hersteller und Gewerke verlangen vom Planer ein detailliertes Wissen über die zugrundeliegende Technik.

Dieses Spezialwissen ist in mehrtägigen Schulungen vermittelbar und muß in der täglichen Praxis gefestigt werden. Vielen Planern und Betreibern fehlt dafür heute der finanzielle und zeitliche Spielraum, so daß sich eine entsprechende Kompetenz nur bei spezialisierten Systemintegratoren findet, der breite Markt der Endanwender jedoch überfordert ist.

2 Lösungsansatz

Wer als Planer aus der Beschäftigung mit LonWorks kein Kerngeschäft machen will, darf bei der ersten Begegnung mit LonWorks nicht mit Spezialwissen „überfallen" werden. Die ihm angebotenen Planungswerkzeuge müssen stattdessen soweit wie möglich auf Vertrautem aufbauen: Ein Planer ist mit dem Umgang von Grundrißplänen vertraut, in die er die benötigten Geräte, Verbindungen und Funktionen hineinskizziert.

Die Rechnerunterstützung für Planung und Inbetriebnahme muß diesem Konzept der grundrißorientierten Planung soweit wie möglich folgen. LonWorks-spezifische Details dürfen zumindest in Standardsituationen auf dieser Ebene gar nicht in Erscheinung treten und nur auf spezielle Anforderung oder für projektspezifische Lösungen bereitstehen, so daß eine schrittweise, bedarfsorientierte Annäherung an LonWorks stattfindet.

3 Umsetzung mit Unilon

3.1 Grundrißorientierte Planung und Inbetriebnahme

Die Teilnehmer eines Lon-Busnetzes sind mit Neurons ausgestattet und verfügen über eigene Intelligenz. Man nennt sie auch *devices*. Sie erhalten während der Inbetriebnahme Informationen über ihre jeweilige Aufgabe und darüber, wie sie mit ihrer Umgebung kommunizieren sollen. Die Festlegung der Kommunikationsstrukturen nennt man auch *binding*. Die Planungsaufgabe besteht im Entwurf dieser Funktionen und der anschließenden Programmierung der Busteilnehmer.

Unilon unterstützt diese Aufgabe grafisch orientiert. Jedes reale Lon-Gerät erhält eine symbolische Entsprechung auf dem Bildschirm. Unterlegt man den Plan mit einem importierten Grundriß, so ist auch eine räumliche Zuordnung möglich. Das erleichtert insbesondere bei späteren Änderungen die Orientierung. Diese Importfunktion unterstützt die gängigen Grafikformate.

3.2 Installation

Als devices stehen die intelligenten Geräte selbst sowie sogenannte *slave devices* zur Verfügung. Das ist z.B. eine Leuchte ohne eigene Intelligenz, die aber an einen intelligenten Lichtregler angeschlossen ist. Sämtliche devices, die zum Einsatz kommen sollen, stehen am Bildschirm in einer „Installationsansicht" auf einem *Toolpanel* zur Verfügung. Den Umfang dieses „Werkzeugkastens" kann der Anwender selbst bestimmen, indem er die zu den Geräten mitgelieferten Softwaremodule hinzuinstalliert oder aber eigene definiert (s. Abschnitt 3.4). Dieses Konzept erlaubt eine flexible und gewerkunabhängige Anpassung an neue Planungsanforderungen oder Geräteneuheiten. Der Planer wählt das entsprechende Gerät auf dem Toolpanel aus und plaziert es anschließend an der gewünschten Stelle auf dem Grundriß. Anschließend verbindet man die Geräte grafisch untereinander so, wie es der Verdrahtung in der Wirklichkeit entspricht.

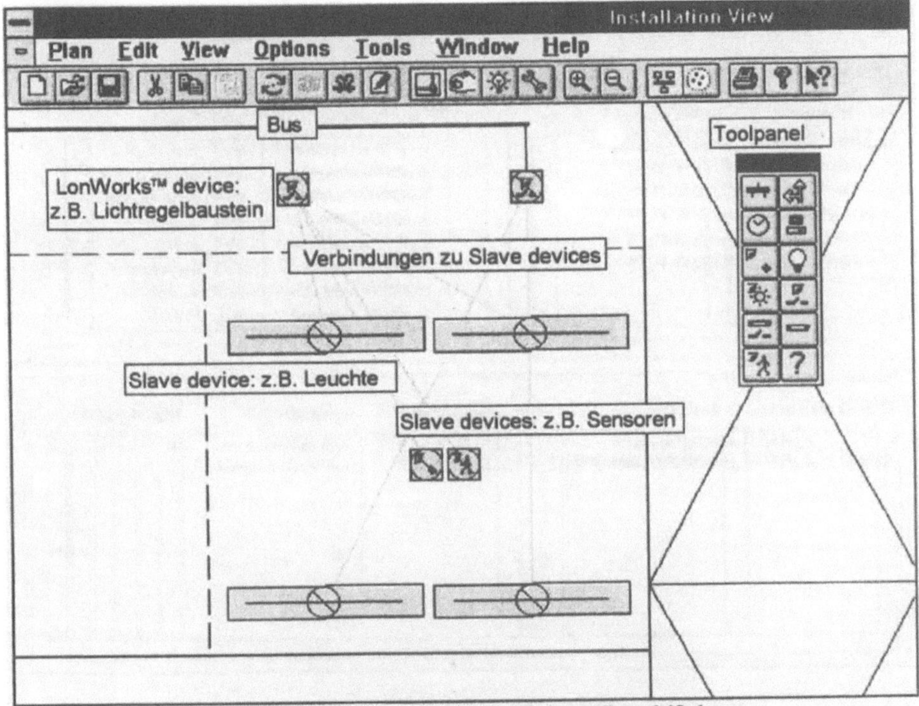

Fig. 1. Installation von devices auf einem Grundrißplan

Der nächste Arbeitsschritt stellt den eindeutigen Bezug zwischen dem realen Netz und der grafischen Nachbildung her: Jedem Neuron ist eine eindeutige, unveränderliche Identifikationsnummer ab Werk „eingebrannt". Diese Nummern werden sukzessive den Bildschirmsymbolen zugeordnet. Damit ist der Bezug hergestellt.

3.3 Konfiguration

Im Anschluß an die Nachbildung der physikalischen Gegebenheiten geschieht die Festlegung der bindings. Dabei gibt es zwei wesentliche Arbeitsschritte:

Wer redet mit wem: In der „Konfigurationsansicht" sind die Funktionalitäten der Geräte als sogenannte *objects* dargestellt. Entsprechend der Vorgehensweise in der Installationsansicht zieht man hier ebenfalls Linien und stellt so das binding her.

Wer macht was: Die bindings können vordefiniert sein. Häufig ist dies jedoch nicht der Fall, zumal sich die Palette der verfügbaren Lon-Geräte ständig ändert und erweitert. In einem solchen Fall bietet Unilon die Neudefinition eines bindings der beiden beteiligten Geräte auf Grundlage der verfügbaren Netzwerkvariablen an. Aus den beiden angebotenen Variablenlisten wählt man diejenigen Variablen aus, die den Informationsaustausch bewerkstelligen sollen. Sie tauchen anschließend in der unteren Tabelle auf und legen so Zeile um Zeile die Funktionalität des bindings fest.

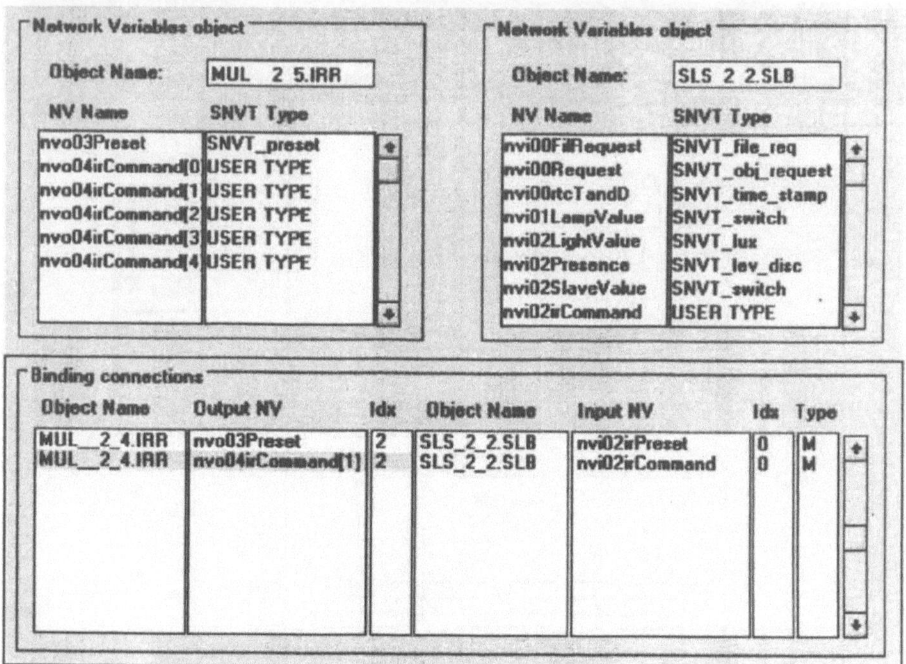

Fig. 2. Definition eines bindings

In gleicher Weise ist auch eine nachträgliche Modifikation eines bestehenden bindings möglich.

Gleichzeitig mit dem Abspeichern des bearbeiteten Planes geschieht automatisch die Programmierung der angeschlossenen Nodes. Eine nachträgliche Änderung von bindings führt zu einer Warnanzeige am Bildschirm, bis ein erneutes Speichern die Funktionalität des realen Netzes wieder anpaßt.

3.4 Eigene Definitionen mit dem Wizard

Aufgrund des umfangreichen Angebotes an Lon-devices ist es vielfach nötig, die entsprechende Software-Repräsentation selbst zu erstellen, da der Gerätehersteller kein entsprechendes Modul mitliefert. Das geschieht mit Hilfe des *Wizard*: Die Eigenschaften eines Nodes sind in zwei Dateien eindeutig abgelegt, dem „.nxe-file" mit den Arbeitsanweisungen des Knotens und dem „.xif-file", das die Netzwerkvariablen für den Datenaustausch mit anderen Busteilnehmern enthält. Unabhängig von einer Applikationssoftware wie z.B. Unilon gilt das für jedes Lon-device. Die Dateien werden vom Hersteller des devices mitgeliefert oder lassen sich mit dem Hilfsprogramm *nodeutil* aus dem betreffenden Node extrahieren. Daraus und mit im Prinzip frei wählbaren bitmaps zur grafischen Darstellung formt man mit dem Wizard ein neues device mit einem oder mehreren zugeordneten functional objects. Dieses device steht anschließend neu im Toolpanel zur Verfügung und ist gleichwertig mit vordefinierten devices.

4 Praxiserfahrungen

Unilon wird für eigene Planungen eingesetzt und ist Bestandteil von Schulungen zum Thema „Bussysteme in der Gebäudesystemtechnik" für Planer und Installateure. Die Erfahrungen lassen sich in der Form zusammenfassen: Eine kurze Einweisung und praktische Demonstration genügt in der Regel, um anschließend z.B. eine einfache Beleuchtungsanlage mit den Funktionen anwesenheitsgesteuertes Licht, tageslichtabhängige Lichtregelung und Infrarotfernsteuerung selbständig zu installieren, zu konfigurieren und in Betrieb zu nehmen. Das funktioniert vollständig auf grafisch-intuitiver Ebene, ohne daß von der Technologie mehr als der Name „LonWorks" bekannt ist.

Die einfache Handhabung schafft Vertrauen und weckt das Interesse nach teifergehenden Informationen, so daß sich im zweiten Schritt z.B. die Integration einer Jalousiesteuerung mit Hilfe des „Wizard" als natürliche Konsequenz anschließt. Ein Schulungstag ist ausreichend zur Heranführung an LonWorks bis hin zur „Wizard"-Programmierung.

Literatur

UNILON Handbook, Philips Lighting Controls, April 1997

Flexible Konfigurierung für die Gebäudeautomation

Joachim Edtmair [1], Herbert Schweinzer [2]

[1] Fa. ATB, Gallneukirchen, A-4210
[2] Inst. f. Elektrische Meßtechnik, TU Wien,· A-1040

Abstract. Die Firma 'ATB Automatisierungstechnik' entwickelte ein Gebäudeautomationssystem, das CAN zur Kommunikation einsetzt. Der Aufbau des Systems ist äußerst flexibel und bietet die Möglichkeit, eine unterschiedliche Anzahl von CAN-Komponenten zu integrieren. Besonderes Augenmerk wurde auf die Inbetriebnahme des Systems gelegt. Eine menügeführte Software ermöglicht dem Inbetriebnehmer die Einstellung der Systemparameter und den Aufbau der Kommunikation. Die Vergabe der Identifier der einzelnen Objekte wird nicht durch Hardwarecodierung, sondern durch einen Programmiervorgang aller am Bus angeschlossenen Komponenten durchgeführt. Jeder Knoten muß für die flexible Vergabe der Identifier vorbereitet sein. Ein zentraler Knoten organisiert das Verteilen der IDs. Zu diesem Zweck übernimmt er für die Dauer der ID-Vergabe die Masterfunktion. Nach Abschluß der Zuteilung ist die Kommunikation entsprechend der Anwendung möglich.
Der Kommunikationsaufbau erfordert ein Tool zur Generierung der IDs entsprechend der Anwenderdaten und ein Tool zur flexiblen Vergabe der IDs.

Abstract. The Austrian company 'ATB Automatisierungstechnik' has developped a new CAN-based system for flexible home automation which cuts down costs for heating and air-conditioning while increasing user-comfort. The system is very flexible and offers the use of a variable amount of CAN-components. Special attention is directed to putting the system into operation. A software which is easy to use makes the setting of system-parameters und the setup of communication possible. The distribution of the identifiers is not carried out by hardware facilities but by a protocol on the CAN-bus. Every node must be prepared for the flexibel distribution of the identifiers which is organized by a master-node. When this process is finished all nodes can be used for their intended application even the master-node.
One tool is needed to generate the IDs in accordance to the application and another one for the distribution.

1 Einleitung

In dezentralen Gebäudeleitsystemen sind räumlich und logisch getrennte digitale Regeleinheiten und I/O-Systeme miteinander verknüpft. CAN bietet dafür durch seine Multimasterfähigkeit und die einfache Prioritätssteuerung der Nachrichten optimale Voraussetzungen.

CAN benutzt eine nachrichtenorientierte Adressierung, das heißt, jeder Nachricht ist ein netzweit eindeutiger Identifier zugeteilt. Die Zuordnung dieser Identifier stellt eine zentrale Aufgabe beim Aufbau eines CAN-Netzwerks dar, insbesondere wenn baugleiche Module in einem System eingesetzt sind. Von besonderer Wichtigkeit erscheint die Möglichkeit, diese Identifier-Zuordnung nicht durch Hardware-codierung, sondern durch einen Programmiervorgang aller am Bus angeschlossenen Komponenten vornehmen zu können. Die aus den Anwenderdaten generierte Zuordnung der Identifier zu den Nachrichtenobjekten liefert die erforderlichen Daten für diesen Programmiervorgang. Die Identifier werden den Modulen einmal im Rahmen der Anlageninbetriebnahme zugewiesen und sind ab diesem Zeitpunkt in jedem Modul versorgungsspannungsunabhängig gespeichert. Nachträgliche Änderungen in einer Anlage (Erweiterungen, Änderungen der Identifier, ...) können ebenfalls über einen Programmiervorgang erfolgen.

Durch entsprechend aufgebaute Module und mit Hilfe geeigneter Engineering-Tools können diese Konfigurierungsaufgaben flexibel und methodisch gelöst werden.

2 Inbetriebnahmephasen

Die Identifier werden grundsätzlich durch einen ausgezeichneten Knoten im System vergeben. Das heißt, daß anders als im „normalen" Betrieb ein Knoten die Master-Funktion übernimmt. Dieser Knoten organisiert sowohl die Zuteilung der Identifier als auch die dazu notwendige Kommunikation.

Eine selbständige Vergabe von Identifiern über den Bus ist nur dann möglich, wenn die angeschlossenen Module voneinander unterschieden werden können. Diese Unterscheidbarkeit ist durch eine Seriennummer möglich. Jedem Gerät werden im Rahmen eines Funktionstests nach der Fertigung diese Seriennummer sowie Funktions- und Kommunikationsparameter (z.B. Abgleichdaten für Sensoren oder Übertragungsraten, ...) zugeteilt.

Nach der *Erstinbetriebnahme* sind die Module vorbereitet für die Phase der „Adreßvergabe". In diesem Zustand werden sie in eine Anlage eingebaut. Durch ein Mastermodul wird das System konfiguriert, d.h. es werden alle für den Normalbetrieb benötigten Identifier dynamisch zugeteilt. Durch spezielle Identifier wird vom Mastermodul zunächst durch Abfrage der Seriennummer die Art und Anzahl der im System installierten Knoten festgestellt. Diese *Identifikation der installierten Knoten* erfolgt in zwei Schritten:

- Bereiche von Seriennummern werden abgefragt:
  Nach einem Suchverfahren wird für Bereiche von Seriennummern abgefragt, ob ein oder mehrere Knoten dieses Seriennummernbereiches im System installiert sind. Alle Bereiche, in denen sich ein oder mehrere Knoten gemeldet haben, werden weiter behandelt.

- Einzelsuche von Seriennummern:
  Die in Schritt 1 erwähnten Bereiche von Seriennummern werden einzeln nach vorhandenen Knoten durchsucht.

Diese Vorgangsweise in zwei Schritten hält die erforderliche Zeit für die Identifikation der Knoten in Grenzen. Das Ergebnis dieser Selbstidentifikation ist eine Liste von Seriennummern der im System installierten Knoten.

Durch die implementierte Funktion der einzelnen Knoten ist festgelegt, welche Daten zu anderen Teilsystem übertragen beziehungsweise von denen empfangen werden. Aus der Projektierung des Gesamtsystems steht also eine *Variablenliste* aller auszutauschenden Daten zur Verfügung. Für die Generierung der CAN-Identifier wird diese Liste durch Angabe von Prioritätsklassen der einzelnen Variablen ergänzt.

Eine im Masterknoten implementierte Software erstellt entsprechend dieser Variablenliste eine CAN-Identifier-Liste, die als Grundlage für die im folgenden durchgeführte *Vergabe der Identifier* für die Variablen der einzelnen Komponenten dient.

Jede Komponente erhält eine der Anzahl der auszutauschenden Objekte entsprechende Menge von Identifiern zugeteilt. Übertragen werden diese Identifier in einer Nachricht in Verbindung mit der Seriennummer, sodaß jeder Knoten durch Vergleich der eigenen Seriennummer mit der übertragenen auswerten kann, ob der empfangene Identifier für ihn relevant ist.

Nachdem alle Identifier vergeben wurden, gehen alle Komponenten in den Normalbetrieb über und sind nur mehr über die zugeteilten Identifier an der Kommunikation beteiligt.

In Abb. 1 sind diese Phasen der Inbetriebnahme dargestellt.

3 Projektierung

Die Aufteilung der Gesamtfunktion in örtlich und vor allem logisch getrennte Teilsysteme, die weitgehend unabhängig voneinander geplant und realisiert werden können, ist ein vielversprechender Ansatz zur Lösung von komplexen Automatisierungsaufgaben. Die markanten Vorteile, die sich dadurch im Engineering und bei der Inbetriebsetzung ergeben, sollen hier nicht weiter diskutiert werden.

Die Vorgangsweise für die Projektierung verteilter Systeme soll im folgenden näher betrachtet werden.

Im ersten Schritt wird für jedes Teilsystem festgelegt, welche Daten von anderen Teilsystemen benötigt werden. Daraus ergibt sich für jedes System eine Liste von Eingangsdaten mit entsprechender Formatangabe und Angabe des Quellsystems, das den Wert liefert. Für das Erstellen dieser Eingangsliste ist noch keine Koordinierung unter den Programmierern der einzelnen Teilsysteme notwendig.

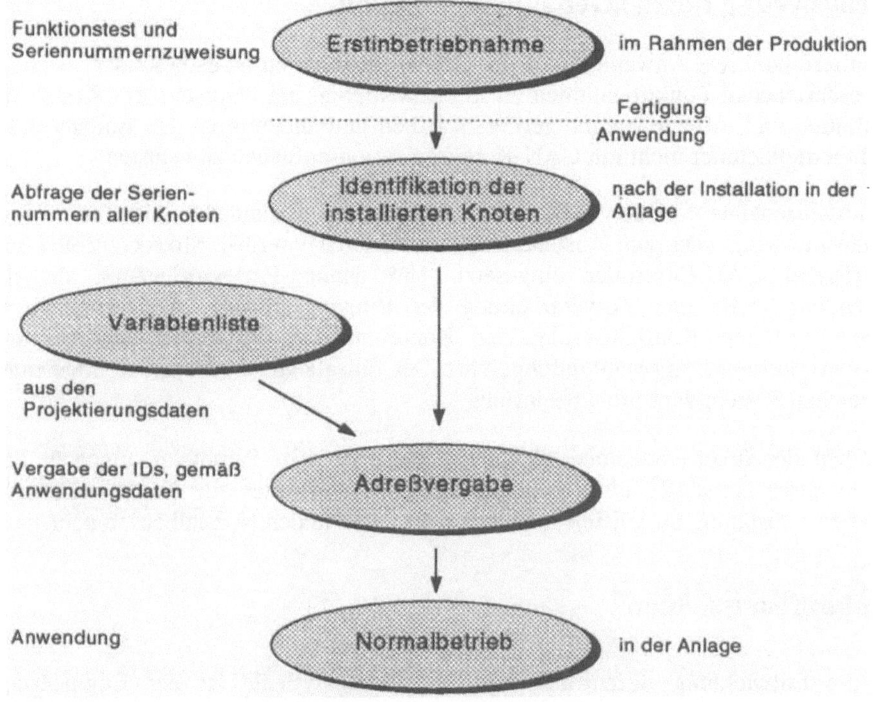

Abb. 1. Inbetriebnahmephasen der Module

Im nächsten Schritt wird eine Gesamtliste erstellt, in der alle Variablen angeführt und nach Prioritätsklassen sortiert werden. Mit diesem Schritt ist die Arbeit der Projektingenieure bereits abgeschlossen. Die weiteren Schritte können weitgehend automatisiert erfolgen.

Ein Softwaretool vergibt nun die Identifier entsprechen der vorhin erstellten Liste und erzeugt ein Include-File mit symbolischen Konstanten für die in C geschriebene Software der einzelnen Knoten. Dieses File wird beim Compilieren der Teilsysteme eingebunden. Es bestehen einige Eingriffsmöglichkeiten:

- Zum Beispiel werden für einfache I/O-Module, deren Identifier mit DIP-Switches fest einzustellen sind, fixe Werte vergeben.

- Einzelbitnachrichten mit der selben Quelle können einer gemeinsamen Nachricht zugeordnet werden.

- Zwischen den Identifiern bleibt Platz für spätere Erweiterungen.

4 Realisierung für die Gebäudeautomation

Für unsere konkrete Anwendung in der Gebäudeautomation ist es besonders wichtig, die beschriebenen Funktionalitäten zu implementieren, um einerseits die Kosten für Installation und Inbetriebnahme gering zu halten und andererseits den Endanwender und Inbetriebnehmer nicht mit CAN-Kommunikationsaufgaben zu belasten.

Für kostensensible Module (z.B. zur Aufnahme von Raumdaten wie Temperatur, Luftfeuchte, etc., oder zur Ansteuerung von Aktoren) werden Mikrocontroller mit integriertem CAN-Controller eingesetzt. Um genug Prozessorleistung für die Anwendung (z.B. eine Vorverarbeitung der Eingangsgrößen) zur Verfügung zu haben, wird die Konfigurierung der Kommunikation (Identifier-Vergabe) mit möglichst geringem Aufwand durchgeführt. Der Einsatz von CANopen Komponenten im gleichen System wird nicht beeinflußt.

Nachdem der Inbetriebnehmer alle anwenderspezifischen Parameter angepaßt hat, werden durch Anwählen eines entsprechenden Menüpunktes alle Kommunikationsaufgaben selbständig initialisiert und das System geht in den Normalbetrieb über.

5 Zukunftsaussichten

Für die Entwicklung dezentraler Automatisierungssysteme ist der Engineering-Aufwand nur dann in Grenzen zu halten, wenn für die Generierung der Identifier aus den Anwendungsdaten Werkzeuge zur Verfügung gestellt werden. In Ergänzung mit der Möglichkeit der flexiblen Vergabe von Identifiern stellen damit Inbetriebnahme, Konfigurierung und Erweiterung, sowie Service und Wartung von Anlagen kein Problem mehr dar.

Voraussetzung und letztendlich entscheidend für den breiten Einsatz dezentraler Systeme als Ersatz für zentrale Strukturen ist die Verfügbarkeit geeigneter Softwarewerkzeuge, die den Ingenieuren eine komfortable Projektierung der gesamten Anlage in ihrer räumlichen Verteilung im Gebäude ermöglichen. Durch massive Unterstützung soll die Akzeptanz der Projektingenieure gesteigert werden, denn ohne sie ist jede wirtschaftlich und technisch noch so vorteilhafte Entwicklung schwer durchsetzbar.

Literatur

1. Produktunterlagen der Fa. ATB Automatisierungstechnik, Tel.: (+43) 07235 / 65040, Fax: (+43) 07235 / 650404, email: wolfgang.bernhard@telecom.at

Easy Installation and Home Management

M. Goossens

The EIB Association (EIBA),
Tinklaan 5, B-1160 Brussels, Belgium

Abstract. Designed for the EIB System for Home and Building Electronics, the EIB Home Management (EHM) concept combines a powerful software framework for PC's (the Home Assistant) with a device model mirrored on EIB Distributed Objects. Its notion of wizard-like Easy Installation realistically matches the aspirations of plug and play with the not always deterministic nature of user expectations. A dedicated Appliance Interface opens up EHM to white and brown goods.
The resulting approach brings management-level control and integration to Home Automation. Thanks to a specially designed touch-screen User Interface, the occupant can unlock the versatility of the underlying EIB installation.

1. Preface

The world of Home & Building Electronic Systems is buzzing with headlines such as *Plug 'n Play, intelligent devices*, etc. Sifting technical reality from marketing hype is not always easy. Still, the prominence of these concepts indicates the underlying need for various forms of *Easy Installation* - particularly for the home market. In contrast with office or commercial buildings, particularly the larger ones, there can be no full-time engineer running about to manage and maintain the building. And for white and brown goods which the consumer installs him- or herself, there is not even a qualified expert around to set up and configure the device... Against this background, EIB Home Management (EHM) was devised as an extension to the EIB system [1].

EHM assumes that an electrical installation based on EIB is present. It is a PC-based, modular and extendible Management system which encompasses all application domains in the home, centred around the *Home Assistant* Framework. Through this platform, the occupants gain access to all available functionality of the underlying system - most of which will otherwise remain out of reach. Typical applications include security, energy management, household appliances, multimedia and communication.

The Home Assistant is already commercially available. Since some features are still on the drawing boards, the information presented in this paper should nevertheless be considered as preliminary.

2. The Home Assistant Framework

2.1 Survey and User Interface

The centrepiece of EHM is the Home Assistant (HA). The HA is an extendible software framework for Windows PC's. (For more on the concept of frameworks: see [2, 3].) With a specially designed touch-screen User Interface, the Home Assistant finally brings the power and flexibility of the underlying HBES to the fingertips of the consumer. It can be used as a console for visualisation and control, available in several rooms. Fig. 1 shows the main menu (German version) with the functions *security, agenda, (tele)communication, appliances, home* and *light & temperature* in the top two rows.

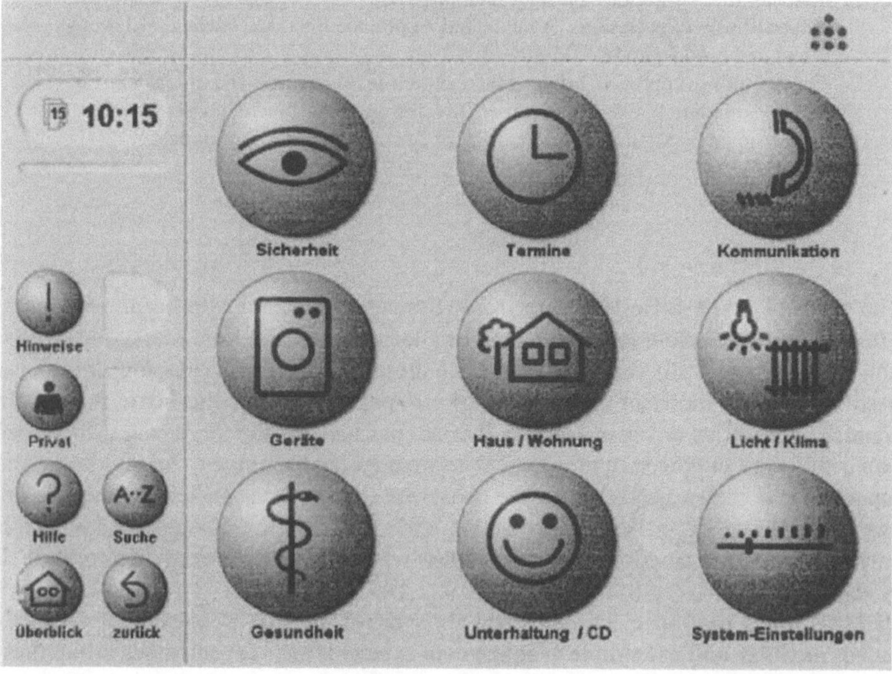

Fig. 1. Touch-screen user interface of the Home Assistant.

2.2 HA Framework Architecture

Fig. 2 illustrates the main plug-in architecture of the HA framework. HA defines a set of API's allowing application or product specific plug-ins access to both kernel and User Interface features of the Home Assistant. Horizontal applications provide general services, dealing with such things as scenario's, telecom and remote services, etc. Vertical applications specialise one particular function domain (such as heating control), or one particular device (such as a household appliance). The Home Assistant Tool Software (HTS) helps 3rd party programmers to create their own HA components and customise OEM features[1].

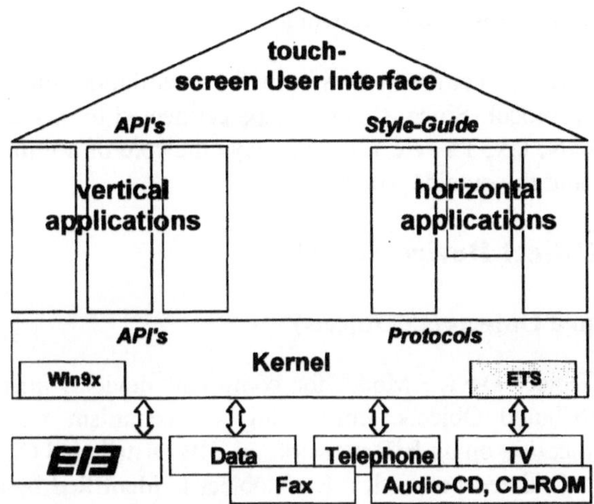

Fig. 2. Architecture of the Home Assistant framework.

2.3 The HA Repository

HA has a Repository, implemented as a relational database, to cope with persistent data. It extends the repository of the general PC software framework for the EIB, known as the EIB Tool Environment (ETE). This vendor-independent infrastructure is used by various tools, including the EIB Tool Software (ETS) for EIB project engineering. ETE contains a database with both (manufacturer specific) product templates as well as (project specific) product instances [4].

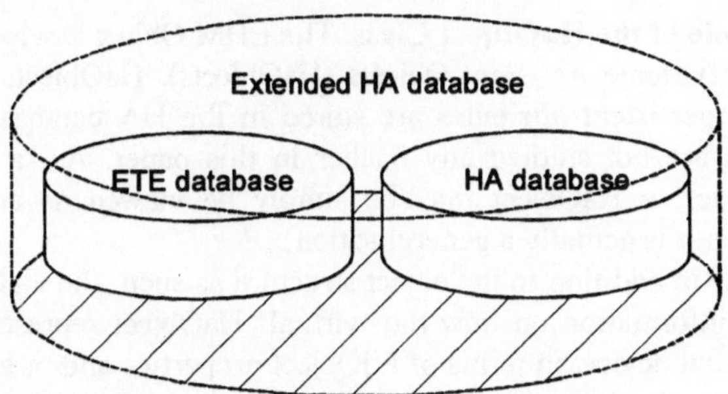

Fig. 3 Joining the ETE and HA databases.

---

[1] Both the Home Assistant framework and the HTS may be licensed from Bosch-Siemens Hausgeräte.

As shown in fig. 3, HA logically extends this ETE repository database, and makes it accessible to HA applications via a HA kernel API.

2.4 Communication on the Home Network

Within the home, the HA communicates with the installation and the EHM appliances via the EIB protocol. Since it can also be connected to any external network (like telephone, ISDN, fax, TV, etc.), it also plays the role of a multimedia gateway between the home and the outside world.

3. The EHM Object Device Model

3.1 EIB Distributed Objects (EdObjects)

EHM defines an Object Device Model for compliant devices, mirroring (and extending) EIB Distributed Objects. Introducing a mechanism for authorised and structured communication on an EIB network, an EIB Distributed Object (EdObject) bundles a number of related properties. Each object is identified by an index in the local device: a communication partner (for instance the Home Assistant) can scan a device's objects and for each object it finds, determine the object type. The object type as well as the type of each property are stored in the device as well. A property is referenced via a Property Identifier (property_id), and may itself be an indexed array. (See [5, 6] for details.)

The fact that the value of property may be declared as a shared variable (for EIB commonly known as a Communication Object), is exploited by EHM. In this case, we say that the property and the shared variable *coincide*. In this way, efficient and transparent run-time access is ensured via multicast (or 'group') addressing [7].

3.2 HaObjects: the Home Assistant's View of an EIB Device

Structure and Role of the HaObject Class. The EHM Object Device Model 'sees' a device as a (set of) Home Assistant Objects (HaObjects). HaObjects live in the HA framework: their persistent attributes are stored in the HA database. Their implementation details are not studied any further in this paper. As 'a coherent set of properties' however, a HaObject may be simply be viewed as an EdObject-like structure (of which it is actually a generalisation).

Fig. 4 depicts how in addition to the object structure as such, the HaObject holds the (device-specific) information on how the 'virtual' HaObject representation is to be mapped to the actual device, in terms of EdObject properties and/or shared variables. In fact, the 'real' device is again replaced by a virtual representative, in this case an ETE device instance, as shown in fig. 4.

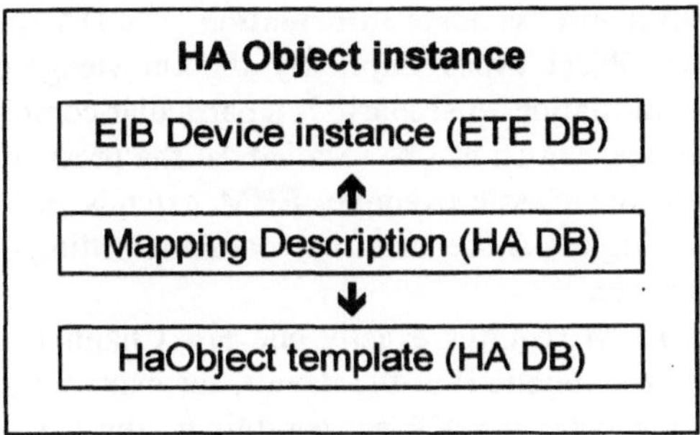

Fig. 4. Logical aggregation of ETE and HA database entities into a HaObject.

Importantly, the HaObjects constitute the full (and only) interface to this device for the HA. They all have to fulfil the following criteria:
1. For each shared variable, there must be a coinciding property.
2. There must be a property which explicitly contains the mapping between the coinciding properties and shared variables.

Actually, the HaObject is a metaclass. The inheritance hierarchy is given in Fig. 5.

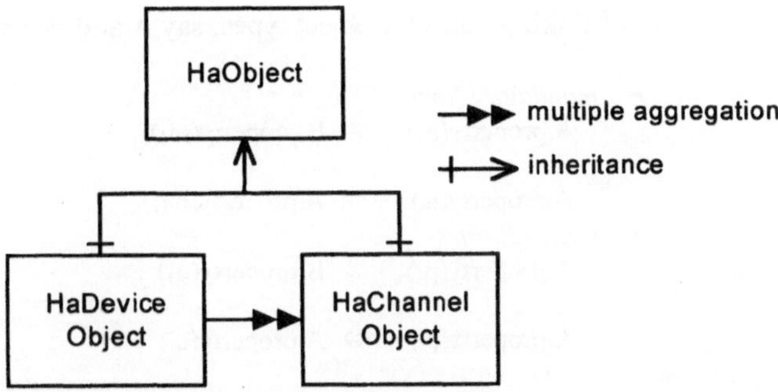

Fig. 5. Inheritance and aggregation for HaObject classes.

The HA Device Object. Each device is modelled as a HaDeviceObject. Essentially, it contains the following properties:
1. *device name* (string)
2. *location* (string): where the device (or its effect) is located
3. *function* (enum): what kind of function the device fulfils, e.g. push-button, sensor, ...
4. *usage* (enum): what kind of application the device is used for, e.g. lighting, temperature control, ...
5. a list of HaChannelObjects, one for each channel of the device (see Fig. 5 and section 3.2.4. below).

HA Device Object Types and Scenario Information. The HA specification defines a whole series of Device Object Types. Implicitly, the knowledge of these types also incorporates *scenario* information. A scenario is a particular combination of settings-values for a series of devices which can be recalled "at the press of a button"; 'mood lighting' scenarios form the classic example. EHM extends this notion across all application domains: a single HA scenario may combine settings for lighting, temperature control, shutters etc.

The Channel Object. EHM requires exactly one HA Channel (or 'Applica-tion') Object for each channel (or function) of the device, for example of an n-fold binary output. In addition to the general HaObject conditions, the Channel Object has to fulfil the following criterion:

- The status of the channel must be explicitly readable via a property. (For example a binary output with an internal delay must contain a property which shows the current status of the output.)

4. Easy Installation

All of this obviously remains invisible to the occupant. Nevertheless, it is precisely the Object Device Model which enables him or her to set up and configure new devices through the Home Assistant. The knowledge of how devices may be linked is embedded in HA as a set of linking-rules for object types, say A and B, as in Table 1.

- *mandatory links*
    $$A.property(n_1) \quad \Leftrightarrow \quad B.property(m_1)$$
    ...
    $$A.property(n_i) \quad \Leftrightarrow \quad A.property(m_i)$$
- *optional links*
    $$A.property(p_1) \quad \Leftrightarrow \quad B.property(q_1)$$
    ...
    $$A.property(p_j) \quad \Leftrightarrow \quad A.property(q_j)$$

Table 1. Linking rules for coinciding shared variable properties.

Importantly, the HA guides the user through the configuration process in a wizard-like way. Thus, the problem of matching the user's expectations with the system's resulting behaviour is elegantly solved. This device-specific procedure is provided by the vendor in a standard HA format on a CD-ROM.

5. The Appliance Interface

For household appliances, a special appliance interface was developed, based on EIB Bus Access Units (BAU) which implement EdObjects. They are available for both the EIB Twisted Pair and Powerline media. As a universal interface, it is not restricted to white goods, however.

One role of the appliance interface is to ensure galvanic separation between appliance and EIB, as well as to provide EMC protection for the EIB network. It also holds the project-specific configuration data (such as addresses) for the device it is connected to. It will also verify whether it is set up for this device, and informs the Home Assistant about the status of the local communication.

Acknowledgements

The author wishes to thank Mr. P. Ferstl (Bosch-Siemens Hausgeräte GmbH) for the numerous yet succinct explanations.

References

1. Goossens, M.: A Survey of the EIB System. In: EIBA Proceedings, EIBA (1997).
2. The Power of Frameworks, Addison-Wesley 1995 (The Taligent Reference Library).
3. Booch, G.: Object-oriented Analysis and Design, Ch. 9. The Benjamin / Cummings Publishing Company, 1994.
4. Goossens, M.: Component-based Project Engineering. In: FeT Proceedings 1997, Springer 1997.
5. The EIB Handbook Issue 3 - Vol.3 (preliminary edition). EIBA 1997.
6. Goossens, M.: Object-based Distributed Application Design. In: FeT Proceedings 1997, Springer 1997.
7. Goossens, M.: Communication and Addressing on EIB. In: EIBA Proceedings, EIBA 1997.

A comprehensive fieldbus testing methodology

Christophe Crocombette

Electricité de France
Direction des Etudes et Recherches, Moret sur Loing, France

Abstract. The purpose of this paper is to present the fieldbus testing methodology which is used in Electricité de France laboratories. Tests are notably achieved to assess interoperability and interwork between products. Their goal is therefore not only to prove that products conform to a specific standard but also to check their ability to interoperate and interwork. Specific tools like a network analyser and a polynomial waveform synthetiser are used to both control and observe communicating product behaviors on the network.

1 Introduction

Electricité de France (EDF) is one of the largest electrical utilities in the world. An integrated public corporation, it carries out the construction, operation and maintenance of French power plants. It is also responsible for the transmission of electricity in France and for its distribution to 29 million customers. Thus, EDF has a broad need of fieldbusses in many areas like process control or home automation and is deeply involved in communicating product testing.

2 Conformance testing limitations

Communicating product testing first relies on conformance testing which is standardized by the ISO-9646 standard [1]. Any implementation under test (IUT) is described thanks to a Protocol Implementation Conformance Statement (PICS). The PICS gives the standard capabilities and options which have been implemented. Four types of conformance testing can be achieved, according to the extent to which they provide an indication of conformance. Basic interconnection tests provide prima facie evidence that the IUT conforms ; capability tests check that the observable capabilities of the IUT conforms to the allowed minimum capabilities of the PICS (static conformance requirements) ; behavior tests check that all the observable behaviors are permitted (dynamic conformance requirements) ; conformance resolution tests probe in depth the conformance of the IUT to particular requirements.

The limitations of conformance testing have various aspects [2, 3, 4] :

1. Exhaustivity of tests : especially, the tests suites do not enable to make sure that two products have enough ressources (memory, speed capacities, etc.) to properly interwork with each other ;

2. Hidden layers : protocol layers may be implemented inside silicon and therefore unaccessible ;

3. The upper tester may only work if the implementation already conforms to the standard, which is precisely what is to be demonstrated [4] ;

4. It does not garantee interworking or interoperability. As mentioned in [1], 'the purpose of conformance testing is to improve the probability that different implementations are able to interwork'.

In brief, conformance testing is only the first step in the global process which aims at proving that any elements of a communication system shall interwork.

3 Interoperability and interworking

Interoperability and interwork are two different concepts which must be referred to when assessing any communicating product.

Interoperability between two implementations is their ability to exchange data between their communication stacks. Interoperability can thus be categorized in different N-interoperability levels, where N is the considered layer. Unfortunately, application layer interoperability (7-interoperability) does not garantee that different products shall interwork properly.

Interwork between two applications (sets of application processes) or products is indeed their ability to exchange data and achieve the specific task they have been built for.

Let us quote an example : if a heat manager requests from a sensor a room temperature in farenheit degrees and if the sensor responses the value in celsius degrees, the products may interoperate but will not interwork ! Thus, a deeper analysis of communication and companion standard is often required to prevent this kind of problems. Interwork tests notably involve both the communication stack and the user layer.

4 Interoperability testing requirements

In conformance testing, a single implementation is confronted with a testing system. On the contrary, interoperability testing, which is not standardized, intrinsequely confronts several implementations with each other and also with a testing system.

The EDF fieldbus testing methodology is based on simulation and observation on both sides of the communication stack [4]. The user layer of any product behaves like a filter which forbides direct access to the communication stack. Node simulation is therefore necessary to provide the most complete covering. Simulation can be achieved using a previously tested communication stack, a library of protocol primitives and a high level language like C. Anyhow, the most flexible and reliable tool for checking communication mechanisms on the medium itself should be a physical layer Protocol Data Unit (PDU) generator. This generator should not depend on any implementation. In fact, this tool can be achieved thanks to a polynomial waveform synthetiser (PWS). With the right interface, it can emulate physical PDU on any medium (twisted pair, powerline carrier, radiofrequency, etc.). Simultaneously, physical observation of the network enables to continually check the emitted frames.

We applied this methodology with many fieldbusses, including PROFIBUS, FIP, BITBUS, CAN for industrial applications and EIBus, Batibus, LON and EHS for home and building automation. The following examples (CAN and EHS networks) illustrate some points which have previously been explained.

5 Examples

5.1 CAN network

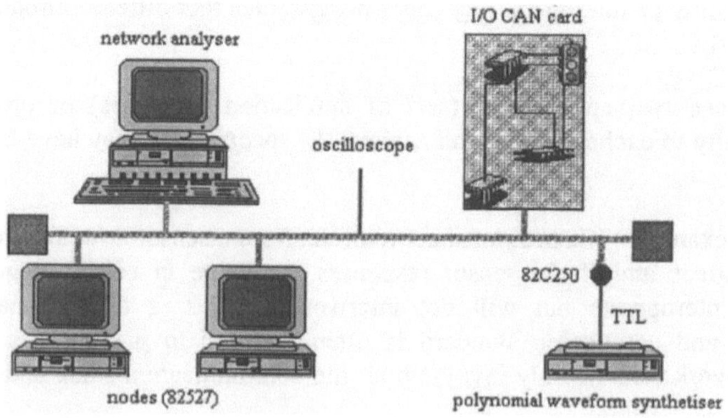

Fig. 1. Test Platform

We assessed [5] a CAN system on an experimental platform which included 3 nodes (a CAN I/0 card based on the NS 57C360 chip, two nodes whose controllers were based on the Intel 82527 chip), a network analyser, an oscilloscope and a PWS. The PWS was connected to the network with a 82C250 chip based interface (figure 1).

MAC sublayer services (MA_DATA.req, MA_DATA.ind, MA_DATA.conf, MA_REMOTE.ind, MA_REMOTE.req, MA_REMOTE.conf, MA_OVLD.req, MA_OVLD.ind, MA_OVLD.conf) were not accessible (limitation 2 of § 2).

The PWS enabled to check CAN error management : for instance, we generated a frame with a wrong CRC to check the CRC error detection and observed the right behavior (in this example, nodes are in the active error state and send an active error frame just after the ack delimiter field, see figure 2).

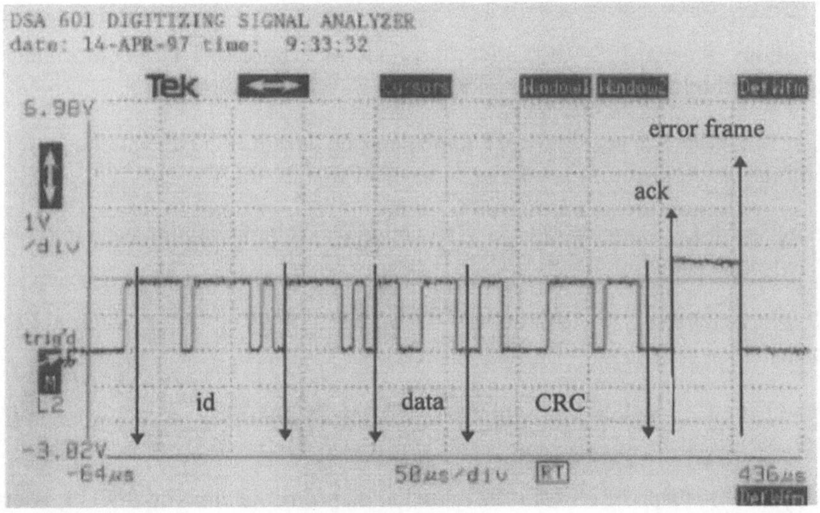

Fig. 2. CRC Error Generation

5.2 EHS (European Home Systems) network

We assessed [6] some EHS prototypes including a light level control system and a heat manager system. A test platform, similar to the CAN test platform (but without PWS) enabled us to analyse EHS exchanges.

The EHS standard has a blocking mechanism which enables a node to send several commands in a single frame. Depending on the size of their reception buffer, we observed that nodes had different blocking capabilities. Our network analyser was itself able to manage 4 command long frame but could not analyse longer frames (limitation 1 of § 2).

We also tried to send a wrong level (more than the maximum level) to a lamp module and observed that this kind of data was not yet secured : the communication stack accepted the command and the lamp had a corrupted behavior.

6 Conclusion

Interoperability and interwork tests are not as expensive as conformance tests and seem to be really relevant when different products from different brands are to communicate. They seem to be complemantary to conformance tests since they focus on a system (assessment of a specific system instead of assessment of a given implementation) and its user layer.

References

1. International Standard ISO/IEC 9646, Information technology - OSI conformance testing methodology and framework, 1989.

2. Siebert M. : L'interopérabilité des équipements de terrain. Vers une méthode de vérification. Application au réseau de terrain FIP, CNAM Thesis, 1993.

3. Benkhellat M. L. : Formalisation et vérification de l'interopérabilité dans les systèmes de communication, Ph. D. Thesis, 1995.

4. Gallon F., Schweitz G. : Méthodologie de test des réseaux de terrain, internal EDF note, 1994.

5. Crocombette C., Joachim S. : Rapport d'essais sur le protocole CAN, internal EDF note, 1997.

6. Labruyère M. : Rapport d'essais d'évaluation du protocole domotique EHS, internal EDF note, 1997.

SIPARK das Parkleitsystem

H.P. Uhl

SIEMENS AG Österreich, IV N, Siemensstr. 88-92, 1210 Wien,
Österreich

Abstract. Am Flughafen in Münster/ Osnabrück in der Bundes-
republik Deutschland wurde erstmals SIPARK das Parkleitsystem von
SIEMENS realisiert. Auf 30 000 m² in fünf Etagen stehen den
Reisenden 1100 Stellplätze zur Verfügung. Über jeden einzelnen
Stellplatz im Parkhaus wird ein Ultraschallsensor montiert, der den
Belegezustand des Parkplatzes ermittelt und mittels Feldbussysteme,
AS- Interface und PROFIBUS DP, an eine Bereichssteuerung weiter-
leitet. Die dadurch gewonnenen Daten werden zur Verkehrsleitung
verwendet.

Abstract. For the first time, SIPARK the carpark guiding system from
SIEMENS was carried out at the airport in Münster/ Osnabrück in the
Federal Republic of Germany. On 30 000 m² at five floors 1100
parkinglots are at the disposal for the travellers. Above every parkinglot there is an Ultrasonic Proximity Switch assembled, which
establiches the occupancy status of the parkinglot and with the help of
two fieldbussystems, AS- Interface and PROFIBUS DP the information is transmitted to the controllunit. The achieved information is
then used for the traffic controllsystem.

1. Systembeschreibung

Das gesamte Parkleitsystem setzt sich aus verschiedenen Komponenten zusammen
(Parkhaussensor, Gateway, Anzeige und Steuerung SIMATIC S7), die sich in
mehrere Hierachieebenen strukturieren lassen.

2. Parkhaussensor

Für die Einzelplatzüberwachung kommt ein berührungs- und kontaktlos arbeitender
Ultraschallnäherungsschalter, ohne mechanische Teile und damit wartungsfrei, zum
Einsatz. Der Parkhaussensor arbeitet nach dem Echo- Laufzeit Verfahren, d.h. es
wird der zeitliche Abstand zwischen Sende- und Echoimpuls ausgewertet. Der
Sensor erkennt durch Ausmessen der Deckenhöhe und Vergleich mit einer Refer-

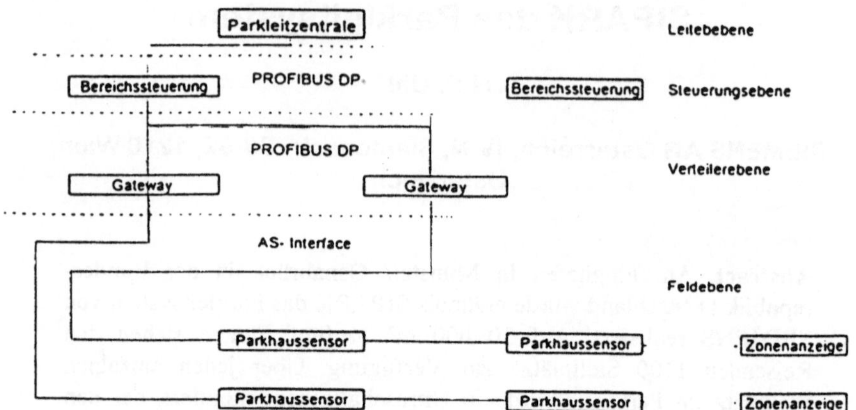

Abb.1. Informationsfluß von den Parkhaussensoren bis zur Parkleitzentrale mit Kennzeichnung der einzelnen Ebenen

enzhöhe, ob ein Parkplatz frei oder belegt ist. Eine Visualisierung „Frei" oder „Belegt" erfolgt durch mehrere, direkt am Sensor angebrachter, LED' s.

Alle Parkhaussensoren werden mit dem Feldbussystem AS- Interface vernetzt. Die Adressierung (1 bis 31) des Parkhaussensors wird einmalig bei der Montage mit einem Adressiergerät durchgeführt. Die Parametrierung (Einstellung vom Sollabstandswert, Meßzyklen und Betriebsart) erfolgt automatisch beim Hochlauf durch die Steuerung mittels spezieller Datenbausteine. Dadurch konnte die Inbetriebnahmezeit um 50% reduziert werden.

Um eine gegenseitige Beeinflussung der Ultraschallsensoren zu vermeiden, müssen die einzelnen Sensoren auf verschiedene Meßkreise aufgeteilt und zusätzlich von der Bereichssteuerung aus getriggert werden. Dadurch wird erreicht, daß benachbarte Stellplätze, in zeitlich voneinander getrennten Intervallen, abgetastet werden.

3. Anzeigen

Neben den Parkhaussensoren bilden die Zonenanzeigen und die Großanzeigen eine Kernkomponente des Parkleitsystems. Die Großanzeigen wurden mit PROFIBUS DP in das Parkleitsystem SIPARK eingebunden. Bei der Einfahrt in das Parkhaus erkennt der Parkplatzsuchende die genaue Anzahl der freien Parklätze in der jeweiligen Etage. Der Parkplatzsuchende wird ab der Einfahrtsschranke durch Zonenanzeigen mit grünen und roten Richtungspfeilen zum nächst gelegenen freien Stellplatz geleitet. Die Zonenanzeigen werden mittels AS- Interface verdrahtet. Die Adressierung wird einmalig bei der Montage durchgeführt. Eine weitere Parametrierung bei der Inbetriebnahme ist nicht erforderlich.

4. Kommunikation

Die vier Hierarchieebenen des Parkhauses kommunizieren mit den beiden unterschiedlichen Feldbussystemen AS- Interface und PROFIBUS DP miteinander. Als Gateway wurde ein DP/ AS- I Link eingesetzt.

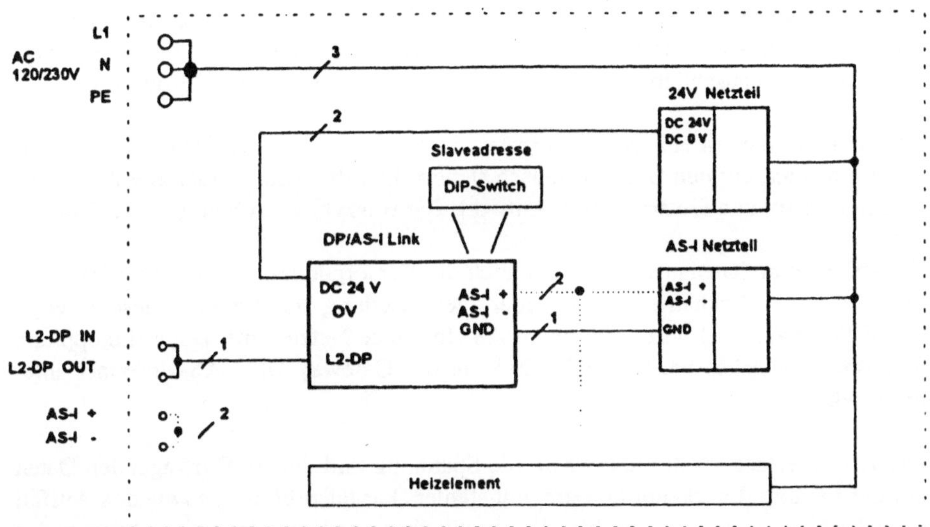

Abb. 2 . Gateway DP/ AS-I Link

4.1 AS- Interface

Das AS- Interface ist ein Vernetzungskonzept für den untersten Feldbereich in der Automatisierungsebene gemäß AS- I Spezifikation nach IEC TG 17 B. AS- Interface ist ein herstellerunabhängiger Standard.

Über die AS- Interfaceleitung kommuniziert der Master (Gateway) mit den Teilnehmern (Parkhaussensoren) im Feid. Zur Vernetzung kann eine einfache Zweidrahtleitung mit einem Querschnitt von 2x 1,5 mm² verwendet werden. Abschirmung, Verdrillung oder Abschlußwiderstände sind nicht notwendig.

Diese Zweidrahtleitung ist profiliert, womit ein verpolungssicherer Anschluß gewährleistet wird. Die Leitung wird mit Hilfe der Durchdringungstechnik kontaktiert. Ein Abschneiden, Abisolieren oder Verschrauben ist nicht erforderlich.

Das Verlegen der AS- Interfaceleitung erfolgt wie bei einer herkömmlichen Elektro-installation. An jeder Stelle der Busleitung kann ein neuer Abzweig begonnen werden. Man kann Stern-, Linien- und Baumstrukturen realisieren. Die maximale Leitungslänge ist 100 m und kann durch die Verwendung von Repeatern bis auf 300 m ausgedehnt werden.

Als Master wurde für SIPARK ein spezielles Gateway zwischen AS- Interface und PROFIBUS DP entwickelt. Dieses Gateway bietet die Möglichkeit bis zu 31 Parkhaussensoren anzuschließen. Jeder Sensor überträgt 4 bit (jeweils 2E und 2A).

Das AS- Interface ist ein „Singel- Master- System", d.h. pro Leitungsstrang existiert nur ein Master, der den Datenaustausch steuert. Er ruft nacheinander alle AS- Interface Slaves auf und erwartet deren Antwort. Die Buszykluszeit beträgt max. 5 ms.

Zusätzlich zum Datenaustausch wird auch die Versorgungsspannung der Elektronik und Sensorik (Parkhaussensoren) über diese Leitung übertragen. Diese Energie (30 V DC max. 7 A) wird von einem AS- Interface Netzteil mit Datenentkopplung eingespeist, welches bei SIPARK direkt in das Gateway DP / AS- Interface integriert wurde.

Der AS- Interface Master überwacht die Spannung und die zu übertragenden Daten auf der Leitung. Er erkennt Übertragungsfehler (Paritätsprüfung), sowie den Ausfall eines Parkhaussensors und meldet dies an die übergeordnete Steuerung (Anzeige des Statusbits 1 im Datenwort des jeweiligen Parkhausensors). Nach dem Austausch wird der neue Parkhaussensor automatisch von der Steuerung durch die beiden Bussysteme neu parametriert. Ein Austausch von Parkhaussensoren ist während des normalen Betriebes möglich und stört die Kommunikation der anderen Teilnehmer nicht.

4.2 PROFIBUS DP

Der PROFIBUS DP zeichnet sich durch seine schnelle (max. 12 Mbit/s), zyklische Kommunikation nach dem Token Ring Prinzip aus. Bei einer Kabellänge von 1000m wird PROFIBUS DP mit 187,5 kBit/s betrieben.

Das elektrische Netz verwendet eine geschirmte, verdrillte Zweidrahtleitung. Die RS 485 Schnittstelle arbeitet mit Spannungs- differenzen und ist daher unempfindlich gegen Störeinflüsse.

Mittels PROFIBUS DP werden die einzelnen Gateways und die Großanzeigen miteinander vernetzt. Weiters wird mittels PROFIBUS DP die Ankopplung an die übergeordnete Steuerung realisiert.

5. Zusammenfassung

SIPARK

+ verbessert den Auslastungsgrad eines Parkhauses
+ sorgt für höchste Stellplatztransparenz
+ garantiert höchste Wirtschaftlichkeit durch hohe Verfügbarkeit der Stellplätze
+ garantiert bei Meldung „Frei" immer einen freien Parkplatz
+ entlastet die Zufahrtswege
+ verringert durch kürzere Zufahrtswege die Schadstoffbelastung und die Unfallgefahr
+ erspart Zeit und Streß bei der Parkplatzsuche

Literatur

1. Werner R. Kriesel, Otto Madelung.: ASI The Actuator - Sensor - Interface for Automation, Hansa 1994, 143-160

2. Forster A.: SIEMENS AG AUT 645, SIPARK.: 04.06.1996 V 0.2, 3 - 13

3. DIN E 19245 T.3, PROFIBUS DP

4. Heger, D. : Konformitätsprüfung offener Kommunikationsprotokolle für die Automatisierungstechnik, Fachberichte MSR, Springer-Verlag Berlin, Heidelberg, New York 1986, Bd.14 S. 658-666

SPARK

- erkennt den Auslastungsgrad eines Softwares
- sorgt für einen Stellplatzumsatz
- garantiert höchste Wirtschaftlichkeit durch hohe Verdichtung an den Stellplätze
- garantiert bei Nutzung ... bei Nutzung der Parkraum
- nützt die Zeitabläufe
- ermöglicht durch klare Zuweisung die Bearbeitung der Stellplätze
- reduziert Zeit und Streß bei der Parkplatzsuche

Literatur

1. Walker R. (Hrsg.): Das Management of the Network. Seminar Internationaler Automaten, Hanau 1994, 241-244

2. Fischer ANSPRACH S AG ALFRED SLE... M/E 1998 V 277 IN

3. DI... SM JESIES 1-5 PROFINUS DP

4. Kögel D.: Kinder-Erziehung von der Kommunität insgesamt Folge für die Automatisierungstechnik. Fachzeitschrift ABK Nummer Verlag Berlin, Heidelberg, New York 1998, Bd. 14 S. 63-690

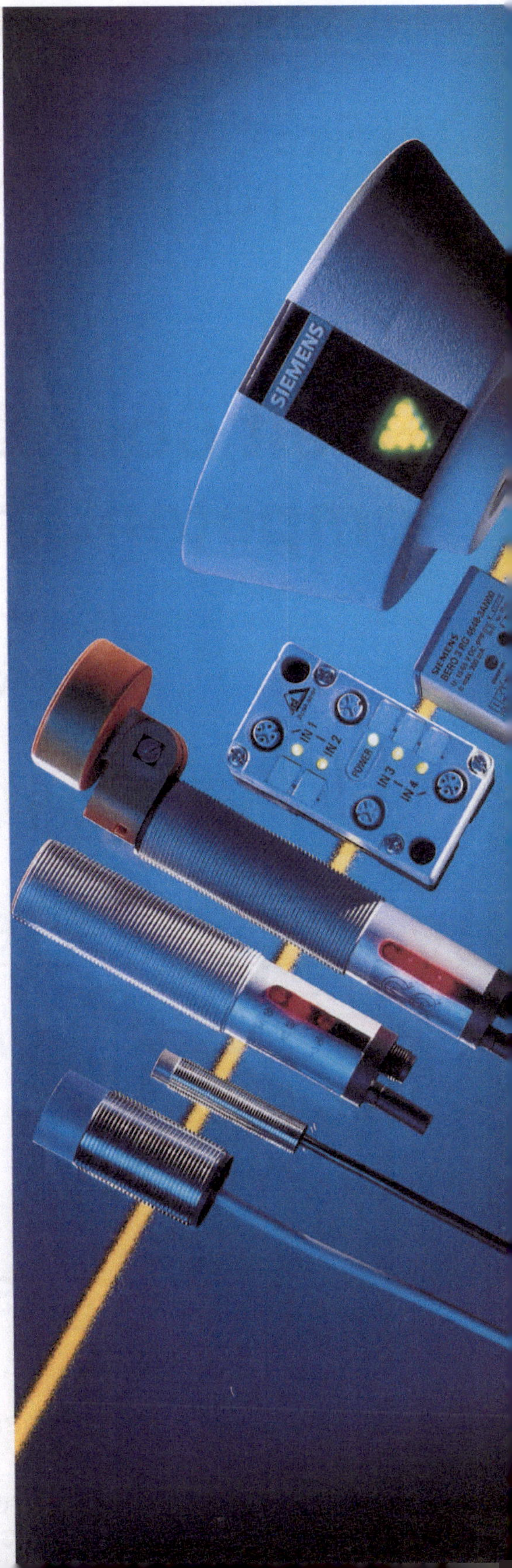

Selection of Fieldbus Systems and the Provision of their Interoperability with IEC 1131

Birgit Scherff, Hagen R. Wenzek, Wolfgang Halang

FernUniversität – University of Hagen
D-58084 Hagen, Germany

Abstract. Open and modular systems architectures are the main requirements for mechanical and process plant engineers when making their selection of components for control and process control technology. In mechanical and process plant engineering a system architecture consists of technology for actuators, electronical controls and process controls. It is built upon a wide scale from PLCs up to real-time applications based on VMEbus and PCs that are interconnected with fieldbusses. To make a solid selection for the appropriate bus system, the project engineer has to consider certain parameters that are different for either mechanical or process plant engineering. The level of decentralisation is even more crucial for the decision towards a certain bus system. Therefore, the actual application needs to be thoroughly analysed with the possible products in mind. However, to get a qualified classification, the next layer of the pyramid of information – the layer of control – needs to be considered as well.

1 System requirements in mechanical and process plant engineering

Mechanical and process plant engineers follow a principle for the selection of components for control and process control technology, that open and modular solutions for system architectures are inherently necessary.

A system architecture in the fields of mechanical and process plant engineering can be divided into technology for either actuators, electronical controls or process controls. Depending on the state of technology, customer and costs requirements and more, a wide range of solutions are available on the current market. These consist of classical PLC technology as well as real-time control systems that are based on VMEbus systems or industrial PCs. The requirements from the technical point of view differ between mechanical and process plant engineering. Therefore it is necessary to classify these differences of parameters. As an overview, this can be done as follows:

Mechanical Engineering	**Process Plant Engineering**
The requirements coming from the machines are:	The character of process engineering leads to:
➢ numerous I/O ➢ fast control programs	➢ low I/O count ➢ highly diversified in the area

Teil 2 Produkt- und anwendungsorientierte Beiträge

Product- and application-oriented contributions

6 Bereichsübergreifende und weitere Beiträge

Interdisciplinary and other contributions

Die Beiträge spiegeln die enorme Themenvielfalt der Feldbustechnologie wider. Die Sicherstellung der Interoperabilität mit Hilfe von IEC 1131 bis hin zur Anwendung von CAN in Kleinsatelliten wird angesprochen.

The papers reflect the enormous variety of topics in fieldbus technology. They address how interoperability can be assured with the aid of IEC 1131 as well as the application of CAN in low earth orbit satellites.

- ➢ exact and quick positioning
- ➢ small measures of modules
- ➢ need for teleservice

- ➢ low requirements on timing for control programs
- ➢ uncritical I/O timing
- ➢ problematic surrounding (climatic, temperature variations)
- ➢ need for teleservice

By analysing the requirements for the parameters at each field, one can find a way to decide on the optimal philosophy for the interconnection between the control modules and the S/A. Using the classification given by the pyramid of information [9] the PLCs and industrial PCs with control programs are located on the control layer. The lower layer is made of the sensors and actuators and is named field layer. This interconnection on the control layer itself and between the control and the field layer can be targeted from the logical and the physical side:

	on control layer	between control and field layer	on the field layer
Physical	industrial networks (e.g. Industrial Ethernet)	fieldbusses (e.g. DeviceNet)	fieldbusses (e.g. Sensor-Loop)
Logical	company specific / open (e.g. STEP 7 / IEC 1131)	IEC 1131 / RACKS CAPI	IEEE P1451

To better communicate on the different fieldbusses whether they are located on the field layer itself or connect the field and the control layer, we define the latter "fieldbus CTRL". The fieldbus that communicates directly with the sensors and actuators on the field layer shall be named "fieldbus SA".

However, as the trends in decentralisation continuos, the requirements for communication with the S/A will shift more and more towards the fieldbus CTRL. This can be expected, because S/As become as intelligent as PLCs when considering timing parameters. Only a small part of the S/As will be connected to a fieldbus SA system in the way as they shuffle bits and bytes in real time on the bus. But as long as this is not the case, the provided differentiation is necessary.

2 Integrating fieldbusses in automation systems

The level of intelligence of the sensors and actuators characterises the state of decentralisation that the automation system is up to. Depending on this state different relationships for the coverage of the fieldbusses can be made:

state of decentralisation	coverage of fieldbusses
low	mainly non-intelligent I/O on fieldbus SA coupled to industrial network backbone (e.g. AS-Interface to SIMATIC NET Industrial Ethernet)

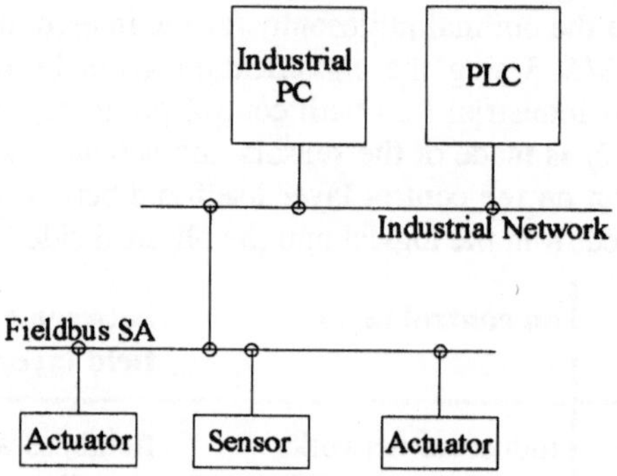

Fig. 1. Low decentralisation

medium	non-intelligent I/O on fieldbus SA coupled to intelligent I/O and PLC on fieldbus CTRL coupled with industrial network (e.g. Sensor-Loop to InterBus-S and Ethernet)

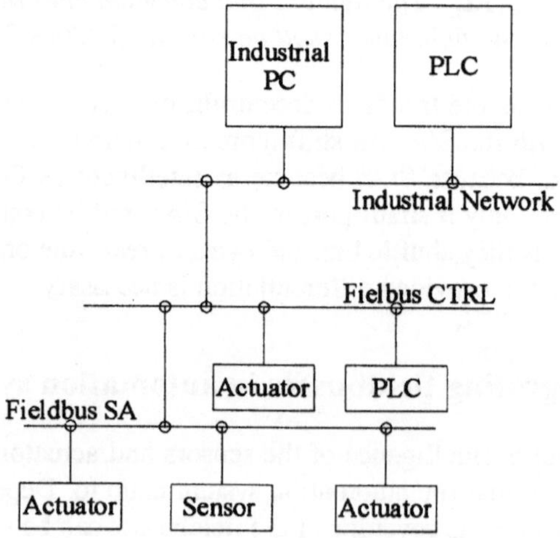

Fig. 2. Medium decentralisation

high

only fieldbus CTRL connecting all kinds of intelligent devices

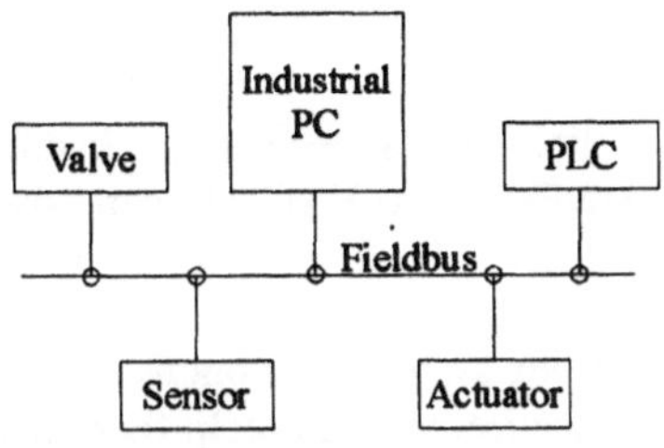

Fig. 3. High decentralisation

Although more decentralisation means that a fieldbus CTRL can cover most of the automation system and therefore build an as homogenous system as possible, the intelligence of the system modules can produce problems in configuration and process control. Whereas the industrial network builds the backbone for the process control and visualisation computers and therefore provides a physical connection to all process plant data, the exchange of information[1] is mostly inhibited by an uncommon interpretation of local data bases.

This leads to the fact that every module of the automation system needs to be configured with its own tool and this tool bases its process information on local data. As soon as a change in the system configuration takes place (like the exchange of a sensor), a chain of database updates arises. This is even the point, when the sensor announces itself with a full set of product definition data to the network. This data can only be understood by the corresponding system, normally a specific fieldbus standard. Although taking one procedure out of the update chain, the main expenses are still in

➢ the connection on the software side,
➢ the hardware planning and management and
➢ the documentation.

The solutions to these problems depend nowadays on the specific fieldbus system, that is implemented in the automation system. One can program several interfaces that interpret the data delivered on the fieldbus for each component that is not directly compatible e.g. as it comes from a different company. However, just to attach e.g. a sensor to a fieldbus is a much smaller task, than to integrate it fully into the whole automation system. With over 60 different sensor network protocols [10], developing interfaces at a single source would be highly unpractical.

---

[1] "Information" in opposition to "data" provides a meaning to the bits and bytes. Pure exchange of data is (nearly) always possible, as long as a physical connection exists. However, to exchange information or gather information out of data, common arrangements like interfaces need to be made.

410

The integration of the fieldbus itself into the automation system needs to take all the above stated parameters into account. However, there are several ways to ease this task either from the logical or the physical side (see also table 1).

2.1 Physical side

When different fieldbusses need to be installed or connected, a common hardware interface like AnyBus from HMS [1] can adapt the physical connection onto a single bus.

To find the right bus system that opens up the best (in terms of cost and effect) solutions, a proper selection tool can help there (see later).

2.2 Logical side

To allow for an interoperability between different fieldbus components and different fieldbusses itself, the IEC 1131 provides the necessary standards.

To add sensors to the standardised programming and configuration interface, the proposed EDS (electronic data sheet) standard proposed for sensors IEEE P1451 can supplement the IEC 1131.

3 Providing Interoperability with IEC 1131

As figure 4 shows, a common database for all services and modules that can be found in an automation system builds the basis for the interoperating system. The common database is fed by the elements on the field and control layer as well as the process control layer. These are connected and interconnected through the different networks provided by fieldbusses, Ethernet etc. The IEC 1131 uses the common database for the direct programming of the PLCs and configuration, management etc. of fieldbusses. This is made possible through the structure of function blocks that is also inherent to IEC 1131 and therefore supported by more and more companies in the automation industry and also subject to a new standard IEC 1499 [5]. These companies provide tools for their fieldbusses using IEC 1131 compatible structures. Unfortunately these tools are still specific to one fieldbus, they only use the programming directives standardised in IEC 1131[2]. An approach to this problem comes from the ESPRIT project RACKS and NOAH [11], by developing an common application programming standard that hides the specific fieldbus protocols from the IEC 1131 programming interface.

However, so far systems for documentation and CAE are not coupled to this common database through a standard. Although company specific protocols do exist or ways

---

[2] E.g. Beckhoff's WinCAT integrates different fieldbusses in one programming and control system. However, to make use of diagnostics delivered by devices on the InterBus-S requires specific tools e.g. from Phoenix Contact.

are found to use the product description standard STEP that is used in IEC 1131 a solution with an automated bi-directional connection needs a careful arrangement of components.

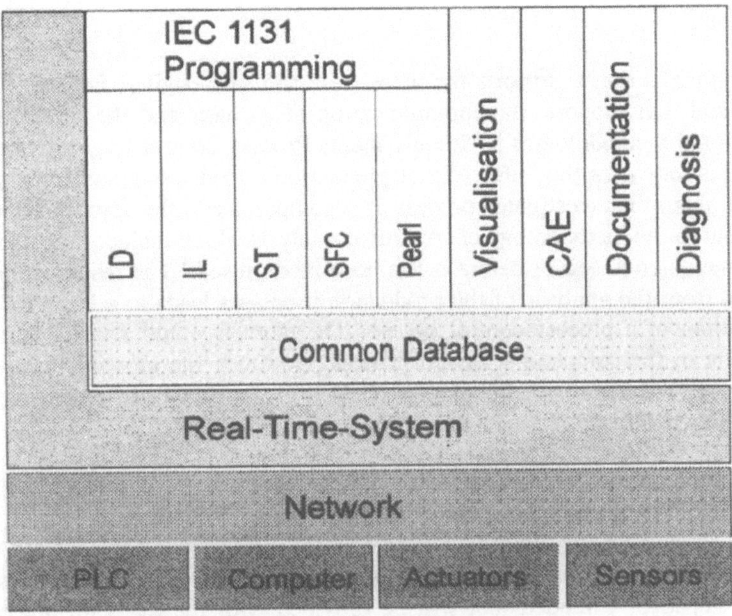

Fig. 4. Integration through common databases

3.1 A tool for the selection of fieldbusses to provide interoperability

As the statements above have shown, the decision for the best fieldbus system can only be made with several cross connecting parameters in mind. By using approaches coming out of the field of software engineering, namely requirements engineering, we propose to develop a tool that takes care of the requirements coming from:

➤ the projected target, i.e. mechanical engineering or process plant engineering,
➤ the level of anticipated decentralisation,
➤ the structure of the upper level networks and
➤ the application the user has in mind

This selection and (later) projection tools directly provide for the interoperation between different systems components that it will have suggested. This directly results from the conceptual works that extend the methodology of requirements engineering to interoperability. The research on the information science side has been done from [6]. By tracing the process flow, the interoperability that is possible or that is necessary builds the basis for deploying the requirements engineering methodology. By following those objectives that lead to a highly user and application driven software program, it should be possible to respectively provide a tool with the same goals for the selection of interoperating fieldbus connected automation systems. This optimistic conclusion comes from the fact, that automation systems are made of a growing percentage of software (some 80% according to Manfred van Raven, Siemens automation technology [4]). Therefore not the classical tools for projection and selection of mainly hardware components provide the best answer, but those that come from the software research area.

Another approach that transports the ideas from software engineering into industrial processes and that supports the communication of the user and the system engineer through formal methodologies is "Requirements Process Control Engineering" by [2]. This methodology uses the "life cycle of production"[3] model as a tool to describe the knowledge about the investigated process. It is comparable to the description tools that can be found in the methodology of structured analysis which includes the former type of requirements engineering before it has been modernised.[4] The resulting functional description from the analysis of the production processes leads to a basis for the further projection of a process control system. Therefore it would be useful to include these results in the complete selection process. However, the growing complexity of using more and different methodologies leads to an artificial as well as a practical restriction to use only one integrated

Tools that are available and support the management of the requirements engineering process, so called requirements management tools (e.g. "RequisitePro" [8] or DOORS [7]), are evaluated if they could have been adapted and extended to conform with the targeted criteria for the selection of fieldbus systems. First a set of libraries for the newly developed or extended tool will be provided with the focus on process plant engineering. These will allow one to follow the process flow with the parameters for process plants stated above in mind, classify the state of decentralisation if necessary and use physical and logical interoperability possibilities to get to a formally documented decision for the proper fieldbus system.

Literature

1. Hasbjer Micro System, Halmstadt, Schweden. http://www.hms.se

2. Lauber, J., "Methode zur funktionalen Beschreibung und Analyse von Produktionsprozessen als Basis zur Realisierung leittechnischer Lösungen". Verlag Mainz, Aachen, 1996.

3. Jarke, M.; "Requirements Engineering: An Integrated View of Representation, Process and Domain".
   ftp://ftp.informatik.rwth-aachen.de/pub/NATURE/NATURE-93-07.ps.Z

4. Happacher, M.; "Ein neues Kapitel – Bill Gates verändert die Automatisierungswelt". Elektronik 6/1997, Franzis-Verlag, Feldkirchen 1997. S.96ff.

---

[3] Original German "Phasenmodell der Produktion" [2]
[4] As is was targeted in the ESPRIT project NATURE (e.g. [3]).

5. Neumann, P. et al.; "Der Weg zu offenen Prozeßleitsystemen". In: VDI/VDE GMA Jahrbuch 1997. VDI-Verlag, Düsseldorf 1997.

6. Pohl, K., B. Nusebeih, A. Finkelstein, J. Kramer; "Process Driven Interoperability". In: B. Kramer, M. Papazoglou "Systems Interoperability", RSP distributed by J. Wiley and Sons Ltd, Sussex, UK 1997.

7. Quality Systems and Software, Ltd. Oxford, England.
http://www.qssinc.com/DOORS/overview.html

8. Rational Software Corporation, Santa Clara, CA, USA
http://www.rational.com/products/reqpro/datasheet.html

9. Scherff, B. et al.; "Automatisierungstechnik im Anlagenbau: Echtzeitsystem und Feldbus – kein Widerspruch". In: Systeme, Franzis-Verlag, Poing, 8(1994)4.

10. Warrior, J.; "Smart Sensor Networks of the Future". Sensors March 1997, Helmers Publishing, Peterborough, NH, USA 1997.

11. Wollschläger, M.; "Einheitlichkeit verbessert Akzeptanz – Die Integration von Feldbus und IEC 1131". Elektrotechnik für die Automatisierung 5 – 2.5.1997. Vogel-Verlag, Würzburg 1997.

OPC-Anschluß an die Automatisierungstechnik

K.H. Deiretsbacher

Siemens AG, AUT GT 14 (Gemeinsame Technik)

Postfach 4848, 90327 Nürnberg

eMail: Karl.Deiretsbacher@nbgm.siemens.de

Abstract.   In heutigen Anlagenlösungen werden vielfach Applikationen von verschiedenen Herstellern eingesetzt (zunehmend auf PCs mit Microsoft Windows). Mit OPC wird ein einheitliches Zugriffsverfahren auf Daten des Prozesses definiert, durch das Geräte und Anwendungskomponenten verschiedener Hersteller einfacher kombinierbar werden. OPC basiert auf dem Komponentenmodell OLE/COM, wodurch die Integration mit anderen Komponenten (bzw. Programmen wie z. B. Visual Basic, Excel) gewährleistet ist und andererseits die Verteilung mit DCOM (**D**istributed **C**omponent **O**bject **M**odel) automatisch unterstützt wird. Der Funktionsumfang von OPC ist in der ersten Version auf effizienten Datentransfer fokussiert. Folgeversionen werden Erweiterungen, z. B. für Alarmbearbeitung und Ereignisaufzeichnungen enthalten. Today manufacturers face the task of integrating plant floor data of different vendors into their business systems (typically PCs with Microsoft Windows). OPC specifies an interface which allows access to this data in a consistent manner. OPC is based on the component object model OLE/COM, providing easy integration with other components (or programs, like Visual Basic, Excel) and distribution with DCOM (**D**istributed **C**omponent **O**bject **M**odel). The first version of OPC focusses on efficient data access. Following versions will add Alarm-Handling, History Data and more.

1 Was ist ”OPC” ?

OPC (OLE for Process Control) ist eine Spezifikation, die eine einheitliche und herstellerunabhängige Softwareschnittstelle zwischen Applikationen und der Automatisierungsebene definiert. Entworfen wurde OPC von einer Task-Force, der einige führende Firmen der Automatisierungsbranche angehören. Die Task-Force wurde beim Entwurf der Spezifikation von der Firma Microsoft unterstützt. Heute werden die Aktivitäten durch die OPC Foundation weitergeführt.

2 Die Motivation für die OPC-Schnittstelle

In vielen Industriebetrieben werden Prozeßdaten, wie z. B. Daten der Produktion, erfaßt und weiterverarbeitet. Diese Vorgänge sind zur Zeit noch kompliziert, da jeder Hersteller von Automatisierungsgeräten eigene Methoden bzw. Protokolle zum Datenaustausch verwendet. Der Aufwand, Geräte von verschiedenen Herstellern in einer Anlage zu einem funktionsfähigen System zu integrieren, ist sehr hoch. Die Folge ist, daß viele Kunden ihr komplettes Automatisierungssystem einschließlich Bedienungssoftware bei einem Hersteller kaufen. Der Kunde bindet sich aus Aufwandsgründen an einen Hersteller und kann sein System nicht individuell, je nach Stärken der einzelnen Anbieter, zusammenstellen.

Die Initiative für OPC wurde gestartet, um basierend auf den Technologien OLE und COM des Betriebssystems Windows ein einheitliches Zugriffsverfahren und eine einheitliche Schnittstelle für Automatisierungsgeräte zu realisieren. Durch OPC ist es möglich, mit Applikationen aus dem industriellen und dem Bürobereich Daten von Prozeßeinheiten zu lesen und weiterzuverarbeiten. Der Entwickler muß sich nicht mehr um die Spezifika eines Herstellers kümmern.

Der Aufbau auf OLE/COM ermöglicht eine einfache Integration von Büroanwendungen in die Prozeßdatenverarbeitung. So kann z. B. ein Excel-Makro mit Hilfe der OPC-Schnittstelle Daten von einer SPS lesen und diese graphisch darstellen.

3 Was ist ein OPC-Server ?

Ein OPC-Server ist eine Realisierung der OPC-Schnittstelle. Applikationen nutzen die Dienste des OPC Servers ausschließlich über die von OPC definierte Schnittstelle. Ein Wissen über Internas ist nicht erforderlich. Die Implementierung eines Servers bildet die Schnittstelle typischerweise auf speziellen Treibern ab (siehe *Abb. 1*).

Die OPC-Spezifikation definiert Syntax und Semantik der Schnittstelle (siehe [3]). Dadurch ist sichergestellt, daß sich OPC-Server unterschiedlicher Hersteller für die Anwendungen gleich verhalten.

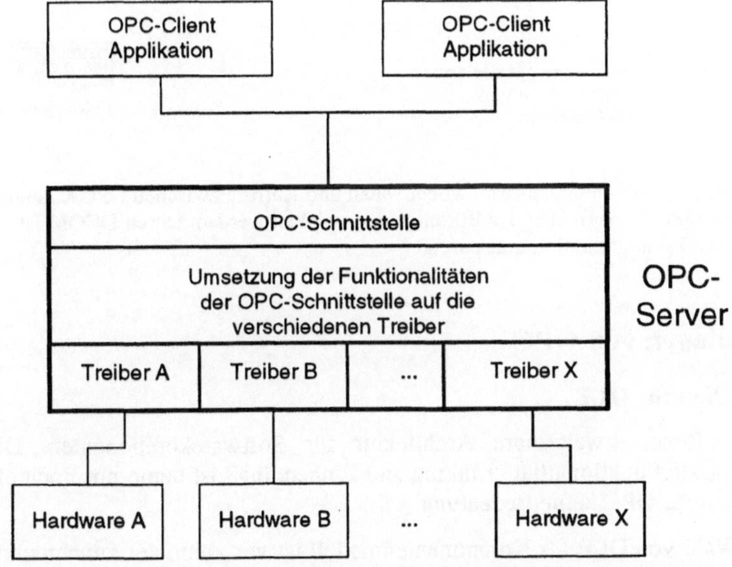

Abb.1: Das Modell eines OPC-Servers, der die OPC Schnittstelle auf die herstellerspezifischen Treiberschnittstellen abbildet.

Ein OPC-Server bietet einem Client folgende Möglichkeiten:
- ♦ den Wert einer oder mehrerer Prozeßvariablen lesen,
- ♦ den Wert einer oder mehrerer Prozeßvariablen ändern,

♦ den Wert einer oder mehrerer Prozeßvariablen überwachen.

Überwacht ein Client eine Prozeßvariable, so prüft der Server innerhalb eines vom Client festgelegten Zeitintervalles, ob sich der Wert der Prozeßvariablen geändert hat. Liegt nach Ablauf des Zeitintervalls ein neuer Wert vor, so meldet der Server dem Client den neuen Wert.

Das Anwendungsszenario von OPC in *Abb. 2* sieht vor, daß Server und Clients auf verschiedenen Rechnern ausgeführt werden können. Die Kommunikation zwischen Server und Clients erfolgt auf Basis der Dienste von OLE (distributed COM).

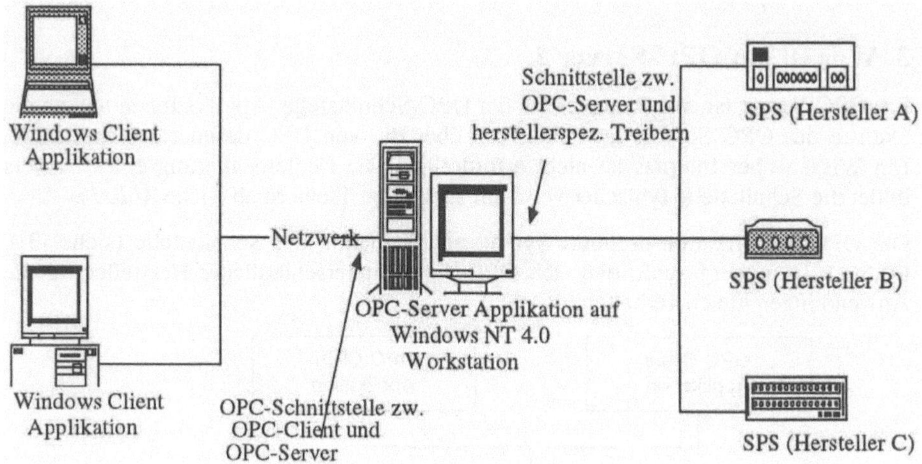

Abb. 2: OPC-Anwendungsszenario, wobei Daten und Aufrufe zwischen OPC-Clients und OPC-Servern über das Büronetz abgewickelt werden. Durch DCOM ist diese Kommunikation transparent.

4 Die Grundlagen von OPC

4.1 Die Motivation für OLE

OLE ist eine offene, erweiterbare Architektur für Softwarekomponenten. Die ursprünglich zentrale Funktionalität "Linking and Embedding" ist heute nur noch ein Teilaspekt und hat für OPC keine Bedeutung.

Grund für die Wahl von OLE als Komponentenmodell ist vor allem der zunehmende Einsatz des Betriebssystems Windows. Dazu kommt, daß im Bereich der Entwicklung von PC-Applikationen der Trend zu Entwicklungsumgebungen wie Visual Basic, Visual C++, Delphi, Power Builder etc. geht. Diese Produkte unterstützen OLE und können somit mit "beliebigen" OLE-Komponenten zusammenarbeiten.

4.2 Das Klassenmodell der OPC-Schnittstellen

Die OPC-Schnittstelle stellt der Client-Anwendung ein Klassenmodell zur Verfügung. Dieses Klassenmodell ist in der *Abb. 3* dargestellt. Für die Darstellung wurde die Notation der Unified Modeling Language (kurz: UML) gewählt. Eine Erklärung der verwendeten Symbole der UML findet sich in [2] bzw. [4].

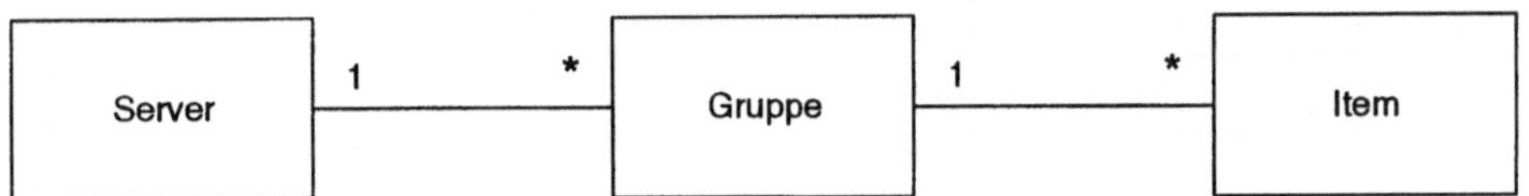

Abb. 3: Das OPC-Klassenmodell, dessen hierarchische Ordnung eine natürliche Aufteilung der Funktionen erlaubt.

Das Klassenmodell besteht aus den Klassen Server, Group und Item. Die Klassen besitzen eine hierarchische Ordnung. Die Klasse Server definiert Attribute über den Status, die Version etc. eines OPC-Servers. Außerdem besitzt sie Methoden, mit denen ein Client Objekte der Klasse Group verwalten kann. Die Klasse Group verwaltet Objekte der Klasse Item. Ein Item-Objekt repräsentiert eine Verbindung zu einer Prozeßvariablen. Dabei ist typischerweise ein schreib- und/oder lesbares Datum der Prozeßperipherie (z. B. die Temperatur eines Kessels) gemeint. Mit Hilfe von mehreren Groups kann ein Client semantisch sinnvolle Einheiten von Items bilden und auf diesen Operationen ausführen. So können sich in einer Group alle Prozeßvariablen der Maschine A befinden und in einer zweiten Group die Temperaturvariablen aller Kessel.

Die Items einer Group können jeweils mit einem Methodenaufruf manipuliert (z. B. gelesen) werden. Man spricht dabei auch von Mengenoperationen, die eine hohe Effizienz der OPC-Schnittstelle gewährleisten.

5 Ein Szenario für die OPC Automation-Schnittstelle

In diesem Abschnitt wird ein Anwendungsszenario mit einem Sequence-Chart und den wesentlichen, dazu erforderlichen VisualBasic Anweisungen vorgestellt.

In diesem Szenario wird eine Gruppe (OPCGroup) erzeugt und aktiviert. Der OPC-Server beobachtet die aktiven Elemente der Gruppe (OPCItems) und verständigt die Anwendung in einem konfigurierten Raster, wenn sich Inhalte dieser Elemente ändern. Man spricht in diesem Zusammenhang auch von einer exception-based connection.

5.1 Konfigurierung

Als Konfigurierung wird die Phase bezeichnet, in der Gruppen und Items erzeugt und eventuell aktiviert werden. Dies kann - muß aber nicht - in einer Anlaufphase der Applikation erfolgen. Variablendeklaration und Fehlerbehandlungsroutinen wurden zur besseren Übersichtlichkeit weggelassen.

Chart 1: Erzeugen einer Gruppe, die von Beginn an aktiv ist

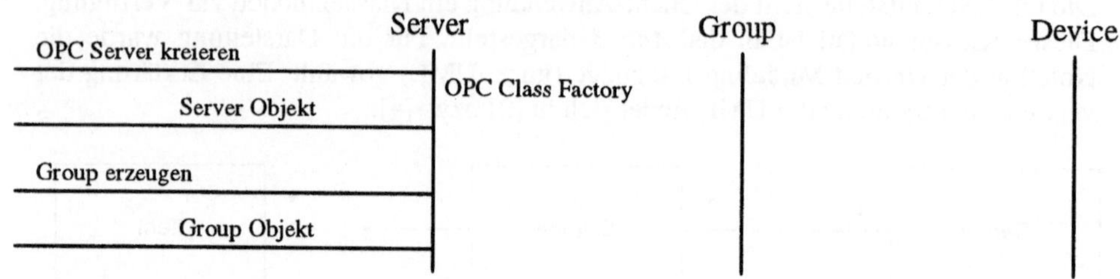


```
` Kreieren eines OPC Servers
SvrObj= CreateObject("OPC.OPCServer")
` Erzeugen einer Gruppe, die von Beginn an aktiv ist
updateRate  = 500      ` 500 ms
Set Grp = SvrObj.AddGroup("MyGroup", active, updateRate)
```

Chart 2: Erzeugen der Items und Registrierung eines 'Callbacks'

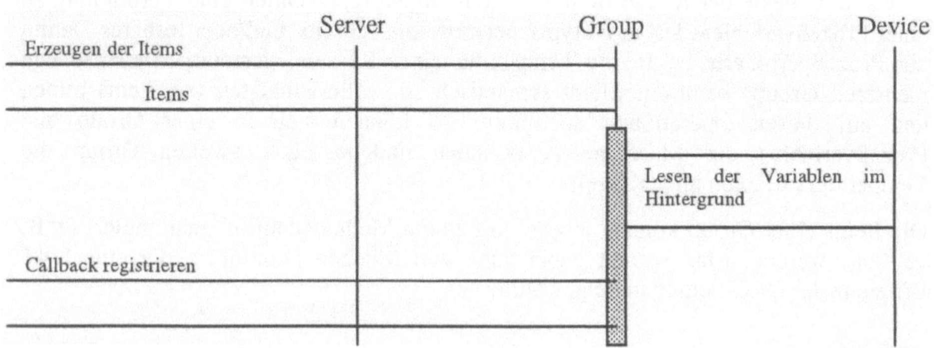


```
` Initialisieren der Iteminformation
ItemIDs(0)      = "[dp:cp_12_1]Slave005EB000"
CltHdls(0)      = 10000
ItemIDs(1)      = "[s7:vessel_control]pressure"
CltHdls(1)      = 10001
ItemIDs(2)      = "[s7:vessel_control]temperature"
CltHdls(2)      = 10002

` Erzeugen der Items
Ecode = Grp.AddItems(3, ItemIDs(0), CltHdls(0), Items(0),...)

` Registrieren des Callbacks
context   = OPC_EXCEPTION | OPC_ASYNCHREAD
Grp.AddCallbackReference(context, CallbackObj)
```

5.2 Bearbeitung zur Laufzeit

Nach Registrierung des 'Callbacks' ist die Konfigurierung abgeschlossen. Die Group ist aktiv und beobachtet, ob sich die Dateninhalte der Variablen ändern. Jeweils nach Ablauf einer `updateRate` werden die veränderten Variablen an die Callback-Methode übergeben. Wenn keine Änderungen stattgefunden haben unterbleibt der Aufruf.

Chart 3: Änderungsmeldungen zur Laufzeit

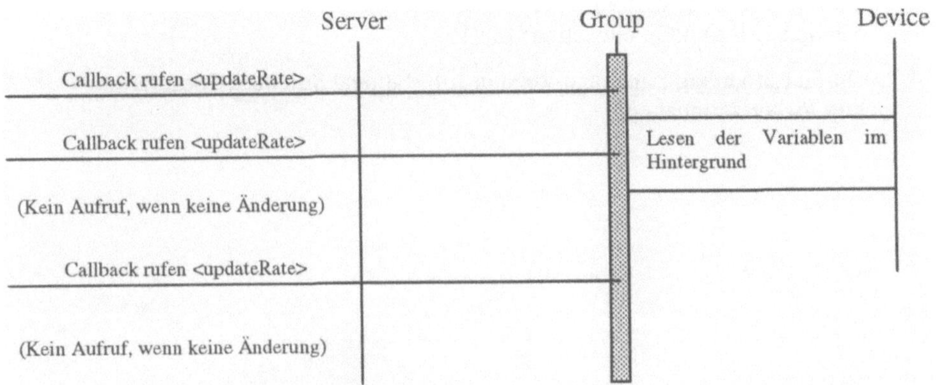

6 OPC Produkte

Um die Funktionsfähigkeit von OPC zu demonstrieren, wurden von den Firmen der OPC-Foundation begleitend zur Spezifikation Prototypen erstellt und getestet. Dadurch war es möglich, bereits kurz nach Veröffentlichung des Standards Produkte anzubieten. Heute gibt es bereits eine Reihe von OPC-Servern und -Anwendungen (Clients).

Als aktives Mitglied der OPC Foundation hat der Bereich Automatisierungstechnik der Siemens AG seine Produkte um die OPC Schnittstelle ergänzt:

OPC-Server für die industrielle Kommunikation mit Simatic NET
> Mit diesem Server ist vom PC über die OPC Schnittstelle der Zugang zu Simatic S7 und S5 möglich. Im ersten Schritt werden die Protokolle DP und S7 unterstützt.

OPC-Kanal DLL für das Prozeßvisualisierungssystem WinCC
> Durch diese Erweiterung wird WinCC ertüchtigt, auf Prozeßdaten über die OPC-Schnittstelle zuzugreifen.

OPC-Server für WinCC
> Durch diese Erweiterung ist ein Zugriff auf alle durch WinCC erfaßten Daten möglich. Dieses Produkt ermöglicht B&B-Hierarchien, in denen ein

untergeordnetes WinCC seine Daten einem übergeordneten WinCC mit Hilfe von OPC zugänglich macht.

Literatur

1. Component Object Model Specification 0.9; o.A.; Microsoft MSDN Library

2. Objektorientierte Softwareentwicklung: Analyse und Design; Oestereich Bernd; Oldenbourg 1997

3. OPCTaskForce: "OLE for Process Control" Final Release Version 1.0; OPC Task-Force; August 1996
Version 1.0A; OPC Foundation; Juli 1997

4. Unified Modeling Language Version 1.0; Rational Software Corporation; http://www.rational.com

ASI–Visualisierung in der Fertigungstechnik

Christian Kral [1], Heinrich Reiter [2]

[1] HTC Kral, 1120 Vienna, Malfattig. 1/2/1, AUSTRIA
[2] Inst. f. Automation, TU Wien, 1040 Vienna, Treitlstr. 1, AUSTRIA

Abstract. Die steigende Anzahl von Sensoren, Aktoren und Feldbussen erhöht die Notwendigkeit adäquater Visualisierungstools. Ein anderer Aspekt ist die aufstrebende Netzwerkstruktur, die ihrerseits die Netzwerkfähigkeit des Visualisierungsteils notwendig macht. Die logische Folge, basierend auf derzeitigen Standards, war die Verwendung des World Wide Web (WWW). Das Hauptziel dieses Systems sind allgemeine Verfügbarkeit, kostenlose oder günstige Software und einfache Handhabung.

As a matter of fact, the growing number of sensors, actors and fieldbuses raised the demand for adequate visualization tools. Another aspect is the upcoming network structure, which in turn brings up the need for network capabilities of the visualization task. The logical solution, based on common standards, was the use of the World Wide Web (WWW). The main goals of this system are common availability, free or inexpensive software and simple handling.

1 Einführung

In den letzten zehn Jahren stieg auf Grund der zunehmenden Anforderungen in der Industrie- und Heimautomatisierung auch die Notwendigkeit einfach handzuhabender und aufzusetzender Kontrollsysteme. Weiters wurden die Prozeßstrukturen zunehmend komplexer und für Einzelpersonen schwerer überschaubar.

Basierend auf dieser Tatsache, wurden Feldbusse in der Automatisierungstechnik immer wichtiger, unabhängig davon, ob die Aufgaben in komplexen Industrieanlagen oder im dynamischen Gebäudemanagement eingesetzt wurden.

Ein besonderes Thema ist dabei die Geschwindigkeit der Visualisierung mit kurzen Antwortzeiten über hochfrequentierte Local oder Wide Area Networks in allen Arten von Applikationen.

2 Systemdesign und Datenakquisition

Am Beginn jedes Automatisierungsprojektes steht als Basis das solide Systemdesign, um später einwandfrei funktionierende Prozeßketten zu erhalten.

Früher durchgeführte Designaufgaben machten sehr umfangreiches Wissen und hochqualifizierte Ingenieure notwendig, die den größten Teil der Arbeit auf Papier erledigten, meist ohne wirklich einen Überblick über das System zu erhalten. Als zweiter Schritt wurden dann Kabel- und Stücklisten sowie Spezifikationen angefertigt, die in einer Art Textverarbeitung eingegeben wurden, um sie für spätere Installations- und Dokumentationszwecke zur Verfügung zu haben. Nach dieser Designarbeit begann die eigentliche Arbeit, nämlich die Programmierung der Steuerung durch Softwarespezialisten.

Heutzutage werden diese Schritte von elektronischen Designsystemen erledigt, die zu jedem Zeitpunkt der Entwicklung einen guten Überblick über das Gesamtsystem ermöglichen, und gleichzeitig sämtliche Systemdaten und Dokumentationen als Nebenprodukt ausgeben. Die Arbeit der Programmierer wird nun über Hochsprachen abgewickelt, wobei alle Kontrollsequenzen aus einfachen Kommandos bestehen. Dies gibt weniger spezialisierten Ingenieuren die Möglichkeit, das System selbst aufzusetzen und vor allem vor Ort jederzeit zu ändern und anzupassen.

3 Aufsetzen des Systems mit einfachen Hilfsmitteln

3.1 Systemdesign, Zuordnen von Namen und Ausgabe der Dokumentation

Die meiste Arbeit kann basierend auf graphischen Editoren gemacht werden, indem Sensoren und Aktoren an ihre entsprechenden Positionen im System eingefügt werden und zwar unabhängig davon, welche Hardware verwendet wird. Der nächste Schritt ist das Zuordnen logischer Namen zu den Komponenten, die dem Bediener helfen, die Baugruppen so handzuhaben, wie sie im System erscheinen; d.h., als Teile logischer Funktionsgruppen.
Ein Beispiel im Bereich der Heimautomatisierung könnte lauten

cluster4.master13.slave3.bit12=window(open/close)

und in Industrieautomatisierungsprojekten

cluster38.master8.slave2.bit3=conveyor belt(go/stop).

Nachdem der passende Sensor oder Aktor aus einem Selektionsbaum ausgewählt wurde, ist nur noch wenig Arbeit notwendig, um die für eine komplette Systemdokumentation benötigten Kabel- und Stücklisten aus dem System zu bekommen.

3.2 Eingabe von Bedingungen

Der nächste Schritt ist die Eingabe von Aktionen, die von den Sensordaten oder Bedienereingaben abhängen. Dies erfolgt durch globale Definitionen wie

if rain-sensor=wet then close(window)

oder

if machineX=error then stop(machineX)
set(alarm)

und erfordert kein spezielles Wissen auf dem Sektor des verwendeten Feldbusses oder eines anderen Datenakquisitionsmediums. Daher kann jeder Anwender sein eigenes System einfach entwerfen, korrigieren und verfeinern.

Es ist dabei nicht einmal notwendig, alle im System vorkommenden Namen zu kennen, da diese aus einem hierarchischen Selektionsbaum ausgewählt werden.

3.3 Graphische Darstellung

Für die graphische Darstellung können von den Softwareherstellern, passend für die jeweilige Anwendung, Symboldatenbanken zur Verfügung gestellt werden. Dies sind beispielsweise Fenster–, Türen– und Kühlschranksymbole für Heimautomatisierungsprojekte und Maschinensymbole für Industrieprojekte. Weitere Symbole kann der User mit jedem beliebigen Graphikeditor erstellen oder in sein System einscannen.

Der User vergibt für jedes Symbol einen Namen, der die Funktion repräsentiert. Dies vervollständigt die Verbindung zwischen Sensoren und Aktoren auf der einen Seite und der Visualisierung auf der anderen. Nach dem Setzen aller graphischen Symbole ist das System betriebsbereit (siehe Abb. 1).

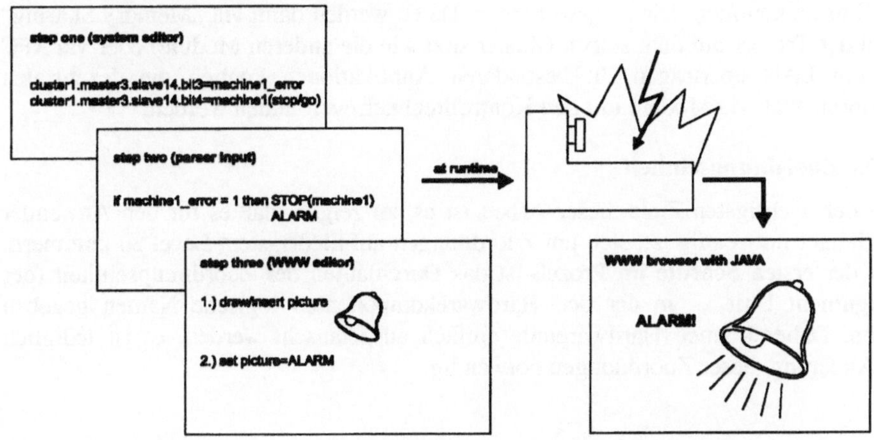

Abb. 1. Einfaches Systemdesign

3.4 Austauschbare Treiber für die Datenakquisition

Die Idee ist, möglichst viel Arbeit des „low level designs" vom Anwender zu nehmen, sodaß (fast) kein Hardwarewissen für das Systemdesign notwendig ist. Ein Systemdesign dieser globalen Struktur kann somit einfach von einem Feldbus- oder Verdrahtungssystem zu einem anderen übernommen werden, indem die Treiber für die Sensoren und Aktoren entsprechend ersetzt werden.

Damit kann der jeweils am besten geeignete Bauteil, abhängig von Preis, Verfügbarkeit, etc. für die jeweilige Anwendung eingesetzt werden, solange entsprechende Treiber verfügbar sind.

4 Laufzeitteile und Datenkontrolle

4.1 Die Aufgabe

Da nun ein Interface zwischen Rechner und Feldbussystem bereitsteht, ist der nächste Schritt die Repräsentation der Feldbusdaten. Der User soll keine Feldbussteuerungen mehr von Grund auf programmieren müssen. Es sollte möglich sein, Konfigurationen oder Kontrollaufgaben mittels graphisch unterstützter Methoden einzugeben, anstatt viele Zeilen Code eintippen zu müssen (siehe [5]). Um aus den Bedienereingaben für den Feldbus verständliche Kommandos zu machen, war es notwendig, einen geeigneten Parser zu entwickeln, der sich um die korrekte Übersetzung der Frontends (WWW-Seite) zum Feldbus bemüht.

4.2 Treiber

Die notwendigen Treiber wurden bereits zuvor besprochen. Diese sind der erste Punkt in der Liste der Laufzeitteile.

Auf jedem Cluster muß zumindest ein für die verwendete Hardware passender Treiber zum Einsatz kommen. Die so gewonnenen Daten werden dann via „Memory Sharing" (wenn der Treiber auf dem selben Cluster sitzt wie die anderen Module) oder via RPC über ein LAN übertragen. In besonderen Applikationen können die dezentralen Strukturen auch via Modem mit den Kontrollrechnern verbunden werden.

4.3 Die Zuordnungseinheit

Eines der wichtigsten Ziele dieser Arbeit ist es, zu zeigen, daß es für den Anwender nicht länger notwendig ist, sich um Zuordnungen auf niedrigstem Level zu kümmern. Einer der ersten Schritte im Prozeß ist das Durchlaufen der Zuordnungseinheit (der „Assignment Unit"), in der den Hardwarekomponenten logische Namen gegeben werden. Daher können Hardwareteile einfach ausgetauscht werden, es ist lediglich eine Änderung dieser Zuordnungen notwendig.

4.4 Der Parser

Der Parser kann als Datenkonzentrator oder Interpreter verstanden werden. Besonders in weit verteilten Systemen müssen Daten vieler Lokationen von einem Punkt aus kontrolliert werden, um dem System die Kontrolle über mehrere Maschineninstanzen geben zu können.

Dieser Prozeß kommuniziert auch mit anderen Anwendungen, sodaß ein Datenaustausch mit Datenbankapplikationen oder Logistikprogrammen erfolgen kann. Daher können Gesamtsteuerungen implementiert werden, anstatt lediglich lokaler Kontrollen auf Feldbusclustern.

Die Struktur der Laufzeitprozesse wird im Abb. 2 dargestellt.

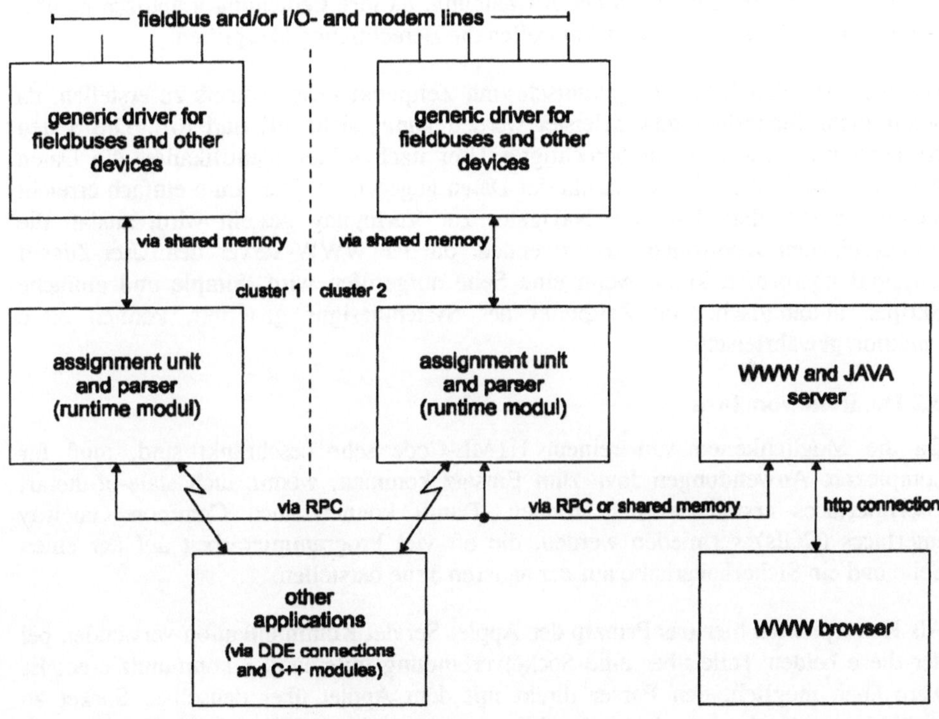

Abb. 2. Struktur der Laufzeitprozesse

5 Das Konzept der Visualisierung via WWW und die Rolle von Java

5.1 Der dynamische HTML-Generator

Die Idee, das WWW als das prinzipielle Visualisierungsmedium einzusetzen, ist, einfache, bekannte und freie oder kostengünstige Software für diese Aufgabe zu verwenden, während der User das System von der ganzen Welt aus kontrollieren kann, unabhängig davon, auf welcher Hard- oder Softwareplattform er arbeitet und welcher WWW-Browser verfügbar ist.

Web-Browser sind mittlerweile überall gängig und es ist fast unmöglich, einen Rechner mit Internetzugang ohne WWW-Browser zu finden. Daher kann der Anwender darauf vertrauen, jederzeit und von überall Zugang zu und Kontrolle über sein System zu haben, unabhängig davon, wo er sich gerade aufhält.

Diese Browser sind einfach in der Anwendung, da ihre Bedienung graphisch geführt und intuitiv ist, während Zutrittskontrollen die Berechtigung überprüfen.

Die Idee ist, HTML-Seiten dynamisch zum Zeitpunkt ihres Aufrufs zu erstellen, da somit mehr Sicherheit gewährleistet werden kann (siehe [4] und [6]). Die Seiten bestehen nur, während ein berechtigter User nach seiner Identifikation die Daten darstellt, womit auch die Aktualität der Daten gegeben ist. Dies kann einfach erreicht werden, indem dem User ein Navigator zur Verfügung gestellt wird, anstatt die browsereigenen Kontrollen zu verwenden, da der WWW-Server den User-Zugriff jedesmal überprüfen kann, wenn eine Seite aufgerufen wird. Simple und einfache Skripts, automatisch zum Zeitpunkt des Systemdesigns generiert, können diese Funktion gewährleisten.

5.2 Die Rolle von Java

Da die Möglichkeiten von reinem HTML-Code sehr beschränkt sind, muß für komplexere Anwendungen Java zum Einsatz kommen, womit auch state-of-the-art Userinterfaces erstellt werden können. Damit können auch Common Gateway Interfaces (CGIs) vermieden werden, die oft viel Programmierarbeit auf der einen Seite und ein Sicherheitsrisiko auf der anderen Seite darstellen.

Als Konzept wird hier das Prinzip der Applet-Servlet Kommunikation verwendet, bei der diese beiden Teile über eine Socketverbindung miteinander kommunizieren. Es wäre auch möglich, den Parser direkt mit dem Applet über denselben Socket zu verbinden, jedoch erfordert ein klares, gekapseltes Systemdesign eine klare Aufgabentrennung, wodurch auch das Implementieren weiterer Features einfacher wird (siehe [2]). Weiters könnte der Javateil mit übergeordneten Datenbanken- oder Kontrollsystemen interagieren (siehe [3]).

Da Java ein C-Interface hat, kann die sogenannte native-method-calling Methode zum Einsatz kommen. Die Methode, C-Funktionen zu implementieren, wurde bereits in [1] ausführlich beschrieben. Da globale Dateninterfaces immer praktikabler, einfacher und damit gebräuchlicher werden, wird die File–Kommunikation immer unbedeutender.

In unserer Applikation wurden Java-Applets für die dynamische, graphische Repräsentation der Daten eingesetzt. Die Idee war, so wenig Daten wie möglich über Netzwerke zu senden, um möglichst gute Antwortzeiten zu erhalten.

Der zweite Punkt war, als Novität für Java–Anwendungen, daß der Server die Verbindung öffnet, sobald Daten zur Visualisierung bereit stehen, anstatt die Applets pollen zu lassen, wodurch es zu einer ständigen Netzwerkbelastung kommt. Damit können, selbst in stark frequentierten Netzwerken, gute Antwortzeiten erreicht werden.

Tests über stark belastete Netzwerkleitungen der Technischen Universität Wien während Spitzenzeiten, die in einer 14,4k Modemleitung in den USA endete, bestätigten diese Hypothese.

In Abb. 3 sehen Sie eine typische Applikation einer verteilten Feldbusanwendung.

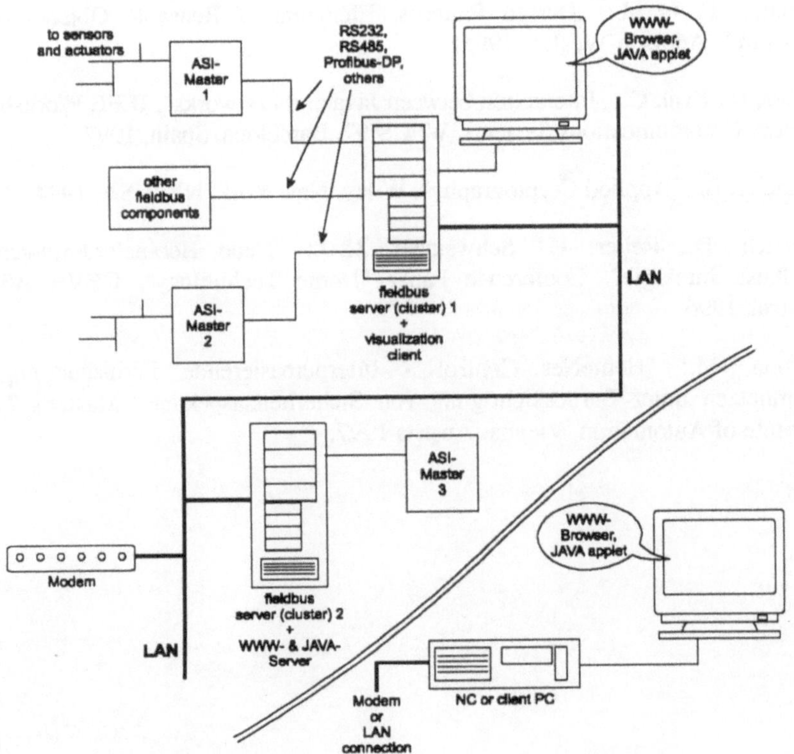

Abb. 3. Typische Anwendung für verteilte Feldbussysteme

6 Schlußwort

Je komplexer die Prozesse werden und umso mehr Komfort sich der User einer Industrie- oder Heimautomatisierungsanlage erwartet, desto mehr Hilfe benötigt der durchschnittlich erfahrene Anwender beim Design und Setup des Systems. Spezialisten werden jedoch immer teurer und die Anlagen werden komplizierter zu warten oder gar zu ändern.

Die Idee der Autoren ist es, jeden potentiellen Anwender im Bereich der Automatisierung mit einer simplen und kostengünstigen Visualisierungs- und Kontrollmöglichkeit für annähernd jede Anwendung zu versorgen, wobei es nebensächlich ist, wo sich der User aufhält und welche Soft- oder Hardwareplattform zur Verfügung steht.

Literatur

1. Bartlett, N. et al.: „JAVA Programming Explorer"; The Coriolis Group, Scottsdale, Arizona, 1996.

2. Gamma, E. et al.: „Design Patterns, Elements of Reusable Object-Oriented Software", Addison-Wesley, 1995.

3. Reiter, H., Kral, C.: „Interaction between Java and Lonworks", IEEE Workshop on Factory Communications Systems, WFCS´97, Barcelona, Spain, 1997.

4. Schneier, B.: "Applied Cryptography", Wiley, New York, NY, USA, 1994.

5. Dietrich, D., Reiter, H., Schweinzer, H.-J.: "Neue Herausforderungen der Feldbustechnologie", Conference Paper "Home Technology", OEVE, Villach, Austria, 1996.

6. Carotta, M.: "HomeNet Control - Internetbasierende Fernsteuerung von Heimnetzen unter Berücksichtigung von Sicherheitsaspekten", Master's Thesis, Institute of Automation, Vienna, Austria 1997.

InterBus-Baustein MIC8052 mit integriertem Mikrocontroller

Michael Bratge

Mikroelektronik Anwendungszentrum GmbH Thüringen

Abstract. Die Mikroelektronik Anwendungszentrum GmbH Thüringen (MAZeT) hat in der Vergangenheit bereits die InterBus Protokollschaltkreise SUPI 3, LPC1 (InterBus Loop) sowie einen skalierbaren InterBus Kern als ASIC-Makro gemeinsam mit der Fa. Phoenix Contact GmbH & Co. entwickelt. MAZeT, Phoenix Contact sowie der Distributor SEI Jermyn bringen gegenwärtig einen Baustein IB8052 [10] auf den Markt, der auf der Basis des SUPI 3-InterBus-Kerns diesen mit einen Mikrocontrollerkern der Fa. Synopsys verbindet. Ein Nachfolger für diesen Baustein (MIC8052) wird einen 2-Kanal Quadraturdecoder, eine EnDat und SSI-Schnittstelle sowie ein spezielles Interface für LED- und LCD-Controller enthalten.

Der Controllerkern des MIC8052 entspricht einem 8052, der einen Befehl in nur 4 Taktzyklen abarbeitet. Der Standard-805x benötigt pro Befehl 12 Taktzyklen. Dadurch erreicht der mit mindestens 16 MHz getaktete Prozessor eine mehrfach höhere Performance, als man von einem herkömmlichen 8051-Derivat gewöhnt ist.

Der Prozessor empfiehlt sich insbesondere für Geräteentwicklungen der InterBus-Gerätegruppen Encoder, Identifikationssysteme, Mensch-Maschine-Kommunikation und die Antriebstechnik.

Abstract. In the past Mikroelektronik Anwendungszentrum GmbH Thüringen (MAZeT) has developed the InterBus protocol circuits SUPI 3, LPC1 (InterBus Loop) and a scaleable InterBus core as ASIC-Makro together with Phoenix Contact GmbH & Co., Blomberg.

At present MAZeT, Phoenix Contact and the Distributor SEI Jermyn bring a new solution IB8052 [10] to the market for InterBus connecting the above mentioned InterBus core with a microcontrollercore compatible to an 8051. This device is suited for applications with middle need of controller Performance and resources. A successor of this ASIC (MIC8052) will contain a dual channel quadrature Encoder, an EnDat /SSI Interface and an interface for LED and LCD controllers.

The controller core corresponds to a 8052, but most instructions are performed within 4 clock cycles only. The original 8052 needs 12 clock cycles. That is why the new device has about the 2,5 Performance over the original controller. The controller can be clocked with 8 or 16 MHz.

The controller especially recommend for devices developed in automation systems for Encoder, Identification Systems, Man-Machine-Communication and also for Technology and Drive control.

1 Übersicht

Für Automatisierungsaufgaben empfehlen sich immer mehr moderne serielle Bussysteme. Das es unter diesen trotzdem noch eine große Vielfalt gibt, zeigt der InterBus, der topologisch eigentlich eine Ringstruktur besitzt. Für den InterBus gibt es universelle Anschaltbausteine, die autark digitale Ein- und Ausgabefunktionen realisieren können. Alternativ können diese Bausteine ein Mikroprozessorinterface bereitstellen, so daß die unterschiedlichsten Mikrocontroller oder -prozessoren an den Bus angeschlossen werden können. In dem Bestreben, einen Baustein zu kreieren, der den Busanschluß mit einem Mikrocontroller verbindet, galt es zunächst, den Mikrocontroller mit der höchsten Akzeptanz im Bereich der Hersteller von Automatisierungstechnik zu finden. Interessanterweise kam hier immer wieder der 8051 trotz seiner inzwischen bescheidenen Leistungsfähigkeit im Vergleich zu jüngeren Entwicklungen zur Sprache. Da es Ziel war, einen leistungsfähigen Prozessor zu integrieren, galt es, einen Controllerkern zu finden, der einerseits codecompatibel zum 8051 ist und der andererseits leistungsfähig genug für moderne Applikationen ist.

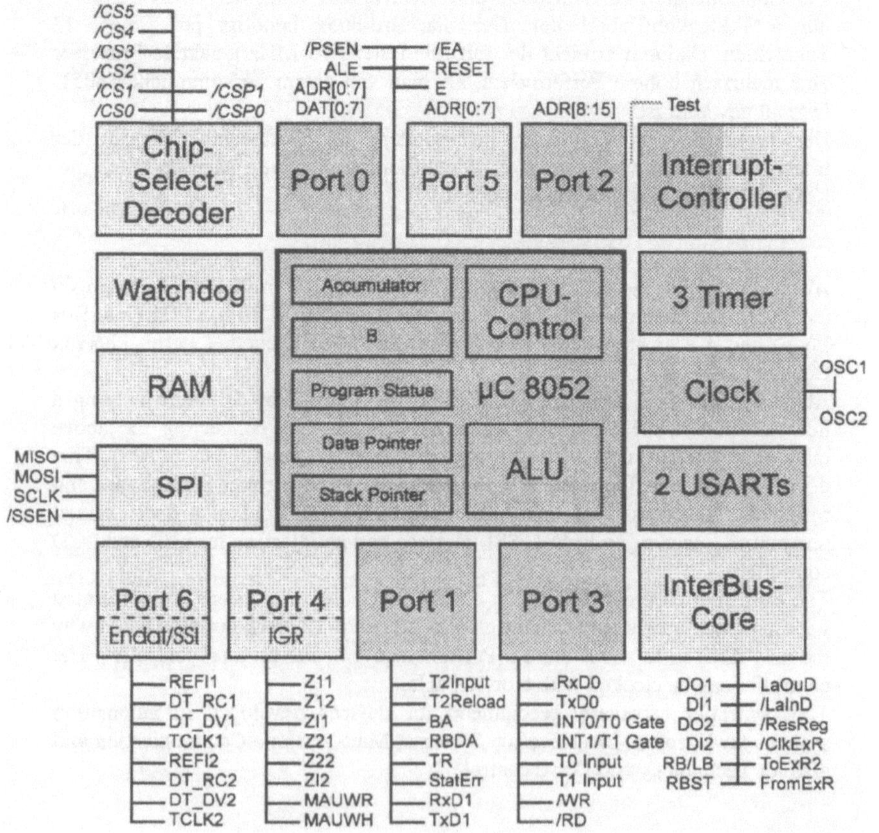

Bild 1: Blockschaltbild des MIC 8052

Der Chip wurde so gestaltet, daß er ein über einen externen Programmspeicher oder per Download frei programmierbarer Mikrocontroller mit IBS-Anschluß ist.

Der Schaltkreis beinhaltet einen Mikroprozessor mit der Funktionalität des 8052 mit

- einer universellen synchronen und asynchronen seriellen Schnittstelle mit der Funktionalität der UART des 8052,
- 6 Parallelports zu je 8 Bit, von denen 5 Ports bitweise adressierbar sind,
- 3 16-bit-Timern mit der Funktionalität der Timer des 8052,
- einer erweiterten Interrupt-Unit zum wahlfreien Zuweisen der vorhandenen 11 Interruptquellen auf 2 unterschiedliche Prioritätsebenen,

sowie zusätzlich:

- einem Watchdog Timer,
- dem integrierten InterBus-Kern,
- 1 KByte integriertem RAM,
- 512 Byte ROM mit einem Urlader,
- einem zweikanaligem Positionsgeberinterface, welches wahlweise als Interface für Inkrementalgeber, SSI oder EnDat Schnittstellen benutzt werden kann, sowie
- einem zusätzlichen synchronen seriellen Port, der sich wie ein SPI-Baustein verhält, zusätzlich jedoch konfigurierbar in seiner Datenlänge ist und die Daten in beiden Richtungen schieben kann (LSB oder MSB first).

Der Interbus-Teil setzt auf der Funktionalität des jüngsten Protokollschaltkreises, dem SUPI 3 auf. Er besitzt darüber hinaus 12 Datenregister a 8 Bit und erweiterte Steuerungsmöglichkeiten durch den Mikroprozessor. Diverse Einstellungen, die beim SUPI 3 hardwaremäßig vorgenommen werden können, werden beim MIC8052 softwaremäßig realisiert. Hierdurch reduziert sich deutlich die Anzahl der benötigten Pins und der Baustein findet in einem 100-poligen QFP-Gehäuse Platz.

Im Unterschied zu einer herkömmlichen Lösung bietet der Baustein die Möglichkeit einer intelligenten InterBus-Anschaltung auf kleinstem Raum, darüber hinaus verbesserte Störsicherheit durch den integrierten Aufbau.

2 Anschluß externer Ressourcen

Beim MIC8052 können 2 Speicherbereiche für externe Programmadressen definiert werden. Beim Zugriff auf diese Bereiche wird ein Chip-Select-Signal /CSP0 bzw. /CSP1 aktiviert. Die Grenze zwischen den Adreßbereichen ist variabel gestaltet und kann an Vielfachen von 1 KByte liegen. Für den Datenspeicher ist die Zahl der Chip-select-Signale erweitert worden. Hier stehen 6 Signale zur Verfügung, die sich den Adreßraum in Schritten von minimal 256 Byte untereinander aufteilen.

Damit ist es möglich, verschiedene Speicher oder Baugruppen mit geringem Aufwand an den Prozessor anzuschließen und den Zugriff darauf zu steuern.

3 Interne Speicher

Der MIC8052 enthält als PROM eine Urladeroutine. Durch die Beschaltung des Pins /EA ist einzustellen, ob das interne Programm oder ein externes Anwenderroutine gestartet wird. Ein High-Pegel am Pin /EA startet das integrierte Urlader-Programm, welches versucht, ein Anwenderprogramm über eine der seriellen Schnittstellen in den internen RAM oder einen angeschlossenen externen Speicher zu laden. Gelingt dies, wird anschließend dieses Programm gestartet. Andernfalls stoppt der Prozessor die Befehlsabarbeitung. Unter anderem kann der Prozessor Programme bis zu 1KByte aus einem seriellen PROM in den internen Datenspeicher laden. Er ist in der Lage, diesen Speicher in den Codespeicherbereich zu mappen und das geladene Programm abzuarbeiten. Auf diesen Weg läßt sich für kleine Applikationen ein externer paralleler Programmspeicher und damit Platz auf der Platine einsparen.

Ein Low-Pegel am Pin /EA startet ein externes Anwenderprogramm. Diese Funktion ist vom 8051 her bekannt. Mit Hilfe eines internen Registers ist es möglich, das vom Pin /EA kommende Signal softwaremäßig zu negieren. Damit kann dynamisch zwischen der integrierten und einer externen Anwenderroutine umgeschaltet werden.

4 Serielle Schnittstellen

Der Schaltkreis enthält zwei asynchrone und synchrone serielle Schnittstellen, die in allen vom 8052 her bekannten Modi arbeiten können.

Die zusätzliche synchrone serielle Schnittstelle wird über das Synchron Serial Control Register oder über ein Taktsignal der Gegenstelle gesteuert. Eine maximale Baudrate von 2 MBd ist möglich. Die Datenübertragung erfolgt immer bidirektional, d. h. es wird mit jeder relevanten Flanke des Clock-Signals sowohl ein Informationsbit in das Schieberegister des Empfangspuffers gelesen als auch ein Informationsbit aus dem Senderegister geschoben. Die synchrone serielle Schnittstelle ist kompatibel zu SPI. darüber hinaus kann sie wahlweise mit dem MSB oder LSB beginnend Daten austauschen. Außerdem ist die Datenlänge zwischen 2 und 8 Bit frei wählbar. Mit dieser Schnittstelle sind sehr bequem A/D-Wandler beliebiger Auflösung, 7-Segmentanzeigen und andere Interfacebausteine ansteuerbar.

5 Postionsgeberinterfaces

Der Baustein hat eine Schnittstelle, die sich wahlweise wie ein Quadraturdecoder MAC4124 [8], wie eine SSI-Schnittstelle [6] oder wie eine EnDat-Schnittstelle [7] verhalten kann. Diese Schnittstelle ist zweikanalig aufgebaut. Als EnDat-Schnittstelle erfüllt sie das Protokoll der Version 2.0 und entspricht dem Baustein MES48 [9]. Die SSI ist eine Untermenge der EnDat-Schnittstelle.

6 Counter / Timer

Die Funktionalität dieser Baugruppe entspricht der entsprechenden Baugruppe des 8052. Die Timer werden in jedem Befehlszyklus, somit alle 4 Systemtakte einmal inkrementiert. Über ein zusätzliches Register können die Zähler so eingestellt werden, daß sie sich verhalten wie die Timer eines originalen 8052, der die Counter alle 12 Takte inkrementiert.

7 Parallele Schnittstellen

Der MIC8052 enthält 7 parallele Ports, die alle mit einer Zweitfunktion belegt sind. Nach RESET sind alle Ports als Eingänge geschaltet.

Wird Port 0 nicht anderweitig benötigt, kann er genutzt werden, um getrennte Adreß- und Datenleitungen für externe Baugruppen am Bus zu verwenden. In diesem Fall wird kein externer Adreßlatch benötigt. Port 5 erfüllt die Funktion des Port 0 aus dem Original 8052: er stellt die Daten und das Low Byte der Adreßsignale für den externen Speicher- und I/O-Bausteinanschluß bereit. Die Ein- und Ausgabefunktion des Port 5 kann nur in Verbindung mit einem in den internen RAM geladenen Programm genutzt werden.

Port 1 hat die gleichen Funktionen wie das 8052-Original, ebenso Port 3, jedoch stellt Port 3 zusätzliche Signale für den InterBus sowie 2 weitere Signale zur Verfügung.

Port 4 wird von der EnDat/SSI - Schnittstelle benutzt und Port 6 für die Signale des Quadraturdecoders sowie für 2 optionale Interbus-Signale.

8 InterBus -Schnittstelle

An den Prozessor ist ein InterBus-Slave-Kern angeschlossen, der das 2-Leiter-Protokoll des InterBus realisiert. Das 8-Leiter-Protokoll realisiert der derzeit aktuelle InterBus-Slave-Baustein SUPI 3 nur noch aus Kompatibilitätsgründen, da der SUPI 3 den Vorgängertypen SUPI 2 aufwärtskompatibel ersetzen kann. Da das 8-Leiter-Protokoll des InterBus stark in den Hintergrund getreten ist, wurde bei den hier vorgestellten Bausteinen auf die Implementierung des 8-Leiter-Protokolls verzichtet.

Über 14 Ein- und Ausgaberegister wird der InterBus-Controller durch den Prozessor bedient und in Typ und Klasse der InterBus-Klassifikation softwaremäßig eingestellt. Diese Einstellmöglichkeit entspricht weitgehend den hierzu vorgesehenen Möglichkeiten im Mikroprozessor-Modus des InterBus-Protokollchips SUPI 3. Auf eine hardwaremäßige Einstellmöglichkeit des InterBus-Betriebsmodus des Bausteins wurde mit Blick auf die Pinzahl des Bausteins und des Gehäuses verzichtet. Diese ist auch nicht notwendig, da beide hier vorgestellten Chips in der Lage sind, sich softwaremäßig am InterBus zu definieren.

Die Datenbreite der übergebenen Nutzdaten beträgt bis zu 12 Byte und ist ggf. über einen externen Registererweiterungsbaustein SRE1 erweiterbar. Die Nutzdaten transportiert der InterBus-Slave-Kern entsprechend dem InterBus-Protokoll weiter. Grundlagen für die InterBus-Schnittstelle können [3] entnommen werden.

Hardwaremäßig sind die folgenden InterBus-Pins verfügbar:

DO1	Datenleitung IN Hinweg
DO2	Datenleitung OUT Hinweg
DI1	Datenleitung OUT Rückweg
DI2	Datenleitung IN Rückweg
RBST	Meldeeingang, ob die weiterführende Schnittstelle abgeschaltet ist

ToExR	InterBus-Ausgang zum externen Register
FromExR	InterBus-Eingang vom externen Register
/ClkExR	Takt für externe Register
/ResReg	Reset für externe Register
LaInD	Übernahmesignal für periphere Eingangsdaten

Optional können über 4 Pins des Port 1 bzw. über 2 Pins des Port 6 folgende InterBus-Signale genutzt werden:

BA	InterBus active	(Diagnosesignal)
RD	Remotebus disabled	(Diagnosesignal)
TR	Transmit/Receive	(Diagnosesignal)
StatErr	Modulfehler	(Diagnosesignal)
MAUWH	Medium Access Warnung Hinweg	
MAUWR	Medium Access Warnung Rückweg	

Diese Signale verhalten sich genauso wie die entsprechenden Signale des SUPI 3. Diagnosesignale können entweder softwaremäßig abgefragt werden oder sie werden wahlweise auf Pins des Parallelports 1 geschaltet. Das Signal ToExR ist softwaremäßig umschaltbar, so daß es sich entweder wie das Signal ToExR1 oder ToExR2 der Interbus-Bausteine SUPI verhalten kann. Die vom SUPI her bekannte Verbindung zwischen den Signalen ToExR und FromExR ist beim IB8052 intern realisiert und kann im Bedarfsfall per Software aufgetrennt werden.

9 Interrupts

Der MIC8052 besitzt Interrupteingänge für alle internen Baugruppen sowie für 2 externe Interruptquellen, die unterschiedlich priorisiert sind. Wie beim 8052 kann

Tabelle 1: Interruptquellen:

Interrupt	Beschreibung	Priorität	Interrupt Vektor
int0	Externer Interrupt. Int0 (P3.2)	1	03h
TF0	Timer 0 overflow	2	0Bh
int1	Externer Interrupt INT1 (P3.3)	3	13h
TF1	Timer 1 overflow	4	1Bh
ti0/ri0	Serial Port 0 transmit or receive	5	23h
TF2	Timer 2 overflow	6	2Bh
ti1/ri1	Serial Port 1 transmit or receive	7	33h
inties1	IGR/EnDat/SSI Kanal 1 Interrupt	8	3Bh
intibs	Interbus Port Interrupt	9	4Bh
int4	Synchron Serial Interface Interrupt	10	53h
inties2	IGR/EnDat/SSI Kanal 2 Interrupt	11	5Bh
wdti	Watchdog Interrupt	12	63h

jeder Interrupt einer von 2 verschiedenen Prioritätsstufen zugeordnet werden. Da es beim MIC8052 mehr Interruptquellen als beim Original 8052 gibt, sind auch die entsprechenden Steuermöglichkeiten dafür durch zusätzliche Interruptregister vorhanden.

10 Takt - Einigkeit macht schnell

Der InterBus-Teil sowie der Mikrocontroller werden beide mit dem gleichen Takt betrieben. Durch die hierdurch erreichte Synchronität sind beide Komponenten in der Lage, ihre Geschwindigkeit optimal auszunutzen. Zur Taktversorgung ist ein Oszillator integriert, der mit einem externen Quarzkristall und minimaler Zusatzbeschaltung die Gesamtschaltung taktet. Alternativ kann auch ein externer Quarzoszillator angeschlossen werden. Die Taktfrequenz wird 16 MHz betragen. Ein zusätzlicher Betrieb bei 8 MHz wird ebenfalls möglich sein. Dazwischenliegende Frequenzen sind ebenfalls denkbar, allerdings ist dann kein Betrieb des Interbus-Interfaces mehr möglich und die Frequenzen aller Baugruppen wie Timer, serielle Schnittstellen, Watchdog-Timer verändern sich entsprechend.

Literatur

1. Feger, Otmar, Die 8051-Mikrocontroller-Familie, Teil 1 und 2, Siemens Aktiengesellschaft, Berlin und München, 1991

2. MCS51 Microcontroller Family User's Manual, Intel Corp. 1994, ON 272383-002

3. IBS SUPI II HB; Anwenderhandbuch für InterBus Slave-Protokollchip SUPI II; Phoenix Contact GmbH & Co., PF 1341, 32819 Blomberg.

4. Anwenderhandbuch für InterBus Evaluation Board IBS EVA Board HB; Phoenix Contact GmbH & Co., PF 1341, 32819 Blomberg.

5. Referenzhandbuch für InterBus Peripheral Communications Protocol IBS PCP RE HB; Phoenix Contact GmbH & Co., PF 1341, 32819 Blomberg.

6. Anordnung zur seriellen Übertragung der Meßwerte wenigstens eines Meßwertwandlers, Patentschrift DE 3445617 C2, Fa. Max Stegmann GmbH.

7. Bieslki, S., Hagl, R, Übertragung absoluter Positionswerte von Längenmeß-systemen, Drehgebern und Winkelmeßsystemen zu Folge-Elektroniken

8. Datenblatt MAC4124, MAZeT GmbH, Jena, 1996

9. Anwenderinformation MES48, MAZeT GmbH, Jena, 1997

10. Bratge, M.; InterBus-Baustein IB8052 mit integriertem Mikrocontroller, http://www.mazet.de/mazet_d/index_df.html, oder über http://www.mazet.de

Untersuchungen zur Anwendung des CAN für Kleinsatelliten

A. Braune, K. Janschek

Institut für Automatisierungstechnik
Technische Universität Dresden, Deutschland

Abstract. Low Earth Orbit (LEO) Satellite Networks are currently established as a novel basis for mobile communication and earth observation including traffic and environmental monitoring. These LEO-satellites are built under stringent commercial constraints and call therefore for new concepts in terms of cost reduction compared to the well established technologies for geostationary telecommunication satellites. The onboard automation plays a key role for the autonomous operation of the satellites within the network. Cost reduction potentials in this area are found through the implementation of functional redundancy to decrease the overall number of equipment and through the use of well proven industrial terrestrial (off-the-shelf) technologies for the onboard realization. Functional redundancy for the onboard automation requires appropriate bus-technologies for an efficient realtime communication with maximum freedom for reconfiguration. Due to its source from automotive applications the CAN bus technology is a very promising candidate in terms of functionality, performance and cost. This paper presents the dimensioning requirements and a first analysis of the applicability of the CAN for LEO-satellites.

1 Motivation

LEO-Satellitennetzwerke (LEO = Low Earth Orbit) sind die Grundlage für zukünftige satellitengestützte Kommunikationssysteme der Mobilkommunikation, der Erdbeobachtung und wissenschaftlicher Untersuchungen zu Verkehr, Umwelt oder Klima. Zur globalen Abdeckung benötigt man dazu eine große Anzahl (40 ... 800) niederfliegender Satelliten (typ. Bahnhöhe 1000 km). Begründet durch die Aufgabenstellung ergeben sich für diese neue Generation von Satelliten völlig neue *Anforderungen* im Vergleich zu den eingeführten geostationären (GEO) Telekommunikationsatelliten, wie z.B.:

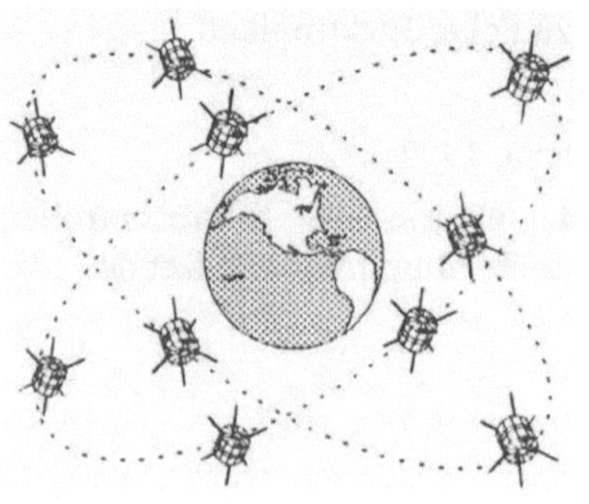

□ kleinere Sendeleistungen und reduzierte Datenkapazität pro Satellit
  ⇨ <u>"kleine" Satelliten</u> (typ. 500 kg)

□ große Anzahl identischer Satelliten
  ⇨ <u>"kostengünstige" Satelliten</u> von der "Stange"

□ verminderte Redundanzanforderungen im Einzelsatelliten durch Systemredundanz des Netzwerkes
  ⇨ <u>"einfache" Satelliten</u>

Abb. 1. LEO Satellitennetzwerk

Wegen der großen Anzahl an Satelliten und der dynamischen Netzwerkstruktur stellt die *Bordautomatisierung* dieser Satelliten eine Schlüsseltechnologie für einen *kommerziell erfolgreichen Betrieb* eines solchen satellitengestützten Kommunikationssystems dar.
Die *Automatisierungsansätze* lassen sich wie folgt zusammenfassen:

□ *Entwurf* der Einzelsatelliten als weitestgehend *autonome Klienten* in einem räumlich verteilten System (Netzwerk) mit den Aufgaben:

- Bereitstellung definierter Satellitenzustände zur Durchführung der Kommunikationsaufgaben (Lage- und Positionsbestimmung, Ausrichtung, Onboard Data Management, etc.)
- bordautonomes Fehlermanagement
- bordautonome Erhaltung der Bahnkonstellation (relative Position im Netzwerk)

□ *Realisierung* der Einzelsatelliten durch *kommerziell attraktive Systemlösungen* ("klein-billig-einfach") wie z.B.

- Ersatz von klassischer Hardwareredundanz durch massiv funktionelle Redundanz, d.h. Mehrfachnutzung von Hardwarekomponenten für unterschiedliche Aufgaben
- Nutzung von Sekundärresourcen - Informationsfusionierung - z.B. zusätzliche Nutzung der Bildinformation einer Erdbeobachtungskamera für Navigationsaufgaben
- Einsatz bzw. Adaption von terrestrischen, industriellen Technologien für Bordaufgaben (z.B. Automobilbau)

An der Technischen Universität Dresden (TUD) werden zur Zeit im Rahmen eines interfakultären Forschungsprojektes (*TUD-Satellitenkolleg*) Schlüsseltechnologien für derartige *LEO-Satellitennetzwerke* der nächsten Generation untersucht. Ausgewählte Ergebnisse dieser Forschungsarbeiten sollen dann mit einem an der TUD entwickelten und gebauten Mikrosatelliten (100 kg) als Technologiedemonstrator im Weltraum erprobt werden [1].
Das *Institut für Automatisierungstechnik* hat dabei die Bearbeitung der folgenden Aufgaben übernommen:

o Bahn- und Lageregelung
o Onboard-Data-Handling
o Fehlermanagement.

In der vorliegenden Arbeit wird über erste Ergebnisse zu Untersuchungen des CAN-Busses als einen möglichen Kandidaten für die Bordkommunikation berichtet. In besonderem Maße interessieren dabei dessen Eigenschaften bezüglich eines Einsatzes in einer dezentralen multifunktionalen Automatisierungsarchitektur.

2 Anforderungen an die Bordkommunikation

Das bordseitige Datenkommunikationssystem verbindet die Komponenten der Systemsteuerung und der Nutzlast(en). Damit erhält es eine zentrale und missionskritische Bedeutung für die Bordautomatisierung und muß Redundanzanforderungen genügen. Abb. 2 zeigt typische informationsverarbeitende Bordfunktionen eines LEO-Satelliten. Zu übertragen sind
-- Systemdaten: z.B. Meß-/Stell-/Status-/Alarmwerte, Systemzeit (10 ... 100 Kbit/s)
-- Nutzlastdaten: z.B. Bilddaten einer Bordkamera (100KBit/s ... 5 Mbit/s)
-- Programmcode zur Reprogrammierung der softwarebasierten Bordfunktionen von der Bodenstation aus über die Telemetrieschnittstelle
System- und Nutzlastdaten unterliegen zum Teil harten Echtzeitforderungen, d.h. zwischen Datenquelle und Datensenke sind maximal zulässige Verzögerungszeiten einzuhalten.

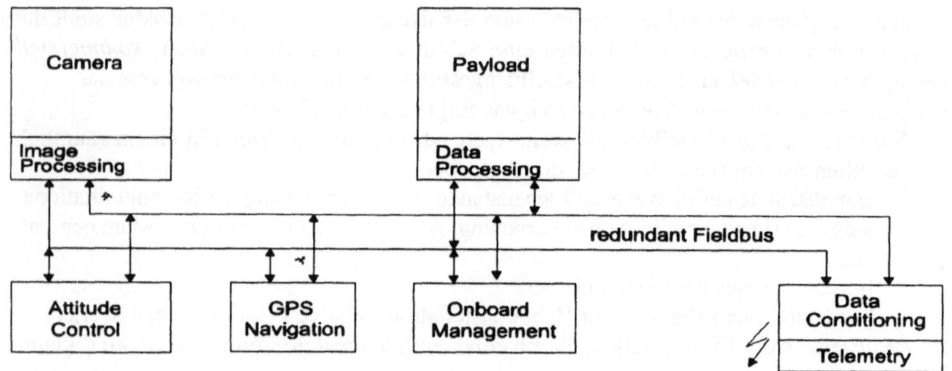

Abb. 2. Onboard-Kommunikationsstruktur eines Kleinsatelliten

Bei Missionsänderungen oder im Fehlerfall können Änderungen der Systemstruktur erforderlich sein. Wenn beispielsweise statt eines ausgefallenen GPS die Informationen der Nutzlast-Kamera durch Landmarkenerkennung für die Lagebestimmung genutzt oder die Aufgaben eines ausgefallenen Bordrechners auf andere Einheiten verteilt werden, muß die Kommunikationsstruktur während des Betriebes einfach änderbar sein. Der Ausfall eines Teilnehmers darf keinen Einfluß auf die Funktionsfähigkeit des Netzes haben.
Der autonome Betriebe unter extremen Umgebungsbedingungen erfordert hohe Übertragungssicherheit. Aus Kosten- und Zuverlässigkeitsgründen sollen standardisierte, industrielle Komponenten verwendet werden. Während der Entwicklungsphase wird auf die Schicht 2 des ISO/OSI-Schichtenmodells aufgesetzt, da die Anforderungen an mögliche Anwendungs- und Nutzerschichten erst im Verlaufe der weiteren Entwicklungsarbeiten spezifiziert werden sollen.

Die dimensionierenden Anforderungen lassen sich folgendermaßen zusammenfassen:
- Nutzung standardisierter Komponenten aus der industriellen Massenproduktion
- Übertragung von Systemdaten (10 Kbit/s ... 100 Kbit/s), von Nutzlastdaten (100 KBit/s ... 5MBit/s) und von Programmen
- z.T. harte Echtzeitanforderungen mit Deadlines im Bereich von 500 μs bis 0,5 ms
- hohe Übertragungssicherheit
- flexible Kommunikationsstrukturen
- zuverlässiger Betrieb unter extremen Umgebungsbedingungen

Als Kandidat für die Onboard-Datenkommunikation bietet sich der CAN-Bus (Controller Area Network) aus der Automobilindustrie an, da dort ähnlich hohe Anforderungen an Robustheit, Übertragungssicherheit und Echtzeitfähigkeit gestellt werden.

3 Eignung des CAN-Bus

3.1 Standardisierung und verfügbare Komponenten
Wegen der hohen Stückzahlen eingesetzter Knoten im Automobilbereich existiert eine große Anzahl kostengünstiger CAN-Controller mit unterschiedlicher Leistungsfähigkeit. CAN-Controller, die unmittelbar im Motorraum eingesetzt sind, entsprechen hinsichtlich ihres

Temperaturbereiches und der mechanischen Belastbarkeit bereits weitgehend den Anforderungen von Kleinsatelliten in erdnahen Bahnen und sind damit für die vorliegenden Anwendung sehr gut geeignet. Bei Einsatz von CMOS-Technologie ist auch die Strahlenbelastung wegen der niedrigen Flughöhe im zulässigen Bereich.

Neueste Entwicklungen stellen bereits integrierte Sensoren plus CAN-Controller zur Verfügung, die beispielweise direkt als Navigationssensoren eingesetzt werden können [2][3].

Die verfügbaren Controller entsprechen weitgehend der Schicht 2 des ISO/OSI-Referenzmodells, so daß diese Ebene zunächst als Experimentierumgebung verwendet wird. Basierend auf Schicht 2 existieren mehrere Schicht 7 - Protokolle und Profile, die teilweise von unterschiedlichen Anwendern entwickelt bzw. für verschiedene Zielgebiete konzipiert wurden (z.B. CAL, SDS, DeviceNet). Weitere Untersuchungen müssen begründen, ob eine höhere Protokollschicht für die Anwendung des CAN in einem Kleinsatelliten geeignet bzw. erforderlich ist.

3.2 Informationsübertragung

Ein CAN-Telegramm überträgt maximal 64 Bit nichtstrukturierter Nutzdaten (Abb. 3). Damit können die Systemdaten eines Gerätes in den überwiegenden Fällen mit einem oder wenigen Telegrammen übertragen werden. Die Nutzlastdaten sind dazu in 64 Bit-Einheiten aufzuteilen. Das CAN-Interface bietet keine Unterstützung bei der Behandlung strukturierter Daten oder zusammengehöriger Datenbereiche.

In höheren Protokollschichten existieren hierfür Objekte und Dienste. Ob die daraus resultierenden längeren Bearbeitungszeiten die Echtzeiteigenschaften wesentlich beeinflussen, muß für konkrete Hard- und Software-Umgebungen untersucht werden.

CAN läßt eine Baudrate von 1 Mbit/s bei bis zu 40m Netzausdehnung zu. Eine maximale Busauslastung von 70% erlaubt in diesem Falle eine Nutzdatenrate von ca. 400 KBit/s. Damit sind die Systemdaten vollständig, die Nutzlastdaten je nach Mission u.U. nur eingeschränkt übertragbar (vgl. Abschnitt 2). Bei bestimmten Fehlersituationen muß der Bus unter Umständen trotz der zu erwartenden kurzen Segmentlängen mit niedrigeren Baudraten betrieben werden. In diesen Fällen sind Datenreduktionen erforderlich.

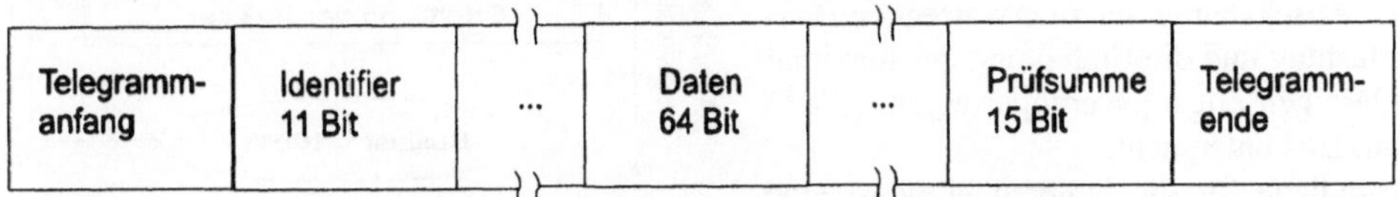

| Telegramm-anfang | Identifier 11 Bit | ... | Daten 64 Bit | ... | Prüfsumme 15 Bit | Telegramm-ende |

Abb. 3. Prinzipieller Aufbau eines CAN-Telegramms

3.3 Multi-Master-Fähigkeit

Der Buszugriff nach einem Master-Slave-Verfahren ist wegen der dann erforderlichen Redundanzen für den Master für die vorliegende Anwendung ungeeignet. Beim CAN erfolgt die Vergabe des Buszugriffsrechts nach einem modifizierten CSMA/CD Verfahren. Jeder Busteilnehmer ist gleichberechtigter Master, besitzt Buszugriffsrechte und ist damit autonom am Busbetrieb beteiligt. Der Ausfall eines Teilnehmers hat damit, wie gefordert, keine Auswirkungen auf den Betrieb des gesamten Netzes.

In einem CAN-Telegramm werden Nachrichten durch einen netzweit eindeutigen Identifier gekennzeichnet. Dieser bestimmt gleichzeitig die Priorität dieser Nachricht. Alle Teilnehmer hören alle Nachrichten auf dem Bus mit und überprüfen deren Relevanz anhand des Identifiers. Damit realisiert bereits das Protokoll die in der vorliegenden Anwendung geforderten flexiblen Kommunikationsstrukturen und Broadcastdienste.

Bei gleichzeitigem Sendewunsch mehrerer Telegramme setzt sich das Telegramm mit der höchsten Priorität durch, so daß eine kurze Reaktionszeit nur für wenige hochpriore Daten garantiert werden kann. Die Einhaltung der Echtzeitanforderungen ist damit nicht in jedem Fall gegeben und muß für die konkrete Anwendung explizit überprüft werden.

3.4 Übertragungssicherheit

Die Vorteile des CAN resultieren weitestgehend aus seinem primären Einsatzgebiet im Automobilbereich und den dort vorliegenden Anwendungsanforderungen. Es existieren vergleichsweise hochentwickelte Verfahren zur Erkennung fehlerhafter Telegramme und zum Erreichen einer hohen Störsicherheit auch unter ungünstigen Umgebungsbedingungen. Darüber hinaus erkennt das Protokoll fehlerverursachende Knoten und schränkt deren Zugang zum Netz ganz oder teilweise ein, wenn bestimmte Fehlerhäufigkeiten überschritten werden [4].

Nachteilig bei CAN ist, daß das Busprotokoll einen vollständigen Teilnehmerausfall nicht erkennt. Hierzu sind zusätzliche, nutzerdefinierte Überwachungsfunktionen zu installieren. Höhere Protokollschichten definieren z.B. Fehlermaster, welche die übrigen Netzknoten zyklisch auf Funktionsfähigkeit überprüfen. Im vorliegendenFall müßte aber auch dieser Fehlermaster wiederum überwacht und redundant ausgeführt werden, so daß eine verteilte Kompetenz zur Fehlererkennung geeigneter erscheint.

4 Entwicklungswerkzeug zur Modellierung des Echtzeitverhaltens

Die Analyse im vorhergehenden Kapitel zeigt einige kritische Aspekte bezüglich des Echtzeitverhaltens auf. Flexible Kommunikationstrukturen bei harten Echtzeitanforderungen müssen bereits in der Entwurfsphase quantitativ analysierbar sein. Dazu wurden verschiedene algorithmische Zugänge zur Abschätzung der zu erwartenden Busbelastung und die Einhaltung der maximal zulässigen Zugangsverzögerungen (ZVZ) zum Bus untersucht.

Grundlage für die Untersuchung ist eine festgelegte Bus-Konfiguration einschließlich der zu erwartenden Telegramme. Vereinfachend wird angenommen, daß alle Telegramme zyklisch auftreten und die Datenlänge während der Betriebszeit konstant ist [5]. Die Telegramme sind definiert durch *Datenlänge, Zykluszeit* und *zulässige Verzögerung*. Für zyklische Nachrichten kann die mittlere Busbelastung ermittelt werden aus dem Verhältnis von Nachrichtenlänge/Zykluszeit für alle auftretenden Nachrichten.

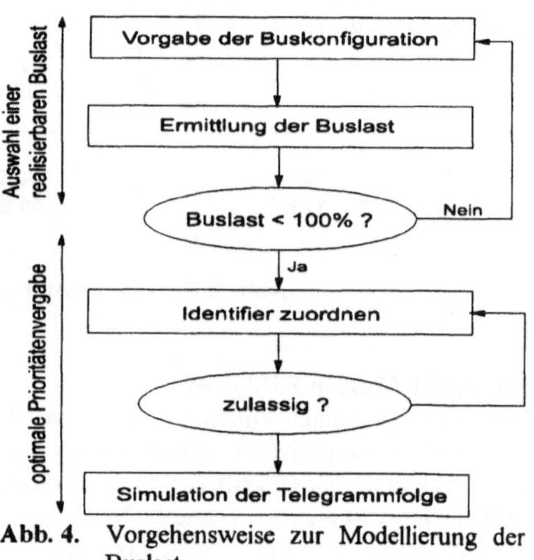

Abb. 4. Vorgehensweise zur Modellierung der Buslast

Ist die Buslast kleiner als 100%, die Konfiguration also prinzipiell lauffähig, wird für jede Nachricht die ungünstigste, d.h. längstmögliche Zugangsverzögerung (worst case) ermittelt. Diese tritt dann auf, wenn ein Telegramm durch das gerade begonnene längste niederpriore

Telegramm verzögert wird und danach alle höherprioren Telegramme ebenfalls Buszugriff fordern. Die Berechnung der Antwortzeiten erfolgt mit dem in [6] vorgeschlagenen *time dilation algorithm*.
Wenn die ermittelte ungünstigste Antwortzeit unter der vorgegebenen zulässigen Antwortzeit liegt, so werden die Echtzeitforderungen als erfüllt angenommen.
Ist das nicht der Fall, wird versucht, geeignetere Prioritätszuordnungen zu finden.
Die Prioritätsvergabe erfolgt nach folgenden Algorithmen [6] :

- *Rate Monotonic*
  vergibt hohe Prioritäten an Nachrichten mit kurzen Zykluszeiten
- *Deadline Monotonic*
  vergibt hohe Prioritäten an Nachrichten mit kurzen Deadlines
- *Last Laxity*
  ermittelt die Differenz zwischen Deadline und eigener Buslaufzeit und bewertet kurze Differenzen mit hohen Prioritäten

Mit einer angeschlossenen Simulation unter dem Systemsimulationswerkzeug *STATEMATE* können die errechneten Ergebnisse überprüft und grafisch dargestellt werden [7,8].

5 Echtzeitverhalten für ein ausgewähltes Beispiel

5.1 Buskonfiguration und Kommunikationsanforderungen

Für erste Untersuchungen der zu erwartenden Echtzeiteigenschaften dient die in Abb. 5 definierte Onboard Referenzstruktur [1]:

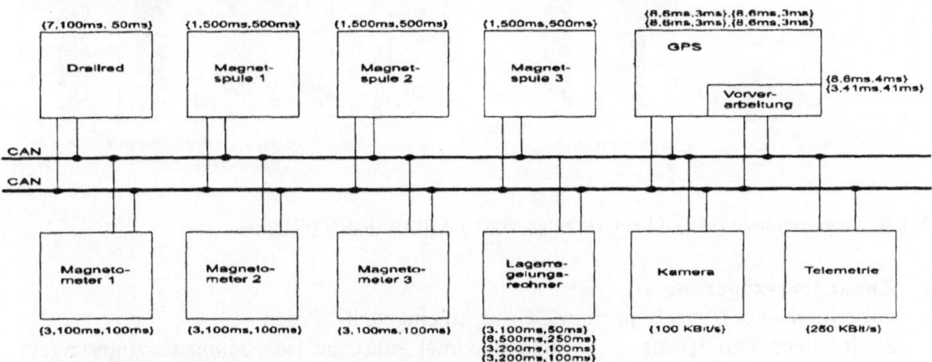

{Datenlänge, Zykluszeit, Deadline} bzw. {Nutzdatenrate} bei Kamera und Telemetrie

Abb. 5: Untersuchte Onboard-Struktur

□ *Redundanter CAN-Bus*

□ *System* (Bahn- und Lageregelung, Onboard Management):
  - *Data Handling & Control* : 3 Magnetometer, 3 Magnetspulen, 1 Drallrad, 1 Lageregelungsrechner
  - *GPS* mit zugehöriger Verarbeitungseinheit als intelligenter Sensor für die Bestimmung der Bahn- und Lagekoordinaten

442

☐   *Nutzlast*:
- Kamera für Aufnahmen mit verschiedenen Auflösungen.
- Funkbasierte Informationen zum Flottenmanagement von Fahrzeugen
Untersucht werden die folgenden Betriebszustände:
- *Nominal*: Beide Busse und alle Komponenten sind funktionsfähig. System und Nutzlast nutzen je einen Bus.
- *Fehler*:
    1. Ausfall der GPS-Verarbeitungseinheit:
       Der Lageregelungsrechner übernimmt die Verarbeitungsfunktionen. Über den "System"-Bus sind die "Rohdaten" der 4 GPS-Antennen zu übertragen.
    2. Ausfall eines Bussystems
       Systemkomponenten und Nutzlast nutzen gemeinsam einen CAN-Bus.
  Doppelfehler:
  Ein Bus und die GPS-Verarbeitungeinheit sind ausgefallen.

5.2   Busbelastung

Abb. 6 zeigt die auftretenden Busbelastungen. Bei 1 Mbit/s Baudrate liegt die Buslast in allen Betriebszuständen unter 70% und wird damit als zulässig angesehen. Bei 500 Kbit/s sind nur die Systemdaten verlustfrei übertragbar, die der Nutzlast nur zu ca. 50% .

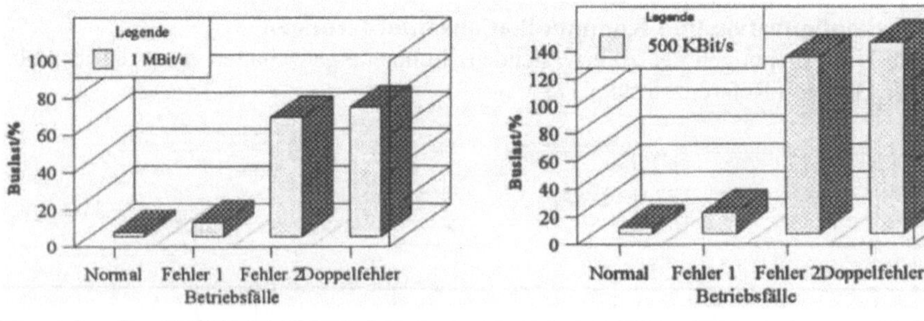

Abb. 6: Zu erwartende Buslast bei Baudraten von 1 MBit/s und 500 KBit/s

5.3   Zugangsverzögerungen

Die zu übertragenden Daten können in 2 Gruppen eingeteilt werden:
1. Zykluszeiten von 10 ms ... 1 s und maximal zulässige Buszugangsverzögerungen (Deadlines) kleiner oder gleich der Zykluszeit. Hierbei handelt es sich meist um Systemdaten u.a. für Regelungs- und Steuerungsaufgaben, so daß keine Deadline-Überschreitung zugelassen wird.
2. Zykluszeiten unter 10 ms. Diese Daten entstehen durch die Aufteilung großer Datenpakete meist der Nutzlasten auf Datentelegramme, z.B. bei einer online-Bildübertragung mit 100 Kbit/s. Die Verzögerung einzelnen Nachrichten auch über die Deadline hinaus ist hier bei entsprechender Zwischenpufferung unkritisch, wenn die Zeitforderungen für das gesamte Datenpaket eingehalten werden.

Die Identifiervergabe erfolgt beispielhaft für die in Abschnitt 4 genannten Algorithmen Bei einer Baudrate von 1 MBit/s passieren alle Nachrichten rechtzeitig den Bus. Es treten keine unzulässigen Verzögerungen auf. Bei 500 Kbit/s und Reduktion der Nutzlastdaten auf 50 %

überschreiten einzelne Bilddatentelegramme kurzzeitig ihre Deadline. Der vorgegebene Zeitrahmen für den gesamten Datenblock wird aber eingehalten.

Wesentliche Unterschiede in der Prioritätenzuordnung durch die verwendeten Algorithmen treten nicht auf, da im vorliegenden Fall die Zykluszeiten und die Deadlines jeweils in der gleichen Größenordnung liegen. Abweichende Ergebnisse sind insbesondere bei der Einbeziehung von Alarmdaten zu erwarten, die azyklisch auftreten, dann aber mit kurzen Verzögerungszeiten übertragen werden müssen.

6 Auswertung, Ausblick

Die Auswertungen zeigen, daß CAN einen geeigneten Kandidat für den Einsatz als Kommunikationssystem in einem Kleinsatelliten darstellt. Seine Stärken liegen besonders bei der Übertragung von Systemdaten und im Vorhandensein preiswerter, umweltrobuster Controller. Nutzlastdaten sind nicht in jedem Falle vollständig bzw. rechtzeitig übertragbar. Hier sollten Bussysteme mit höheren Baudraten untersucht werden. Wegen der z.T. harten Echtzeitanforderungen und der notwendigen Redundanz zur Systemdatenübertragung werden weitere Feldbusse, wie z.B. der Profibus DP in Betracht gezogen.

Weitere Untersuchungen sind erforderlich zur Nutzung höherer Protokollschichten und zum Test weiterer Konfigurationen. Die unter *STATEMATE* entwickelte Simulations- und Testumgebung hat sich während des Entwurfes als sehr hilfreich erwiesen.

Literatur

[1] Janschek, K. et.al.: TUD-Satellit Demonstrationsmission "Satellitengestütztes Monitoring mobiler Objekte", Förderantrag Vorentwurfsphase zu Projekt 1203/68501-0 gefördert durch Sächsiches Staatsministerium für Wissenschaft und Kunst, 1997.

[2] Ulbricht, S; Budde, W.; Gottfried-Gottfried, R.; Sauer, B.; Wende, U.; A monolithically integrated two-axis magnetic field sensor system, ESSCIRC,1996, Proceedings S 132-135.

[3] Bender, S.; Fischer, W.J.; Heinig, A.; Köhler, C.; Rietz,S.; Schneider,B.: Multi-sensor controller with CAN bus interface, Micro system technologies, 1996, Potsdam, Proceedings S. 151-155.

[4] Etschberger, K.: CAN Controller Area Network, Hanser, 1994.

[5] Staub, M.: Mit dem Worst Case kalkuliert - Zugriffsgarantie bei CAN-Netzen, elektronik 1995, Heft 12.

[6] Stümpfle, M.: Ober sticht unter - Prioritätenvergabe in Echtzeitsystemen, elektronik 1995, Heft 22.

[7] Hedrich, M.: Entwicklung eines CAN-Simulators, TU Dresden, Fak. Elektrotechnik, Institut für Automatisierungstechnik, Studienarbeit, 1996.

[8] Glaß, R.: Entwicklung und Aufbau einer CAN-Testumgebung, TU Dresden, Fak. Elektrotechnik, Institut für Automatisierungstechnik, Diplomarbeit 1996.